SELECTED PAPERS ON
Frequency Modulation

SELECTED PAPERS ON

Frequency Modulation

EDITED BY

Jacob Klapper

Associate Professor of Electrical Engineering
Newark College of Engineering

Dover Publications, Inc., New York

Published in Canada by General Publishing
Company, Ltd., 30 Lesmill Road, Don Mills,
Toronto, Ontario.

Published in the United Kingdom by Constable
and Company, Ltd., 10 Orange Street, London
WC 2.

This Dover edition, first published in 1970, is a
selection of papers which appear here in collected
form for the first time. Authors' corrections have
been made in seven papers—Nos. 3, 4, 5, 9, 13,
14, and 18. With these exceptions the papers are
reproduced without change from the original
journals.

The editor and publisher are indebted to the
authors and original publishers for permission to
reprint these papers, and for supplying copies of
papers for reproduction.

Standard Book Number: 486-62136-7
Library of Congress Catalog Card Number: 71–95223

Manufactured in the United States of America
Dover Publications, Inc.
180 Varick Street
New York, N.Y. 10014

INTRODUCTION

Frequency modulation (FM), now a major method of transmitting intelligence, continues to gain ground over the older and more established amplitude modulation (AM). There is a growing interest in FM radio for entertainment; the transmittal of sound in television is via FM; commercial and military communication systems favor FM; artificial satellites, our newest path of communication, use FM; and, finally, the growing volume of digital intelligence transmission is largely via digital FM.

What has been and is responsible for the popularity of FM? Its major advantage is the simple tradeoff it permits between noise immunity and bandwidth occupancy. This accounts for the popularity of the FM radio and the use of FM in satellite communication systems. Also favoring the widespread use of FM is the efficiency with which it can be generated at appropriate levels of power. Since the FM carrier is of constant amplitude, it may be passed through nonlinear stages for efficient power amplification. Military and space communications find this property especially useful. Finally, the widespread use of digital FM is due to its simplicity and superior performance.

The development of FM has an interesting history. About half a century ago, engineers in their search for ways to reduce the bandwidth required for information transmittal proposed the method of FM. It was hypothesized that if the frequency of a carrier were varied by only small amounts, then the frequency spread of this carrier would be smaller than for AM. Furthermore, it was thought that a reduced signal bandwidth, permitting a narrower filter, would yield greater immunity to noise. Carson (see reference 1 on page 415) disproved this hypothesis in 1922 and showed that, on the contrary, a frequency modulated signal occupies more spectrum than an AM signal. Carson's findings discouraged much further effort in the utilization of FM for more than a decade.

It was only in 1936 that Armstrong first showed that the advantages of FM lay exactly in the larger frequency spread. He stipulated and experimentally verified that the larger frequency spread of FM differentiates the signal from typical noise, thereby giving it greater noise immunity. Thus the modern era of FM began.

Selected Papers on Frequency Modulation is divided into four sections: I. General FM Theory and Basic Experiments; II. FM Circuit Theory; III. FM Threshold Reduction; and IV. Digital FM. Most of the papers will be recognized as the classics in FM development. They are arranged chronologically in each section, with a few exceptions made for the sake of better continuity or pedagogical reasons. Following the papers is an annotated Selected Bibliography for Further Reading. The book is intended as a reference work for the practitioner, as a guide for those interested in entering the field, and as a textbook in FM principles.

The collection opens with the now famous paper of Armstrong in which he for the first time announced to the world the successful utilization of FM. Here he relates the major property and advantage of FM, namely its greater immunity to noise interference, and describes a practical FM system still popular today. The second paper, the very readable classic by Crosby published soon after Armstrong's, presents basic theory and experiments on noise immunity properties of FM. Corrington (paper 3) discusses another very important consideration in FM, the required bandwidth for its faithful transmission.

The noise immunity of FM does not hold, however, under high noise-level conditions. Additive noise enters into the FM demodulation process in a nonlinear manner, rendering the exact analysis of the demodulated signal-to-noise ratio a very difficult problem. The region where the demodulated signal-to-noise ratio begins to deviate and fall off more rapidly than predicted by linear analysis is referred to as the "FM Improvement Threshold," or simply "threshold." The region below the threshold is marked by a rapid deterioration of the demodulated output, nullifying the noise immunity property of FM. The problem of finding the exact demodulated noise values at and below threshold is discussed by Rice and Stumpers (papers 4 and 5) using different analytical approaches. In 1963 Rice published a very useful approximate approach to the problem (paper 7) based on the notion of click noise. Common- and adjacent-channel interference is discussed very lucidly by Corrington (paper 6).

Section II (FM Circuit Theory) presents methods for dealing with the effect of circuitry on frequency modulated signals. This is an important and intriguing problem, generally involving approximations of one sort or another. The two classical methods for quasi-stationary signals are those of "Carson and Fry" and "van der Pol and Stumpers" (papers 8, 9, and 10). The paper by Weiner and Leon (paper 11) gives a useful quasi-stationary solution with the correction term in closed form. A further useful approximate method is given in the recent paper of Bedrosian and Rice (paper 12).

Since FM noise immunity is limited by the threshold effect, much research effort has recently been expended, especially in connection with space and military communication, to reduce the threshold. It has been found that through the use of FM and/or phase feedback, demodulators with thresholds at considerably lower received carrier-to-noise ratios are possible. These are the so-called *phase-locked* and *FM-feedback loops*. Section III (FM Threshold Reduction) is devoted to this topic. Enloe's paper (paper 13) features the "two threshold" theory which is widely used in the design of FM feedback demodulators. Develet (paper 14) presents an approximate nonlinear analysis of the phase-locked loop applicable when both the noise and the signal are of a Gaussian distribution. He also gives the limit of threshold performance for optimum demodulators. Viterbi (paper 15) gives an exact nonlinear analysis for the first-order phase-locked loop in the presence of white Gaussian noise interference.

The last section deals with digital FM. This area is already prominent and is gaining further impetus from the trend toward digitalization of information. The salient consideration here is the probability of mistaking a "mark" for a "space," or vice versa, for a given received signal and noise power level. The paper by Bennett and Salz (paper 16) is a comprehensive treatment of system performance for cases where postdemodulation filtering can be ignored. The editor's paper (paper 17) includes the effect of postdemodulation filtering through an approximate approach, and at the same time links the theories of analog and digital FM. Concluding the collection is the paper by Pelchat (paper 18) on the power spectrum of PCM/FM.

I am grateful to Dr. T. T. N. Bucher for comments and suggestions, and to A. Newton and J. Frankle for many fruitful discussions. Roxana Klapper is always at my side.

JACOB KLAPPER

Newark College of Engineering
Newark, New Jersey
December, 1969

CONTENTS

III. FM Threshold Reduction

IV. Digital FM

I

General FM Theory and Basic Experiments

PAPER NO. 1

Reprinted from *Proc. IRE*, Vol. 24, No. 5, pp. 689–740, May 1936

A METHOD OF REDUCING DISTURBANCES IN RADIO SIGNALING BY A SYSTEM OF FREQUENCY MODULATION*

BY

EDWIN H. ARMSTRONG

(Department of Electrical Engineering, Columbia University, New York City)

Summary—*A new method of reducing the effects of all kinds of disturbances is described. The transmitting and receiving arrangements of the system, which makes use of frequency modulation, are shown in detail. The theory of the process by which noise reduction is obtained is discussed and an account is given of the practical realization of it in transmissions during the past year from the National Broadcasting Company's experimental station on the Empire State Building in New York City to Westhampton, Long Island, and Haddonfield, New Jersey. Finally, methods of multiplexing and the results obtained in these tests are reported.*

PART I

I T IS the purpose of this paper to describe some recent developments in the art of transmitting and receiving intelligence by the modulation of the frequency of the transmitted wave. It is the further purpose of the paper to describe a new method of reducing interference in radio signaling and to show how these developments may be utilized to produce a very great reduction in the effects of the various disturbances to which radio signaling is subject.

HISTORICAL

The subject of frequency modulation is a very old one. While there are some vague suggestions of an earlier date, it appears to have had its origin shortly after the invention of the Poulsen arc, when the inability to key the arc in accordance with the practice of the spark transmitter forced a new method of modulation into existence. The expedient of signaling (telegraphically) by altering the frequency of the transmitter and utilizing the selectivity of the receiver to separate the signaling wave from the idle wave led to the proposal to apply the principle to telephony. It was proposed to effect this at the transmitter by varying the wave length in accordance with the modulations of the voice, and the proposals ranged from the use of an electrostatic micro-

* Decimal classification: R400×R430. Original manuscript received by the Institute, January 15, 1936. Presented before New York meeting, November 6, 1935.

3

phone associated with the oscillating circuit to the use of an inductance therein whose value could be controlled by some electromagnetic means. At the receiver it was proposed to cause the variations in frequency of the received wave to create amplitude variations by the use of mistuned receiving circuits so that as the incoming variable frequency current came closer into or receded farther from the resonant frequency of the receiver circuits, the amplitude of the currents therein would be correspondingly varied and so could be detected by the usual rectifying means. No practical success came from these proposals and amplitude modulation remained the accepted method of modulating the arc. The various arrangements which were tried will be found in the patent records of the times and subsequently in some of the leading textbooks.[1] The textbooks testify unanimously to the superiority of amplitude modulation.

Some time after the introduction of the vacuum tube oscillator attempts were again made to modulate the frequency and again the verdict of the art was rendered against the method. A new element however, had entered into the objective of the experiments. The quantitative relation between the width of the band of frequencies required in amplitude modulation and the frequency of the modulating current being now well understood, it was proposed to narrow this band by the use of frequency modulation in which the deviation of the frequency was to be held below some low limit; for example, a fraction of the highest frequency of the modulating current. By this means an economy in the use of the frequency spectrum was to be obtained. The fallacy of this was exposed by Carson[2] in 1922 in the first mathematical treatment of the problem, wherein it was shown that the width of the band required was at least double the value of the highest modulating frequency. The subject of frequency modulation seemed forever closed with Carson's final judgment, rendered after a thorough consideration of the matter, that "Consequently this method of modulation inherently distorts without any compensating advantages whatsoever."

Following Carson a number of years later the subject was again examined in a number of mathematical treatments by writers whose results concerning the width of the band which was required confirmed those arrived at by Carson, and whose conclusions, when any were expressed, were uniformly adverse to frequency modulation.

[1] Zenneck, "Lehrbuch der drahtlosen Telegraphy," (1912).
　Eccles, "Wireless Telegraphy and Telephony," (1916).
　Goldsmith, "Radio Telephony," (1918).
[2] "Notes on the theory of modulation," Proc. I.R.E., vol. 10, pp. 57–82; February, (1922).

In 1929 Roder[3] confirmed the results of Carson and commented adversely on the use of frequency modulation.

In 1930 van der Pol[4] treated the subject and reduced his results to an excellent form for use by the engineer. He drew no conclusions regarding the utility of the method.

In 1931, in a mathematical treatment of amplitude, phase, and frequency modulation, taking into account the practical aspect of the increase of efficiency at the transmitter which is possible when the frequency is modulated, Roder[5] concluded that the advantages gained over amplitude modulation at that point were lost in the receiver.

In 1932 Andrew[6] compared the effectiveness of receivers for frequency modulated signals with amplitude modulated ones and arrived at the conclusion that with the tuned circuit method of translating the variations in frequency into amplitude variations, the frequency modulated signal produced less than one tenth the power of one which was amplitude modulated.

While the consensus based on academic treatment of the problem is thus heavily against the use of frequency modulation it is to the field of practical application that one must go to realize the full extent of the difficulties peculiar to this type of signaling.

Problems Involved

The conditions which must be fulfilled to place a frequency modulation system upon a comparative basis with an amplitude modulated one are the following:

1. It is essential that the frequency deviation shall be about a fixed point. That is, during modulation there shall be a symmetrical change in frequency with respect to this point and over periods of time there shall be no drift from it.

2. The frequency deviation of the transmitted wave should be independent of the frequency of the modulating current and directly proportional to the amplitude of that current.

3. The receiving system must have such characteristics that it responds only to changes in frequency and that for the maximum change of frequency at the transmitter (full modulation) the selective characteristic of the system responsive to frequency changes shall be such that substantially complete modulation of the current therein will be produced.

[3] "Ueber Frequenzmodulation," *Telefunken-Zeitung* no. 53, p. 48, (1929).
[4] "Frequency modulation," Proc. I.R.E., vol. 18, pp. 1194–1205; July, (1930).
[5] "Amplitude, phase, and frequency modulation," Proc. I.R.E., vol. 19, pp. 2145–2176; December, (1931).
[6] "The reception of frequency modulated radio signals," Proc. I.R.E., vol. 20, pp. 835–840; May, (1932).

4. The amplitude of the rectified or detected current should be directly proportional to the change in frequency of the transmitted wave and independent of the rate of change thereof.

5. All the foregoing operations should be carried out by the use of aperiodic means.

The Transmitting System

An extensive experience with the various known methods of modulating the frequency convinced the writer as indeed it would anyone who has tried to work with this method of modulation at a high frequency that some new system must be evolved. During the course of this work there was evolved a method which, it is believed, is a complete solution of the transmitter problem. It consists in employing the modulating current to shift the phase of a current derived from a source of fixed phase and frequency by an amount which is directly proportional to the amplitude of the modulating current and inversely proportional to its frequency. The resulting phase shift is then put through a sufficient number of frequency multiplications to insure 100 per cent modulation for the highest frequency of the modulating current. By keeping the initial phase shift below thirty degrees substantial linearity can be obtained.

The means employed to produce the phase shift consisted of a source of fixed frequency, a balanced modulator excited by this source, and arrangements for selecting the side frequencies from the modulator output and combining them in the proper phase with an unmodulated current derived from the initial source. The phase relations which must exist where the combination of the modulated and unmodulated currents takes place are that at the moment the upper and lower side frequencies produced by the balanced modulator are in phase with each other, the phase of the current of the master oscillator frequency with which they are combined shall differ therefrom by ninety degrees.

The schematic and diagrammatic arrangements of the circuits may be visualized by reference to Figs. 1 and 2, and their operation understood from the following explanation. The master oscillator shown in these diagrams may be of the order of fifty to one hundred thousand or more cycles per second, depending upon the frequency of the modulating current. An electromotive force derived from this oscillator is applied in like phase to the grid of an amplifier and both grids of a balanced modulator. The plate circuits of the modulator tubes are made nonreactive for the frequency applied to their grids by balancing out the reactance of the transformer primaries as shown. The plate currents are therefore in phase with the electromotive force applied to the grid. The succeeding amplifier is coupled to the output transformer by a coil whose natural period is high compared to the frequency of the master oscillator and the electromotive force applied to the grid of this amplifier when the modulator tubes are unbalanced by a modulating voltage applied to the screen grids is therefore shifted in phase ninety

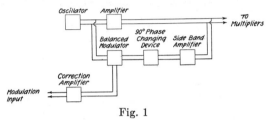

Fig. 1

degrees (or 270 degrees) with respect to the phase of the electromotive force applied to the grids of the balanced modulators. Hence it follows that the phase of the currents existing in the plate circuit of the amplifier of the output of the balanced modulator (at the peak of the modulation voltage) is either ninety degrees or 270 degrees apart from the phase of the current existing in the plate circuit of the amplifier of the

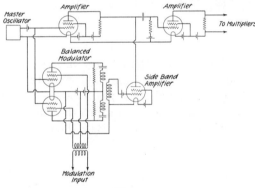

Fig. 2

unmodulated master oscillator current. Therefore the voltages which they develop across the common resistance load will be ninety degrees apart.

The resulting effect on the phase of the voltage developed across the resistance in the plate circuits of these two amplifiers when modulation is applied, compared to the phase of the voltage which would exist there in the absence of modulation will appear from Fig. 3. It will be observed from the vector diagrams that the phase of the voltage across the common resistance load is alternately advanced and retarded by the combination of the modulated and unmodulated components and that the maximum phase shift is given by an angle whose tangent is the sum of the peak values of the two side frequencies divided by the peak value of the unmodulated component. By keeping this angle

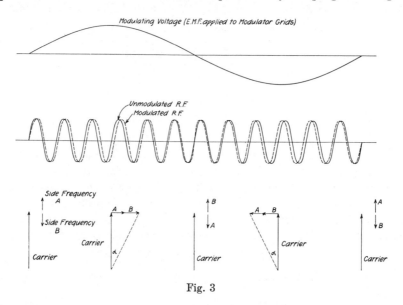

Fig. 3

sufficiently small (not greater than thirty degrees) it may be made substantially proportional to the amplitude of the two side frequencies and hence to the amplitude of the initial modulating current.[7] It will be observed that if the angle through which the phase is shifted be the same for all frequencies of modulation then the rate of increase or de-

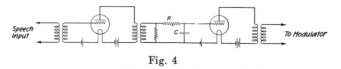

Fig. 4

crease of the angle will be proportional to the frequency of modulation and hence the deviation in frequency of the transmitted wave will be proportional to the frequency of the modulating current. In order to insure a frequency deviation which is independent of the modulation

---

[7] For the large angular displacements there will be an appreciable change in amplitude of the combined currents at double the frequency of the modulating current. This variation in amplitude is not of primary importance and is removed subsequently by a limiting process.

frequency it is necessary that, for a constant impressed modulating electromotive force, the angle through which the phase is shifted be made inversely proportional to the frequency of the modulating current. This is accomplished by making the amplification of the input amplifier inversely proportional to frequency by means of the correction network shown in Fig. 4. The network consists of a high resistance in series with a capacity whose impedence for the lowest frequency of modulation is relatively small with respect to the series resistance. The voltage developed across the capacity which excites the succeeding amplifier stage is therefore inversely proportional to frequency and hence it follows that the angle through which the current is advanced or retarded becomes directly proportional to the amplitude of the modulating current and inversely proportional to its frequency. The resulting phase shift must be multiplied a great many times before a frequency modulated current which can be usefully employed is produced. This will be clear from an examination of the requirements of a circuit over which it is desired to transmit a frequency range from thirty to 10,000 cycles. Since the lowest frequency is limited to a phase shift of thirty degrees it follows that for 10,000 cycles the phase shift will be but 0.09 degree. The minimum phase shift for 100 per cent modulation of the transmitted wave is roughly forty-five degrees. A frequency multiplication of 500 times is required, therefore, to produce a wave which is fully modulated[8] and capable of being effectively handled by the receiver in the presence of disturbing currents.

Under ordinary conditions this multiplication of frequency can be realized without loss of linearity by a series of doublers and triplers operated at saturation provided the correct linkage circuits between the tubes are employed. Where however the wide band frequency swing which will be described subsequently in this paper is employed unexpected difficulties arise. These also will be dealt with subsequently.

From the foregoing description it will be seen that this method of obtaining frequency modulation consists in producing initially phase modulation in which the phase shift is inversely proportional to the frequency of modulation and converting the phase modulated current into a frequency modulated one by successive multiplications of the phase shift. The frequency stability, of course, is the stability attainable by a crystal controlled oscillator and the symmetry of the deviation may be made substantially perfect by compensating such asymmetrical action in the system as may occur. With the method of phase

---

[8] One in which the side frequencies are sufficiently large with respect to the carrier to make it possible to produce at the receiver 100 per cent modulation in amplitude, without the use of expedients which affect unfavorably the signal-to-noise ratio.

shifting shown in Fig. 2 there is an asymmetry which is of importance when the frequency of modulation is high compared to the master oscillator frequency. It occurs in the plate transformer of the balanced modulator. The plate circuits of these tubes are substantially aperiodic and consequently the amplitudes of the upper and lower side frequencies are approximately equal and from this it follows that the electromotive forces induced in the secondary are directly proportional to the values of these frequencies. Where the master oscillator frequency is 50,000 cycles and a frequency of modulation of 10,000 cycles is applied, the upper side frequency may be fifty per cent greater than the lower. This inequality may be compensated by a resistance-capacity network introduced subsequent to the point at which the combination of carrier and side frequencies is effected but prior to any point at which loss of linearity of amplitude occurs. The level in the amplifiers ahead of the compensating network must be kept sufficiently low so that the operation of the system is linear. After the side frequencies are equalized amplitude linearity ceases to be of importance.

The performance of transmitters operating on this principle has been in complete accord with expectations. While the arrangements may seem complex and require a large amount of apparatus the complexity is merely that of design, not of operation. The complete arrangement, up to the last few multiplifier stages may be carried out most effectively with receiving type tubes, these last multiplier stages consisting of power type pentodes for raising the level to that necessary to excite the usual power amplifiers.

The Receiving System

The most difficult operation in the receiving system is the translation of the changes in the frequency of the received signal into a current which is a reproduction of the original modulating current. This is particularly true in the case of the transmission of high fidelity broadcasting. It is, of course, essential that the translation be made linearly to prevent the generation of harmonics but it must also be accomplished in such a manner that the signaling current is not placed at a disadvantage with respect to the various types of disturbances to which radio reception is subject. In the particular type of translation developed for this purpose which employs the method of causing the changes in frequency to effect changes in amplitude which are then rectified by linear detectors, it is essential that for the maximum deviation of the transmitted frequency there shall be a substantial amplitude modulation of the received wave. At first sight it might appear that 100 per cent or complete modulation would be the ideal, but there are objections to approaching this limit too closely. It will, however, be clear that where the translation is such that only a few per cent amplitude modulation results from the maximum deviation of the frequency of the transmitted wave the receiver is hopelessly handicapped with respect to amplitude disturbances. This is true because even when the level of the voltage applied to the conversion system is kept constant by a current limiting device or automatic volume control there still remains those intervals wherein the incoming disturbances arrive in the proper phase to neutralize the signaling current in the detector, effecting thereby substantially complete modulation of the rectified current or the intervals wherein the disturbing currents themselves effect greater amplitude changes than the signal itself by cross modulation of its frequency.

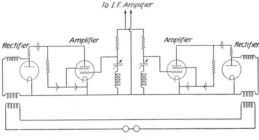

Fig. 5

An arrangement in which linear conversion can be effected without handicapping the system with respect to amplitude disturbances is illustrated diagrammatically in Fig. 5. Two branch circuits each containing resistance, capacity, and inductance in series as shown are connected to the intermediate-frequency amplifier of a superheterodyne at some suitable frequency. One capacity and inductance combination is made nonreactive for one extreme of the frequency band which the signal current traverses and the other capacity and inductance combination is made nonreactive for the other end of the band. The resistances are chosen sufficiently high to maintain the current constant over the frequency range of the band; in fact, sufficiently high to make each branch substantially aperiodic. The reactance characteristics taken across each capacity and inductance combination will be as illustrated in Fig. 6 by curves A and B. Since the resistances in series with the reactance combinations are sufficient to keep the current constant throughout the frequency band it follows that the voltages developed across each of the two combinations will be proportional to their reactances as is illustrated in curves A' and B'. The two voltages are

11

applied respectively to the two equal aperiodic amplifiers, each of which is connected to a linear rectifier. The rectifiers are in series with equal output transformers whose secondaries are so poled that changes in the rectifier currents resulting from a change in the frequency of the received signal produce additive electromotive forces in their secondaries. Since amplifiers and rectifiers are linear the output currents will follow the amplitude variations created by the action of the capacity-inductance combinations. While the variation in reactance is not linear with respect to the change of frequency, particularly where the width of the band is a substantial percentage of the frequency at which the operation takes place, as a practical matter, by the proper choice of values together with shunts of high resistance or reactance

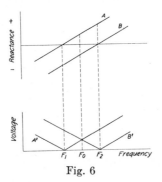

Fig. 6

these characteristics may be rendered sufficiently straight within the working range to meet the severest requirements of high fidelity broadcasting. The operation of the system is aperiodic and capable of effecting 100 per cent modulation if desired, this last depending on the separation of the two nonreactive points with respect to the frequency swing. Generally the setting of the nonreactive frequency points should be somewhat beyond the range through which the frequency is swung.

There is shown in Fig. 7 an alternative arrangement of deriving the signal from the changes in frequency of the received wave which has certain advantages of symmetry over the method just described. In this arrangement a single capacity-inductance combination with the nonreactive point in the center of the frequency band is used and the rectifiers are polarized by a current of constant amplitude derived from the received current. In this way, by properly phasing the polarizing current, which is in effect a synchronous heterodyne, differential rectifying action can be obtained. In Fig. 7 the amplified output of the receiver is applied across the single series circuit consisting of resistance R, capacity C, and inductance L. The reactance of C and L are equal for the mid-frequency point of the band and the reactance curve is as illustrated in *A* of Fig. 8. At frequencies above the nonreactive

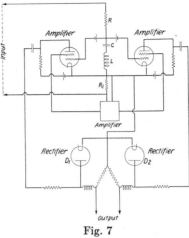

Fig. 7

point the combination acts as an inductance; at frequencies below the nonreactive point as a capacity and the phase of the voltage developed across the combination with respect to the current through it

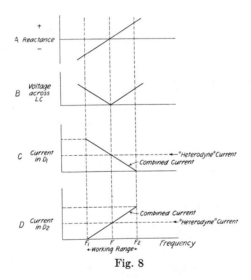

Fig. 8

differs, therefore, by 180 degrees above and below the nonreactive point. Since the current through the circuit is maintained constant over the working range by the resistance *R* and since the resistance of the

13

capacity C and inductance L may be made very low the electromotive force developed across C and L is of the form shown in curve B. This curve likewise represents the variation in voltage with variation in frequency which is applied to the grids of the amplifiers and eventually to the two rectifiers D_1 and D_2.

The heterodyning or polarizing voltage is obtained by taking the drop across the resistance R_1, amplifying it, changing its phase through ninety degrees and applying the amplified voltage to the screen grids of the amplifiers in opposite phase. The characteristic of this amplifying and phase changing system must be flat over the working range. Under these conditions the signaling and heterodyning voltages are exactly in phase in one rectifier and 180 degrees out of phase in the other, and hence for a variable signaling frequency the rectifying characteristics are as shown in curves C and D the detector outputs being cumulatively combined for frequency changes. Adjustment of the relative amplitudes of the signaling and polarizing voltages in the rectifier controls the degree of amplitude modulation produced from 100 per cent down to any desired value.

PART II

With the foregoing description of the instrumentalities for transmitting and receiving frequency modulated waves it is now in order to consider the main object of the paper; the method of reducing disturbances and the practical results obtained by its use.

METHOD OF REDUCING DISTURBANCES

The basis of the method consists in introducing into the transmitted wave a characteristic which cannot be reproduced in disturbances of natural origin and utilizing a receiving means which is substantially not responsive to the currents resulting from the ordinary types of disturbances and fully responsive only to the type of wave which has the special characteristic.

The method to be described utilizes a new principle in radio signaling the application of which furnishes an interesting conflict with one which has been a guide in the art for many years; i.e., the belief that the narrower the band of transmission the better the signal-to-noise ratio. That principle is not of general application. In the present method an opposite rule applies.

It appears that the origin of the belief that the energy of the disturbance created in a receiving system by random interference depended on the band width goes back almost to the beginning of radio. In the days of spark telegraphy it was observed that "loose coupling" of the conventional transmitter and receiver circuits produced a "sharper wave" and that interference from lightning discharges, the principle "static" of those days of insensitive and nonamplifying receivers was decreased. Further reduction in interference of this sort occurred when continuous-wave transmitters displaced the spark and when regeneration narrowed the band width of the receiving system. It was observed, however, that "excessive resonance" must not be employed either in telegraphic or more particularly in telephonic signaling or the keying and speech would become distorted. It was concluded in a qualitative way that there was a certain "selectivity" which gave the best results.

In 1925 the matter was placed on a quantitative basis by Carson[9] where in a mathematical treatment of the behavior of selective circuits when subjected to irregular and random interference (with particular reference to "static"), on the basis of certain assumptions, the proposition was established that "if the signaling system requires the transmission of the band of frequencies corresponding to the interval $\omega_2 - \omega_1$ and if the selective circuit is efficiently designed to this end, then the mean square interference current is proportional to the frequency band width $(\omega_2 - \omega_1)/2\pi$.

Hazeltine[10] pointed out that when a detector was added to such a system and a carrier of greater level than the interference currents was present, that for aural reception only those components of the interfering current lying within audible range of the carrier frequency were of importance and that Carson's theory should be supplemented by the use of a factor equal to the relative sensitivity of the ear at different frequencies.

With the discovery of shot effect and thermal agitation noises and the study of their effect on the limit of amplification quantitative relations akin to those enunciated by Carson with respect to static were found to exist.

Johnson,[11] reporting the discovery of the electromotive force due to thermal agitation and considering the problem of reducing the noise in amplifiers caused thereby, points out that for this type of disturbance the theory indicates, as in the Carson theory, that the frequency range of the system should be made no greater than is essential for the proper transmission of the applied input voltage, that where a voltage of constant frequency and amplitude is used one may go to extremes in

[9] J. R. Carson, "Selective circuits and static interference," *Bell Sys. Tech. Jour.*, vol. 4, p. 265, (1925).

[10] L. A. Hazeltine, Discussion on "The shielded neutrodyne receiver," PROC. I.R.E., vol. 14, pp. 408, 409; June, (1926).

[11] J. B. Johnson, "Thermal agitation of electricity in conductors, *Phys. Rev.*, vol. 32, no. 1, July, (1926).

making the system selective and thereby proportionately reducing the noise, but that when the applied voltage varies in frequency or amplitude the system must have a frequency range which takes care of these variations and the presence of a certain amount of noise must be accepted.

Ballantine[12] in a classical paper discussing the random interference created in radio receivers by shot and thermal effects obtained a complete expression for the noise output.[13]

Johnson and Llewellyn,[14] in a paper dealing generally with the limits to amplification, point out that in a properly designed amplifier the limit resides in thermal agitation in the input circuit to the amplifier, that the power of the disturbance in the output of the amplifier is proportional to its frequency range and that this, the only controllable factor in the noise equation, should be no greater than is needed for the transmission of the signal. A similar conclusion is reached in the case of a detector connected to the output of a radio-frequency amplifier and supplied with a signal carrier.

It is now of interest to consider what happens in a linear detector connected to the output of a wide band amplifier which amplifies uniformly the range from 300 to 500 kilocycles. Assume that the amplification be sufficiently great to raise the voltage due to thermal agitation and shot effect to a point sufficient to produce straight-line rectification and that no signal is being received. Under these conditions the frequencies from all parts of the spectrum between 300 and 500 kilocycles beat together to contribute in the output of the detector to the rough hissing tone with which the art is familiar. The spectrum of frequencies in the rectified output runs from some very low value which is due to adjacent components throughout the range beating with one another to the high value of 200 kilocycles caused by the interferences of the extremes of the band.

It is important to note that all parts of the 300- to 500-kilocycle spectrum contribute to the production in the detector output of those frequencies with which we are particularly interested—those lying within the audible range.

[12] Stuart Ballantine, "Fluctuation noise in radio receivers," Proc. I.R.E., vol. 18, pp. 1377–1387; August, (1930).

[13] Ballantine expressed his result as follows: "In a radio receiver employing a square-law detector and with a carrier voltage impressed upon the detector, the audio-frequency noise, as measured by an instrument indicating the average value of the square of the voltage (or current), is proportional to the area under the curve representing the square of the over-all transimpedance (or of the transmission) from the radio-frequency branch in which the disturbance originates to the measuring instrument as a function of frequency and proportional to the square of the carrier voltage."

[14] J. B. Johnson and F. B. Llewellyn, "Limits to amplification," Trans. A.I.E.E., vol. 53, no. 11, November, (1934).

Assume now that an unmodulated signal carrier is received of, for example, 400 kilocycles and that its amplitude is greater than that of the disturbing currents. Under these circumstances an entirely new set of conditions arise. The presence of the 400-kilocycle current stops the rectification of the beats which occur between the various components of the spectrum within the 300- to 500-kilocycle band and forces all rectification to take place in conjunction with the 400-kilocycle carrier. Hence in the output of the rectifier there is produced a series of frequencies running from some low value up to 100 kilocycles. The lowest frequency is produced by those components of the spectrum which lie adjacent to the 400-kilocycle current, the highest by those frequencies[15,16] which lie at the extremity of the band; i.e., 300 and 500 kilocycles, respectively.

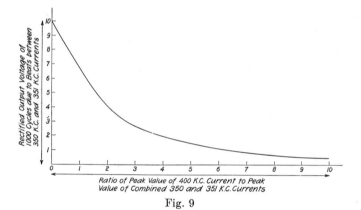

Fig. 9

The characteristics of the rectifiers and the magnitude of some of the effects involved in the above-described action may be visualized by reference to the succeeding figures. The actual demodulation of the beats occurring between adjacent frequency components by the presence of the 400-kilocycle current is shown by the characteristic of Fig. 9, which illustrates what happens to the output voltage of a rectifier produced by beating together two equal currents of 350 and 351 kilocycles, respectively, when a 400-kilocycle current is introduced in the same rectifier and its amplitude progressively increased with respect

[15] It has been pointed out by Ballantine[16] that it is improper to speak of the amplitude of a single component of definite frequency and that the proper unit is the noise per frequency interval. This is, of course, correct, but to facilitate the physical conception of what occurs in this system the liberty is taken of referring to the noise components as though they were of continuous sine wave form. The behavior of the system may be checked by actually introducing from a local generator such components.

[16] "Fluctuation noise in radio receivers," Proc. I.R.E., vol. 18, pp. 1377–1387; August, (1930).

to the amplitude of these two currents. The characteristic was obtained
with the arrangement shown in Fig. 10, in which two oscillators of 350
and 351 kilocycles produced currents of equal strength in a linear
rectifier, this rectifier consisting of a diode in series with 10,000 ohms
resistance. The output of the rectifier is put through a low-pass filter,
a voltage divider, and an amplifier. The 400-kilocycle current is intro-

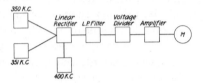

Fig. 10

duced into the rectifier without disturbing the voltage relations of the
other two oscillators and the effect on the rectified output voltage observed as the 400-kilocycle current is increased. The purpose of the
low-pass filter is to prevent the indicating instrument from responding
to the 49- or 50-kilocycle currents produced by the interaction of the
350- and 351-kilocycle currents with the current of 400 kilocycles. The

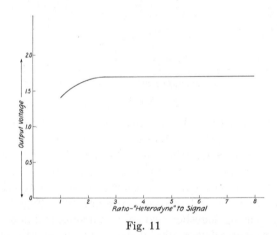

Fig. 11

linearity characteristic of the rectifier is shown in Fig. 11 where the
voltage produced by the beats between a current of constant amplitude
and one whose amplitude is raised from equality with, to many times
the value of, the first current is plotted against the ratio of the two.
The linearity of the rectifier is such that after the ratio of the current
becomes two to one no further increase in rectifier output voltage
results. In fact with the levels used in these measurements when the

two currents are equal there is an efficiency of rectification of only about twenty per cent less than the maximum obtained.

It is important to note here that the only frequencies in the spectrum which contribute to the production of currents of audible frequency in the detector output circuit are those lying within audible range of the signal carrier. We may assume this range as roughly from 390 to 410 kilocycles. The frequencies lying beyond these limits beat against the 400-kilocycle carrier and of course are rectified by the detector but the rectified currents which are produced are of frequencies which lie beyond the audible range and produce therefore no effect which is apparent to the ear. It follows that if the signal carrier is somewhat greater in amplitude than the disturbing currents the signal-to-noise ratio for a receiver whose band of admittance covers twice the audible range will be the same as for one whose band width is many times that value. (There are, of course, certain second order effects, but they are of such minor importance that the ear cannot detect them.) The amplitude of the disturbances in the detector output, will vary in accordance as the components of the disturbing currents come into or out of phase with the signal carrier, the rectified or detector output current increasing above and decreasing below the level of the rectified carrier current by an amount proportional to the amplitude of the components of the 300–500-kilocycle band. The reasons for the independence of the signal-to-noise ratio of the band width under the circumstances which have been described should now be apparent. In any event, it can be readily demonstrated experimentally.

It is now in order to consider what happens when a current limiting device is introduced between the output of the amplifier and the detector input. (Assume signal level still above peak noise level.) Two effects will occur. One of the effects will be to suppress in the output circuit of the limiter all components of the disturbing currents which are in phase with, or opposite in phase to, the 400-kilocycle carrier. The other effect will be to permit the passage of all components of the disturbing currents which are in quadrature with the 400-kilocycle current.

Both the above effects are brought about by a curious process which takes place in the limiter. Each component within the band creates an image lying on the opposite side of the 400-kilocycle point whose frequency difference from the 400-kilocycle current is equal to the frequency difference between that current and the original component. The relative phase of the original current in question, the 400-kilocycle current and the image current is that of phase modulation—that is, at the instant when the original component and its

19

image are in phase with each other, the 400-kilocycle current will be in quadrature with them both and at the instant that the 400-kilocycle current is in phase with one of these two frequencies, it will be out of phase with the other.

The relation (obtained experimentally) between the amplitudes of the original current and the image is illustrated by the curve of Fig. 12,

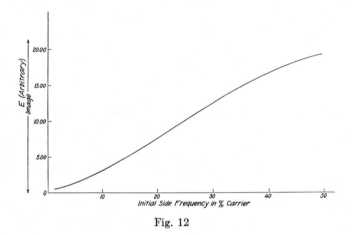

Fig. 12

which shows the relation between the amplitude of a 390-kilocycle current introduced into a limiter along with the 400-kilocycle current and the resulting 410-kilocycle image in terms of percentage amplitude of the 400-kilocycle current. It will be obvious from the curve that in the region which is of interest—that is, where the side frequencies are smaller than the mid-frequency—that the effect is substantially linear.

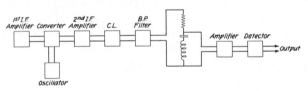

Fig. 13

With the above understanding of what takes place in the current limiter it is now in order to consider what happens when a selective system as illustrated in Fig. 13 is interposed between the limiter and the detector. (The band-pass filter is for the purpose of removing limiter harmonics.) A rough picture of what occurs may be had by considering a single component of the interference spectrum. Suppose

this component to be at 390 kilocycles and that by the action already explained it has created its image at 410 kilocycles. These two frequencies are equal in amplitude and so phased with respect to each other and with respect to the 400-kilocycle carrier that no amplitude change results.

Assume now that the selective system has the characteristic *MN* which as shown in Fig. 14 is designed to give complete modulation for a ten-kilocycle deviation of frequency. Since at 390 kilocycles the reactance across the capacity-inductance combination is zero and at 410 kilocycles double what it is at 400 kilocycles it follows that the 390-kilocycle component becomes equal to zero but the ratio of the 410-kilocycle component to the 400-kilocycle carrier is doubled; that

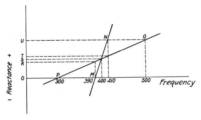

Fig. 14

is, it is twice as great as is the ratio in the circuits preceding the selective system. The change in amplitude, therefore, becomes proportional to *OU*. Therefore in combination with the 400-kilocycle carrier a variation in amplitude is produced which is substantially identical with that which would be obtained were the current limiter removed and the selective system replaced by an aperiodic coupling of such value that the same detector level were preserved.

Now consider what occurs when a selective system having the characteristic such as *PQ* and requiring a deviation of 100 kilocycles to produce full modulation is employed instead of one such as *MN*, where a ten-kilocycle deviation only is required. Assume the same conditions of interference as before. The 400-kilocycle voltage applied to the rectifier will be the same as before, but the *relative amplitudes of the 390- and 410-kilocycle voltages will only be slightly changed.* The 410-kilocycle voltage will be increased from a value which is proportional to *OS* to one which is proportional to *OT* and the 390-kilocycle voltage will be reduced from a value proportional to *OS* to one proportional to *OR*. The difference in value of the two frequencies will be proportional to the difference between *OS* and *OT* or *RT*, and the change in amplitude produced by their interaction with the 400-

kilocycle current will be likewise proportional to RT. The reduction in the amplitude of the disturbance as measured in the detector output by the use of a 200-kilocycle wide selective system as compared to the use of one only twenty kilocycles wide is therefore the ratio RT/OU. In this case it is ten per cent. The power ratio is the square of this or one per cent.

The above reasoning holds equally well if a balanced rectifying system is used where the characteristics of the selective system are as shown in Fig. 15. The output of the system insofar as voltages resulting from changes in frequency are concerned is the sum of outputs of the two sides of the balance.

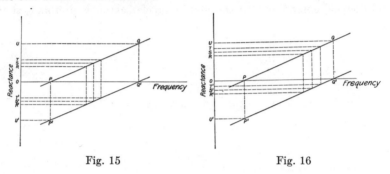

Fig. 15   Fig. 16

It is of course clear that disturbing currents lying farther from the 400-kilocycle point than the ten-kilocycle limit will, by interaction with the 400-kilocycle current, produce larger values of rectified current than those lying within that band. *But the rectified currents produced in the detector output by those components of frequency which lie at a greater than audible frequency distance from the 400-kilocycle current will be beyond the audible range and hence will produce no disturbance which is audible.* (It is generally advisable to eliminate them from the audio amplifier by a low-pass filter to prevent some incidental rectification in the amplifier making their variations in amplitude audible.)

It remains only to consider what happens when the frequency of the 400-kilocycle current is varied in accordance with modulation at the transmitter. It is clear from Fig. 14 that when the selective system has the characteristic MN that a deviation of 10,000 cycles will produce complete modulation of the signal or a change in amplitude proportional to OU. Similarly, when the characteristic is according to the curve PQ it is clear that a 100,000-cycle deviation is required to produce complete modulation, which is likewise proportional to the same value OU. As the signal current is swung back and forth over the range of frequencies between 300 and 500 kilocycles the band of frequencies from which the audible interference is derived continually changes, the band progressively lying about ten kilocycles above and ten kilocycles below what we may call the instantaneous value of the frequency of the signal. The effect is illustrated by Fig. 16 and from this it will be seen that the amplitude of the disturbances in the output circuit of the rectifiers, which is proportional to the sum of RT and $R'T'$ will be constant. This will be true where the ratio of the amplitude of the signal to the disturbing currents is sufficiently large—where this condition does not exist then there are certain other effects which modify the results, but these effects will only be of importance at the limits of the practical working range.

COMPARISON OF NOISE RATIOS OF AMPLITUDE AND FREQUENCY MODULATION SYSTEMS

From the foregoing description it will be clear that as between two frequency modulation systems of different band widths the signal-to-noise power ratio in the rectified output will vary directly as the square of the band width (provided the noise voltage at the current limiter is less than the signaling voltage). Thus doubling the band width produces an improvement of 4 to 1 and increasing it tenfold an improvement of 100 to 1.

The comparison of relative noise ratios of amplitude and frequency modulation systems cannot be made on so simple a basis as there are a number of new factors which enter, particularly when the comparison is viewed from the very practical aspect of how much greater power must be used with an amplitude modulated transmitter than with a frequency modulated one. If the academic comparison be made between a frequency modulated system having a deviation of ten kilocycles and an amplitude modulated one of similar band width and the *same carrier* level (also same fidelity), it will be found that the signal-to-noise voltage ratio as measured by a root-mean-square meter will favor the frequency modulation system by about 1.7 to 1, and that the corresponding power ratio will be about 3 to 1. This improvement is due to the fact that in the frequency modulation receiver it is only those noise components which lie at the extremes of the band; viz., ten kilocycles away from the carrier which, by interaction with the carrier (when unmodulated) can produce the same amplitude of rectified current as will be produced by the corresponding noise component in the amplitude modulated receiver.

Those components which lie closer to the carrier than ten kilocycles will produce a smaller rectified voltage, the value of this depending on their relative distance from the carrier. Hence the distribution of energy in the rectified current will not be uniform with respect to frequency but will increase from zero at zero frequency up to a maximum at the limit of the width of the receiver, which is ten kilocycles in the present case. The root-mean-square value of the voltage under such a distribution is approximately 0.6 of the value produced with the uniform distribution of the amplitude receiver.

Similarly in comparing an amplitude modulation system arranged to receive ten-kilocycle modulations and having, of course, a band width of twenty kilocycles, with a 100-kilocycle deviation frequency

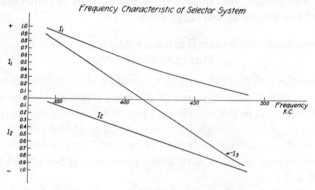

Fig. 17

modulation system (same carrier level and same fidelity) there will be an improvement in noise voltage ratio of

$$1.7 \times \frac{\text{deviation}}{\text{audio-frequency range}} \quad \text{or} \quad 1.7 \times \frac{100}{10} = 17.$$

The above comparisons have been made on the basis of equal carrier. The practical basis of comparison between the two is that of half carrier for the amplitude modulation and full carrier for the frequency modulation system. This results in about the equivalent amount of power being drawn from the mains by the two systems. On this basis the voltage improvement becomes thirty-four and the signal-to-noise power ratio 1156. Where the signal level is sufficiently large with respect to the noise it has been found possible to realize improvements of this order.

The relative output signal-to-noise ratios of an amplitude modulation system fifteen kilocycles wide (7.5-kilocycle modulation frequency) and a frequency modulation system 150 kilocycles wide (75-kilocycle deviation) operating on forty-one megacycles have been compared on the basis of equal fidelity and half carrier for amplitude modulation.

The characteristic of the selective system for converting frequency changes to amplitude changes, which was used, is shown in Fig. 17. The variation of the output signal-to-noise ratio with respect to the corresponding radio-frequency voltage ratio is illustrated in Fig. 18. The curves show that where the radio-frequency peak voltage of the noise measured at the current limiter is less than ten per cent of the

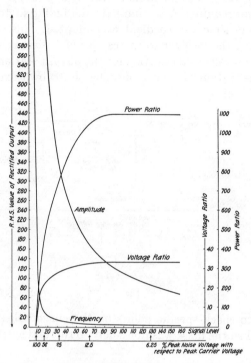

Fig. 18

signal peak voltage then the energy of the disturbance in the rectified output will be reduced by a factor which is approximately 1100 to 1. When the peak radio-frequency noise voltage is twenty-five per cent of the signal peak voltage then the energy of the disturbance in the rectified output has been reduced to about 700 to 1, and when it is fifty per cent the reduction of the disturbance drops below 500 to 1. Finally when the noise and signal peak voltages become substantially equal the improvement drops to some very low value. While it is unfortunate, of course, that the nature of the effect is such that the amount of noise reduction becomes less as the noise level rises with respect to the signal, nevertheless this failing is not nearly so important as it would appear. In the field of high fidelity broadcasting a signal-to-noise voltage ratio of at least 100 to 1 is required for satisfactory reception. It is just within those ranges of noise ratios which can be reduced to this low level that the system is most effective.

The arrangements employed in obtaining these characteristics and the precautions which must be observed may perhaps be of interest. As it was obviously impracticable to vary the power of a transmitter over the ranges required or to eliminate the fading factor except over short periods of time an expedient was adopted. This expedient consisted in tuning the receiver to the carrier of a distant station, determining levels and then substituting for the distant station a local signal generator, the distant station remaining shut down except as it was

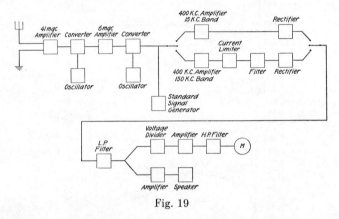

Fig. 19

called upon to check specific points on the curve. Observations were taken only when the noise was due solely to thermal agitation and shot effect.

Fig. 19 shows the arrangement of apparatus. The receiver was a two-intermediate-frequency superheterodyne with provision for using either a narrow band second intermediate amplifier with the amplitude modulation system or a wide band amplifier with the frequency modulation system. The band width of the amplitude modulation system was fifteen kilocycles or twice the modulation frequency range. The band width of the frequency modulation receiver was 150 kilocycles or twice the frequency deviation. Provision was made for shifting from one intermediate amplifier to the other without disturbing the remainder of the system. The forty-one-megacycle circuits and the first intermediate amplifier circuits were wide enough to pass the frequency swing of 150 kilocycles. Identical detection systems were used, the frequency modulation detector being preceded by a selective system for translating changes in frequency into changes in amplitude. The output circuits of the detectors were arranged to be connected alternately to a 7500-cycle low-pass filter with a voltage divider across its output. An amplifier with a flat characteristic over the audible range and a root-mean-square meter connected through a high-pass, 500-cycle filter provided the visual indication.

The standard signal was introduced into the input of the two branches of the second intermediate-frequency stage at 400 kilocycles. As long as the receiver is linear between the antenna and the point at which the standard signal is introduced it is immaterial whether the signal be of forty-one megacycles, six megacycles, or 400 kilocycles. This has been checked experimentally but 400 kilocycles was chosen on account of the greater accuracy of the signal generator on low frequencies.

The relative noise levels to be compared varied over such ranges that lack of linearity had to be guarded against and readings were made by bringing the output meter to the same point on the scale each time by adjustment of the voltage divider, and obtaining the relative voltages directly from the divider.

Two other precautions are essential. The absolute value of the noise voltage on the frequency modulation system becomes very low for high signal levels. If the voltages due to thermal agitation and shot effect are to be measured rather than those due to the power supply system the output meter must be protected by a high-pass filter of high attentuation for the frequencies produced by the power system. The cutoff point should be kept as low as possible since because of the difference in the distribution of energy in the rectified outputs of frequency and amplitude modulation receivers already referred to there is a certain error introduced by this filter which is small if the band width excluded by the filter is small but which can become appreciable if too much of the low-frequency part of the modulation frequency range be suppressed.

A second precaution is the use of a low-pass filter to cut off frequencies above the modulation range. Because of the wide band passed by the amplifiers of the frequency modulation part of the system there exists in the detector output rectified currents of frequencies up to seventy-five kilocycles. The amplitude of these higher frequencies is much greater than those lying within the audible range. The average detector output transformer will readily pass a substantial part of these superaudible frequencies which then register their effect upon the output meter although they in no way contribute to the audible disturbance.

The procedure which was followed in making the measurements we are considering consisted in tuning the receiver to the distant transmitter and adjusting the two detector levels to the same value for the respective carrier levels to be employed. This was done by cutting the carrier in half at the transmitter when the amplitude modulation detector level was being set and using full carrier for the adjustment of the frequency modulation detector. Each system was then modulated seventy-five per cent and output voltages checked against each other. If they were equal the modulation was removed and the relative noise voltages measured for the respective carrier levels. This gave the first point on the curve. The transmitter was then shut down and a local carrier introduced which gave the same level in the 400-kilocycle intermediate amplifier circuits as the half carrier distant signal. This level was directly ascertainable from the rectified detector current in the amplitude modulation system. From this point on the procedure was entirely within the control of the receiving station. The noise ratios could be compared at any signal level by adjusting the voltage introduced by the signal generator to any fraction of that of the distant signal, bringing the level in the amplitude modulation detector up to the same original value by adjustment of the amplification of the second intermediate amplifier (the frequency modulation detector stays at its point of reference because of the current limiter) and comparing the two output voltages. The level of the detector in the amplitude modulation receiver was of course set with the half carrier value of the signal generator and the output voltage measured at that level. The output voltage of the frequency modulation system was measured when twice that voltage was applied.

It is important to keep in mind just what quantities have been measured and what the curves show. The results are a comparison between the relative noise levels in the two systems (root-mean-square values) *when they are unmodulated.* In both an amplitude and in a frequency modulation receiver the noise during modulation may be greater than that obtained without modulation. In the frequency modulation receiver two principal sources may contribute to this increase, one of which is of importance only where the band for which the receiver is designed is narrow, the other of which is common to all band widths. If the total band width of the receiver is twenty kilocycles and if the deviation is, for example, ten kilocycles, then as the carrier frequency swings off to one side of the band, it approaches close to the limit of the filtering system of the receiver. Since the sides of the filter are normally much steeper than the selective system employed to convert the changes in frequency into amplitude variations and since the frequency of the signaling current will have approached to within the range of good audibility of the side of the filter a considerable increase in both audibility and amplitude of the disturbance may occur, caused by the sides of the filter acting as the translating device. This effect is obviously not of importance where a wider frequency swing is employed.

The other source of noise which may occur when the signal frequency swings over the full range is found in systems of all band widths. It was first observed on an unmodulated signal when it was noted that swinging the intermediate frequency from the mid-point to one side or the other by adjustment of the frequency of the first heterodyne produced an increase in the amplitude and a change in the character of the noise. The effect was noted on a balanced detector system and at first it was attributed to the destruction of the amplitude balance as one detector current became greater than the other. Subsequently when it was noted that the increase in the noise was produced by the detector with the smaller current and that the effect was most pronounced when the signal level was relatively low, the explanation became apparent. As long as the signal frequency was set at the mid-point of the band its level in the detector was sufficiently large to prevent the production of audible beats between the noise components lying respectively at the two ends of the band where the reactance of the selective systems is a maximum.

When however the signal frequency moves over to one side of the band the amplitude of the voltage applied to one of the detectors progressively decreases, approaching zero as the frequency coincides with the zero reactance point of the selective system. The demodulating effect of the signaling current therefore disappears and the noise components throughout the band, particularly those at the other side of it, are therefore free to beat with each other. The noise produced is the characteristic one obtained when the high-frequency currents caused by thermal agitation and shot effect are rectified in a detector without presence of a carrier. The effect is not of any great importance on the ordinary working levels for simplex operation, although it may become so in multiplex operation. It indicates, however, that where the signal-to-noise level is low, complete modulation of the received signal by the conversion system is not desirable and that an adjustment of the degree of modulation for various relative noise levels is advantageous.

In the course of a long series of comparisons between the two systems a physiological effect of considerable importance was noted. It was observed that while a root-mean-square meter might show the same reading for two sources of noise, one derived from an amplitude modulation, and the other from a frequency modulation receiver (both of the same fidelity) that the disturbance perceived by the ear was more annoying on the amplitude modulation system. The reason for this is the difference in the distribution of the noise voltage with respect to frequency in the rectified output currents of the two systems, the distribution being substantially uniform in the amplitude system but proportional to frequency in the frequency modulation system. Hence in the latter there is a marked absence of those frequencies which lie in the range to which the ear is the most sensitive. With most observers this difference results in their appraising a disturbance produced in the speaker by an amplitude modulation system as the equivalent of one produced therein by a frequency modulation system of about 1.5 times the root-mean-square voltage although of course the factor varies considerably with the frequency range under consideration and the characteristic of the individual's aural system.

On account of this difference in distribution of energy the correct method of procedure in making the comparison is that given in the article by Ballantine,[16] but lack of facilities for such determinations made necessary the use of a root-mean-square meter for the simultaneous measurement of the entire noise frequency range. The increase in noise voltage per frequency interval with the frequency may be readily demonstrated by means of the ordinary harmonic analyzer of the type now so generally used for the measurement of distortion. Because of the extremely narrow frequency interval of these instruments it is not possible to obtain sufficient integration to produce stable meter readings and apparatus having a wider frequency interval than the crystal filter type of analyzer must be used. The observation of the action of one of these analyzers will furnish convincing proof that peak voltmeter methods must not be used in comparing the rectified output currents in frequency and amplitude modulation receivers.

All the measurements which have been heretofore discussed were taken under conditions in which the disturbing currents had their origin in either thermal agitation or shot effect, as the irregularity of atmospheric disturbances or those due to automobile ignition systems were too irregular to permit reproducible results. The curves apply generally to other types of disturbances provided the disturbing voltage is not greater than that of the signal. When that occurs a different situation exists and will be considered in detail later.

There are numerous second order effects produced, but as they are of no great importance consideration of them will not be undertaken in the present paper.

THE NEW YORK-WESTHAMPTON AND HADDONFIELD TESTS

The years of research required before field tests could even be considered were carried out in the Marcellus Hartley Research Laboratory at Columbia University. Of necessity both ends of the circuit had to be under observation simultaneously and a locally generated signal was used. The source of signal ultimately employed consisted of a standard signal generator based upon the principle of modulation already described and capable of giving 150,000 cycles swing on forty-four megacycles. The generator was also arranged to give amplitude modulated signals. Suitable switching arrangements for changing rapidly from frequency to amplitude modulation at either full or half carrier were set up and a characteristic similar to that of Fig. 18 ultimately obtained.

A complete receiving system was constructed and during the Winter of 1933–1934 a series of demonstrations were made to the executives and engineers of the Radio Corporation of America. That wholly justifiable suspicion with which all laboratory demonstrations of "static eliminators" should be properly regarded was relieved when C. W. Horn of the National Broadcasting Company placed at the writer's disposal a transmitter in that company's experimental station located on top of the Empire State Building in New York City. The transmitter used for the sight channel of the television system delivered about two kilowatts of power at forty-four megacycles to the antenna and it was the one selected for use. This offer of Mr. Horn's greatly facilitated the practical application of the system as it eliminated the necessity of transmitter construction in a difficult field and furnished the highly skilled assistance of R. E. Shelby and T. J. Buzalski, the active staff of the station at that time. Numerous difficulties, real and imaginary, required much careful measurement to ascertain their presence or absence and the relative importance of those actually existing. The most troublesome was due to the position of the transmitter, which is located on the eighty-fifth floor of the building and is connected by a concentric transmission line approximately 275 feet long with a vertical dipole antenna about 1250 feet above ground. Investigation of the characteristics of this link between transmitter and antenna showed it to be so poorly matched to the antenna that the resulting standing waves attained very large amplitude. The problem of termination afforded peculiar difficulties because of the severe structural requirements of the antenna above the roof and of the transmission line below it. It was however completely solved by P. S. Carter of the R.C.A. Communications Company in a very beautiful manner, the standing waves being practically eliminated and the antenna broadened beyond all requirements of the modulating system contemplated. With the transmitter circuits no difficulty was encountered at this time. The frequency of the system was ordinarily controlled by a master oscillator operating at 1733 kilocycles which was multiplied by a series of doublers and a tripler to forty-four megacycles. The multiplier and amplifier circuits were found to be sufficiently broad for the purposes of the initial tests.

The crystal control oscillator was replaced by the output of the modulation system shown in Fig. 20 in which an initial frequency of 57.33 kilocycles was multiplied by a series of doublers up to the input frequency of the transmitter of 1733 kilocycles. It was found possible to operate this apparatus as it is shown installed in the shielded room

Fig. 20

of the television studio at the Empire State station as the shielding furnished ample protection against the effects of the high power stages of the transmitter located some seventy-five feet away.

The receiving site selected was at the home of George E. Burghard at Westhampton Beach, Long Island, one of the original pioneers of amateur radio, where a modern amateur station with all facilities, including those for rigging directive antennas, were at hand. Westhampton is about sixty-five miles from New York and 800 or 900 feet below line of sight.

The installation is illustrated in Figs. 21 and 22 which show both frequency and amplitude modulation receivers and some of the measuring equipment for comparing them. The frequency modulation receiver consisted of three stages of radio-frequency amplification (at forty-one megacycles) giving a gain in voltage of about 100. This frequency was heterodyned down to six megacycles where an amplification of about 2000 was available and this frequency was in turn heterodyned down to 400 kilocycles where an amplification of about 1000 could be realized. Two current limiting systems in cascade each with a separate amplifier were used. At the time the photograph was taken the first two radio-frequency stages had been discarded.

Fig. 21

The initial tests in the early part of June surpassed all expectations. Reception was perfect on any of the antennas employed, a ten-foot wire furnishing sufficient pickup to eliminate all background noises. Suc-

Fig. 22

cessive reductions of power at the transmitter culminated at a level subsequently determined as approximately twenty watts. This gave a signal comparable to that received from the regular New York broadcast stations (except WEAF, a fifty-kilowatt station approximately forty miles away).

The margin of superiority of the frequency modulation system over amplitude modulation at forty-one megacycles was so great that it was at once obvious that comparisons of the two were principally of academic interest.

The real question of great engineering and economic importance was the comparison of the ultra-short-wave frequency modulation system with the existing broadcast service and the determination of the question of whether the service area of the existing stations could not be more effectively covered than at present. The remainder of the month was devoted to such a comparison. With the Empire State transmitter operating with approximately two kilowatts in the antenna, at all times and under all conditions the service was superior to that provided by the existing fifty-kilowatt stations, this including station WEAF. During thunderstorms, unless lightning was striking within a few miles of Westhampton, no disturbance at all would appear on the system, while all programs on the regular broadcast system would be in a hopeless condition. Background noise due to thermal agitation and tube hiss were likewise much less than on the regular broadcast system.

The work at Westhampton demonstrated that in comparing this method of transmission with existing methods two classes of services and two bases of comparisons must be used. It was found that the only type of disturbance of the slightest importance was that caused by the ignition systems of automobiles, where the peak voltage developed by the interference was greater than the carrier level. In point-to-point communication this difficulty can be readily guarded against by proper location of the receiving system, and then thermal agitation and shot effect are the principal sources of disturbance; lightning, unless in the immediate vicinity, rarely producing voltages in excess of the carrier level which would normally be employed to suppress the thermal and shot effects. Under these conditions the full effect of noise suppression is realized and comparisons can be made with precision by means of the method already described in this paper. An illustration of the practical accomplishment of this occurred at Arney's Mount, the television relay point between New York and Camden of the Radio Corporation of America. This station is located about sixty miles from the Empire State Building and the top of the tower is only a few feet below line of sight. It is in an isolated spot and the noise level is almost entirely that due to the thermal and shot effects. It was noted by C. M. Burrill of the RCA Manufacturing Company who made the observations at Arney's Mount that with fifty watts in the antenna frequency modulated (produced by a pair of UX 852 tubes), a signal-to-noise ratio of the same value as the two-kilowatt amplitude modulation transmitter (eight-kilowatt peaks) was obtained.

The power amplifier and the intermediate power amplifier of the frequency modulation transmitter is shown in Fig. 23. The signal with fifty watts output would undoubtedly have had a better noise ratio than the two-kilowatt amplitude modulation system had full deviation of seventy-five kilocycles been employed, but on the occasion it was not possible to use a deviation of greater than twenty-five kilocycles. It was also observed at the same time that when the plate voltage on the power amplifier was raised to give a power of the order of 200 watts in the antenna a better signal-to-noise ratio was obtained than

Fig. 23

that which could be produced by the two-kilowatt amplitude modulation. A casual comparison of the power amplifier stages of the frequency modulation transmitter shown in Fig. 23 with the water-cooled power amplifier and modulation stages of the Empire State transmitter is more eloquent than any curves which may be shown herein.

In the broadcast service no such choice of location is possible and a widely variable set of conditions must be met. Depending on the power at the transmitter, the elevation of the antenna, the contour of the intervening country, and the intensity of the interference there will be a certain distance at which peaks of ignition noise become greater than the carrier. The irregularity and difficulty of reproduction of these disturbances require a different method of comparison which will be hereinafter described.

As the site at Westhampton, which was on a section of the beach remote from man-made static, was obviously too favorable a site, a new one was selected in Haddonfield, New Jersey, and about the end of June the receiving apparatus was moved there and erected at the home of Harry Sadenwater. Haddonfield is located about eighty-five miles from New York in the vicinity of Camden, New Jersey, and is over 1000 feet below line of sight of the top of the Empire State Building in New York. Although the field strength at Haddonfield was considerably below that at Westhampton Beach, good reception was obtained almost immediately, the sole source of noise heard being ignition noise

Fig. 24

from a few types of cars in the immediate vicinity of the antenna, or lightning striking within a few miles of the station. At this distance fading made its appearance for the first time, a rapid flutter varying in amplitude three- or four-to-one being frequently observable on the meters. The effect of it was not that of the selective fading so well known in present-day broadcasting. Very violent variations as indicated by the meters occurred without a trace of distortion being heard in the speaker. During a period of over a year in which observations have been made at Haddonfield, but two short periods of fading have been observed where the signal sank to a level sufficient to bring in objectionable noise, one of these occurring prior to an insulation failure at the transmitter.

It is a curious fact that the distant fading, pronounced though it may be at times, is not so violent as that which may be encountered at a receiving station located within the city limits of New York. The effect, which appears to be caused by moving objects in the vicinity of the receiving antenna, causes fluctuations of great violence. In was ap-

Fig. 25

parently first observed by L. F. Jones of the RCA Manufacturing Company within a distance of half a mile of the Empire State transmitter. It occurs continually at Columbia University located about four miles from the Empire State transmitter but no injurious effect on the quality of transmission has ever been noted.

While at first, because of the lower field strength at Haddonfield and the greater prevalence of ignition disturbances, the superiority over the regular broadcast service was not so marked as at Westhampton Beach, the subsequent improvements which were instituted at both transmitting and receiving ends of the circuit have more than offset the lower signal level. Some idea of their extent may be gained by comparison of the initial and final antenna structures. Fig. 24 shows the original antenna during course of erection, a sixty-five foot mast bearing in the direction of New York permitting the use of an eight-wave length sloping wire of very useful directive properties. Fig. 25

37

shows the final form on which the results are now much better than were originally obtained with the directional wire.

During the past summer, which was marked by thunderstorms of great severity in the vicinity of Philadelphia, it was the exception when it was agreeable or even possible to listen to the nightly programs of the regular broadcast service from the fifty-kilowatt New York stations. In some of the heaviest storms when lightning was striking within the immediate vicinity of the antenna, so close in fact that the lead-in was sparking to a near-by water pipe, perfectly understandable speech could be received on the frequency modulation system, although the disturbance was sufficient to cause annoyance on a musical program; but these periods seldom lasted more than fifteen minutes when the circuit would again become quiet. On numerous occasions the Empire State signal was better than that of the fifty-kilowatt Philadelphia station WCAU located at a distance of twenty miles from Haddonfield. Likewise during periods of severe selective side-band fading in the broadcast band which occurs even from station WJZ at Bound Brook, New Jersey, some sixty miles away, no signs of this difficulty would appear on the ultra-high-frequency wave.

Some of the changes which contributed to the improvement during the past year may be of interest. The introduction of the Thompson-Rose tube permitted the radio-frequency amplification required at forty-one megacycles to be accomplished with one stage and with considerable improvement of signal-to-noise ratio. It had a further interesting result. The tubes previously used for amplifying at this frequency were those developed by the Radio Corporation for the ultra-short-wave interisland communication system in the Hawaiian Islands. On account of the relatively low amplification factor of these tubes the shot effect in the plate circuit of the first tube exceeded the disturbances due to thermal agitation in the input circuit of that tube by a considerable amount. With the acorn type tube, however, the situation is reversed, the thermal noise contributing about seventy-five per cent of the rectified output voltage.

It should be noted here by those who may have occasion to make this measurement on a frequency modulation system that it cannot be made in the ordinary way by simply mis-tuning the input circuit to the first tube. To do so would remove the carrier from the current limiter and be followed by a roar of noise. The measurement must be made with a local signal of the proper strength introduced into one of the intermediate-frequency amplifiers. Under these conditions the antenna may be mis-tuned without interfering with the normal action of the limiter and the relative amounts of noise due to the two sources may readily be segregated.

Considerable trouble was caused during the early stages of the experiments by an order of the Federal Radio Commission requiring the changing of the frequency of the Empire State transmitter from forty-four to forty-one megacycles; this necessitating the realignment of the large number of interstage transformers in the modulating equipment shown in Fig. 20 and also the retermination of the antenna. It, however, led to the application of the idea inherent in superheterodyne design.

Fig. 26

While the circuits of the old modulator were temporarily modified and work carried on, a new modulation system was designed standardizing on an initial frequency of 100 kilocycles which was then multiplied by a series of doublers up to 12,800 kilocycles. By means of a local oscillator this frequency was heterodyned down to 1708 kilocycles, the new value of input frequency to the transmitter required to produce forty-one megacycles in the antenna. Any future changes in wave length can be made by merely changing the frequency of this second oscillator. The frequencies chosen were such that a deviation of 100 kilocycles could be obtained without difficulty, because of the extra number of frequency multiplications introduced. Fig. 26 shows the two modulation systems during the process of reconstruction with arrangements for making the necessary step-by-step comparisons between them.

Much attention was paid during the year to the frequency characteristic of the transmitter, which was made substantially flat from thirty to 20,000 cycles. This required careful attention to the characteristics of the doubler and amplifier circuits of the transmitter, and to John Evans of the RCA Manufacturing Company and to T. J. Buzalski I am indebted for its accomplishment. Continuous improvement of the transmitter and antenna efficiency was effected throughout the year, but of this phase of the development R. M. Morris of the National Broadcasting Company, under whose direction the work was carried on, is better qualified to speak. As the final step, the lines connecting the transmitter with the control board of the National Broadcasting Company at Radio City, from which the test programs were usually supplied, were equalized to about 13,000 cycles, and when this had been done the quality of reception at Haddonfield was far better than that obtainable from any of the regular broadcast stations.

Interference and Fading

Reference has heretofore been made to the difficulty of comparing the amounts of interference produced in amplitude and frequency modulation systems by the transient type of disturbance, particularly when, as in ignition noise, the peaks are greater in amplitude than the signal carrier. The best method of comparison seems to be that of observing how much greater signal level from the standard signal generator must be introduced into the receiving system when it is arranged to receive amplitude modulation than is required for the same signal-to-noise ratio on a frequency modulated system. The experimental procedure of making such comparison is to change the connection of the speaker rapidly from one receiver to the other, simultaneously changing the level of the local generator until the two disturbances as perceived by the ear are equal. At all times, of course, the amplification in the amplitude modulation receiver is correspondingly changed as the signal generator level is varied to apply the same voltage to the amplitude as to the frequency modulation detector so that the audio-frequency signal level which will be produced by the two systems is the same. The square of the ratio of the two voltages of the signal generator gives the factor by which the *carrier* power of the amplitude modulated transmitter must be increased to give equal performance. While the measurement is difficult to make, the following approximations may give some idea of the magnitudes involved.

If the peak voltage of the ignition noise is twice the carrier level of the frequency modulation system, about 150 to 200 times the power must be used in the carrier of the amplitude modulation system to

40

reduce the disturbance level to the same value. When the peak voltage is five times as great, about 35 to 40 times the power in the amplitude modulation carrier is sufficient to produce equality.[17] These observations have been checked aurally and by the oscilloscope. The results of measurements where the disturbances are due solely to the thermal and shot effects have been compared to those obtained with the method previously described and are found to check with it. The chief value of this method of measurement, however, lies in the ability to predict with certainty the signal level required to suppress all ignition noise. An experimental determination made at Haddonfield shows that a signal introduced from the local generator which produces at the current limiter ten times the voltage of the Empire State signal is sufficient to suppress the disturbance caused by the worst offender among the various cars tested. These cars were located as closely as possible to the doublet antenna shown in Fig. 25, the distance being about forty feet. The increase in field strength necessary to produce this result can be readily obtained by an increase in the transmitter power to twenty or twenty-five kilowatts and the use of a horizontally directional antenna array. An increase in the field strength of three or four to one by means of an array is within the bounds of engineering design so that the practical solution of the problem of this type of interference is certainly at hand up to distances of one hundred miles.

So also is the solution of the problem at its source. It has been determined experimentally that the introduction of 10,000 ohms (a value of resistance which is not injurious to motor performance) into the spark plug and distributor leads of the car referred to eliminates the interference with the Empire State signal.

Since active steps are now being taken by the manufacturers of motor cars to solve the more difficult general problem, the particular one of interference with sets located in the home will thus automatically disappear. The problem of eliminating the disturbance caused by an automobile ignition system in a receiving set whose antenna is a minimum of fifty feet away from the car is obviously a much simpler one than that of eliminating the interference in a receiver located in the car or in another car a few feet away.

During the course of the experimental work in the laboratory a very striking phenomenon was observed in the interference characteristics between frequency modulation systems operating within the same wave band. The immunity of a frequency modulation system from interference created by another frequency modulated transmission is of the

[17] Linear detection was used in the amplitude modulation receiver but no limiting was employed.

same order of magnitude as the immunity with regard to tube noises. This property merits the most careful study in the setting up of a broadcast system at those wave lengths at which the question of inter-station interference is a major factor. It is well known that when the carriers of two amplitude modulated transmitters are sufficiently close in frequency to produce an audible beat that the service range of each of them is limited to that distance at which the field strength of the distant station becomes approximately equal to one per cent of the field strength of the local station. As a consequence of this, the service area of each station is very greatly restricted; in fact the service area of the two combined is but a small percentage of the area which is rendered useless for that frequency due to the presence thereon of the two interfering stations. With the wide band frequency modulation system, however, interference between two transmissions does not appear until the field strength of the interfering station rises to a level in the vicinity of fifty per cent of the field strength of the local one. The reason for this lies in the fact, that while the interfering signal in beating with the current of the local station under such conditions may be producing a fifty per cent change in the voltage applied to the current limiter, the system is substantially immune to such variations in amplitude. The only way in which the interfering signal can make its presence manifest is by cross modulation of the frequency of the local signal. Since, under the conditions, this cross modulation produces less than a thirty-degree phase shift and since the characteristics of the wide band receiver are such that, at least within the range of good audibility, thousands of degrees of phase shift are necessary to produce full modulation, it is clear that a thirty-degree phase shift will not produce very much of a rectified output. For example, assuming two unmodulated carriers are being received, that their amplitudes have a ratio of two to one, and that their frequencies differ by 1000 cycles, then for a system having a wide band (of the order of 150,000 cycles) the modulation produced by the interaction of the two carriers would be of the order of one per cent of that produced by full modulation of the stronger carrier. This example, however, represents perhaps the worst possible condition as during modulation of either station, with the proper type of conversion system, the aural effect of the disturbance is greatly reduced. The whole problem of interference between unmodulated carriers may, however, be entirely avoided by separating them in frequency by an amount beyond the audible range. Hence it follows that with two wide band frequency modulated transmitters occupying the same frequency band that only the small area located midway between the two wherein the field strength of one station is less than twice the field strength of the other will be rendered useless for reception of either station. This area may well be less than ten per cent of the total area. Even in this area reception may be effected as a receiving station located within it has only to erect directional aerials having a directivity of two to one to receive either station. The two-to-one ratio of field strength which has been referred to as the ratio at which interference appears is not by any means the limit but rather one which can be realized under practically all conditions. Better ratios than this have been observed, but the matter is not of any great importance since by the use of the directional antennas referred to it becomes possible to cover the sum of the areas which may be effectively covered by each station operating alone, subject only to the limitations of the noise level. The problem of the interference due to overlapping has been completely wiped out. One precaution only should be observed—the unmodulated carriers should be offset in frequency by an amount beyond the audible limit.

In the above analysis it has been assumed, of course, that the distance between stations has been selected so that the "no-mans land" between stations is not sufficiently distant from either one to be within the zone where any large amount of fading occurs. If the distance between stations is such that the signal strength varies appreciably with time then the directivity of the receiving antennas must be greater than two to one.

DIFFICULTIES AND PRECAUTIONS

The principles which have been described herein were successfully applied only after a long period of laboratory investigation in which a series of parasitic effects that prevented the operation of the system were isolated and suppressed. The more important of these effects, which will be of interest to those who may undertake work in this field, will be referred to briefly.

It was observed in the early work in the laboratory that it was at times impossible to secure a balance in the detector system, and that the amplitudes of the currents in the rectifiers varied in very erratic fashion as the frequency of the first heterodyne was changed. Under these conditions it was not possible to produce any appreciable noise suppression. The effect varied from day to day and the cause defied detection for a long period of time. Ultimately the presence of two side frequencies in the detector circuits was discovered, one of these frequencies lying above and the other below the unmodulated intermediate frequency by an amount equal to the initial crystal frequency of the transmitter. It was then discovered that the trouble had its origin in the transmitting system and that a current having the fundamental frequency of the crystal, (in the present case 57.33 kilocycles), passed through the first doubler circuits in such phase relation to the doubled frequency as to modulate the doubled frequency at a rate corresponding to 57.33 kilocycles per second. This modulation of frequency then passed through all the transmitter doubler stages, increasing in extent with each frequency multiplication and appearing finally in the forty-four-megacycle output as a fifty-seven-kilocycle frequency modulation of considerable magnitude. In the first doubler tank circuit of the transmitter a very slight change in the adjustment of the tuning of the circuit produced a very great change in the magnitude of this effect. A few degrees shift in the tuning of the first doubler tank condenser, so small that an almost unnoticeable change in the plate current of the doubler occurred, would increase the degree of the modulation to such extent as to make the first upper and lower side frequencies in the forty-four-megacycle current greater than the carrier or mid-frequency current (when no audio modulation was applied). Under such conditions the proper functioning of the receiving system was impossible.

The delay in uncovering this trouble lay in the fact that it was obscured by the direct effect of harmonics from the transmitter doubler stages which had to be set up in an adjoining room and by the numerous beats which can occur in a double intermediate-frequency superheterodyne. To these effects were added an additional complication caused by the presence of harmonics in the circuits of the selective system resulting from the action of the limiter which the filtering arrangements did not entirely remove. The coincidence of one of these harmonics with the natural period of one of the inductances in the branch circuits likewise interfered with the effectiveness of the noise suppression. The causes of all these spurious effects were finally located and necessary steps taken to eliminate them.

With the removal of these troubles a new one of a different kind came to light, and for a time it appeared that there might be a very serious fundamental limitation in the phase shifting method of generating frequency modulation currents. There was found to be in the output of the transmitter at forty-four megacycles a frequency modulation which produced a noise in the receiver similar to the usual tube hiss. The origin of it was traced to the input of the first doubler or the output of the crystal oscillator where a small deviation of the initial frequency was produced by disturbances originating in these circuits. While the frequency shift in this stage must have been very small, yet on account of the great.amount of frequency multiplication (of the order of 800 times) it became extremely annoying in the receiver; in fact for low levels of receiver noise that noise which originated in the transmitted wave was by far the worse. For a time it seemed as though the amount of frequency multiplication which could be used in the transmitter was limited by an inherent modulation of the frequency of the oscillator by disturbances arising in the tube itself. The proper proportioning of the constants of the circuits, however, reduced this type of disturbance to a point where it was no longer of importance and frequency multiplications as high as 10,000 have since been effectively used. On account of the very large amount of frequency multiplication, any troubles in these low-frequency circuits caused by noisy grid leaks, improper bypassing of power supply circuits, or reaction of one circuit upon another become very much more important than they would normally be. Difficulties of all these kinds were encountered, segregated, and eliminated.

Another source of trouble was discovered in the correction system. Because of the range in frequency required, particularly in multiplex work where thirty to 30,000 cycles was frequently used, the output voltage of the correction system at the higher frequencies became very much less than the input voltage, hence any leakage or feed-forward effect due to coupling through the power supply circuits developed a voltage across the output much higher than that required by the inverse frequency amplification factor as determined by the correction network. Hence, the frequency swing for the upper frequencies of modulation would frequently be several hundred per cent greater than it should be. Likewise, at the lower frequency end of the scale various reactions through the power supply were very troublesome. All these effects, however, were overcome and the correction system designed so that its accuracy was within a few per cent of the proper value.

From the foregoing it might be assumed that the transmitting and receiving apparatus of this system are inherently subject to so many new troubles and complications that their operation becomes impracticable for ordinary commercial applications. Such is not the case. The difficulties are simply those of design, not of operation. Once the proper precautions are taken in the original design these difficulties never occur, except as occasioned by mechanical or electrical failure of material. During the period of over a year in which the Empire State transmitter was operated, only two failures chargeable to the modulating system occurred. Both were caused by broken connections. Even the design problems are not serious as methods are now available for detecting the presence of any one of the troubles which have been here enumerated.

These troubles were serious only when unsegregated and en masse they masked the true effects and made one wonder whether even the laws of electrical phenomena had not been temporarily suspended.

Multiplex Operation

During the past year, two systems of multiplexing have been operated successfully between New York and Haddenfield and it has been

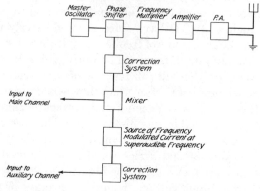

Fig. 27

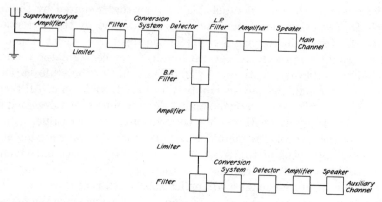

Fig. 28

found possible to transmit simultaneously the red and blue network programs of the National Broadcasting Company, or to transmit simultaneously on the two channels the same program. This last is much the simpler thing to accomplish as the cross-talk problem is not a serious one. The importance of multiplexing in point-to-point communication services has long been recognized. In broadcasting there are several applications which, while their practical application may be long deferred, are clearly within view.

Two general types of multiplexing were used. In one type a current of superaudible frequency is caused to modulate the frequency of the transmitted wave. The frequency at which the transmitted wave is caused to deviate is the frequency of this current and the extent of the deviation is varied in accordance with modulation of the amplitude of the superaudible frequency current. At the receiver detection is accom-

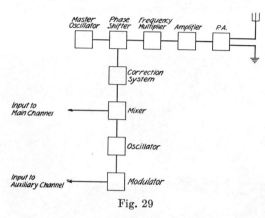

Fig. 29

plished by separating the superaudible current and its component modulations from the rectified audible frequency currents of the main channel and reproducing the original modulating current from them by a second rectification. The general outline of the system is illustrated in Figs. 27 and 28. The setting of the levels of the main and auxiliary channels must be made in this system of modulation with due regard

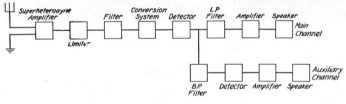

Fig. 30

to the fact that the deviation of the transmitted wave produced by the superaudible frequency current of the second channel is a variable one and changes between the limits of zero and double the unmodulated deviation.

In the second method of multiplexing a superaudible current produces a frequency modulation of the transmitted wave of constant deviation, the rate of the deviation being varied in accordance with the frequency of the superaudible current and modulation being produced by varying the frequency of this auxiliary current and thereby the rate at which the superimposed modulation of frequency of the transmitted wave changes. The operations which must be carried out at the receiver are the following: After suitable amplification, limiting, and filtering, an initial conversion and rectification produces in the output of the detector the audible frequencies of the main channel and a super-audible constant amplitude variable frequency current. This last is selected by means of a band-pass filter, passed through a second con-

Fig. 31

version system to translate the changes in the frequency into variations of amplitude, and then rectified to recreate the initial modulating current of the auxiliary channel. The general arrangment of the system is illustrated in Figs. 29 and 30. This latter method of multiplexing has obvious advantages in the reduction of cross modulation between the channels and in the fact that the deviation of the transmitted wave produced by the second channel is constant in extent, an advantage being gained thereby which is somewhat akin to that obtained by frequency, as compared to amplitude, modulation in simplex operation. The subject of the behavior of these systems with respect to interference of various sorts is quite involved and will be reserved for future treatment as it is beyond the scope of the present paper.

The final arrangement of the modulating equipment installed at the Empire State station is illustrated in Figs. 31 and 32. The main channel apparatus is shown on the five tables located on the right side of the room. The vertical rack in the left center contains three channels for transmitting facsimile by means of the amplitude modulation method of multiplexing already described. In Fig. 32, located on the four tables on the left of the room is shown the auxiliary channel of the frequency modulation type already described. The comparatively

Fig. 32

low frequency of this channel was obtained by the regular method of phase shifting and frequency multiplication, the frequency multiplication being carried to a high order and the resultant frequency modulated current heterodyned down to twenty-five kilocycles (mid-frequency). A deviation up to ten kilocycles was obtainable at this frequency.

The receiving apparatus located at Haddonfield is illustrated in Figs. 33 and 34. Fig. 33 shows the modified Westhampton receiver and Fig. 34 the multiplex channels of the receiver. The vertical rack to the right holds a three-channel receiver of the amplitude modulation type. The two panels in the foreground constitute the frequency modulation type of auxiliary channel.

Some of the practical results may be of interest. It was suggested by C. J. Young of the RCA Manufacturing Company that it might be possible to transmit simultaneously a facsimile service at the same time that a high quality broadcast program was being transmitted. With the assistance of Mr. Young and Maurice Artzt this was accomplished

Fig. 33

over a year ago between New York and Haddonfield, New Jersey, the two services operating without interference or appreciable loss of efficiency at the distance involved. Two additional channels, a synchronizing channel for the facsimile and a telegraph channel, were also operated. The character of the transmission is illustrated in Fig.

Fig. 34

35, which shows a section of the front page of the *New York Times*, This particular sheet was transmitted under considerable handicap at the transmitter as due to a failure of the antenna insulator on the forty-one-megacycle antenna it had become necessary to make use of the sixty-megacycle antenna for the forty-one-megacycle transmission. It is an interesting comment on the stability of the circuits that all four were kept in operation at the transmitter by one man, Mr. Buzalski,

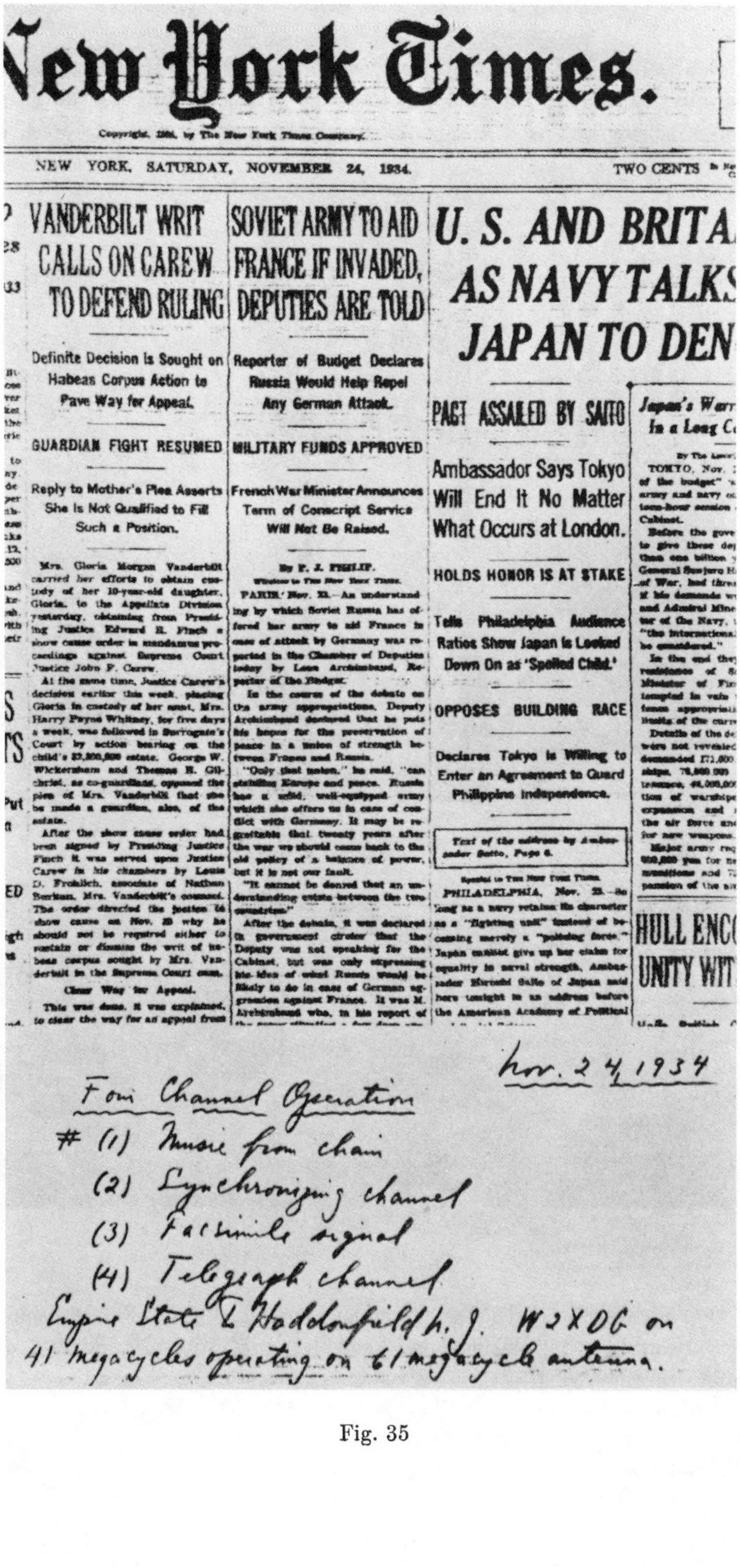

nov. 24 1934

Four Channel Operation
(1) Music from chain
(2) Synchronizing channel
(3) Facsimile signal
(4) Telegraph channel
Empire State & Haddonfield N. J. W2XDG on
41 megacycles operating on 61 megacycle antenna.

Fig. 35

who was alone in the station on that day. The combined sound and facsimile transmission has been in successful operation for about a year, practically perfect copy being obtained throughout the period of the

Fig. 36

Fig. 37

severe atmospheric disturbances of the past Summer. The subject of this work and its possibilities can best be handled by Mr. Young, who is most familiar with it.

Acknowledgment

On account of the ramifications into which this development entered with the commencement of the field tests many men assisted in this work. To some reference has already been made.

I want to make further acknowledgment and express my indebtedness as follows:

To the staff of the National Broadcasting Company's station W2XDG for their help in the long series of field tests and the conducting of a large number of demonstrations, many of great complexity, without the occurrence of a single failure;

To Mr. Harry Sadenwater of the RCA Manufacturing Company for the facilities which made possible the Haddonfield tests and for his help with the signal-to-noise ratio measurements herein recorded;

To Mr. Wendell Carlson for the design of many of the transformers used in the modulating equipment;

To Mr. M. C. Eatsel and Mr. O. B. Gunby of the RCA Manufacturing Company for the sound film records showing the comparison, at Haddonfield, of the Empire State transmission with that of the regular broadcast service furnished by the New York stations;

To Mr. C. R. Runyon for his development of the two-and-one-half-meter transmitters and for the solution of the many difficult problems involved in the application of these principles of modulation thereto;

To Mr. T. J. Styles and particularly to Mr. J. F. Shaughnessy, my assistants, whose help during the many years devoted to this research has been invaluable.

Conclusion

The conclusion is inescapable that it is technically possible to furnish a broadcast service over the primary areas of the stations of the present-day broadcast system which is very greatly superior to that now rendered by these stations. This superiority will increase as methods of dealing with ignition noise, either at its source or at the receiver, are improved.

Appendix

Since the work which has been reported in this paper on forty-one megacycles was completed attention has been paid to higher frequencies. On the occasion of the delivery of the paper a demonstration of transmission on 110 megacycles from Yonkers to the Engineering Societies Building in New York City was given by C. R. Runyon, who described over the circuit the transmitting apparatus which was used. A brief description of this transmitter is reproduced here.

The power delivered to the antenna was approximately 100 watts at 110 megacycles and the deviation (one half total swing) used during the demonstration was under 100 kilocycles. Fig. 36 illustrates the modulating equipment for this transmitter and the low power frequency multiplication stages. Fig. 37 shows the higher power frequency multiplier and power amplifier stages of the transmitter.

The rack shown in Fig. 36 consists of six panels. Panel number one at the top contains the correction system. Panel number two contains the master oscillator of 100 kilocycles and the modulator circuits. Panel number three contains a pair of doublers for multiplying the 100-kilocycle frequency to 400 kilocycles and the necessary filtering means for avoiding the modulation of the currents in the succeeding doubler stages by the 100-kilocycle oscillator current. Panel number four contains the doubling apparatus for raising the frequency to 3200 kilocycle and panel number five the multipliers for raising it to 12,800 kilocycles. Panel number five also contains a heterodyning and conversion system for beating the 12,800 kilocycles down to 2292 kilocycles. Panel number six contains a doubler for raising this to 4584 kilocycles and an amplifier for increasing the level sufficiently to drive the succeeding power stage. The output of this amplifier is fed through a transmission line to the metal box at the extreme right of Fig. 36 which contains a series of doublers and amplifiers for increasing the level and raising the frequency to 36,672 kilocycles. Adjacent to this box is a second box which contains a fifty-watt amplifier. This amplifier drives a tripler located in the third box and the tripler in turn drives the power amplifier located at the extreme left at 110 megacycles. The transmitter circuits were designed for total frequency swing of 500 kilocycles and may be effectively so operated. Because of the limitation of the receiver available at that time the demonstration was carried out with a swing of 200 kilocycles.

PAPER NO. 2

Reprinted from *Proc. IRE*, Vol. 25, No. 4, pp. 472–514, April 1937

FREQUENCY MODULATION NOISE CHARACTERISTICS*

By

MURRAY G. CROSBY

(RCA Communications, Inc., Riverhead, L. I., New York)

Summary.—Theory and experimental data are given which show the improvements in signal-noise ratio effected by frequency modulation over amplitude modulation. It is shown that above a certain carrier-noise ratio in the frequency modulation receiver which is called the "improvement threshold," the frequency modulation signal-noise ratio is greater than the amplitude modulation signal-noise ratio by a factor equal to the product of a constant and the deviation ratio (the deviation ratio is equal to the ratio between the maximum frequency deviation and the audio modulation band width). The constant depends upon the type of noise, being slightly greater for impulse than for fluctuation noise. In frequency modulation systems with high deviation ratios, a higher carrier level is required to reach the improvement threshold than is required in systems with low deviation ratios; this carrier level is higher for impulse than for fluctuation noise. At carrier-noise ratios below the improvement threshold, the peak signal-noise ratio characteristics of the frequency modulation receiver are approximately the same as those of the amplitude modulation receiver, but the energy content of the frequency modulation noise is reduced.

An effect which is called "frequency limiting" is pointed out in which the peak value of the noise is limited to a value not greater than the peak value of the signal. With impulse noise this phenomenon effects a noise suppression in a manner similar to that in the recent circuits for reducing impulse noise which is stronger than the carrier in amplitude modulation reception.

When the power gain obtainable in certain types of transmitters by the use of frequency modulation is taken into account, the frequency modulation improvement factors are increased and the improvement threshold is lowered with respect to the carrier-noise ratio existing in a reference amplitude modulation system.

INTRODUCTION

IN A previously published paper,[1] the propagation characteristics of frequency modulation were considered. Prior to, and during these propagation tests, signal-noise ratio improvements effected by frequency modulation were observed. These observations were made at an early stage of the development work and were investigated by experimental and theoretical methods.

It is the purpose of this paper to consider that phase of the frequency modulation development work by RCA Communications,

---

* Decimal classification: R148×R270. Original manuscript received by the Institute, November 23, 1936. Presented as part of a paper on "Propagation and characteristics of frequency modulated waves," before New York meeting, January 8, 1936. Revised and presented in full before Chicago Section, September 11, 1936.

[1] Murray G. Crosby, "Frequency modulation propagation characteristics," PROC. I.R.E., vol. 24, pp. 898–913; June, (1936).

Inc., in which the signal-noise characteristics of frequency modulation are studied. The theory and experimental work consider the known systems of frequency modulation including that independently developed by E. H. Armstrong.[2]

TABLE OF SYMBOLS

C = carrier peak voltage.

C/N = theory: Ratio between peak voltage of carrier and instantaneous peak voltage of the noise in the frequency modulation receiver. Experiment: Ratio between peak voltage of carrier and *maximum* instantaneous peak voltage of the noise.

C/n = ratio between the peak voltage of the carrier and the peak voltage of the noise component.

C_a/N_a = carrier-noise ratio in the amplitude modulation receiver.

F_a = maximum audio frequency of modulation band.

F_c = carrier frequency.

F_d = peak frequency deviation due to applied modulation.

F_{dn} = peak frequency deviation of the noise.

F_1 = intermediate-frequency channel width.

F_m = modulation frequency.

F_n = frequency of noise resultant or component.

F_d/F_a = deviation ratio.

K = slope filter conversion efficiency.

M = modulation factor of the amplitude modulated carrier.

M_f = modulation factor at the output of the sloping filter.

M_{fn} = modulation factor at the output of the sloping filter when noise modulates the carrier.

N = instantaneous peak voltage of the noise.

n = peak voltage of the noise component.

N_a = noise peak or root-mean-square voltage at amplitude modulation receiver output.

N_f = noise peak or root-mean-square voltage at frequency modulation receiver output.

$p = 2\pi F_m$.

S_a = signal peak or root-mean-square voltage at amplitude modulation receiver output.

S_f = signal peak or root-mean-square voltage at frequency modulation receiver ouput.

$\omega = 2\pi F_c$.

[2] Edwin H. Armstrong, "A method of reducing disturbances in radio signaling by a system of frequency modulation," PROC. I.R.E., vol. 24, pp. 689–740; May, (1936).

$$\omega_n = 2\pi F_n.$$
$$\omega_{na} = (\omega - \omega_n) = 2\pi(F_c - F_n) = 2\pi F_{na}.$$
$$Z = C/n + n/C.$$
$\phi(t) =$ phase variation of noise resultant as a function of time.

Theory

In the following analysis, frequency modulation is studied by comparing it with the familiar system of amplitude modulation. In order to do this, the characteristics of frequency modulation reception are analyzed so as to make possible the calculation of the signal-noise ratio improvement effected by frequency modulation over amplitude modulation at various carrier-noise ratios.[3] The amplitude modulation standard of comparison consists of a double side-band system having the same audio modulation band as the frequency modulation system and producing the same carrier at the receiver. Differences in transmitter power gain due to frequency modulation are then considered separately. The frequency modulation reception process is analyzed by first considering the components of the receiver and the manner in which they convert the frequency modulated signal and noise spectrum into an output signal-noise ratio.

The Frequency Modulation Receiver

The customary circuit arrangement used for the reception of frequency modulation is shown in the block diagram of Fig. 1. The intermediate-frequency output of a superheterodyne receiver is fed through a limiter to a slope filter or conversion circuit which converts

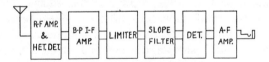

Fig. 1—Block diagram of a frequency modulation receiver.

the frequency modulation into amplitude modulation. This amplitude modulation is then detected in the conventional amplitude modulation manner. The audio-frequency amplifier is designed to amplify only the modulation frequencies; hence it acts as a low-pass filter which rejects noise frequencies higher than the maximum modulation frequency.

[3] Throughout this paper, carrier-noise ratio will refer to the ratio measured at the output of the intermediate-frequency channel. Signal-noise ratio will refer to that measured at the output of the receiver and will depend upon the depth of modulation as well as upon the carrier strength.

The purpose of the limiter is to remove unwanted amplitude modulation so that only the frequency modulation component of the signal will be received. It may take the form of an overloaded amplifier tube whose output cannot rise above a certain level regardless of the input. Care must also be exercised to insure that the output of the overloaded amplifier does not fall off as the input is increased since this would introduce amplitude modulation of reverse phase, but of equally undesirable character.

The main requirement of the conversion circuit for converting the frequency modulation into amplitude modulation is that it slope linearly from a low value of output at one side of the intermediate-frequency channel to a high value at the other side of the channel. To do this, an off-tuned resonant circuit or a portion of the characteristic of

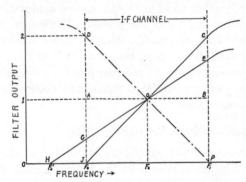

Fig. 2—Ideal sloping filter characteristics.

one of the many forms of wave filters may be utilized. The ideal slope filter would be one which gave zero output at one side of the channel, an output of one voltage unit at carrier frequency, and an output of two units at the other side of the channel. Such a characteristic is given by the curve JOC of Fig. 2. From this curve it is easily seen that if the frequency is swung between the limits F_0 and F_1, about the mean frequency F_c, the output of the filter will have an amplitude modulation factor of unity. The modulation factor for a frequency deviation, F_d, will be given by

$$M_f = \frac{F_d}{(F_1 - F_c)} = \frac{F_d}{(F_c - F_0)} = \frac{2F_d}{F_i} \tag{1}$$

where F_i = intermediate-frequency channel width.

When the converting filter departs from the ideal characteristic in the manner of the filter of curve $HGOE$ of Fig. 2, the modulation factor produced by a given frequency deviation is reduced by a factor equal

58

to the ratio between the distances AG and AJ or BE and BC. A convenient term for this reduction factor of the filter is "conversion efficiency" of the filter. Taking into account this conversion efficiency, the modulation factor for a frequency deviation F_d becomes

$$M_f = \frac{2KF_d}{F_i} \qquad (2)$$

where $K = AG/AJ = BE/BC = $ conversion efficiency of the filter.

A low conversion efficiency may be used as long as the degree of limiting is high enough to reduce the amplitude modulation well below the level of the converted frequency modulation. This is true since lowering the conversion efficiency reduces the output of the noise in the same proportion as the signal as long as no amplitude modulation is present. Hence the signal-noise *ratio* is unimpaired and the only effect is a reduction of the audio gain by the factor K. If insufficient limiting is applied so that the output of the limiter contains appreciable amplitude modulation, a filter with a high conversion efficiency is desirable so that the amplitude modulation noise will not become comparable to the frequency modulation noise and thereby increase the resultant noise.

A push-pull, or "back-to-back" receiver may be arranged by providing two filters of opposite slope and separately detecting and combining the detected outputs in push-pull so as to combine the audio outputs in phase. Another slope filter having a characteristic as shown by the dot-dash line DOP in Fig. 2 would then be required.

A further type of receiver in which amplitude modulation is also balanced out may be arranged by making one of the slope filter circuits of the above-mentioned back-to-back type of receiver a flat-top circuit for the detection of amplitude modulation only. The sloping filter channel then detects both frequency and amplitude modulation; the flat-top channel detects only amplitude modulation. When these two detected outputs are combined in push-pull, the amplitude modulation is balanced out and the frequency modulation is received. This type of detection, as well as that in which opposite slope filters are used, has the limitation that the balance is partially destroyed as modulation is applied. However, if a limiter is used, the amplitude modulation is sufficiently reduced before the energy reaches the slope filters; consequently, for purposes of removing amplitude modulation, the balancing feature is not a necessity.

Noise Spectrum Analysis

The first step in the procedure to be followed here in determining the noise characteristics of the frequency modulation receiver will be

59

to determine mathematically the fidelity with which the noise is transmitted from the radio-frequency branch, in which it originates, to the measuring instrument as a function of frequency. To do this, the waves present at the receiver input will be assumed to be the frequency modulated carrier and the spectrum of noise voltages. This wave and spectrum will be combined into a single resultant whose amplitude and phase are functions of the constants of the component waves. The resultant will then be "mathematically" passed through the limiter to remove the amplitude modulation. From a determination of the instantaneous frequency of the resultant, the peak frequency deviation effected by the noise will be found. A single noise component of arbitrary frequency will then be substituted for the resultant of the noise spectrum, and the modulation factor at the output of the converting filter will be obtained. This noise component will then be varied in frequency to determine the over-all transmission of the receiver in terms of the modulation factor at the sloping filter output. The area under the curve representing the square of this over-all transmission will then be determined. By comparing this area with the corresponding area for an amplitude modulation receiver under equivalent conditions, and taking into consideration the pass band of the intermediate- and audio-frequency channels, a comparison will be obtained between the average noise powers, or the average root-mean-square noise voltages from the two receivers.[4]

The peak voltage characteristics of the two receivers will be compared for fluctuation noise by a correlation of known crest factors with the root-mean-square characteristics. (Crest factor = ratio between the peak and root-mean-square voltages.) The peak voltage characteristics of impulse noise will be determined by a separate consideration of the effect of the frequency modulation over-all transmissions upon the peak voltage of this type of noise.

After a comparison between the noise output voltages from the frequency and amplitude modulation receivers has been obtained, the respective signal output voltages will be taken into consideration so that the improvement in signal-noise ratio may be determined.

In the process of determining the over-all transmission of the noise, the frequency modulated wave may be expressed by

$$e_s = C \sin \left\{ \omega t + (F_d/F_m) \cos pt \right\} \tag{3}$$

[4] Stuart Ballantine, "Fluctuation noise in radio receivers," Proc. I.R.E., vol. 18, pp. 1377–1387; August, (1930). In this paper, Ballantine shows that the average value of the square of the noise voltage " . . . is proportional to the area under the curve representing the square of the over-all transimpedance (or of the transmission) from the radio-frequency branch in which the disturbance originates to the measuring instrument as a function of frequency . . . ."

where $C=$ carrier peak voltage, $\omega=2\pi F_c$, $F_c=$ carrier frequency, $F_d=$ applied frequency deviation, $p=2\pi F_m$, and $F_m=$ modulation frequency. The noise spectrum may be expressed by its resultant,[5]

$$e_n = N \sin (\omega_n t + \phi(t)) \tag{4}$$

where $N=$ instantaneous peak voltage of the noise (a function of time). $\phi(t)$ takes into account the fact that the noise resultant is phase modulated, as would be the case with the resultant of a spectrum of many noise voltages. $\omega_n=2\pi F_n$, $F_n=$ frequency of the noise resultant.

The signal voltage given by (3) and the noise voltage given by (4) may be combined by vector addition to give

$$e = \sqrt{C^2+N^2+2CN \cos\left\{(\omega-\omega_n)t-\phi(t)+\frac{F_d}{F_m}\cos pt\right\}}$$

$$\sin\left[\omega t+\frac{F_d}{F_m}\cos pt+\tan^{-1}\frac{\sin\left\{(\omega-\omega_n)t-\phi(t)+\frac{F_d}{F_m}\cos pt\right\}}{\frac{C}{N}+\cos\left\{(\omega-\omega_n)t-\phi(t)+\frac{F_d}{F_m}\cos pt\right\}}\right]. \tag{5}$$

When the resultant wave given by (5) is passed through the limiter in the frequency modulation receiver, the amplitude modulation is removed. Hence the amplitude term is reduced to a constant and the only part of consequence is the phase angle of the wave. The rate of change of this phase angle, or its first derivative, is the instantaneous frequency of the wave. Taking the first derivative and dividing by 2π to change from radians per second to cycles per second gives

$$\frac{d\left[\omega t+\frac{F_d}{F_m}\cos pt+\tan^{-1}\dfrac{\sin\left\{\omega_{na}t-\phi(t)+\frac{F_d}{F_m}\cos pt\right\}}{\frac{C}{N}+\cos\left\{\omega_{na}t-\phi(t)+\frac{F_d}{F_m}\cos pt\right\}}\right]}{dt}\times\frac{1}{2\pi}$$

$$=f=F_c-F_d \sin pt-\frac{\left(F_{na}-\dfrac{1}{2\pi}\dfrac{d\phi(t)}{dt}-F_d \sin pt\right)}{\dfrac{\frac{C}{N}+\cos\left\{\omega_{na}t-\phi(t)+\frac{F_d}{F_m}\cos pt\right\}}{\frac{N}{C}+\cos\left\{\omega_{na}t-\phi(t)+\frac{F_d}{F_n}\cos pt\right\}}+1} \tag{6}$$

[5] John R. Carson, "The reduction of atmospheric disturbances," Proc. I.R.E., vol. 16, pp. 967–975; July, (1928).

in which $\omega_{na} = (\omega - \omega_n) = 2\pi(F_c - F_n) = 2\pi F_{na}$.

Equation (6) gives the instantaneous frequency of the resultant wave consisting of the signal wave and the noise resultant voltage. From this equation the signal and noise frequency deviations may be obtained. In order to determine the over-all transmission with respect to the various components in the noise spectrum, a single component of noise, with constant amplitude and variable frequency, will be substituted for the resultant noise voltage given by (4). This makes N equal to n, which is not a function of time, and $\phi(t)$ equal to zero. Making these changes in (6) gives

$$f = F_c - F_d \sin pt - \frac{(F_{na} - F_d \sin pt)}{\dfrac{\dfrac{C}{n} + \cos\left\{\omega_{na}t + \dfrac{F_d}{F_m}\cos pt\right\}}{\dfrac{n}{C} + \cos\left\{\omega_{na}t + \dfrac{F_d}{F_m}\cos pt\right\}} + 1}. \qquad (7)$$

The equations for the instantaneous frequency, given by (6) and (7), show the manner in which the noise combines with the incoming carrier to produce a frequency modulation of the carrier. From these equations the frequency deviations of the noise may be determined, and from the frequency deviations the modulation factor at the output of the sloping filter may be found. Hence the over-all transmission may be obtained in terms of the modulation factor at the output of the sloping filter for a given carrier-noise ratio. When the carrier-noise ratio is high, (6) and (7) simplify so that calculations are fairly easy. When the carrier-noise ratio is low, the equations become involved to a degree which discourages quantitative calculations.

High Carrier-Noise Ratios

When C/n is large compared to unity, and the applied modulation on the frequency modulated wave is reduced to zero ($F_d = 0$), (7) reduces to

$$f = F_c - \frac{n^2}{C^2} F_{na} - \frac{n}{C} F_{na} \cos \omega_{na}t. \qquad (8)$$

From (8) the effective peak frequency deviation of a single noise component of the spectrum is

$$F_{dn} = \frac{nF_{na}}{C}\left(\frac{n}{C} + 1\right). \qquad (9)$$

But, since n/C is negligible compared to unity,

$$F_{dn} = \frac{nF_{na}}{C}.$$ (10)

When this value of frequency deviation is substituted in (1) to find the modulation factor[6] of the energy at the output of the sloping filter, the following results:

$$M_{fn} = \frac{n}{C} \times \frac{2F_{na}}{F_i}.$$ (11)

Equation (11) shows that the modulation factor of the noise is inversely proportional to the carrier-noise ratio and directly proportional to the ratio between the noise audio frequency and one half the intermediate-frequency channel width. When this equation is plotted with the noise audio frequency, F_{na}, as a variable and the modulation factor as the ordinate, the audio spectrum obtained for the detector output is like that of the triangular spectrum OBA in Fig. 3. Such a spectrum would be produced by varying F_n through the range between the upper and lower cutoff frequencies of the intermediate-frequency channel. The noise amplitude would be greatest at a noise audio frequency equal to one half the intermediate-frequency channel width. At this noise audio frequency, the ratio $2F_{na}/F_i$ is equal to unity and the modulation factor becomes equal to n/C. If the detector output is passed through an audio system having a cutoff frequency F_a, the maximum frequency of the audio channel governs the maximum amplitude of the spectrum. The maximum amplitude of the detector output is therefore reduced by the ratio $F_i/2 : F_a$. This can be seen by a comparison of the spectrum OBA for the detector output and the spectrum ODH for the audio channel output.

When the amplitude modulation reception process is analyzed with a carrier and noise spectrum present at the receiver input, the modulation factor of the energy fed to the detector is found to be equal to the reciprocal of the carrier-noise ratio for all of the noise frequencies in the spectrum. That is to say, the receiver transmission for amplitude modulation will be constant for all of the frequencies in the spectrum. Normally the upper cutoff frequency of the audio amplifier is equal to one half the intermediate-frequency channel width ($F_a = F_i/2$). Consequently the audio spectrum of the amplitude modulated noise fed to

---

[6] The ideal filter is used in this case since the use of a filter with a low conversion efficiency would merely require the addition of audio gain to put the frequency modulation receiver on an equivalent basis with the amplitude modulation receiver. The audio gain necessary would be equal to the reciprocal of the conversion efficiency of the sloping filter.

the detector will be the same as that at the receiver output and will be as portrayed by the rectangle *OCEH*.

The spectra of Fig. 3 show the manner in which the frequency modulation receiver produces a greater signal-noise ratio than the amplitude modulation receiver. The noise at the output of the detector of the frequency modulation receiver consists of frequencies which extend out to an audio frequency equal to one half the intermediate-frequency channel width, and the amplitudes of these components are proportional to their audio frequency. Hence in passing through the audio channel the noise is reduced not only in range of frequencies, but also in amplitude. On the other hand, the components of the signal wave are properly disposed to produce detected signal frequencies which fit into the audio channel. In the case of the amplitude modulation receiver, the amplitude of the components at the output of the audio

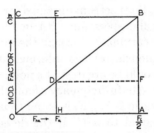

Fig. 3—Amplitude and frequency modulation receiver noise spectra. *OBA* = frequency modulation detector output. *ODH* = frequency modulation receiver output. *OCEH* = amplitude modulation receiver output.

channel is the same as that at the output of the detector since the spectrum is rectangular. Thus the frequency modulation signal-noise ratio is greater than the amplitude modulation signal-noise ratio by a factor which depends upon the relative magnitudes of the spectra *OCEH* and *ODH*. The magnitudes which are of concern are the root-mean-square and peak values of the voltage due to the spectra.

Root-Mean-Square Noise Considerations

The average noise power or average root-mean-square voltage ratio between the rectangular amplitude modulation spectrum *OCEH* and the trangular frequency modulation spectrum *ODH*, of Fig. 3, may be found by a comparison of the squared-ordinate areas of the two spectra. Thus,

$$\frac{W_a}{W_f} = \frac{\text{area } OCEH \text{ (ordinates)}^2}{\text{area } ODH \text{ (ordinates)}^2}$$

$$= \frac{\left(\dfrac{n}{C}\right)^2 F_a}{\displaystyle\int_0^{F_a} \left(\dfrac{n}{C} \times \dfrac{2F_{na}}{F_i}\right)^2 dF_{na}} = 3 \left(\frac{F_i}{2F_a}\right)^2 \qquad (12)$$

where W_a/W_f is the ratio between the amplitude modulation average noise power and the frequency modulation average noise power at the receiver outputs. The root-mean-square noise voltage ratio will be

$$\frac{N_a}{N_f} \text{ (r-m-s fluctuation)} = \sqrt{\frac{W_a}{W_f}} = \sqrt{3}\,\frac{F_i}{2F_a}. \qquad (13)$$

Equation (13) gives the root-mean-square noise voltage ratio for equal carriers applied to the two receivers. The modulation factor of the frequency modulated signal due to the applied frequency deviation, F_d, is, from (3), equal to $2F_d/F_i$. The modulation factor of the amplitude modulated signal may be designated by M and has a maximum value of 1.0. Thus the ratio between the two signals will be given by

$$\frac{S_a}{S_f} \text{ (peak or r-m-s values)} = \frac{F_i M}{2F_d} = \frac{F_i}{2F_d} \text{ (for } M = 1.0). \quad (14)$$

Dividing (13) by (14), to find the ratio between the signal-noise ratios at the outputs of the two receivers, gives

$$\frac{S_f/N_f}{S_a/N_a} \text{ (r-m-s values)} = \sqrt{3}\,\frac{F_d}{F_a}. \qquad (15)$$

It is apparent that the ratio between the frequency deviation and the audio channel, F_d/F_a, is an important factor in determining the signal-noise ratio gain effected by frequency modulation. A convenient term for this ratio is the "deviation ratio" and it will be designated as such hereinafter.

Equation (15) gives the factor by which the amplitude modulation root-mean-square signal-noise ratio is multiplied in order to find the equivalent frequency modulation signal-noise ratio. Since this factor is used so frequently hereinafter, it will be designated by the word "improvement." The improvement given by (15) has been developed under the assumption of zero applied frequency deviation (no modulation) and a carrier which is strong compared to the noise. However, as will be shown later, as long as the carrier is strong compared to the noise, this equation also holds true for the case where modulation is present.

Peak Noise Considerations

In the ultimate application of signal-noise ratios, peak voltages are of prime importance since it is the peak of the noise voltage which seems to produce the annoyance. This is especially true in the case of impulse noise such as ignition where the crest factor of the noise may be very high. Thus the energy content of a short duration impulse might be very small in comparison with the energy content of the signal, but the peak voltage of the impulse might exceed the signal peak voltage and become very annoying. The degree of this annoyance would of course depend upon the type of service and will not be gone into here. In view of this importance of peak noise considerations, the final judgment in the comparison between the systems of frequency and amplitude modulation treated here will be based upon peak signal-noise ratios.

When the peak voltage or current ratio of the frequency and amplitude modulation spectra is to be determined, the characteristics of the different types of noise must be taken into consideration. There seem to be two general types of noise which require consideration. The first of these is fluctuation noise, such as thermal agitation and shot effect, which is characterized by a random relation between the various frequencies in the spectrum. The second is impulse noise, such as ignition or any other type of noise having a spectrum produced by a sudden rise of voltage, which is characterized by an orderly phase and amplitude relation between the individual frequencies in the spectrum.

Experimental data taken by the author have shown that the fluctuation noise crest factor is constant, independent of band width, when the carrier is strong compared to the noise. Thus the peak voltage of fluctuation noise varies with band width in the same manner that root-mean-square voltage does, namely, as the square-root of the band width ratio. Consequently, for the strong-carrier condition, the peak voltage characteristics of fluctuation noise may be determined by applying the experimentally determined crest factor to the root-mean-square characteristics. Hence, in the case of fluctuation noise, (15) applies for peak noise improvement as well as for average root-mean-square noise improvement.

Impulse Noise Characteristics

A simple way of visualizing the manner in which impulse noise produces its peak radio-frequency voltage is to consider the case of a recurrent impulse. It is well known that a recurrent impulse, such as square-wave-form dots, may be expressed by a Fourier series which consists of a fundamental and an infinite array of harmonics. The amplitudes of these harmonics are inversely proportional to their frequencies. The components of the single impulse will be similar to those of the recurrent impulse since the single impulse may be considered as a recurrent impulse with a very low rate of recurrence. The part of this impulse spectrum that is received on a radio receiver is a small band of the very high order harmonics. Since the frequency difference between the highest and lowest frequencies of this band is small compared to the mid-frequency of the band, all of the frequencies received are of practically equal amplitude. These harmonics are so related to each other by virtue of their relation to a common fundamental that they are all in phase at the instant the impulse starts or stops. Hence, for the interval at the start or stop of the impulse, all of the voltages in the band add up arithmetically and the peak voltage of the combination is directly proportional to the number of individual voltages. Since the individual voltages of the spectrum are equally spaced throughout the band, the number of voltages included in a given band is proportional to the band width. Consequently, the peak voltage of the resultant of the components in the spectrum is directly proportional to the band width. Thus impulse noise varies, not as the square root of the band width, as fluctuation noise does, but directly as the band width.[7] Since the voltages in the spectrum add arithmetically, their peak amplitude is proportional to their average ordinate as well as proportional to the band width. This makes the peak voltage of impulse noise, not proportional to the square root of the ratio between the *squared-ordinates areas*, as is the case with root-mean-square noise, but proportional to the ratio between the *areas* of the two spectra. Hence, (referring to Fig. 3)

$$\frac{N_a}{N_f} \text{ (peak values, impulse)} = \frac{\text{area } OCEH}{\text{area } ODH}$$

$$= \frac{(n/C) \times F_a}{F_a \times \frac{1}{2} \times \frac{2F_a}{F_i} \times \frac{n}{C}} = \frac{F_i}{F_a}. \quad (16)$$

[7] The fact that the peak voltage of impulse noise varies directly with the band width was first pointed out to the author by V. D. Landon of the RCA Manufacturing Company. The results of his work were later presented by him as a paper entitled "A study of noise characteristics," before the Eleventh Annual Convention, Cleveland, Ohio, May 13, 1936; published in the Proc. I.R.E., vol. 24, pp. 1514–1521; November, (1936).

Dividing (16) by (14) to obtain the ratio between the frequency and amplitude modulation output signal-noise ratios gives

$$\frac{S_f/N_f}{S_a/N_a} \text{ (peak values, impulse)} = 2\,\frac{F_d}{F_a}. \tag{17}$$

Equation (17) shows that the frequency modulation peak voltage improvement with respect to impulse noise is equal to twice the deviation ratio or about 1.16 times more improvement than is produced on fluctuation noise. The peak power gain would be equal to the square of the peak voltage gain or four times the square of the deviation ratio.

Low Carrier-Noise Ratios

When the expression for the instantaneous frequency of the wave modulated by the noise component and signal, given by (7), is resolved into its components by the use of the binomial theorem, the following is the result:

$$f = F_c - F_d \sin pt - \frac{(F_{na} - F_d \sin pt)}{Z}\left[\frac{n}{C} - \left(1 - \frac{2n}{ZC}\right)\left\{\frac{K_0}{Z}\right.\right.$$

$$- K_1 \cos (\omega_{na}t - F_d \sin pt) + K_2 \cos 2(\omega_{na}t - F_d \sin pt)$$

$$\left.\left.+ K_3 \cos 3(\omega_{na}t - F_d \sin pt) \cdots \right\}\right] \tag{18}$$

in which $Z = \dfrac{C}{n} + \dfrac{n}{C}$ and

$$K_0 = K_1 = \left(1 + \frac{3}{Z^2} + \frac{10}{Z^4} + \frac{35}{Z^6} + \frac{126}{Z^8} + \frac{462}{Z^{10}} + \frac{1716}{Z^{12}} + \cdots\right) \tag{19}$$

$$K_2 = \left(\frac{1}{Z} - \frac{4}{Z^3} - \frac{15}{Z^5} - \frac{56}{Z^7} - \frac{210}{Z^9} - \frac{792}{Z^{11}} - \frac{3003}{Z^{13}} - \cdots\right) \tag{20}$$

$$K_3 = \left(\frac{1}{Z^2} + \frac{5}{Z^4} + \frac{21}{Z^6} + \frac{84}{Z^8} + \frac{330}{Z^{10}} + \frac{1287}{Z^{12}} + \frac{5005}{Z^{14}} + \cdots\right). \tag{21}$$

Additional terms of the series of (19), (20), and (21), as well as higher order series, may be found with the aid of a table of binomial coefficients.

Equation (18) shows that, as the carrier-noise ratio approaches unity, the effective signal-noise ratio at the receiver output is no longer directly proportional to the carrier-noise ratio. The effective frequency deviation produced by the noise has harmonics introduced and a constant frequency shift added. The effect of the harmonics and constant shift is to make the wave form of a single noise component very peaked and of the nature of an impulse. Because of the selectivity of the audio channel, none of the harmonics are present for the noise frequencies in the upper half of the audio spectrum. As the frequency of the noise voltage is lowered, more and more harmonics are passed by the audio channel and as a consequence, the peak frequency deviation due to the noise is increased. This can be more easily understood from the following calculation of the wave form produced by the instantaneous frequency deviation of the single noise component.

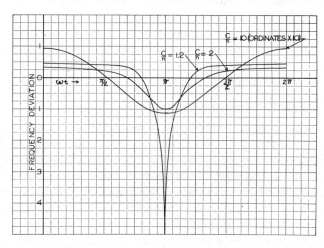

Fig. 4—Calculated wave forms showing the distortion produced on the instantaneous frequency deviation of the wave composed of the combination of the carrier and a single noise component. C/n = ratio between the peak voltage of the carrier and the peak voltage of the noise component.

The curves of Fig. 4 have been calculated from (7) and show how the instantaneous frequency deviation varies with time or the phase angle of the wave. A wave with the instantaneous frequency given by these curves would produce voltages in the output of the detector of the frequency modulation receiver which are proportional to the frequency deviations. It can be seen from these curves that, as the carrier-noise ratio approaches unity, the wave form becomes more and more peaked. The harmonics which enter in to make up this peaked wave form are given by (18) and are completely present for all noise frequencies only in the absence of audio selectivity.

In the presence of audio selectivity, the condition portrayed by (18) is approached as the audio frequency of the noise approaches zero. Thus the wave form of the noise is sinusoidal at a noise frequency high enough to have its harmonics eliminated by the audio selectivity, but becomes more peaked as the frequency is made lower so that more harmonics are included. This effect tends to increase the peak voltage of the low-frequency noise voltages which have a large number of harmonics present. Thus, as the carrier-noise ratio approaches unity, the triangular audio spectrum is distorted by an increase in the amplitude of the lower noise frequencies.

The above gives a qualitative and partially quantitative description of the noise spectrum which results at the lower carrier-noise ratios. Further development would undoubtedly make possible the exact calculation of the peak and root-mean-square signal-noise ratio at the receiver output when the carrier-noise ratio at the receiver input is close to unity, but, because of the laborious nature of the calculations involved in evaluating the terms of (18), and pressure of other work, the author is relying upon experimental determinations for these data.

Noise Crest Factor Characteristics

The crest factor characteristics of the noise can be studied to an approximate extent by a study of (6). This equation portrays the resultant peak frequency deviation of the wave at the output of the limiter. From it, the crest factor characteristics of the output of the detector may be determined since in the frequency modulation receiver frequency deviations are linearly converted into detector output voltages. However, the crest factor characteristics of the receiver output are different from those at the detector output due to the effect of the selectivity of the audio channel. This is especially true in the case of the frequency modulation receiver with a deviation ratio greater than unity, that is, where the audio channel is less than one half the intermediate-frequency channel. Consequently, in order to obtain the final results, the effect of the application of the audio selectivity must be applied to the results determined from a study of (6).

From the curves of Fig. 4, it can be seen that the peak frequency deviation of the wave given by (7) occurs at a phase angle equal to 180 degrees. From the similarity of (6) and (7), it can be seen that the peak frequency deviation of (6) would also occur at a phase angle of 180 degrees. At this phase angle the noise peak frequency deviation from (6) is

$$f_{dn}(\text{peak}) = \frac{\left(F_{na} - \dfrac{1}{2\pi} \dfrac{d\phi(t)}{dt} - F_d \sin pt \right)}{\dfrac{(C/N) - 1}{(N/C) - 1} + 1}$$

$$= \frac{- \left(F_{na} - \dfrac{1}{2\pi} \dfrac{d\phi(t)}{dt} - F_d \sin pt \right)}{(C/N) - 1}. \tag{22}$$

Equation (22) shows that the peak frequency deviation of the noise, for any value of carrier-noise ratio, C/N, is proportional to the noise instantaneous audio frequency given by the quantity

$\left(F_{na} - \dfrac{1}{2\pi} \dfrac{d\phi(t)}{dt} - F_d \sin pt \right)$, and to the quantity $1/\{(C/N) - 1)\}$.

C/N is the resultant instantaneous peak carrier-noise ratio which is present in the output of the frequency modulation intermediate-frequency channel. It is apparent that when this carrier-noise ratio is high, the peak frequency deviation of the noise is proportional to N/C. When the carrier-noise ratio is equal to unity, the peak frequency deviation becomes infinite and it is evident that the frequency modulation improvement, which is based on a high carrier-noise ratio, would be lost at this point. The term "improvement threshold" will be employed hereinafter to designate this point below which the frequency modulation improvement is lost and above which the improvement is realized. Theoretically this term would refer to the condition where the instantaneous peak voltage of the noise is equal to the peak voltage of the carrier. However, in the practical case, where only *maximum* peak values of the noise are measured, the improvement threshold will refer to the condition of equality of the *maximum* instantaneous peak voltage of the noise and the peak voltage of the carrier.

As the experimental characteristics will show, this increase in peak frequency deviation of the noise is manifested in an increase in crest factor of the noise. The crest factor cannot rise to infinity, however, due to the limitations imposed by the upper and lower cutoff frequencies of the intermediate-frequency channel. This selectivity limits the peak frequency deviation of the resultant of the noise and applied modulation to a value not greater than one half the intermediate-frequency channel width. Hence, in the absence of applied frequency modulation, the peak voltage of the noise at the detector output may rise to a value equal to the peak voltage due to the applied frequency modulation with maximum frequency deviation. In the presence of the applied frequency modulation, the total peak frequency deviation is limited so that the noise peaks depress the signal, that is, they punch holes in the signal, but do not rise above it. Thus a phenomenon which might be termed "frequency limiting" takes place. This frequency

71

limiting limits frequency deviations in the same manner that amplitude limiting limits amplitude deviations. The resulting effect is the same as though an amplitude limiter were placed at the detector output to limit the output so that the peak voltage of the noise or signal, or their resultant, cannot rise above a voltage corresponding to that produced by the signal alone at full modulation.

Since the frequency limiting limits the noise so that its maximum amplitude cannot rise above the maximum amplitude of the applied modulation, a noise suppression effect is present which is similar to that effected by the recent noise suppression circuits[8,9] used for reducing impulse noise which is stronger than the amplitude modulated carrier being received. The result of such limiting is a considerable reduction of the annoyance produced by an intermittent noise, such as ignition, where the duration of the impulses is short and the rate of recurrence is low. With such noise, the depression of the signal for the duration of the impulse reduces the presence of the signal for only a small percentage of the time; the resultant effect is a considerable improvement over the condition where the peaks of the noise are stronger than the signal. On the other hand, for steady noise such as fluctuation noise, as the carrier-noise ratio is made less than unity, the signal is depressed more and more of the time so that it is gradually smothered in the noise.

When the effect of the audio selectivity is considered in conjunction with the frequency limiting, it is found that the noise suppression effect is somewhat improved for the case of a deviation ratio greater than unity. The reason for this is as follows: The frequency limiting holds the peak voltage of the noise at the output of the detector so that it cannot rise above the maximum value of the signal. However, in passing through the audio channel, the noise is still further reduced by elimination of higher frequency components whereas the signal passes through without reduction. Consequently the over-all limiting effect is such that the noise is limited to a value which is *less* than the maximum value of the signal. The amount that it is less depends upon the difference between the noise spectra existing at the output of the detector and the output of the audio selectivity.

Experimental determinations, which will be shown later, point out that as unity carrier-noise ratio is approached, the frequency

[8] Leland E. Thompson, "A detector circuit for reducing noise interference in C.W. reception," *QST*, vol. 19, p. 38; April, (1935). A similar circuit for telephony reception is described by the same author in *QST*, vol. 20, pp. 44–45; February, (1936).

[9] James J. Lamb, "A noise-silencing I.F. circuit for superhet receivers," *QST*, vol. 20, pp. 11–14; February, (1936).

modulation audio noise spectrum changes from its triangular shape to a somewhat rectangular shape. Hence the noise spectrum at the output of the detector when frequency limiting is taking place would be approximately as given by $OCBA$ of Fig. 3. When the audio selectivity is applied, the spectrum would be reduced to $OCEH$ and the band width of the noise would be reduced by a ratio equal to the deviation ratio. This would reduce the peak voltage of fluctuation noise by a ratio equal to the square root of the deviation ratio and that of impulse noise by a ratio equal to the deviation ratio. Thus, the resultant effect of the frequency limiting is that the fluctuation noise output is limited to a value equal to the maximum peak voltage of the signal divided by the square root of the deviation ratio. The corresponding value of impulse noise is limited to a value equal to the maximum peak voltage of the signal divided by the deviation ratio. Consequently, with fluctuation noise, when the noise and signal are measured in the absence of each other, the signal-noise ratio cannot go below a value equal to the square root of the deviation ratio; the corresponding signal-noise ratio impulse noise cannot go below a value equal to the deviation ratio. However, these minimum signal-noise ratios are only those which exist when the noise is measured in the absence of the applied frequency modulation. When the applied modulation and the noise are simultaneously present, the noise causes the signal to be depressed. When this depressed signal, with its depression caused by noise composed of a wide band of frequencies, is passed through the audio selectivity, the degree of depression is reduced. The amount of the reduction will be different for the two kinds of noise. The determination of the actual magnitude of this reduction of the signal depression, as effected by the audio selectivity, will be left for experimental evaluation.

In comparing frequency modulation systems with different deviation ratios at the *low carrier-noise ratios*, the wider intermediate-frequency channel necessary for the high deviation ratio receiver gives that receiver a disadvantage with respect to the low deviation ratio receiver. Since this wider channel accepts more noise than the narrower intermediate-frequency channel of the low deviation ratio receiver, when equal carriers are fed to both such receivers equality of carrier and noise occurs at a higher carrier level in the high deviation ratio receiver. As a result, a higher carrier voltage is required to reach the improvement threshold in the case of the high deviation ratio system. Thus at certain low carrier levels, the carrier-noise ratio could be above the improvement threshold in the low deviation ratio system, but below in the high deviation ratio system; at this carrier level the low deviation ratio system would be capable of producing a better output signal-noise ratio than the high deviation ratio system.

The difference between the improvement thresholds of receivers with different deviation ratios may be investigated by a determination of the carrier-noise ratio which exists in the reference amplitude modulation receiver when the improvement threshold exists in the frequency modulation receiver. This carrier-noise ratio may be found by a consideration of the relative band widths of the intermediate-frequency channels of the receivers. Thus, when the deviation ratio is unity, and the intermediate-frequency channel of the frequency modulation receiver is of the same width as that of the amplitude modulation receiver,[10] the two receivers would have the same carrier-noise ratio in the intermediate-frequency channels. When the deviation ratio is greater than unity, and the intermediate-frequency channel of the frequency modulation receiver is broader than that of the amplitude modulation receiver, the carrier-noise ratio in the frequency modulation receiver is less than that in the amplitude modulation receiver. For the case of fluctuation noise, where the peak values vary as the square root of the ratio between the two band widths concerned, the carrier-noise ratio in the frequency modulation intermediate-frequency channel would be less than that in the amplitude modulation intermediate-frequency channel by a ratio equal to the square root of the deviation ratio. Thus, when equal carrier voltage is fed to both receivers,

$$C_a/N_a = (C/N)\sqrt{F_d/F_a} \text{ (fluctuation noise, peak or r-m-s values)} \quad (23)$$

in which C_a/N_a = carrier-noise ratio in the amplitude modulation intermediate-frequency channel and C/N = corresponding ratio in the frequency modulation intermediate-frequency channel.

In the case of impulse noise, where the peak values of the noise

---

[10] In order to assume that the frequency modulation receiver with a deviation ratio of unity has the same intermediate-frequency channel width as the corresponding amplitude modulation receiver, the assumption would also have to be made that the peak frequency deviation due to the applied frequency modulation is equal to one half the intermediate-frequency channel width. In the ideal receiver with a square-topped selectivity characteristic, this amount of frequency deviation would produce considerable out-of-channel interference and would introduce distortion in the form of a reduction of the amplitudes of the higher modulation of frequencies during the intervals of high peak frequency deviation. However, under actual conditions, where the corners of the selectivity characteristic are rounded, it has been found that the frequency deviation may be made practically equal to one half the normal selectivity used in amplitude modulation practice without serious distortion. Receivers with high deviation ratios are less susceptible to this distortion due to the natural distribution of the side bands for the high values of F_d/F_m which are encountered with such receivers.

vary directly with the band-width ratio, the carrier-noise ratios in the two receivers are related by

$$C_a/N_a = \frac{C}{N} \frac{F_d}{F_a} \text{ (impulse noise, peak values).} \qquad (24)$$

From (23), it can be seen that, with fluctuation noise, a carrier-noise ratio equal to the square root of the deviation ratio would exist in the amplitude modulation intermediate-frequency channel when the carrier-noise ratio is at the improvement threshold $(C/N=1)$ in the frequency modulation intermediate-frequency channel. Likewise, from (24), with impulse noise, the frequency modulation improvement threshold occurs at a peak carrier-noise ratio in the amplitude modulation intermediate-frequency channel which is equal to the deviation ratio.

Effect of Application of the Modulation

For the condition of a carrier which is strong compared to the noise, the equation for the instantaneous frequency of the wave modulated by the noise and signal, given by (7), may be reduced to the following:

$$f = F_c - F_d \sin pt - \frac{n^2}{C^2} (F_{na} - F_d \sin pt)$$

$$- \frac{n}{C} (F_{na} - F_d \sin pt) \cos \left\{ \omega_{na}t + \frac{F_d}{F_m} \cos pt \right\}. \qquad (25)$$

By neglecting the inconsequential term proportional to n^2/C^2, applying the sine and cosine addition formulas, the Bessel function expansions, and the Bessel function recurrence formulas, (25) may be resolved into

$$f = F_c - (n/C) \left[J_0\left(\frac{F_d}{F_m}\right) F_{na} \cos \omega_{na}t \right.$$

$$- J_1\left(\frac{F_d}{F_m}\right) \{ (F_{na}+F_m) \sin (\omega_{na}t+pt) + (F_{na}-F_m) \sin (\omega_{na}t-pt) \}$$

$$- J_2\left(\frac{F_d}{F_m}\right) \{ (F_{na}+2F_m) \cos (\omega_{na}t+2pt) + (F_{na}-2F_m) \cos (\omega_{na}t-2pt) \}$$

$$+ J_3\left(\frac{F_d}{F_m}\right) \{ (F_{na}+3F_m) \sin (\omega_{na}t+3pt) + (F_{na}-3F_m) \sin (\omega_{na}t-3pt) \}$$

$$\left. + J_4 \cdots \right]. \qquad (26)$$

This resolution shows that the application of frequency modulation to the carrier divides the over-all transmission of the receiver into components due to the carrier and each side frequency. The amplitudes of these components are proportional to the frequency difference between the noise voltage and the side frequency producing the component. The frequency of the audio noise voltage in each one of these component spectra is equal to the difference between the side frequency and the noise radio frequency. Thus the application of the modulation changes the noise from a single triangular spectrum due to the carrier, into a summation of triangular spectra due to the carrier and side frequencies. In the absence of selectivity, the total root-mean-square noise would be unchanged by the application of the modulation since the root-mean-square summation of the frequency modulation carrier and side frequencies is constant; hence the root-mean-square summation of noise spectra whose amplitudes are proportional to the strength of the carrier and side frequencies would be constant. However, since selectivity is present, the noise is reduced somewhat. This can be seen by considering the noise spectrum associated with one of the higher side frequencies. The noise spectrum associated with this side frequency, which acts as a "carrier" for its noise spectrum, is curtailed at the high-frequency end by the upper cutoff of the intermediate-frequency channel. The region of noise below the side frequency is correspondingly increased in range, but yields high-frequency noise voltages which are eliminated by the audio-frequency selectivity. Consequently when modulation is applied, the noise is slightly reduced. The amount of this reduction may be calculated by a root-mean-square summation of the individual noise spectra due to the carrier and side frequencies. For the case of a deviation ratio of unity, an actual summation of the various spectra for full applied modulation has shown the root-mean-square reduction to be between two and three decibels depending upon the audio frequency of the noise. The same sort of summations also shows that the reduction becomes less as the deviation ratio is increased.

The weak-carrier root-mean-square noise characteristics in the presence of applied frequency modulation do not lend themselves to such straightforward calculation as the corresponding strong-carrier characteristics and will not be gone into here. The same can be said for the peak-noise characteristics in the presence of applied frequency modulation.

Transmitter Frequency Modulation Power Gain

The above considerations, which are based upon the equivalent conditions of equal carrier amplitude at the input of the amplitude and frequency modulation receivers, do not take into account the power gain effected by the use of frequency modulation at the transmitter. Since the power in a frequency modulated wave is constant, the radio-frequency amplifier tubes in the transmitter may be operated in the class C condition instead of the class B condition as is required for a low level modulated amplitude modulation system. In changing from the class B to the class C condition, the output voltage of the amplifier may be doubled. Consequently a four-to-one power gain may be realized by the use of frequency modulation when the amplitude modulation transmitter uses low-level modulation. On the other hand, when the amplitude modulation transmitter uses high level modulation—that is, when the final amplifier stage is modulated, the power gain is not so great. However, for the purpose of showing the effect of a transmitter power gain, the amplitude modulation transmitter will be assumed to be modulated at low levels.

As this paper is in the final stages of preparation, systems of amplifying amplitude modulation have been announced wherein plate efficiencies of linear amplifiers have been increased practically to equal the class C efficiencies.[11,12] Since these systems are not in general use as yet, it will suffice to say that such improvements in amplitude modulation transmission will tend to remove the frequency modulation transmitter gain in accordance with these improvements. Hence the overall frequency modulation gain will more nearly approach that due to the receiver[13] alone.

With a four-to-one power gain at the transmitter, a frequency modulation system would deliver twice the carrier voltage to its receiver that an amplitude modulation system would with the same transmitter output stage. Hence (15) and (17), and (23) and (24) become, respectively,

$$\frac{S_f/N_f}{S_a/N_a} \text{ (peak values, fluctuation noise)} = 2\sqrt{3}F_d/F_a \qquad (27)$$

$$\frac{S_f/N_f}{S_a/N_a} \text{ (peak values, impulse noise)} = 4F_d/F_a \qquad (28)$$

---

[11] W. H. Doherty, "A new high efficiency power amplifier for modulated waves," presented before Eleventh Annual Convention, Cleveland, Ohio, May 13, (1936); published in Proc. I.R.E., vol. 24, pp. 1163–1182; September, (1936).

[12] J. N. A. Hawkins, "A new, high-efficiency linear amplifier," *Radio*, no. 209, pp. 8–14, 74–76; May, (1936).

[13] The receiver and transmitter gain are mentioned rather loosely when they are separated in this way. However, it will be understood that the receiver gain could not be realized without providing a transmitter to match the requirements of the receiver.

$C_a/N = (C/2N)\sqrt{F_d/F_a}$ (fluctuation noise, r-m-s or peak values) (29)

$C_a/N = (C/2N)F_d/F_a$ (impulse noise, peak values). (30)

These equations show that this increase in carrier fed to the frequency modulation receiver not only increases the frequency modulation improvement, but also lowers the carrier-noise ratio received on the amplitude modulation receiver when the improvement threshold exists in the frequency modulation receiver.

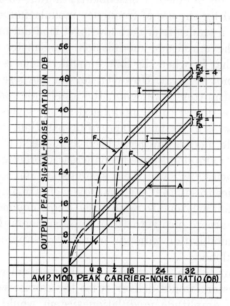

Fig. 5—Theoretical signal-noise ratio characteristics of frequency and amplitude modulation without the transmitter gain taken into account. Curve A = amplitude modulation receiver. The curves marked with I and F show the characteristics of the frequency modulation receivers for impulse and fluctuation noise, respectively. F_d/F_a = deviation ratio.

Theoretical Conclusions

The curves of Fig. 5 and 6 summarize the theoretical conclusions by means of an example in which receivers with deviation ratios of four and one are compared with each other and with an amplitude modulation receiver at various carrier-noise ratios. Fig. 5 shows the receiver gain only, whereas Fig. 6 takes into consideration a transmitter power gain of four to one. The curves are plotted with peak carrier-noise ratio in the amplitude modulation selectivity channel as a standard of comparison. Thus the curve for the amplitude modulation receiver is a straight line with a slope of forty-five degrees. The curves for the frequency modulation receivers show output signal-noise ratios which are greater or less than those obtained from the amplitude modulation receiver depending upon the carrier-noise ratio.

For Fig. 5, (15) and (17) were used to obtain the strong-carrier frequency modulation improvement factors. Hence the frequency modulation output signal-noise ratios were obtained by multiplying the amplitude modulation signal-noise ratios by the frequency modulation improvement factors. The carrier-noise ratios which exist in the amplitude modulation receiver when the improvement threshold exists

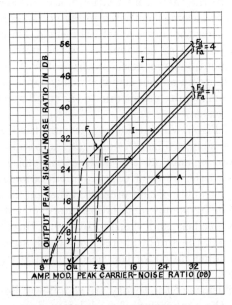

Fig. 6—Theoretical signal-noise ratio characteristics of frequency and amplitude modulation receivers with the transmitter gain taken into account.

in the frequency modulation receiver were determined by substituting a value of unity carrier-noise ratio in (23) and (24). The improvement thresholds are designated in both Figs. 5 and 6 by the points u and z for fluctuation and impulse noise, respectively. Since the theory does not permit actual calculation of signal-noise ratios in the region between high ratios and the improvement threshold, that part of the curves has been sketched in with a dashed line.

The part of the impulse-noise curve, for the deviation ratio of four represented by the line x-y shows the characteristic which would be obtained if the noise and signal were measured in the absence of each other. Because of frequency limiting, the noise is limited to equality with the signal at the output of the detector and is then reduced in peak voltage by the audio selectivity. The amount of this reduction for impulse noise would be a ratio equal to the deviation ratio or, in this case, twelve decibels. In the case of fluctuation noise, the reduction of the noise, which is present in the absence of modulation, would be equal to the square root of the deviation ratio, or six decibels, and the corresponding curve is shown by the line *v-w*. However, these lines do not portray the actual signal-noise ratio characteristics since the noise depresses the signal when the carrier-noise ratios go below the improvement threshold. In the case of fluctuation noise this signal depression causes the signal to become smothered in the noise as the carrier-noise ratio is lowered below the improvement threshold. On the other hand, with impulse noise such as ignition, where the pulses are short and relatively infrequent, carrier-noise ratios below the improvement threshold will present an output signal which is depressed by the noise impulses, but which is quite usable due to the small percentage of time that the impulse exists.

The curves of Fig. 6, which take into account the frequency modulation transmitter gain, utilize (27) and (28) to obtain the frequency modulation improvements at the high carrier-noise ratios. These curves assume a carrier at the frequency modulation receiver inputs which is twice the strength of that present at the amplitude modulation receiver input. The frequency modulation improvements are therefore increased by six decibels and the improvement thresholds occur at signal-noise ratios in the amplitude receiver which are six decibels below the corresponding ratios for the case where the transmitter gain is not taken into account.

Further conclusions of the theory are as follows: For the high carrier-noise ratios, the application of modulation does not increase the root-mean-square value of the noise above its unmodulated value. Also, in the case of the low deviation ratio receivers, the root-mean-square value of the noise will be slightly reduced as the modulation is applied.

Experiment

In the experimental work it was desired to obtain a set of data from which curves could be plotted showing the frequency modulation characteristics in the same manner as the theoretical curves of Fig. 5. To do this it was necessary to have an amplitude modulation reference system and frequency modulation receivers with deviation ratios of unity and greater than unity. Equal carrier voltages and noise spectra could then be fed to these receivers and the output signal-noise ratios measured while the carrier-noise ratio was varied. Since it was not convenient to measure the carrier-noise ratio at intermediate frequency, the output signal-noise ratios of the amplitude modulation receiver were measured instead and were plotted as abscissas in place of the carrier-noise ratios. This gives an abscissa scale which is practically the same as that which would be obtained by plotting carrier-noise ratios. The validity of this last statement was checked by measuring the linearity with which the output signal-noise ratio of the amplitude modulation receiver varied from high to low values as the carrier-noise ratio was varied by attenuating the carrier in known amounts in the presence of a constant noise. At the very low root-mean-square ratios the inclusion of the beats between the individual noise frequencies in the spectrum increases the apparent value of the

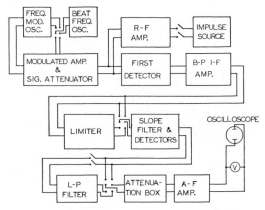

Fig. 7—Block diagram of experimental setup.

root-mean-square resultant of the noise voltages about two or three decibels. Thus, except for this small error at the low root-mean-square carrier-noise ratios, the amplitude modulation signal-noise ratio can be assumed equal to the carrier-noise ratio.

The block diagram of Fig. 7 shows the arrangement of apparatus used in obtaining the experimental data. The frequency modulated oscillator employed a circuit which was similar to that used in the previously mentioned propagation tests.[1] The modulated amplifier consisted of a signal generator which was capable of being amplitude modulated, but whose master oscillator energy was supplied from the frequency modulated oscillator. Thus a signal generator was available which was capable of being either frequency or amplitude modulated. A two-stage radio-frequency amplifier, tuned to the carrier frequency, but with no signal at its input, was used as the source of fluctuation noise. For the impulse noise measurements, the radio-frequency output of a square-wave multivibrator was fed to the input of this radio-frequency amplifier.

In order to make available frequency modulation receivers with different deviation ratios, a method was devised which made possible the use of a single intermediate-frequency channel and detection system for all receivers. The method consisted in the insertion of a low-pass filter in the audio output of the receiver so as to reduce the width of the audio channel and thereby increase the deviation ratio of the receiver. This procedure is not that which might be normally followed since to increase the deviation ratio, the audio channel would normally be left constant and the intermediate-frequency channel increased.

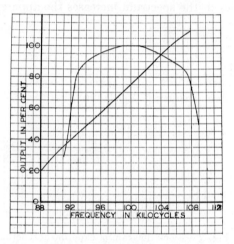

Fig. 8—Band-pass characteristic of receiver intermediate-frequency amplifier, and characteristic of sloping filter.

However, since it is only the *ratio* between the intermediate- and audio-frequency channels which governs the frequency modulation improvement, such an expedient is permissible for the purpose of the experiments.

The band-pass filter of the receiver intermediate-frequency amplifier was adapted from broadcast components and gave an output which was about one decibel down at 6500 cycles off from mid-band frequency. (See Fig. 8.) Hence maximum frequency deviation was limited to 6500 cycles. The audio channel of the receiver cut off at 6500 cycles and the low-pass filter cut off at 1600 cycles. Thus the following four different types of receivers were available: Number one, a frequency modulation receiver with a deviation ratio of unity which would receive a 6500-cycle modulation band. Number two, an amplitude modulation receiver which would receive a 6500-cycle modulation band. Number three, a frequency modulation receiver with a deviation ratio of about four (6500÷1600) which would receive a 1600-cycle modulation band. Number four, an amplitude modulation receiver which would receive a 1600-cycle modulation band.

With these four receivers, a comparison between number two and number one would produce a comparison between amplitude modulation reception and frequency modulation reception with a deviation ratio of unity. A comparison between receivers number four and number three would produce a comparison between amplitude modulation reception and frequency modulation reception with a deviation ratio of four. Thus both frequency modulation receivers had as a standard of comparison an amplitude modulation receiver with an audio channel equal to that of the frequency modulation receiver.

The limiter of the frequency modulation receiver consisted of four stages of intermediate-frequency amplification arranged alternately to amplify and limit. The sloping filter detectors utilized the same circuit as used in the propagation tests[1] except that only one sloping filter was used in conjunction with a flat-top circuit as described in the theoretical section of this paper. Thus a balanced detector type of receiver was available which would also receive amplitude modulation by switching off the frequency modulation detector and receiving the detected output of the flat-top circuit. The characteristic of the sloping filter is shown in Fig. 8.

The output of the detectors was fed to a switching system which connected either to a low-pass filter and attenuator or directly to the attenuator. The output of the attenuator passed to an audio-frequency amplifier having an upper cut-off frequency of 6500 cycles. The indicating instruments were connected to the amplifier output terminals. For the root-mean-square fluctuation noise measurements, a copper-oxide-rectifier type meter was used.[14] A cathode-ray oscilloscope was used for all peak voltage measurements.

In the procedure used to obtain the data, the carrier-noise ratio was varied over a wide range of values and the receiver output signal-noise ratios were measured at each value of carrier-noise ratio. To do this, the output of the noise source was held constant while the carrier was

[14] In the preliminary measurements, a thermocouple meter was connected in parallel with the copper-oxide-rectifier meter in order to be sure that no particular condition of the fluctuation noise wave form would cause the rectifier meter to deviate from its property of reading root-mean-square values on this type of noise. It was found that the rectifier type of instrument could be relied upon to indicate correctly so that the remainder of the measurements of root-mean-square fluctuation noise were made using the more convenient rectifier type of instrument.

varied by means of the signal generator attenuator. The output peak signal-noise ratios were obtained by first measuring the peak voltage of the tone output with the noise source shut off and then measuring the peak voltage of the noise with the tone shut off. The *maximum* peak voltage of the noise was read for its peak voltage. The root-mean-square signal-noise ratios were measured by reading the root-mean-square voltage of the tone in the presence of the noise and then reading the voltage of the noise alone. The signal was then separated from the noise by equating the measured signal-plus-noise voltage to $\sqrt{S^2 + N^2}$, substituting the measured noise voltage for N, and solving for the signal, S. In these measurements, a 1000-cycle tone was used to modulate

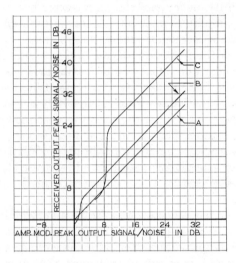

Fig. 9—Measured peak signal-noise ratio characteristics for fluctuation noise. Curve A = amplitude modulation receiver. Curve B = frequency modulation receiver with deviation ratio equal to unity. Curve C = frequency modulation receiver with deviation ratio equal to four.

at fifty per cent the amplitude modulator or to produce one-half frequency deviation (3250 cycles) on the frequency modulator. The output signal-noise ratios were corrected to a 100 per cent, or full modulation, basis by multiplying them by two. The radio frequency used was ten megacycles.

Fluctuation Noise Characteristics

The curves of Fig. 9 show the fluctuation noise characteristics, in which peak signal-noise ratios were measured. These curves check the theoretical curves of Fig. 5 as nearly as such measurements can be expected to check. With the deviation ratio of four (low-pass filter in), the theoretical strong-carrier improvement should be $4 \times 1.73 = 6.9$ or 16.8 decibels; the measured improvement from Fig. 9 is about 14 decibels. With the deviation ratio of unity (low-pass filter out), the measured improvement was about 3.5 decibels as compared to the 4.76-decibel theoretical figure. The full frequency modulation improvement is seen to be obtained down to carrier-noise ratios about two or three decibels above the improvement threshold (equality of peak carrier and noise). The fact that the frequency modulation improvement threshold occurs at a higher carrier-noise ratio in the case of the receiver with a deviation ratio of four than in the case of the receiver with a deviation ratio of unity, also checks the theoretical predictions. In this case of fluctuation noise, the improvement threshold for the receiver with the deviation ratio of four should occur at a carrier-noise ratio in the amplitude modulation intermediate-frequency channel which was twice the corresponding ratio for the receiver with a deviation ratio of unity. The curves show these two points to be about seven decibels apart or within one decibel of the theoretical figure of six decibels.

The data for the curves of Fig. 9 were obtained by measuring the peak value of the noise alone and signal alone and taking the ratio of these two values as the signal-noise ratio. Hence the signal depressing effect, occurring for carrier-noise ratios below the improvement threshold, does not show up on the curves. In order to obtain an approximate idea as to the order of magnitude of this effect, observations were made in which the carrier-noise ratio was lowered below the improvement threshold while the tone modulation output (100 per cent modulation in the case of the amplitude modulation observation and full frequency deviation in the case of the frequency modulation observation) was being monitored by ear and oscilloscope observation. It was found that the fluctuating nature of the instantaneous peak voltage of the fluctuation noise had considerable bearing upon the effects observed. Due to the fact that the instantaneous value of the peak voltage is sometimes far below the maximum instantaneous value, frequency modulation improvement is obtained to reduce still further the peak voltage of these intervals of noise having instantaneous peak voltages lower than the maximum value. This effect seems to produce a signal at the output of the frequency modulation receiver which sounds "cleaner," but which has the same maximum peak voltage characteristics as the corresponding amplitude modulation receiver. Thus, as far as maximum peak voltage of the noise is concerned, the frequency modulation receiver produces about the same output as the amplitude modulation receiver for carrier-noise ratios below the improvement threshold. The reduction of the peak voltage of the noise during the intervals of lower instantaneous peak value reduces the energy content of the noise in the output; hence some idea of the magnitude of this effect can be obtained from the root-mean-square characteristics of the noise.

The curves of Fig. 10 are similar to those of Fig. 9 except that the root-mean-square signal-noise ratios are plotted as ordinates. Since the crest factor of the signal is three decibels and that of fluctuation noise is about thirteen decibels (as later curves will show), the root-mean-square signal-noise ratios are ten decibels higher than the corresponding peak ratios. It can be seen that the root-mean-square characteris-

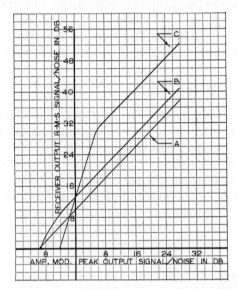

Fig. 10—Measured root-mean-square signal-noise ratio characteristics for fluctuation noise. Curve A = amplitude modulation receiver. Curve B = frequency modulation receiver with deviation ratio equal to unity. Curve C = frequency modulation receiver with deviation ratio equal to four.

tics differ from the peak characteristics in the range of carrier-noise ratios below the improvement threshold; above the improvement threshold, the characteristics are similar.

Since the root-mean-square and peak signal-noise ratios display different characteristics below the improvement threshold, it is quite evident that the crest factor of the noise changes as the carrier-noise ratio is lowered below this point. The crest factor can be obtained from the curves of Fig. 9 and 10 as follows: By adding three decibels to the ordinates of Fig. 10 they will be converted to peak signal to root-mean-square noise ratios. Hence by subtracting from these ratios the corresponding ordinates of Fig. 9, the crest factor of the noise is obtained. The results of such a procedure are shown in Fig. 11.

In the case of the frequency modulation receiver with a deviation ratio of four, Fig. 11 shows that the crest factor increases by about 14.5 decibels at the improvement threshold. Hence the frequency modulation improvement, which is about fourteen decibels by measurement and sixteen by calculation, is counteracted by an increase in crest factor. This same situation exists in the case of the receiver with a deviation ratio of unity. Here the increase in crest factor is about four decibels; the measured frequency modulation improvement is about 3.5 decibels and the calculated value 4.76 decibels.

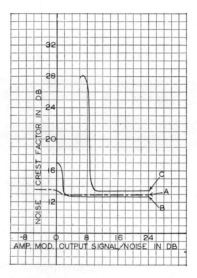

Fig. 11—Crest factor characteristics of frequency and amplitude modulation receivers. Curve A =amplitude modulation receiver with 6500-cycle audio channel. Curve B =frequency modulation receiver with deviation ratio equal to unity. Curve C =frequency modulation receiver with deviation ratio equal to four.

Curve A of Fig. 11 shows the crest factor characteristics of the amplitude modulation receiver. It is seen that this crest factor is about equal to that for frequency modulation above the improvement threshold. The average value of the crest factor for both amplitude and frequency modulation in this region is thirteen decibels or about 4.5 to one. This value checks previous measurements of crest factor where a slide-back vacuum tube voltmeter was used in place of an oscilloscope to measure the peak voltage and a thermocouple was used to measure the root-mean-square values.

The point where the crest factor of the noise increases, which occurs at the frequency modulation improvement threshold, has a rather distinctive sound to the ear. When fluctuation noise is being observed, as this point is approached the quality of the hiss takes on a more intermittent character, somewhat like that of ignition. This point has been termed by the author the "sputter point," and since it coincides with the improvement threshold it is a good indicator for locating the improvement threshold. It is caused by the fact that the fluctuation noise voltage has a highly variable instantaneous peak voltage so that there are certain intervals during which the instantaneous peak voltage of the noise is higher than it is during other intervals. Consequently, as the maximum peak value of the noise approaches the peak value of the signal, the higher instantaneous peaks will have their crest factor increased to a greater degree than the lower instantaneous peaks. Fig. 12 shows oscillograms taken on the fluctuation noise output of the

F   A

Fig. 12—Wave form of the fluctuation noise output at unity carrier-noise ratio in the frequency modulation receiver. *F* =frequency modulation receiver. *A* =amplitude modulation receiver.

frequency and amplitude modulation receivers with the 1600-cycle low-pass filter in the audio circuit and with the signal-noise ratio adjusted to the sputter point. These oscillograms also tend to show how the frequency modulation signal would sound "cleaner" than the amplitude modulation signal when the carrier-noise ratio is below the improvement threshold.

Data were also taken to show the fluctuation noise characteristics as frequency modulation is applied. These data were taken by inserting low-pass or high-pass filters in the audio system and then applying a modulation frequency to the frequency modulated oscillator which would fall outside the pass band of the filters. The low-pass filter cut off at 1600 cycles so that modulating frequencies higher than 1600 cycles were applied. The output of the filter contained only noise in the range from zero to 1600 cycles and the change of noise versus frequency deviation of the applied modulation could be measured. The high-pass filter also cut off at 1600 cycles so that measurements of the noise in the range from 1600 to 6500 cycles were made while applying modulation frequencies below 1600 cycles. In the case of the high-pass filter, the harmonics of the modulating frequencies appeared at the filter output in addition to the noise. Consequently, a separate measurement of the harmonics in the absence of the noise was made so that the noise could be separated from the harmonics by the quadrature relations. The results with the low-pass filter are shown in Fig. 13. The results with the high-pass filter are shown in Fig. 14.

The curves of Fig. 13 are representative of a system with a deviation ratio of four. They point out the fact that when the peak carrier-noise ratio in the frequency modulation intermediate-frequency channel

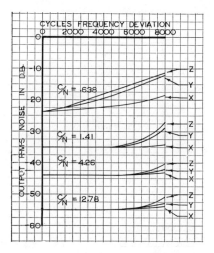

Fig. 13—Variation of frequency modulation receiver output noise as frequency modulation is applied. 1600-cycle low-pass filter in audio output. Modulation frequency: for curve $X = 6000$ cycles, $Y = 3000$ cycles, and $Z = 2000$ cycles. C/N = peak carrier-noise ratio in the output of intermediate-frequency channel.

is greater than unity, the root-mean-square noise is substantially unchanged due to the application of modulation. The one curve for a carrier-noise ratio less than unity shows a gradual increase of the noise, which would effect a decrease of the signal-noise ratio as the modulation is applied; this increase in the noise is displayed to a greater extent on the lower modulation frequency of 2000 cycles than on the higher modulation frequencies of 3000 and 6000 cycles.

Fig. 14 is approximately representative of a receiver with a deviation ratio of unity. This is because the range of noise frequencies from zero to 1600 cycles, which were eliminated by the high-pass filter, were a small part of the total range extending out to 6500 cycles. At the highest carrier-noise ratio, the noise is decreased as the modulation is applied. This is in accordance with the deductions of the theory in the section *Effect of Application of Modulation*. As the carrier-noise ratio is lowered this tendency is eliminated.

Data similar to that for Fig. 13, with the low-pass filter in the audio circuit, were taken measuring the output *peak* voltage of the noise. The characteristics obtained were identical to those obtained with root-mean-square measurements.

Since the harmonics of the tone present in the output of the high-pass filter could not readily be separated from the noise for the peak

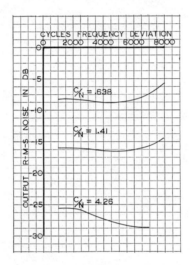

Fig. 14—Variation of frequency modulation receiver output noise as frequency modulation is applied. 1600-cycle high-pass filter in audio output. Modulation frequency = 1000 cycles.

voltage measurements, the high-pass filter data were taken by root-mean-square measurements only.

Measurements were also made to determine how much the audio selectivity reduced the degree of signal depression present at the output of the detector of the frequency modulation receiver. The carrier-noise ratio was set so that the maximum peak voltage of the fluctuation noise was equal to the peak voltage of the carrier. At this carrier-noise ratio the maximum noise peaks depressed the signal down to zero at the output of the detector. At the output of the 1600-cycle low-pass filter, the maximum noise peaks depressed the signal five decibels. Thus, without the audio selectivity, the signal was depressed by an amount equal to its total amplitude; with the audio selectivity, the signal was depressed to five decibels below full amplitude or down to an amplitude of 56 per cent. Hence the reduction of the depth of the signal depression was from a 100 per cent depression to a depression of $(100-56)=44$ per cent or a reduction of about seven decibels. The theoretical reduction of the fluctuation noise in the absence of the modulation would be equal to the square root of the ratio of band widths or six decibels. Thus the reduction of the signal depression is, for all practical purposes, the same as the reduction in the peak voltage of the noise alone.

Impulse Noise Measurements

The first measurements on impulse noise were made using an automobile ignition system driven by an electric motor. However the output from this generator proved to be unsteady and did not allow a reasonable measurement accuracy. Consequently a square-wave multivibrator was set up. This type of impulse noise generator proved to be even more stable than the fluctuation noise source and allowed accurate data to be obtained. On the other hand, the output of the receiver being fed by this noise generator was not as steady as would be expected. In the absence of the carrier the output was steady, but as the carrier was introduced the output peak voltage started to fluctuate. Apparently the phase relation between the components of the noise spectrum and the carrier varies in such a manner as to form a resultant wave which varies between amplitude modulation and phase or frequency modulation. Hence the output of a receiver which is adjusted to receive either type of modulation alone will fluctuate depending upon the probability considerations of the phase of combination of the carrier and noise voltages.

The preliminary impulse noise measurements were made on an amplitude modulation receiver by comparing the peak voltage ratio between the two available band widths of 6500 and 1600 cycles. The 6500-cycle channel was fed to one set of oscilloscope plates and the 1600-cycle channel to the other. Thus, when the peak voltages at the outputs of the two channels were equal the oscilloscope diagram took a symmetrical shape somewhat like a plus sign. The two channel levels were equalized by means of a tone. Hence, when the noise voltage was substituted for the tone, the amount of attenuation that had to be inserted in the wider band to produce a symmetrical diagram on the oscilloscope was taken as the ratio of the peak voltages of the two band widths. In this manner a series of readings was taken which definitely proved that the peak voltage ratio of the two band widths was proportional to the band width ratio. These readings were taken on both the ignition system noise generator and the multivibrator generator. As a check, readings on fluctuation noise were also taken which showed that the peak voltage of fluctuation noise varies as the square root of the band width.

The final measurements on impulse noise were made using the same procedure followed for the fluctuation noise measurements of Fig. 9. Only peak voltage measurements were made on this type of noise. The curves are shown in Fig. 15. It can be seen that the peak voltage characteristics of impulse noise are similar to those of fluctuation noise except for the location of the improvement threshold. For the receiver

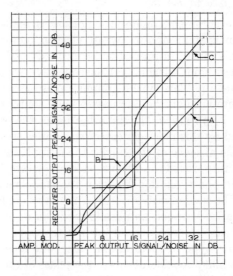

Fig. 15—Measured peak signal-noise ratio characteristics of impulse noise. Curve *A* = amplitude modulation receiver. Curve *B* = frequency modulation receiver with deviation ratio of unity. Curve *C* = frequency modulation receiver with deviation ratio of four.

with a deviation ratio of four, the improvement threshold occurs at a carrier-noise ratio slightly above sixteen decibels as compared with slightly above eight decibels for fluctuation noise. The difference between the improvement thresholds for the two frequency modulation receivers is about fourteen decibels; the corresponding theoretical figure, which is equal to the ratio of the two deviation ratios, is twelve decibels. The theoretical difference between the strong-carrier frequency modulation improvements for impulse and fluctuation noises, as indicated by the difference between the factors two and the square root of three respectively, is too small to be measurable with such variable quantities as these noise voltages.

Since the signal-noise ratios for the curves of Fig. 15 were obtained by measuring the noise and signal in the absence of each other, the signal-depressing effect of the noise does not show up. However, in the case of impulse noise, these curves are more representative of the actual situation existing, because the noise depresses the signal for only a small percentage of the time. In the listening and oscilloscope observations conducted with carrier-noise ratios below the improvement threshold, it was observed that at unity carrier-noise ratio the noise peaks depressed the amplitude of the signal to zero at the output of the detector. When the low-pass filter was inserted in the audio circuit, the impulse noise peaks depressed the signal about 2.5 decibels or reduced the amplitude from 100 per cent to 75 per cent. The effective signal-noise ratio is then increased from unity to $100/(100-75)=4$ or 12

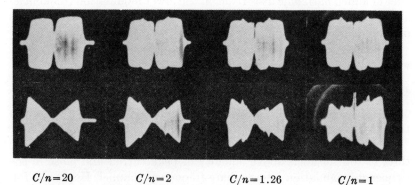

$C/n=20$ $C/n=2$ $C/n=1.26$ $C/n=1$

Fig. 16—Over-all transmission oscillograms of the frequency and amplitude modulation receivers. 1600-cycle low-pass filter in audio output. Top row = amplitude modulation, bottom row = frequency modulation. C/n = ratio between the carrier and the variable-frequency heterodyning voltages.

decibels. This is equal to the theoretical reduction in peak voltage of impulse noise which would be effected by this four-to-one band width ratio. It is then evident that the reduction of the depth of the signal depression caused by the impulse noise is of the same magnitude as the reduction of the peak voltage of the noise alone.

Over-all Transmissions

The oscillograms of Fig. 16 show the over-all transmissions of the amplitude and frequency modulation receivers at various carrier-noise ratios. These oscillograms were taken by tuning the receiver to a carrier, and then, to simulate the noise, manually tuning a heterodyning signal across the intermediate-frequency channel. The audio beat output of the receiver was applied through the low-pass filter to the vertical plates of the oscilloscope. A bias proportional to the frequency change of the heterodyning voltage was applied to the other set of oscilloscope plates. Consequently the spectra obtained are those which would be produced by the combination of a single noise component of variable frequency and the carrier. At the higher carrier-noise ratios, the spectrum is rectangular for amplitude modulation and triangular for frequency modulation. The dip in the middle of the amplitude modulation spectrum is where the audio output is near zero beat. As the carrier-noise ratio is decreased, the frequency modulation spectrum deviates from its triangular shape and the wave form of the receiver output has increased harmonic content at the lower audio frequencies where the audio selectivity does not eliminate the harmonics.

The amplitude modulation spectra of Fig. 16 also show the presence of added harmonic distortion on the lower modulation frequencies and lower carrier-noise ratios. However, the effect is so small that it is of little consequence.

The spectra of Fig. 16 allow a better understanding of the situation which is theoretically portrayed by (7) of the theory.

Experimental Conclusions

It can be concluded that the experimental data in general confirm the theory and point out the following additional information:

The improvement threshold starts at a carrier-noise ratio about three or four decibels above equality of peak carrier and noise in the frequency modulation intermediate-frequency channel. Hence the full frequency modulation improvement may be obtained down to a peak carrier-noise ratio in the frequency modulation receiver of three or four decibels.

The root-mean-square fluctuation noise characteristics differ from the peak fluctuation noise characteristics for carrier-noise ratios below the improvement threshold. The improvement threshold starts at about the same peak carrier-noise ratio, but the improvement does not fall off as sharply as it does for peak signal-noise ratios. Thus, for carrier-noise ratios below the improvement threshold the energy content of the frequency modulation noise is reduced, but the peak characteristics are approximately the same as those of the amplitude modulation receiver. The characteristics are not exactly the same due to the frequency limiting which allows the noise peaks to depress the signal, but does not allow them to rise above the signal.

The crest factor of the fluctuation noise at the outputs of the frequency and amplitude modulation receivers is about thirteen decibels or 4.5 to one for the strong-carrier condition. The crest factor of amplitude modulation fluctuation noise remains fairly constant regardless of the carrier-noise ratio. At equality of peak carrier and peak noise in the frequency modulation intermediate-frequency channel, the crest factor of the noise in the output of the frequency modulation receiver rises to a value which counteracts the peak signal-noise ratio improvement over amplitude modulation; the improvement threshold manifests itself in this manner.

At the improvement threshold, the application of the audio selectivity reduces the signal depression due to a noise peak by the same ratio that it reduces the noise in the absence of the signal. Thus the depth of a noise depression in the signal is reduced by a ratio equal to the square root of the deviation ratio in the case of fluctuation noise, and equal to the deviation ratio in the case of impulse noise.

General Conclusions

The theory and experimental data point out the following conclusions:

A frequency modulation system offers a signal-noise ratio improvement over an equivalent amplitude modulation system when the carrier-noise ratio is high enough. For fluctuation noise this improvement is equal to the square root of three times the deviation ratio for both peak and root-mean-square values. For impulse noise the corresponding peak signal-noise ratio improvement is equal to twice the deviation ratio. When the carrier-noise ratio is about three or four decibels above equality of peak carrier and peak noise in the frequency modulation intermediate-frequency channel, the peak improvement for either type of noise starts to decrease and becomes zero at a carrier-noise ratio about equal to unity. Below this "improvement threshold," the peak characteristics of the frequency modulation receiver are approximately the same as those of the equivalent amplitude modulation receiver. The root-mean-square characteristics of the frequency modulation noise show a reduction of the energy content of the noise for carrier-noise ratios below the improvement threshold; this is evidenced by the fact that the improvement threshold is not as sharp for root-mean-square values as for peak values.

At the lower carrier-noise ratios, frequency modulation systems with lower deviation ratios have an advantage over systems with higher deviation ratios. Since the high deviation ratio system has a wider intermediate-frequency channel, more noise is accepted by that channel so that the improvement threshold occurs at a higher carrier level in the high deviation ratio system than in the low. Hence the low deviation ratio systems retain their frequency modulation improvement down to lower carrier levels.

The peak voltage of fluctuation noise varies with band width in the same manner as the root-mean-square voltage, namely, as the square root of the band width. The peak voltage of impulse noise varies directly as the band width. In frequency modulation systems with a deviation ratio greater than unity, this difference in the variation with band width makes the improvement threshold occur at a higher carrier level with impulse noise than with fluctuation noise. Hence frequency modulation systems with higher deviation ratios are more susceptible to impulse noise interference.

Because of a phenomenon called "frequency limiting" the peak frequency deviations of the noise or the noise-plus-signal are limited so that the peak value cannot rise above the maximum peak value of the signal at the output of the detector. The application of audio selectivity reduces this maximum value of the noise so that fluctuation noise cannot rise to a value higher than the maximum value of the signal divided by the square root of the deviation ratio; the corresponding value of impulse noise cannot rise to a value higher than the maximum peak voltage of the signal divided by the deviation ratio. Inherent with this limiting effect is a signal-depressing effect which causes the fluctuation noise gradually to smother the signal as the carrier-noise ratio is lowered below the improvement threshold. However in the case of impulse noise, the signal depression is not as troublesome, and a noise-suppression effect is created which is similar to that effected in the recent circuits for suppressing impulse noise which is stronger than the carrier in an amplitude modulation system. When the deviation ratio is greater than unity, this frequency limiting is more effective than the corresponding amplitude modulation noise-suppression circuits; this is caused by the audio selectivity reducing the maximum peak value of the noise so that it is less than the peak value of the signal.

For carrier-noise ratios greater than unity, the application of frequency modulation to the carrier does not increase the noise above its value in the absence of applied frequency modulation.

At the transmitter, a four-to-one power gain is obtained by the use of class C radio-frequency amplification for frequency modulation instead of the customary class B amplification as is used for low level amplitude modulation. Therefore, for the same transmitter power input, a frequency modulation system will produce at its receiver a carrier which is twice as strong as that produced at the receiver of an amplitude modulation system. This results in two effects: first, the frequency modulation improvement is doubled for carrier-noise ratios above the improvement threshold; second, when the improvement threshold occurs in the frequency modulation receiver, the carrier-noise ratio existing in the amplitude modulation receiver is one half of what it would have been without the transmitter power gain.

ACKNOWLEDGMENT

The author thanks Mr. H. H. Beverage and Mr. H. O. Peterson for the careful guidance and helpful suggestions received from them during the course of this work. The assistance of Mr. R. E. Schock in the experimental work is also appreciated.

PAPER NO. 3

Reprinted from *Proc. IRE*, Vol. 35, No. 10, pp. 1013–1020, Oct. 1947

Variation of Bandwidth with Modulation Index in Frequency Modulation*

MURLAN S. CORRINGTON†

Summary—Equations are derived for the carrier and side-frequency amplitudes which are obtained when a carrier wave is frequency-modulated by a complex audio signal. The bandwidth occupied by such a frequency-modulated wave is defined as the distance between the two frequencies beyond which none of the side frequencies is greater than 1 per cent of the carrier amplitude obtained when the modulation is removed.

Curves are given to show the amount this bandwidth exceeds the extremes of deviation for a range of modulation indexes from 0.1 to 10,000, for sinusoidal, square, rectangular, and triangular modulation. For more precise definitions of bandwidth, curves are also given for side-frequency amplitude limits of 0.1 per cent and 0.01 per cent of the carrier-wave amplitude. For complex modulation the total bandwidth can be estimated by computing the bandwidth that would be required by each audio-frequency component, if it were on separately, and adding the results.

INTRODUCTION

IF A CARRIER wave is frequency-modulated with a sinusoidal audio voltage, an infinite number of side frequencies is produced. The carrier amplitude is reduced when the modulating voltage is applied, and may even become zero. If the deviation is increased, additional side frequencies are produced in both sidebands, and the distribution of the intensities of the previous ones is changed. For a single audio tone, the distance between side frequencies is always equal to the audio frequency. When two or more modulating tones are used simultaneously, side frequencies are produced which are separated from the carrier by all possible combination frequencies which can be obtained from sums and differences of harmonics of the modulating frequencies.[1,2] Thus, if there are two audio tones of frequencies μ_1 and μ_2, the side frequencies are separated from the carrier by $\pm r\mu_1 \pm s\mu_2$ where r and s are positive integers or zero.

Although, theoretically, an infinite number of side frequencies is produced, in practice the ones separated from the carrier by a frequency greater than the deviation decrease rapidly toward zero, so the bandwidth always exceeds the total frequency excursion, but nevertheless is limited. For large modulation indexes and a sinusoidal modulating voltage, the bandwidth approaches, and is only slightly greater than, the total frequency excursion.

To show how the bandwidth changes with modulation index, exact mathematical expressions for the spectrum will now be obtained.

THE SPECTRUM OF A CARRIER WAVE WHICH IS FREQUENCY-MODULATED WITH A SINUSOIDAL SIGNAL

When a carrier wave is frequency-modulated with a single audio tone, the equation for the voltage is

$$e = E \sin\left(\omega t + \frac{D}{\mu}\sin 2\pi\mu t\right) \qquad (1)$$

where

$E =$ amplitude of the wave
$\omega =$ angular frequency of the carrier, radians per second
$D =$ deviation, cycles per second
$\mu =$ audio frequency, cycles per second
$t =$ time in seconds
$D/\mu =$ modulation index.

This expression can be expanded in a spectrum consisting of a carrier and side frequencies, in accord with the result[3]

$$e = E \sum_{n=-\infty}^{\infty} J_n(D/\mu) \sin(\omega t + 2\pi n\mu t) \qquad (2)$$

where $J_n(D/\mu)$ is a Bessel function of the first kind of order n and argument D/μ.

Graphs of the Bessel Functions

To plot the spectrum of a frequency-modulated wave for a given modulation index D/μ, use a table of Bessel

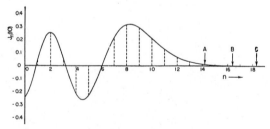

Fig. 1—Graph of $J_n(10)$.

functions[4,5] to obtain the amplitudes of the carrier and the side frequencies in the upper sideband. The odd-order side frequencies in the lower sideband will have

* Decimal classification: R148.12. Original manuscript received by the Institute, May 27, 1946; revised manuscript received, February 5, 1947.
† RCA Victor Division, Radio Corporation of America, Camden, N. J.

1 Murray G. Crosby, "Carrier and side-frequency relations with multitone frequency or phase modulation," *RCA Rev.*, vol. 3, pp. 103–106; July, 1938.
2 M. Kulp, "Spektra und Klirrfaktoren frequenz- und amplituden-modulierter Schwingungen," Part I. *Elek. Nach.-Tech.*, vol. 19, pp. 72–84; May, 1942.

3 A. Bloch, "Modulation theory," *Jour. I.E.E.*, (London), Part III, vol. 91, pp. 31–42; March, 1944.
4 E. Jahnke and F. Emde, "Tables of functions with formulae and curves," Dover Publications, New York, N. Y., 1943, p. 171.
5 August Hund, "Frequency Modulation," McGraw-Hill Book Co., New York, N. Y., 1942; Table VI, p. 352.

signs opposite to those in the upper sideband, and the even-order side frequencies will have the same sign. Fig. 1 is a graph of $J_n(10)$. If the ordinates are drawn for each integer, as shown by the dotted lines, the side frequencies in the upper sideband will be proportional to these ordinates and the carrier will be proportional to the ordinate at $n = 0$.

If the modulation index is increased to 1000, the part of the curve for n nearly equal to the modulation index is similar, but is reduced in amplitude.[6] Fig. 2 shows the variation of the side frequencies near the

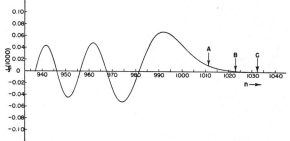

Fig. 2—Graph of $J_n(1000)$.

upper edge of the band. The curve oscillates with gradually increasing amplitude and slowly increasing period all the way from the origin to the last maximum, which is also the absolute maximum, and then decreases rapidly toward zero. The maximum energy occurs at a point in the band just inside the frequencies which correspond to the ends of the swings. When the deviation increases and the modulating frequency remains constant, the total energy of the spectrum is spread over a greater bandwidth, and the average amplitude of the side frequencies must decrease uniformly to maintain constant total energy in the modulated carrier wave.

The absolute maximum value of the Bessel function, for positive values of m, is shown by Fig. 3. For a given

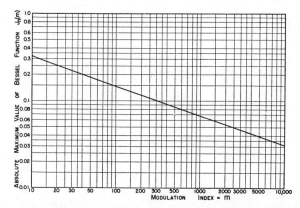

Fig. 3—Absolute maximum value of Bessel function $J_n(m)$.

modulation index, this maximum occurs for a value of the order n slightly less than the modulation index m. For example, for a modulation index of 1000 the

[6] Murlan S. Corrington, "Tables of Bessel functions $J_n(1000)$," *Jour. Math. Phys.* (M.I.T.), vol. 24, pp. 144–147; November, 1945.

maximum occurs at $n = 991.91$ and equals 0.06756. If the modulation index is 10, the maximum occurs at $n = 8.23$ and equals 0.3210. The curve of Fig. 3 shows this maximum value for a range of modulation indexes from 10 to 10,000. It was computed from the formulas of Meissel[7] which state that the Bessel function $J_n(k)$ reaches its absolute maximum

$$J_n(k) = \frac{0.6748\ 8509\ 6430}{\sqrt[3]{k}} + \frac{0.0727\ 6309\ 8182}{k}$$
$$+ \frac{0.0199\ 5975\ 0328}{\sqrt[3]{k^5}} + \cdots \quad (3)$$

for the value

$$n = k - 0.8086\ 1651\ 7466\sqrt[3]{k} - \frac{0.0606\ 4998\ 7910}{\sqrt[3]{k}}$$
$$- \frac{0.0316\ 7351\ 0263}{k} - \cdots . \quad (4)$$

A family of curves for modulation indexes from one to twenty is shown by Fig. 4. The vertical scale represents the amplitude of the given side frequency for each modulation index. The curve of Fig. 1 can be obtained by cutting a section through the surface for a modulation index of 10. Contour line A is for the constant value of the Bessel function $J_n(D/\mu) = 0.01$. Similarly, the contour B corresponds to $J_n(D/\mu) = 0.001$, and contour C is drawn for $J_n(D/\mu) = 0.0001$.

Curve D is shown for the order of the Bessel function equal to the argument. If the bandwidth of a frequency-modulated carrier wave were just equal to twice the deviation, the side frequencies would not extend beyond curve D. It is evident that for a given modulation index the bandwidth extends beyond curve D (say to curve A), but that the intensities of the side frequencies beyond curve D are decreasing rapidly.

Curve E is drawn along the top of the first crest and gives the absolute maximum value of the envelope of the side frequencies for each modulation index. This curve is also given by Fig. 3. The curves F, G, H, I, J, and K show where the surface goes through zero, i.e., the zeros of the Bessel functions.

Definition of Bandwidth

Theoretically, there is an infinite number of side frequencies in the spectrum of a frequency-modulated carrier wave, but the amplitudes decrease very rapidly beyond the last maximum. Point A on Fig. 1 corresponds to the value $J_n(10) = 0.01$ and will be defined as the edge of the band. Point B is shown for $J_n(10) = 0.001$, and point C for $J_n(10) = 0.0001$. These latter two points can be used for a more precise definition of bandwidth, but point A is to be taken as the usual limit for practical purposes.

[7] E. Meissel, "Beitrag zur Theorie der Bessel'schen Functionen," *Astronom. Nach.*, vol. 128, cols. 435–438; 1891.

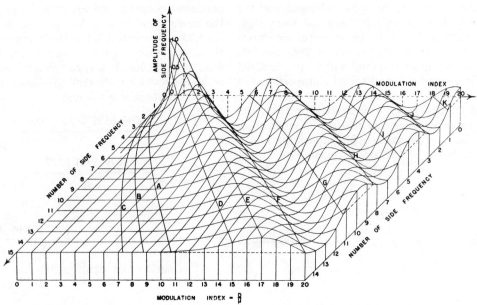

Fig. 4—Side-frequency amplitudes.

The curves of Fig. 5 show the variation of the bandwidth as the modulation index is changed.[8]

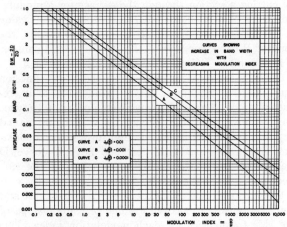

Fig. 5—Variation of bandwidth with modulation index.

Example: Let the deviation be ± 50 kilocycles and the audio frequency be 5 kilocycles. Find the bandwidth. The modulation index is $50/5 = 10$. From curve A of Fig. 5, the increase in bandwidth is approximately 0.42 or 42 per cent, so the bandwidth is approximately $2(50)(1+0.42) = 142$ kilocycles.

[8] A rather simple method for computing the argument for $J_n(x) = 0.01, 0.001, 0.0001$ is to use the approximate formula

$$J_\nu(\nu \operatorname{sech} \alpha) = \frac{\tanh \alpha}{\pi \sqrt{3}} \exp \left\{ \nu(\tanh \alpha + \tfrac{1}{3}\tanh^3 \alpha - \alpha) \right\} K_{1/3}\left(\frac{\nu}{3}\tanh^3 \alpha\right)$$

given in G. N. Watson, "A treatise on the theory of Bessel functions," The Macmillan Company, New York, N. Y., Second Edition, 1944, p. 250, where $K_{1/3}(x)$ is a modified Bessel function of the second kind of order $\tfrac{1}{3}$ and argument x. A series of values of α was chosen and the corresponding Bessel functions computed. These values were plotted on semilog paper and the arguments corresponding to the ordinates 0.01, 0.001, and 0.0001 were read off. The curves of Fig. 5 were obtained directly from these readings.

Bandwidth Required for Complex Modulation

If several modulating tones are present simultaneously, the carrier wave can be expressed as

$$e = E \sin \left\{ \omega t + \sum_{s=1}^{S} \frac{D_s}{\mu_s} \sin (2\pi\mu_s t + \epsilon_s) \right\} \quad (5)$$

where E is the amplitude of the wave, ω is the carrier angular frequency, D_s is the deviation corresponding to the audio frequency μ_s, t is the time, and ϵ_s is the phase angle corresponding to μ_s. This modulated carrier wave can be represented by a spectrum[3,9–11]

$$e = E \sum_{k_s=-\infty}^{\infty} \left\{ \prod_{s=1}^{S} J_{k_s}(m_s) \right\} \sin \left(\omega t + \sum_{s=1}^{S} k_s \theta_s \right) \quad (6)$$

where

$$m_s = D_s/\mu_s \quad \text{and} \quad \theta_s = 2\pi\mu_s t + \epsilon_s.$$

In the case of two-tone modulation this becomes[1]

$$E \sin \left\{ \omega t + D_1/\mu_1 \sin 2\pi\mu_1 t + D_2/\mu_2 \sin 2\pi\mu_2 t \right\}$$

$$= E \sum_{m=-\infty}^{\infty} \sum_{n=-\infty}^{\infty} J_m(D_1/\mu_1) J_n(D_2/\mu_2) \sin (\omega + 2\pi m\mu_1 + 2\pi n\mu_2)t. \quad (7)$$

This result shows that the spectrum is now much more complicated than for a single modulating tone,

[9] E. C. Cherry and R. S. Rivlin, "Non-linear distortion, with particular reference to the theory of frequency modulated waves, Part I," *Phil. Mag.*, vol. 32, pp. 265–281; October, 1941.

[10] A. S. Gladwin, "Energy distribution in the spectrum of a frequency modulated wave, Part I," *Phil. Mag.*, vol. 35, pp. 787–802; December, 1944.

[11] K. R. Sturley, "Frequency modulation," *Jour. I.E.E.* (London), vol. 92, Part III, pp. 197–218, September, 1945.

and that side frequencies will be produced at spacings from the carrier given by all the possible combinations $\pm m\mu_1 \pm n\mu_2$. The amplitude of each side frequency will be proportional to the product of the two Bessel functions. Just as the maximum deviation occurs when D_1 and D_2 are in phase, the maximum bandwidth is given approximately by the sum of the two bandwidths that would be obtained with the two modulating tones used one at a time.

The graph of Fig. 6 shows the spectrum obtained when two tones are present simultaneously, in accord with (7). The side frequencies are no longer symmetrical

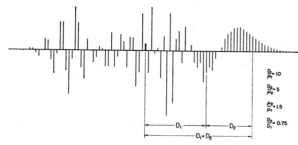

Fig. 6—Spectrum for complex modulation.

about the carrier and, when they are separated from the carrier by a frequency greater than D_1+D_2, decrease rapidly toward zero. The upper sideband contains 57.9 per cent of the power, the lower sideband 42.0 per cent, and the carrier 0.1 per cent.

General Method for Computing Side-Frequency Amplitudes

If the modulating signal is given, the variations of the phase angle will be proportional to the integral of the signal. This integration can be done directly, or by numerical integration, and the constant of integration should be chosen so the average value of the phase angle, over a complete cycle, is zero. If the phase angle is $S(t)$ the frequency-modulated carrier wave can be expressed as

$$e = E \sin \{\omega t + S(t)\}$$
$$= E \sin \omega t \cos S(t) + E \cos \omega t \sin S(t). \quad (8)$$

Expand in the Fourier series

$$\cos S(t) = \sum_{n=0}^{\infty} \{a_n \sin n\theta + b_n \cos n\theta\} \quad (9)$$

$$\sin S(t) = \sum_{n=0}^{\infty} \{c_n \sin n\theta + d_n \cos n\theta\} \quad (10)$$

where $\theta = 2\pi\mu t$ and μ is the repetition rate of the signal in cycles per second. This expansion can be done by direct integration of the integrals for the Fourier coeffici-

ents or by one of the numerical methods for harmonic analysis.[12,13]

Then

$$e = E \sin \omega t \sum_{n=0}^{\infty} \{a_n \sin n\theta + b_n \cos n\theta\}$$

$$+ E \cos \omega t \sum_{n=0}^{\infty} \{c_n \sin n\theta + d_n \cos n\theta\}$$

$$= E \sum_{n=0}^{\infty} \{\tfrac{1}{2}(a_n + d_n) \cos (\omega - 2\pi n\mu)t$$

$$- \tfrac{1}{2}(a_n - d_n) \cos (\omega + 2\pi n\mu)t$$

$$+ \tfrac{1}{2}(b_n - c_n) \sin (\omega - 2\pi n\mu)t$$

$$+ \tfrac{1}{2}(b_n + c_n) \sin (\omega + 2\pi n\mu)t\} \quad (11)$$

which gives the side-frequency amplitudes directly.

The results of numerical computation can be checked by taking the sum of the squares of the carrier and each of the side-frequency amplitudes; they should add up to E^2.

THE SPECTRUM OF A CARRIER WAVE WHICH IS FREQUENCY-MODULATED WITH A RECTANGULAR SIGNAL

When the signal is a rectangular or square wave, as in frequency-modulated telegraphy, or in television video and synchronizing signals, the carrier wave can be analyzed into a spectrum in a similar manner. If the

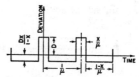

Fig. 7—Modulating signal.

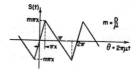

Fig. 8—Variation of phase angle.

modulating signal is as shown by Fig. 7, the phase angle $S(t)$ will be 2π times the integral of this curve, as shown by Fig. 8, where $m = D/\mu$. The equation for the frequency-modulated wave becomes

$$e = E \sin \{\omega t + S(t)\}$$
$$= E \sin \omega t \cos S(t) + E \cos \omega t \sin S(t). \quad (12)$$

Since $\cos S(t)$ is symmetrical about the origin, it can be expanded in a cosine series. Similarly, $\sin S(t)$ is

[12] C. Runge and H. König, "Vorlesungen Über Numerisches Rechnen," Julius Springer, Berlin, 1924, pp. 211–231.
[13] R. P. G. Denman, "36 and 72 ordinate schedules for general harmonic analysis," *Electronics*, vol. 15, pp. 44–47, September, 1942. In addition to the corrections listed in vol. 16, pp. 214–215, April, 1943, change the correction in column one, p. 215, from "Column for B_{11} and B_{25}, line for $\alpha = 20°$, for Δ_2, read Δ_2," to read "... for Δ_2, read $-\Delta_2$."

skew symmetric about the origin and can be expanded in a sine series. From Fig. 8,

$$\cos S(t) = \cos m\theta \qquad\qquad 0 \leqq \theta \leqq \pi x$$

$$= \cos \frac{mx(\pi - \theta)}{1 - x} \qquad \pi x \leqq \theta \leqq \pi$$

$$= \sum_{n=0}^{\infty} b_n \cos 2\pi n\mu t \qquad (13)$$

where

$$b_n = \frac{2}{\pi} \int_0^{\pi x} \cos m\theta \cos n\theta d\theta$$

$$+ \frac{2}{\pi} \int_{\pi x}^{\pi} \cos \frac{mx}{1 - x} (\pi - \theta) \cos n\theta d\theta \qquad (14)$$

$$b_0 = \frac{1}{\pi m x} \sin \pi m x. \qquad (15)$$

Similarly,

$$\sin S(t) = \sin m\theta \qquad\qquad 0 \leqq \theta \leqq \pi x$$

$$= \sin \frac{mx(\pi - \theta)}{1 - x} \qquad \pi x \leqq \theta \leqq \pi$$

$$= \sum_{n=0}^{\infty} c_n \sin 2\pi n\mu t \qquad (16)$$

where

$$c_n = \frac{2}{\pi} \int_0^{\pi x} \sin m\theta \sin n\theta d\theta$$

$$+ \frac{2}{\pi} \int_{\pi x}^{\pi} \sin \frac{mx}{1 - x} (\pi - \theta) \sin n\theta d\theta. \qquad (17)$$

When these results are substituted into (11), the spectrum is given by

$$e = \sum_{n=-\infty}^{\infty} \frac{mE}{\pi(m - n)(mx - nx + n)} \sin \pi x(m - n)$$
$$\cdot \sin (\omega + 2\pi n\mu)t \qquad (18)$$

where

E = amplitude of the wave
m = modulation index = D/μ
D = maximum frequency deviation, cycles per second
μ = repetition rate, cycles per second
x = fraction of the time the frequency is at the extreme deviation D
ω = angular frequency of the carrier, radians per second
t = time in seconds.

When $n = 0$, the carrier amplitude is given by

$$\text{Carrier} = \frac{E}{\pi m x} \sin m\pi x \sin \omega t. \qquad (19)$$

The side frequencies adjacent to the carrier are given by $n = \pm 1$ and are separated from the carrier by an amount equal to the audio frequency. They are given by:

First upper side frequency

$$= \frac{mE}{\pi(m-1)(mx-x+1)} \sin \pi x(m-1) \sin (\omega+2\pi\mu)t. \qquad (20)$$

First lower side frequency

$$= \frac{mE}{\pi(m+1)(mx+x-1)} \sin \pi x(m+1) \sin (\omega-2\pi\mu)t. \qquad (21)$$

The other side frequencies can be determined by assigning appropriate values to n in (18).

The indeterminate cases must be evaluated separately:

Case I $\qquad m = n, \qquad \frac{1}{2}(b_n + c_n) = x \qquad (22)$

Case II $\quad \dfrac{mx}{1-x} = -n, \quad \frac{1}{2}(b_n+c_n) = \dfrac{1}{\pi(m-n)} \sin \pi x(m-n)$
$$\qquad\qquad\qquad\qquad + (-)^n(1-x) \qquad (23)$$

Case III $\qquad m = -n, \qquad \frac{1}{2}(b_n - c_n) = x \qquad (24)$

Case IV $\quad \dfrac{mx}{1-x} = n, \qquad \frac{1}{2}(b_n-c_n) = \dfrac{1}{\pi(m+n)} \sin \pi x(m+n)$
$$\qquad\qquad\qquad\qquad + (-)^n(1-x). \qquad (25)$$

The case of square-wave modulation is obtained by setting $x = \frac{1}{2}$. This gives the result

$$e = \sum_{n=-\infty}^{\infty} \frac{2mE}{\pi(m^2-n^2)} \sin (m-n)\frac{\pi}{2} \sin (\omega+2\pi n\mu)t$$

$$= \frac{2E}{\pi m} \sin m\frac{\pi}{2} \sin \omega t$$

$$+ \frac{2mE}{\pi(m^2-1^2)} \cos \frac{m\pi}{2} \left\{ \sin (\omega-2\pi\mu)t - \sin (\omega+2\pi\mu)t \right\}$$

$$- \frac{2mE}{\pi(m^2-2^2)} \sin \frac{m\pi}{2} \left\{ \sin (\omega-4\pi\mu)t + \sin (\omega+4\pi\mu)t \right\}$$

$$- \frac{2mE}{\pi(m^2-3^2)} \cos \frac{m\pi}{2} \left\{ \sin (\omega-6\pi\mu)t - \sin (\omega+6\pi\mu)t \right\}$$

$$+ \cdots . \qquad (26)$$

This result, for $x = \frac{1}{2}$, agrees with that previously obtained by van der Pol.[14]

The limits for the amplitudes of the side frequencies can be determined from the coefficients of (18). Thus, if $m = D/\mu = 5$, and if $x = \frac{1}{4}$, the limit of the amplitudes becomes

$$\text{Amplitude limit} = \frac{m}{\pi(m - n)(mx - nx + n)}$$

$$= \frac{20}{\pi(5 - n)(5 + 3n)}. \qquad (27)$$

[14] Balth. van der Pol., "Frequency modulation," Proc. I.R.E., vol. 18, pp. 1194–1205; July, 1930.

This curve is shown by Fig. 9. Actually, most of the side-frequency amplitudes will be less than this because of the first sinusoidal term of (18). As shown by

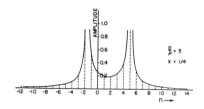

Fig. 9—Limits for side-frequency amplitudes.

Fig. 10, the amplitudes oscillate within the limits of the curve of Fig. 9. It may be easily seen that most of the

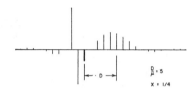

Fig. 10—Spectrum for rectangular modulation.

energy of the spectrum is concentrated about the frequencies that correspond to the two limits of the deviation.

Bandwidth Required for Rectangular Modulation

Equation 18 shows that there is an infinite number of side frequencies in the spectrum of a frequency-modu-

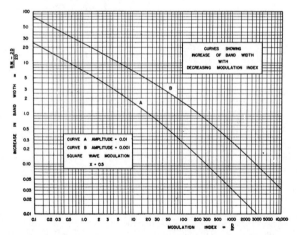

Fig. 11—Variation of bandwidth with modulation index.

lated wave with rectangular modulation. As shown by Figs. 9 and 10, the amplitudes of these side frequencies

decrease uniformly beyond the limits of the deviations. If the edges of the band are defined as the points corresponding to a limiting amplitude of $0.01E$, the bandwidth can be computed directly from (18). For the case of square-wave modulation, $x = 0.5$, and the increase in bandwidth with decreasing modulation index will be as shown by Fig. 11. If a more strict definition of bandwidth is required, curve B shows the width for the limiting amplitude $0.001E$. Curve A is an accurate enough limit for most practical cases.

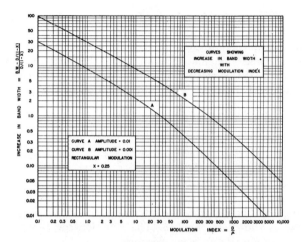

Fig. 12—Variation of bandwidth with modulation index.

If the maximum deviation is for one-fourth the time, $x = 0.25$, the curves of Fig. 12 show the corresponding limits of the bandwidth. Other sets of curves, for other values of x, can be computed from (18).

It will be noted that the band does not end as abruptly with rectangular modulation as it did with sinusoidal modulation. The curves of Figs. 11 and 12 are much farther apart than the corresponding curves of Fig. 5.

THE SPECTRUM OF A CARRIER WAVE WHICH IS FREQUENCY-MODULATED WITH A TRIANGULAR SIGNAL

When a uniformly spaced series of parallel bars, each one unit wide, is scanned at a uniform rate with a

Fig. 13—Scanning of picture element.

rectangular aperture of unit width, as shown by Fig. 13, the resulting signal is proportional to the area of the bar covered by the aperture. The signal will have a triangular wave form, as shown by Fig. 14. During the time the aperture is between the bars, the output will be constant. As the aperture starts to cover a bar, the

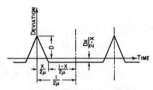

Fig. 14—Modulating signal.

output increases linearly until the aperture covers the entire bar. As the aperture moves on, the signal decreases linearly until it reaches the previous constant value, and remains constant until the next bar is reached. If this wave form is used to modulate the frequency of a carrier wave, the variation of the phase

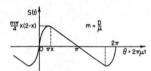

Fig. 15—Variation of phase angle.

angle will be 2π times the integral of this curve of Fig. 14, as shown by Fig. 15. The equation for this frequency-modulated wave becomes

$$e = E \sin \left\{ \omega t + S(t) \right\}$$
$$= E \sin \omega t \cos S(t) + E \cos \omega t \sin S(t) \quad (28)$$

$$S(t) = \frac{D}{\mu} \left\{ \frac{\pi x(2-x)\theta - \theta^2}{\pi x(2-x)} \right\} \quad 0 \leq \theta \leq \pi x \quad (29)$$

$$= \frac{D}{\mu} \left\{ \frac{x(\pi - \theta)}{2 - x} \right\} \quad \pi x \leq \theta \leq \pi. \quad (30)$$

When $S(t)$ is expanded in a Fourier series[15] and (11) is used, the

---

[15] The integrals can be evaluated by the following process:

$$\int_0^{\pi x} \cos (\alpha\theta^2 + \beta\theta)d\theta = \int_0^{\pi x} \cos \left\{ \alpha(\theta + \beta/2\alpha)^2 - \beta^2/4\alpha \right\} d\theta$$

$$= \cos \frac{\beta^2}{4\alpha} \int_0^{\pi x} \cos \left\{ \alpha(\theta + \beta/2\alpha)^2 \right\} d\theta$$

$$+ \sin \frac{\beta^2}{4\alpha} \int_0^{\pi x} \sin \left\{ \alpha(\theta + \beta/2\alpha)^2 \right\} d\theta.$$

Let

$$\sqrt{\alpha}(\theta + \beta/2\alpha) = \pm \sqrt{v} \quad \text{and} \quad \sqrt{\alpha} \, d\theta = \pm \frac{dv}{2\sqrt{v}}.$$

Then

$$\int_0^{\pi x} \cos \left\{ \alpha(\theta + \beta/2\alpha)^2 \right\} d\theta = \sqrt{\frac{\pi}{2\alpha}} \int_{v_1}^{v_2} \frac{\cos v \, dv}{\sqrt{2\pi v}}$$

where

$$v_1 = \text{sgn} \, \beta \, \frac{\beta^2}{4\alpha}; \quad v_2 = \text{sgn} \, \gamma \, \frac{\gamma^2}{4\alpha}.$$

The same transformation can be used on the second integral.

---

amplitude of the nth side frequency

$$= \frac{1}{\sqrt{2\pi\alpha}} \cos \frac{\beta^2}{4\alpha} \left[\text{sgn} \, \gamma \, C \left\{ \frac{\gamma^2}{4\alpha} \right\} - \text{sgn} \, \beta \, C \left\{ \frac{\beta^2}{4\alpha} \right\} \right]$$

$$+ \frac{1}{\sqrt{2\pi\alpha}} \sin \frac{\beta^2}{4\alpha} \left[\text{sgn} \, \gamma \, S \left\{ \frac{\gamma^2}{4\alpha} \right\} - \text{sgn} \, \beta \, S \left\{ \frac{\beta^2}{4\alpha} \right\} \right]$$

$$- \frac{1}{\pi\gamma} \sin (\pi\gamma x + \epsilon) \quad (31)$$

and the amplitude of carrier

$$= \frac{1}{\sqrt{2\pi\alpha}} \cos \frac{\beta^2}{4\alpha} \left[C \left\{ \frac{\alpha\pi^2 x^4}{4} \right\} + C \left\{ \frac{\beta^2}{4\alpha} \right\} \right]$$

$$+ \frac{1}{\sqrt{2\pi\alpha}} \sin \frac{\beta^2}{4\alpha} \left[S \left\{ \frac{\alpha\pi^2 x^4}{4} \right\} + S \left\{ \frac{\beta^2}{4\alpha} \right\} \right]$$

$$+ \frac{1}{\epsilon} \sin (\pi\gamma x + \epsilon) \quad (32)$$

where

$$\alpha = \frac{m}{\pi x(2 - x)} \qquad \gamma = \frac{mx}{2 - x} + n$$

$$\beta = n - m \qquad \epsilon = \frac{-\pi m x}{2 - x}.$$

sgn β means the algebraic sign of β, and the C and S functions are the Fresnel integrals

$$C(z) = \frac{1}{\sqrt{2\pi}} \int_0^z \frac{\cos t \, dt}{\sqrt{t}} = \frac{1}{2} \int_0^z J_{-1/2}(t)dt \quad (33)$$

$$S(z) = \frac{1}{\sqrt{2\pi}} \int_0^z \frac{\sin t \, dt}{\sqrt{t}} = \frac{1}{2} \int_0^z J_{1/2}(t)dt. \quad (34)$$

These integrals are tabulated over a considerable range.[16],[17]

The vertical lines of Fig. 16 show the spectrum for triangular modulation with a modulation index of 10. The dotted line is the Bessel function $J_n(10)$; it gives the amplitudes of the side frequencies for the corresponding sine-wave signal. During triangular modula-

[16] See Table V, pp. 744–745, of footnote reference 8. Tables of $C(x)$ and $S(x)$, $x=0.02(0.02)1.00$; 7D, and $x=0.5(0.5)50.0$; 6D. For list of errors, see J. W. Wrench, Jr., "Mathematical tables—Errata," Mathematical Tables and other Aids to Computation, vol. 1, pp. 366–367; January, 1945.
[17] J. R. Airey, Sec'y., "Fresnel's Integrals, $S(x)$ and $C(x)$," British Association for the Advancement of Science, Report of the Ninety-fourth Meeting, 1926, pp. 273–275. Tables of $C(x)$ and $S(x)$, $x=0.0(0.1)20.0$; 6D.

tion, $x = 1$, the frequency varies linearly from one extreme of the frequency excursion to the other, while for

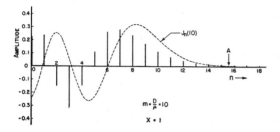

Fig. 16—Spectrum for triangular modulation.

sinusoidal modulation the frequency is near the extremes of frequency a greater portion of the time. As might be expected, more of the energy in the spectrum is near the ends of the swing for sine-wave modulation than for triangular modulation.

Bandwidth Required for Triangular Modulation

If the bandwidth is defined as the extremes of frequency beyond which none of the side-frequency amplitudes are greater than 1 per cent of the carrier amplitude that would be obtained if the modulation were removed, the variation of bandwidth with modulation index can be computed from the equations for the side-frequency amplitudes. Curve A of Fig. 17 shows how the bandwidth increases as the repetition rate is increased. For a more precise definition of bandwidth, either curve B or curve C can be used.

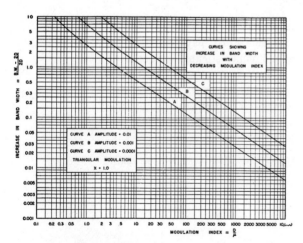

Fig. 17—Variation of bandwidth with modulation index.

If x is reduced to 0.1, the signal becomes a series of triangular pulses with blank spaces between. Most of the sideband energy will occur near the frequency which the carrier wave has between pulses, but the pulses

will cause energy to be distributed on both sides of this frequency. Fig. 18 shows the spectrum for a modula-

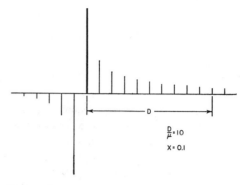

Fig. 18—Spectrum for triangular pulse modulation.

tion index of 10. The amplitudes decrease much more slowly than in the case of triangular or sinusoidal modulation.

CONCLUSIONS

When a carrier wave is modulated in frequency, an infinite number of side frequencies is produced. As the modulation index is changed, the amplitudes of the side frequencies change and the carrier is likewise reduced and may even become zero. Although the bandwidth is theoretically infinite, in practice the side frequencies gradually decrease in amplitude for frequencies beyond the extremes of the total frequency excursions. The bandwidth can be defined as the extremes of frequency beyond which none of the side-frequency amplitudes are greater than 1 per cent of the carrier voltage obtained when the modulation is removed.

The bandwidth so defined always exceeds the total frequency excursion, but is nevertheless limited. For large modulation indexes, i.e., the deviation much greater than the repetition rate, the bandwidth approaches the actual variation in frequency and is only slightly greater. For small modulation indexes, the bandwidth may be several times the actual frequency excursion. Curves are given to show the bandwidth for modulation indexes from 0.1 to 10,000 for sinusoidal, square, rectangular, and triangular modulation. For a more precise definition of bandwidth, curves are also given for amplitude limits of $0.001E$ and $0.0001E$.

When several modulating tones are on simultaneously, the side frequencies are produced at frequencies separated from the carrier by all combination frequencies that can be obtained by taking sums and differences of all the harmonics of the tone frequencies. The same curves can be used to determine the bandwidth when several audio tones are used simultaneously, since the bandwidth will be equal approximately to the sum of the bandwidths for each tone separately.

PAPER NO. 4

Reprinted from *BSTJ*, Vol. 27, No. 3, pp. 109–157, Jan. 1948

Statistical Properties of a Sine Wave Plus Random Noise

By S. O. RICE

INTRODUCTION

IN SOME technical problems we are concerned with a current which consists of a sinusoidal component plus a random noise component. A number of statistical properties of such a current are given here. The present paper may be regarded as an extension of Section 3.10 of an earlier paper,[1] "Mathematical Analysis of Random Noise", where some of the simpler properties of a sine wave plus random noise are discussed.

The current in which we are interested may be written as

$$
\begin{aligned}
I &= Q\cos qt + I_N \\
&= R\cos (qt + \theta)
\end{aligned}
\tag{3.4}
$$

where Q and q are constants, t is time, and I_N is a random (in the sense of Section 2.8 of Reference A) noise current. When the second expression involving the envelope R and the phase angle θ is used, the power spectrum of I_N is assumed to be confined to a relatively narrow band in the neighborhood of the sine wave frequency $f_q = q/(2\pi)$. This makes R and θ relatively slowly (usually) varying functions of time.

In Section 1, the probability density and cumulative distribution of I are discussed. In Section 2, the upward "crossings" of I (i.e., the expected number of times, per second, I increases through a given value I_1), are examined.

The probability density and the cumulative distribution of R are given in Section 3.10 of Reference A. The crossings of R are examined in Section 4 of the present paper.

The statistical properties of θ', the time derivative of the phase angle θ, are of interest because the instantaneous frequency of I may be defined to be $f_q + \theta'/(2\pi)$. The probability density of θ' is investigated in Section 5 and the crossings of θ' in Section 6. θ' is a function of time which behaves somewhat like a noise current and may accordingly be considered to consist of an infinite number of sinusoidal components. The problem of determining the "power spectrum" $W(f)$ of θ', i.e., the distribution of the mean square value of the components as a function of frequency, is attacked in

---

[1] *B.S.T.J.* 23 (1944), 282–332 and 24 (1945), 46–156. This paper will be called "Reference A".

Sections 7 and 8. The correlation function of θ' is expressed in terms of exponential integrals in Section 7, the power spectrum of I_N being assumed symmetrical and centered on f_q. In Section 8, values of $W(f)$ are obtained for the special case in which the power spectrum of I_N is centered on f_q and is of the normal-law type.

It is believed that some of the material presented here may find a use in the study of the effect of noise in frequency modulation systems. For example, the curves in Section 8 yield information regarding the noise power spectrum in the output of a primitive type of system. Also, the procedure employed to obtain the expression (5.7) for $\bar{\theta}'$ may be used to show that if

$$Q\cos[(A/\omega_0) \cos \omega_0 t + qt] + I_N = R\cos (qt + \theta)$$

the sinusoidal component of $d\theta/dt$ is[2]

$$-A (1 - e^{-\rho}) \sin \omega_0 t$$

where ρ is the ratio $Q^2/(2 \overline{I_N^2})$. This illustrates the "crowding effect" of the noise. The statistical analysis associated with R and θ of equations (3.4) (when the sine wave is absent) is similar to that used in the examination of the current returned to the sending end of a transmission line by reflections from many small irregularities distributed along the line. This suggests another application of the results.

ACKNOWLEDGMENT

I am indebted to a number of my associates for helpful discussions on the questions studied here. In particular, I wish to thank Mr. H. E. Curtis for his suggestions regarding this subject. As in Reference A, all of the computations for the curves and tables have been performed by Miss M. Darville. This work has been quite heavy and I gratefully acknowledge her contribution to this paper.

1. PROBABILITY DISTRIBUTION OF A SINE WAVE PLUS RANDOM NOISE

A current consisting of a sine wave plus random noise may be represented as

$$I = Q\cos qt + I_N \tag{1.1}$$

where Q and q are constants, t is the time, and I_N is a random noise current. The frequency, in cycles per second, of the sine wave is $f_q = q/(2\pi)$. In all

---

[2] The first person to obtain this result was, I believe, W. R. Young who gave it in an unpublished memorandum written early in 1945. He took the output of a frequency modulation limiter and discriminator to be proportional to either the signal frequency or to the instantaneous frequency (assumed to be distributed uniformly over the input band) of I_N according to whether Q is greater or less than the envelope of I_N. His memorandum also contains results which agree well with several obtained in this paper.

2

our work we denote the power spectrum of I_N by $w(f)$ and its correlation function by $\psi(\tau)$. The mean square value of I_N is denoted by ψ_0.

The study of the probability distribution of I is essentially a study of the integral[3]

$$p(I) = \frac{1}{\pi\sqrt{\psi_0}} \int_0^\pi \varphi\left[\frac{I - Q\cos\theta}{\sqrt{\psi_0}}\right] d\theta \qquad (1.2)$$

where

$$\varphi(x) = \frac{1}{\sqrt{2\pi}} e^{-x^2/2} \qquad (1.3)$$

and $p(I)$ is the probability density of I, i.e. $p(I)dI$ is the probability that a value of current selected at random will lie in the interval $I, I + dI$. Another expression for $p(I)$ is given by equation (3.10–6) of Reference A, namely

$$p(I) = \frac{1}{2\pi} \int_{-\infty}^{+\infty} e^{-izI - \psi_0 z^2/2} J_0(Qz) \, dz \qquad (1.4)$$

where $J_0(Qz)$ denotes the Bessel function of order zero.

The substitutions

$$y = \frac{I}{\sqrt{\psi_0}}, \qquad a = \frac{Q}{\sqrt{\psi_0}} \qquad (1.5)$$

enable us to write (1.2) as

$$p_1(y) = \sqrt{\psi_0}\, p(I) = \frac{1}{\pi} \int_0^\pi \varphi(y - a\cos\theta)\, d\theta, \qquad (1.6)$$

where $p_1(y)$ denotes the probability density of y. This is the expression actually studied. Curves showing $p_1(y)$ and the cumulative distribution function

$$\int_{-\infty}^{I} p(I_1)dI_1 = \int_{-\infty}^{y} p_1(y_1)dy_1$$

$$= \frac{1}{\pi} \int_0^\pi \varphi_{-1}(y - a\cos\theta)\, d\theta, \qquad (1.7)$$

where

$$\varphi_{-1}(x) = \int_{-\infty}^{x} \varphi(x_1)dx_1 = \tfrac{1}{2} + \tfrac{1}{2}\, erf\,(x/\sqrt{2}) \qquad (1.8)$$

[3] W. R. Bennett, "Response of a Linear Rectifier to Signal and Noise," *Jour. Acous. Soc. Amer.* Vol. 15 (1944), 164–172, and *B.S.T.J.* Vol. 23 (1944), 97–113.

3

are shown in Figs. 1 and 2.   The curves for $a = 10$ and $a = \sqrt{10}$ were computed by Simpson's rule from (1.6) and (1.7), and the curves for $a = 1$ were computed from the series (1.10) given below.   Since both $\varphi(x)$ and

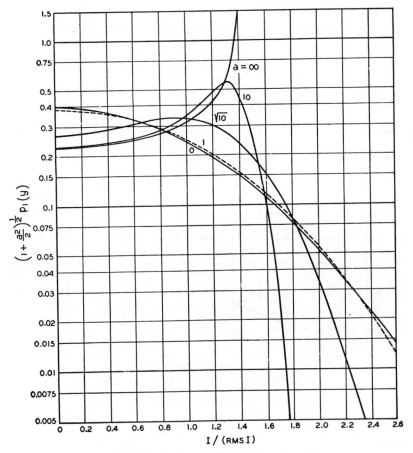

Fig. 1—Probability density of sine wave plus noise.

$I = Q\cos qt + I_N$, $a = Q/\sqrt{\psi_0}$, $y = I/\sqrt{\psi_0}$, $\sqrt{\psi_0} =$ rms I_N
$p_1(y)\, dI/\sqrt{\psi_0} =$ probability total current lies between I and $I + dI$
$y(1 + a^2/2)^{-1/2} = I/(\text{rms } I)$.   Curves are symmetrical about $y = 0$.

$\varphi_{-1}(x)$ are tabulated[4] functions the integrals in (1.6) and (1.7) are well suited to numerical evaluation.

[4] $\varphi(x)$ is given directly and $\varphi_{-1}(x)$ may be readily obtained from W.P.A., "Tables of Probability Functions," Vol. II, New York (1942).   The functions $\varphi^{(m)}(y)$ are tabulated in Table V of "Probability and its Engineering Uses" by T. C. Fry (D. Van Nostrand Co., 1928).

4

The form assumed by $p_1(y)$ as the parameter a becomes large is examined in the latter portion (from equation (1.12) onwards) of the section.

Series which converge for all values of a but which are especially suited for calculation when $a \leq 1$ may be obtained by inserting the Taylor's series (in powers of x) for $\varphi(y + x)$ and $\varphi_{-1}(y + x)$, $x = -a \cos \theta$, in (1.6) and (1.7) and integrating termwise. When we introduce the notation[4]

$$\varphi^{(m)}(y) = \frac{d^m}{dy^m} \varphi(y) = \frac{1}{\sqrt{2\pi}} \frac{d^m}{dy^m} e^{-y^2/2} \tag{1.9}$$

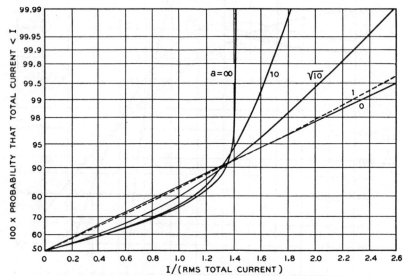

Fig. 2—Cumulative distribution of sine wave plus noise.

Ordinate $= 100 \displaystyle\int_{-\infty}^{y} p_1(y_1) \, dy_1$. See Fig. 1 for notation.

we obtain

$$p_1(y) = \sum_{n=0}^{\infty} \frac{1}{n!n!} \left(\frac{a}{2}\right)^{2n} \varphi^{(2n)}(y)$$

$$\int_{-\infty}^{y} p_1(y_1) dy_1 = \varphi_{-1}(y) + \sum_{n=1}^{\infty} \frac{1}{n!n!} \left(\frac{a}{2}\right)^{2n} \varphi^{(2n-1)}(y) \tag{1.10}$$

The second equation of (1.10) may be shown to be valid by breaking the interval $(-\infty, y)$ into $(-\infty, 0)$ and $(0, y)$. In the first part,

$$\int_{-\infty}^{0} p_1(y_1) \, dy_1 = \varphi_{-1}(0)$$

since both sides have the value 1/2. In the second, term by term integration is valid since the series integrated are uniformly convergent as may be seen from the inequality

$$| \varphi^{(n)}(y) | \leq \left(\frac{n!}{2\pi}\right)^{1/2} \left(\frac{2}{\pi n}\right)^{1/4} [1 + 0(n^{-1}) + 0(y^2 n^{-1})], \quad (1.11)$$

in which we suppose that y remains finite as $n \to \infty$. This may be obtained by using the known behavior of Hermite polynomials of large order.[5]

When $Q \gg rms\ I_n$ so that a is very large the distribution approaches that of a sine wave, namely

$$p_1(y) \sim \begin{cases} 0, & |y| > a \\ (a^2 - y^2)^{-1/2}/\pi, & |y| < a \end{cases} \quad (1.12)$$

$$\int_{-\infty}^{y} p_1(y_1) dy_1 \sim \frac{1}{2} + \frac{1}{\pi} \arcsin \frac{y}{a}, \quad |y| < a$$

In order to study the manner in which the limiting expressions (1.12) are approached it is convenient to make the change of variable

$$x = y - a \cos \theta, \qquad d\theta = [a^2 - (y - x)^2]^{-1/2} dx$$

$$z = x - y + a$$

in (1.6). We obtain

$$p_1(y) = \frac{1}{\pi} \int_{y-a}^{y+a} \varphi(x) [a^2 - (y - x)^2]^{-1/2} dx$$

$$= \frac{1}{\pi} \int_0^{2a} \varphi(z + y - a)[z(2a - z)]^{-1/2} dz. \quad (1.13)$$

An asymptotic (as a becomes large) expression for $p_1(y)$ suitable for the middle portion of the distribution where $a - |y| \gg 1$ may be obtained from the first integral in (1.13). Since the principal contributions to the value of the integral come from the region around $x = 0$ we are led to expand the radical in powers of x and integrate termwise. Legendre polynomials enter naturally since they are sometimes defined as the coefficients in such an expansion. Replacing the limits of integration $y + a$ and $y - a$ by $+\infty$ and $-\infty$, respectively and integrating termwise gives

$$p_1(y) \sim \frac{(a^2 - y^2)^{-1/2}}{\pi} \left[1 + \sum_{n=1}^{\infty} (-)^n \frac{1.3.5. (2n - 1)}{(a^2 - y^2)^n} P_{2n}(it^{1/2}) \right]$$

$$= \frac{(a^2 - y^2)^{-1/2}}{\pi} \left[1 + \frac{3t + 1}{2(a^2 - y^2)} + \frac{3(35t^2 + 30t + 3)}{8(a^2 - y^2)^2} + \cdots \right] \quad (1.14)$$

[5] A suitable asymptotic formula is given in Orthogonal Polynomials, by G. Szegö, *Am. Math. Soc. Colloquium, Pub.*, Vol. 23, (1939), p. 195.

6

where $t = y^2/(a^2 - y^2)$ and $P_{2n}(\ )$ denotes the Legendre polynomial of order $2n$. We have written this as an asymptotic expansion because it obviously is one when y, and hence t, is zero in which case

$$P_{2n}(0) = (-)^n \frac{1.3.5\cdots(2n-1)}{2.4.\ldots 2n}$$

When y is near a or is greater than a, a suitable asymptotic expansion may be obtained from the second integral in (1.13) by expanding $(2a - z)^{-1/2}$ in powers of $z/(2a)$ and integrating termwise. The upper limit of integration, $2a$, may be replaced by ∞ since $\varphi(z + y - a)$ may be assumed to be negligibly small when z exceeds 2a. We thus obtain

$$
\begin{aligned}
p_1(y) &\sim \frac{1}{\pi} \sum_{n=0}^{\infty} \frac{(\frac{1}{2})_n}{n!} \left(\frac{1}{2a}\right)^{n+1/2} \int_0^{\infty} \varphi(z + y - a) z^{n-1/2}\, dz \\
&= \frac{\varphi(y-a)}{\pi} \sum_{n=0}^{\infty} \frac{(\frac{1}{2})_n}{n!} \left(\frac{1}{2a}\right)^{n+1/2} \int_0^{\infty} e^{-z(y-a)-(z^2/2)} z^{n-1/2}\, dz
\end{aligned}
\tag{1.15}
$$

where we have used the notation

$$(\alpha)_0 = 1, \qquad (\alpha)_n = \alpha(\alpha + 1) \cdots (\alpha + n - 1).$$

The integrals occurring in (1.15) are related to the parabolic cylinder function[6] $D_m(x)$. Their properties may be obtained from the known properties of these functions or may be obtained by working directly with the integrals.

Suppose now that a is very large so that only the leading term in the series (1.15) for $p_1(y)$ need be retained. Then

$$p_1(y) \sim a^{-1/2} F(y - a) \tag{1.16}$$

where

$$F(s) = \pi^{-1} 2^{-1/2} \int_0^{\infty} \varphi(z + s) z^{-1/2}\, dz \tag{1.17}$$

By writing out $\varphi(z + s)$, expanding $exp\,(-zs)$ in a power series, and integrating termwise we see that

$$
\begin{aligned}
F(s) &= \frac{\varphi(s) 2^{-5/4}}{\pi} \sum_{\ell=0}^{\infty} \frac{\Gamma\left(\dfrac{\ell}{2} + \dfrac{1}{4}\right)}{\ell!} (-s\sqrt{2})^\ell \\
&= (2\pi)^{-1} s^{1/2} \varphi(s/\sqrt{2}) K_{\frac{1}{4}}(s^2/4)
\end{aligned}
\tag{1.18}
$$

where K denotes a modified Bessel function.[7] The relation (1.18) may also

[6] Whittaker and Watson, "Modern Analysis," 4th ed. (1927), 347–351.

[7] A table of $K_{\frac{1}{4}}(x)$ is given by H. Carsten and N. McKerrow, Phil. Mag. S7, Vol. 35 (1944), 812–818.

7

be obtained from pair 923.1 of "Fourier Integrals for Practical Applications," by G. A. Campbell and R. M. Foster.[8]

A curve showing $F(y - a)$ plotted as a function of $y - a$ is given in Fig. 3. It was obtained from the relation

$$F(s) = 2^{1/4}\pi^{-3/2}\chi(-s/\sqrt{2})$$

where

$$\chi(x) = \int_0^\infty e^{-(x-w^2)^2}\, dw$$

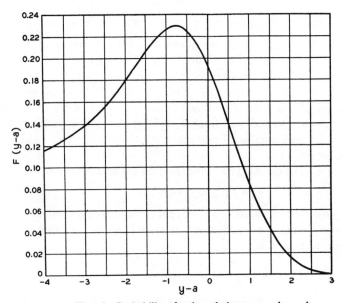

Fig. 3—Probability density of sine wave plus noise.
When rms $I_N \ll Q$ and I is near Q, $p_1(y) \sim a^{-1/2}F(y - a)$, $y - a = (I - Q)/(\text{rms } I_N)$.
See Fig. 1 for notation.

This function has been tabulated by Hartree and Johnston.[9]

The probability that I exceeds Q, or that y exceeds a, is, integrating the second of expressions (1.13),

$$\int_a^\infty p_1(y)\, dy = \frac{1}{\pi} \int_0^{2a} \frac{dz}{\sqrt{z(2a - z)}} \int_z^\infty \varphi(x)\, dx.$$

An asymptotic expansion may be obtained by expanding $(2a - z)^{-1/2}$ as in the derivation of (1.15) but we shall be content with the leading term.

[8] *Bell Telephone System Monograph* B-584.
[9] *Manchester Lit. and Phil. Soc. Memoirs*, v. 83, 183–188, Aug., 1939.

8

113

Using

$$\int_0^\infty z^{-1/2}\,dz \int_z^\infty \varphi(x)\,dz = \int_0^\infty \varphi(x)\,dx \int_0^x z^{-1/2}\,dz = 2^{1/4}\pi^{-1/2}\Gamma(\tfrac{3}{4})$$

we obtain

$$\int_a^\infty p_1(y)\,dy \sim 2^{-1/4}\pi^{-3/2}\Gamma(\tfrac{3}{4})a^{-1/2} = 0.185\cdots a^{-1/2} \tag{1.19}$$

For use in computations we list the following values

$$\Gamma(\tfrac{1}{4}) = 3.62561, \qquad \Gamma(\tfrac{3}{4}) = 1.22542, \qquad \Gamma(\tfrac{5}{4}) = 0.90640$$

2. Expected Number of Crossings of I per Second

In this section, we shall study two questions. First, what is the probability $P(I_1,\ t_1)dt$ of I increasing through the value I_1 (i.e. of I passing through the value I_1 with positive slope) during the infinitesimal interval $t_1,\ t_1 + dt$? Second, what is the expected number $N(I_1)$ of times per second I increases through the value I_1. When I_1 is zero, $2N(0)$ is the expected number of zeros per second, and when I_1 is large $N(I_1)$ is approximately equal to the expected number of maxima lying above the value I_1 in an interval one second long.

We start on the first question by considering the random function

$$z = F(a_1, a_2, \cdots a_N;\ t)$$

where the a's are random variables. The probability that the random curve obtained by plotting z as a function of t increases through the value $z = z_1$ in the interval $t_1,\ t_1 + dt$ is[10]

$$dt \int_0^\infty \eta p(z_1,\ \eta;\ t_1)\,d\eta \tag{2.1}$$

where $p(\xi,\ \eta;\ t_1)$ denotes the probability density of the random variables

$$\xi = F(a_1, a_2, \cdots, a_N;\ t_1)$$

$$\eta = \left[\frac{\partial F}{\partial t}\right]_{t=t_1}.$$

In our case z becomes the current I defined by equation (1.1). The method used to obtain equation (3.3–9) of Reference A may also be used to show that the quantity $p(I_1,\ \eta,\ t_1)$ (which now appears in (2.1)) is given by

$$p(I_1,\ \eta,\ t_1) = \frac{\pi N_0}{-\psi_0''}\,\varphi(y - a\cos qt_1)\varphi(x + b\sin qt_1) \tag{2.2}$$

[10] This result is a straightforward generalization of expression (3.3-5) in Section 3.3 of Reference A where references to related results by M. Kac are given. A formula equivalent to (2.1) has also been given by Mr. H. Bondi in an unpublished paper written in 1944. He applies his formula to the problem studied in Section 4.

9

114

where $\varphi(\ )$ denotes the normal law function defined by equation (1.3) and

$$-\psi_0'' = 4\pi^2 \int_0^\infty w(f)f^2 \, df, \qquad N_0 = \frac{1}{\pi} \sqrt{\frac{-\psi_0''}{\psi_0}}, \qquad y = \frac{I_1}{\sqrt{\psi_0}},$$

$$a = \frac{Q}{\sqrt{\psi_0}}, \qquad x = \frac{\eta}{\sqrt{-\psi_0''}}, \qquad b = \frac{Qq}{\sqrt{-\psi_0''}} = \frac{2af_q}{N_0}. \tag{2.3}$$

Equation (3.3–11) of Reference A shows that N_0 is the expected number of zeros per second which I_N would have if it were to flow alone.

Let $P(I_1, t_1)dt$ be the probability that I will increase through the value I_1 during the interval $t_1, t_1 + dt$. Then (2.1) and (2.2) give

$$P(I_1, t_1) = \int_0^\infty \eta p(I_1, \eta, t_1) \, d\eta$$

$$= \pi N_0 \varphi(y - a \cos qt_1) \int_0^\infty x\varphi(x + b \sin qt_1) \, dx. \tag{2.4}$$

The integral in (2.4) is of the form

$$\int_0^\infty x\varphi(x + v) \, dx = \varphi(v) - v \int_v^\infty \varphi(x) \, dx$$

$$= -\frac{v}{2} + \varphi(v) + v \int_0^v \varphi(x) \, dx$$

$$= -v + \varphi(v) + v\varphi_{-1}(v) \tag{2.5}$$

$$= -\frac{v}{2} + (2\pi)^{-1/2} {}_1F_1\left(-\frac{1}{2}; \frac{1}{2}; -\frac{v^2}{2}\right)$$

where v replaces $b \sin qt_1$ and ${}_1F_1$ denotes a confluent hypergeometric function.

The distribution of the crossings at various portions of the cycle (of the sine wave) may be obtained by giving special values to qt_1 in (2.4).

The expected number of times I increases through the value I_1 in one second is

$$N(I_1) = \operatorname*{Limit}_{T \to \infty} \frac{1}{T} \int_0^T P(I_1, t_1) \, dt_1$$

$$= N_0 \int_0^\pi \varphi(y - a \cos \theta) \left[\varphi(b \sin \theta) + b \sin \theta \int_0^{b\sin\theta} \varphi(x) \, dx \right] d\theta \tag{2.6}$$

where we have used (2.4) and the second equation of (2.5). The integrand in (2.6) is composed of tabulated functions and is of a form suited to numerical integration. Expanding $\varphi(y - a \cos \theta)$ in (2.6) as in the derivation

10

115

of (1.10), replacing the quantity within the brackets by the series shown in the last equation of (2.5) ,and integrating termwise leads to

$$N(I_1) = N_0(\pi/2)^{1/2} \sum_{n=0}^{\infty} \frac{\varphi^{(2n)}(y)}{n!\,n!} \left(\frac{a}{2}\right)^{2n} {}_1F_1\left(-\tfrac{1}{2}; n + 1; -\frac{b^2}{2}\right) \qquad (2.7)$$

The series (2.7) converges for all values of a, y, and b. This follows from the inequality (1.11) which may be applied to $\varphi^{(2n)}(y)$, and from the fact that the ${}_1F_1$ is less than $exp\,(b^2/2)$ as may be seen by comparing corresponding terms in their expansions.

The expected number of zeros, per second, of I is $2N(0)$ where we have set I_1, and hence y, equal to zero. In this case the integral in (2.6) may be simplified somewhat and we obtain

$$2N(0) = N_0\left[e^{-\alpha} I_0(\beta) + \frac{b^2}{2\alpha} Ie\left(\frac{\beta}{\alpha}, \alpha\right)\right] \qquad (2.8)$$

where $I_0(\beta)$ is the Bessel function of order zero and imaginary argument and

$$\alpha = \frac{a^2 + b^2}{4}, \qquad \beta = \frac{a^2 - b^2}{4}$$

$$Ie(k, x) = \int_0^x e^{-u} I_0(ku)\, du.$$

The integral $Ie(k, x)$ is tabulated in Appendix I.

I have been unable to obtain a simple derivation of (2.8). It was originally obtained from the following integral

$$N(I_1) = \frac{N_0}{2} \int_{-\pi}^{\pi} d\theta\, \varphi(y - a \cos \theta) \int_0^{\infty} x\varphi(x + b \sin \theta)\, dx \qquad (2.9)$$

which may be derived from the second equation of (2.4) and the first of (2.6). Setting I_1 and y equal to zero and writing out the φ's gives

$$2N(0) = \frac{N_0}{2\pi} \int_{-\pi}^{\pi} d\theta \int_0^{\infty} dx$$

$$x \exp\left[-\tfrac{1}{2}(x^2 + 2bx \sin \theta + a^2 \cos^2 \theta + b^2 \sin^2 \theta)\right].$$

Equation (2.8) was obtained by applying the method of Appendix III to this expression.

3. Definitions and Simple Properties of R and Θ

The remaining portion of this paper is concerned with the envelope R and the corresponding phase angle θ. These quantities are introduced and some of their simpler properties discussed in this section.

Suppose that the frequency band associated with I_N is relatively narrow

11

and contains the sine wave frequency f_q. The noise current may be resolved into two components, one "in phase" and the other "in quadrature" with $Q \cos qt$. Using the representation (2.8-6) of reference A and proceeding as in Section 3.7 of that paper:

$$I_N = \sum_{n=1}^{M} c_n \cos (\omega_n t - \varphi_n) \tag{3.1}$$

$$= \sum_{n=1}^{M} c_n \cos [(\omega_n - q)t - \varphi_n + qt]$$

$$= I_c \cos qt - I_s \sin qt \tag{3.2}$$

where

$$I_c = \sum_{n=1}^{M} c_n \cos [(\omega_n - q)t - \varphi_n]$$

$$\tag{3.3}$$

$$I_s = \sum_{n=1}^{M} c_n \sin [(\omega_n - q)t - \varphi_n]$$

$$\omega_n = 2\pi f_n, \qquad f_n = n\Delta f, \qquad c_n^2 = 2w(f_n)\Delta f$$

$w(f)$ denotes the power spectrum of I_N and the φ_n's are random variables distributed uniformly over the interval $(0, 2\pi)$.

The total current I may be written as

$$I = Q \cos qt + I_N$$

$$= (Q + I_c) \cos qt - I_s \sin qt$$

$$= R \cos \theta \cos qt - R \sin \theta \sin qt \tag{3.4}$$

$$= R \cos (qt + \theta)$$

where we have introduced the envelope function R and the phase angle θ by means of

$$R \cos \theta = Q + I_c$$

$$R \sin \theta = I_s \tag{3.5}$$

Since I_c and I_s are functions of t whose variations are relatively slow in comparison with those of $\cos qt$, the same is true of R and (usually) θ.

A graphical illustration of equations (3.4) and (3.5) which is often used is shown in Fig. 4.

In accordance with the usual convention used in alternating current theory, the vector OQ is supposed to be rotating about the origin O with angular velocity q. If I_N happened to have the frequency $q/2\pi$, its vector representation QT would be fixed relative to OQ. In general, however, the

12

117

length and inclination of QT will change due to the random fluctuations of I_N. Thus the point T will wander around on the plane of the figure. If rms I_N is much less than Q, T will be close to the point Q most of the time. In this case

$$R = [(Q + I_c)^2 + I_s^2]^{1/2} \sim Q + I_c$$

$$\theta = \tan^{-1} \frac{I_s}{Q + I_c} \sim \frac{I_s}{Q} \tag{3.6}$$

$$\frac{d\theta}{dt} \sim \frac{d}{dt} \frac{I_s}{Q} = \frac{I_s'}{Q}$$

and a number of statistical properties of R and θ may be obtained from the corresponding properties of noise alone when we note that I_c, I_s, and I_s' behave like noise currents whose power spectra are concentrated in the lower portion of the frequency spectrum.

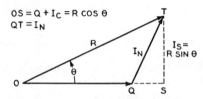

$$OS = Q + I_C = R \cos \theta$$
$$QT = I_N$$

Fig. 4—Graphical representation of $I = Q\cos qt + I_N$.

By squaring both sides of equations (3.1) and (3.3) and then averaging with respect to t and the φ_n's we may show that I_c, I_s, and I_N all have the same rms value, namely $\psi_0^{1/2}$.

It may be seen from (3.3) that the power spectra of I_c and I_s are both given by

$$w(f_q + f) + w(f_q - f) \tag{3.7}$$

where it is assumed that $0 \leq f \ll f_q$. Likewise the power spectrum of the time derivative I_s' of I_s is

$$4\pi^2 f^2 [w(f_q + f) + w(f_q - f)] \tag{3.8}$$

This follows from the representation of I_s' obtained by differentiating the expression (3.3) for I_s with respect to t, the procedure being the same as in the derivation of equation (7.2) in Section 7. The power spectra shown in Table 1 were computed from equations (3.7) and (3.8).

The correlation function for I_c, and hence also for I_s, is, from equations (A2–1) and (A2–3) of Appendix II,

$$\overline{I_c(t)I_c(t + \tau)} = g = \int_0^\infty w(f) \cos 2\pi(f - f_q)\tau \, df$$

13

where the bar denotes an average with respect to t and g is a function of τ. From (A2-3) the correlation function for I'_s is $-g''$ where the double prime on g denotes the second derivative with respect to τ.

Attention is sometimes fixed upon the variation in distance between successive zeros of I. The time between two successive zeros of I at, say, t_0 and t_1 is the time taken for $qt + \theta$, as appearing in $R \cos (qt + \theta)$, to increase by π. This assumes that the envelope R does not vanish in the interval. For the moment we write θ as $\theta(t)$ in order to indicate its dependence on the time t. Then t_0 and t_1 must satisfy the relation

$$qt_1 + \theta(t_1) - qt_0 - \theta(t_0) = \pi \tag{3.9}$$

Since $\theta(t)$ is a relatively slowly varying function we write

$$\theta(t_1) - \theta(t_0) = (t_1 - t_0)\theta'(t_0) + (t_1 - t_0)^2\theta''(t_0)/2 + \cdots$$

TABLE 1

POWER SPECTRA OF I_N, I_c, I_s, AND I'_s

I_N	I_c and I_s	I'_s
$w(f) = w_0 = \psi_0/\beta$ for $f_q - \beta/2 < f < f_q + \beta/2$ $w(f) = 0$ elsewhere $f_q = $ mid-band frequency	$2w_0$ for $0 < f < \beta/2$ 0 elsewhere	$8\pi^2 f^2 w_0$ for $0 < f < \beta/2$ 0 elsewhere
$w(f) = w_0 = \psi_0/\beta$ for $f_q - \beta < f < f_q$ $w(f) = 0$ elsewhere $f_q = $ top frequency	w_0 for $0 < f < \beta$ 0 elsewhere	$4\pi^2 f^2 w_0$ for $0 < f < \beta$ 0 elsewhere
$w(f) = \dfrac{\psi_0}{\sigma\sqrt{2\pi}} e^{-(f-f_q)^2/(2\sigma^2)}$	$\dfrac{2\psi_0}{\sigma\sqrt{2\pi}} e^{-f^2/(2\sigma^2)}$	$\dfrac{8\pi^2 f^2 \psi_0}{\sigma\sqrt{2\pi}} e^{-f^2/(2\sigma^2)}$

where the primes denote differentiation with respect to t. When this is placed in (3.9) and terms of order $(t_1 - t_0)^2$ neglected, we obtain

$$\frac{1}{2(t_1 - t_0)} = \frac{q}{2\pi} + \frac{1}{2\pi}\theta'(t_0) \tag{3.10}$$

which relates the interval between successive zeros to θ'.

The expression on the right hand side of (3.10) may be defined as the instantaneous frequency:

$$\text{Instantaneous frequency} = f_q + \frac{1}{2\pi}\frac{d\theta}{dt} \tag{3.11}$$

This definition is suggested when $\cos 2\pi f t$ is compared with $\cos (qt + \theta)$ and also by (3.10) when we note that the period of the instantaneous fre-

14

quency is approximately equal to twice the distance between two successive zeros which is $2(t_1 - t_0)$.

The probability density of R is[11]

$$\frac{R}{\psi_0} \exp\left[-\frac{R^2 + Q^2}{2\psi_0} \right] I_0(RQ/\psi_0) \tag{3.12}$$

where $I_0 (RQ/\psi_0)$ denotes the Bessel function of order zero with imaginary argument. In Section 3.10 of Reference A, it is shown that the average value of R^n is[*]

$$\overline{R^n} = (2\psi_0)^{n/2} \Gamma\left(\frac{n}{2} + 1\right) {}_1F_1\left(-\frac{n}{2}; \quad 1; \quad -\rho\right), \tag{3.13}$$

where $\rho = Q^2/(2\psi_0)$, of which special cases are

$$\begin{aligned}
\bar{R} &= e^{-\rho/2}(\pi\psi_0/2)^{1/2}\left[(1 + \rho)I_0(\rho/2) + \rho I_1(\rho/2)\right] \\
\overline{R^2} &= Q^2 + 2\psi_0
\end{aligned} \tag{3.14}$$

Curves showing the distribution of R are also given there.

4. Expected Number of Crossings of R per Second

Here we shall obtain expressions for the expected number N_R of times, per second, the envelope passes through the value R with positive slope. When R is large, N_R is approximately equal to the expected number of maxima of the envelope per second exceeding R and when R is small N_R is approximately equal to the expected number of minima less than R. For the special case in which the noise band is symmetrical and is centered on the sine wave frequency f_q N_R is given by the relatively simple expression (4.8).

The probability that the envelope passes through the value R during the interval $t, t + dt$ with positive slope is, from (2.1),

$$dt \int_0^\infty R'p(R, R', t)\, dR' \tag{4.1}$$

where $p(R, R', t)$ denotes the probability density of R and its time derivative R', t being regarded as a parameter.

An expression for $p(R, R', t)$ may be obtained from the probability density of I_c, I_s, I'_c, I'_s. From our representation of a noise current and the central limit theorem it may be shown (as is done for similar cases in Part III of Reference A) that the probability distribution of these four variables is

[11] In equation (60-A) of an unpublished appendix to his paper appearing in the *B.S.T.J.* Vol. 12 (1933), 35–75, Ray S. Hoyt gives an integral, obtained by integrating (3.12) with respect to R, for the cumulative distribution of R.

[*]The correlation function for the envelope of a signal plus noise, together with associated probability densities of the envelope and phase, is given by D. Middleton in a paper appearing soon in the Quart. Jl. of Appl. Math.

15

normal in four dimensions. If the variables be taken in the order given above the moment matrix is, from equations (A2–2) of Appendix II,

$$M = \begin{bmatrix} b_0 & 0 & 0 & b_1 \\ 0 & b_0 & -b_1 & 0 \\ 0 & -b_1 & b_2 & 0 \\ b_1 & 0 & 0 & b_2 \end{bmatrix} \tag{4.2}$$

where the b's are defined by the integrals in equations (A2–1). The inverse matrix is

$$M^{-1} = \frac{1}{B} \begin{bmatrix} b_2 & 0 & 0 & -b_1 \\ 0 & b_2 & b_1 & 0 \\ 0 & b_1 & b_0 & 0 \\ -b_1 & 0 & 0 & b_0 \end{bmatrix}, \quad B = b_0 b_2 - b_1^2 \tag{4.3}$$

which may be readily verified by matrix multiplication, and the determinant $| M |$ is B^2. The normal distribution may be written down at once when use is made of the formulas given in Section 2.9 of Reference A. The substitutions

$$\begin{aligned} I_c &= R\cos\theta - Q, & I_c' &= R'\cos\theta - R\sin\theta\,\theta' \\ I_s &= R\sin\theta, & I_s' &= R'\sin\theta + R\cos\theta\,\theta' \end{aligned} \tag{4.4}$$

$$dI_c dI_s dI_c' dI_s' = R^2 dR dR' d\theta d\theta'$$

enable us to write

$$\begin{aligned} b_2(I_c^2 &+ I_s^2) + b_0(I_c'^2 + I_s'^2) \\ &- 2b_1(I_c I_s' - I_s I_c') = b_2(R^2 - 2QR\cos\theta + Q^2) \\ &\qquad\qquad + b_0(R'^2 + R^2\theta'^2) \\ &\qquad\qquad - 2b_1 R^2\theta' + 2b_1 Q(R'\sin\theta + R\theta'\cos\theta). \end{aligned}$$

Consequently the probability density of R, R', θ, θ' is

$$\begin{aligned} p(R, R', \theta, \theta') &= \frac{R^2}{4\pi^2 B} \exp\Big\{ -\frac{1}{2B} [b_2(R^2 - 2QR\cos\theta + Q^2) \\ &\quad + b_0(R'^2 + R^2\theta'^2) - 2b_1 R^2\theta' + 2b_1 Q(R'\sin\theta + R\theta'\cos\theta)] \Big\} \end{aligned} \tag{4.5}$$

In this expression R ranges from 0 to ∞, θ from $-\pi$ to π, and R' and θ' from $-\infty$ to $+\infty$. The probability density for R and R' is obtained by

16

integrating (4.5) with respect to θ and θ' over their respective ranges. The integration with respect to θ' may be performed at once giving

$$p(R, R', t) = \frac{R(2\pi)^{-3/2}}{\sqrt{Bb_0}} \int_{-\pi}^{\pi} d\theta$$
$$\exp\left\{-\frac{1}{2Bb_0}[B(R^2 - 2RQ\cos\theta + Q^2) + (b_0R' + b_1Q\sin\theta)^2]\right\} \tag{4.6}$$

From (4.1) and (4.6) it follows that the expected number N_R of times per second the envelope passes through R with positive slope is

$$N_R = \frac{R(2\pi)^{-3/2}}{\sqrt{Bb_0}} \int_{-\pi}^{\pi} d\theta \int_0^\infty R'dR'$$
$$\exp\left\{-\frac{1}{2Bb_0}[B(R^2 - 2RQ\cos\theta + Q^2) + (b_0R' + b_1Q\sin\theta)^2]\right\} \tag{4.7}$$

When the power spectrum $w(f)$ of the noise current I_N is symmetrical about the sine wave frequency f_q, b_1 is zero and B is equal to b_0b_2. In this case the integrations in (4.7) may be performed. We obtain

$$N_R = \left(\frac{b_2}{2\pi}\right)^{1/2} \frac{R}{b_0} I_0\left(\frac{RQ}{b_0}\right) \exp\left(-\frac{R^2 + Q^2}{2b_0}\right)$$
$$= \left(\frac{b_2}{2\pi}\right)^{1/2} \times \begin{bmatrix} \text{Probability density of} \\ \text{envelope at the value } R \end{bmatrix} \tag{4.8}$$

where the second line is obtained from expression (3.12). As will be seen from its definition (A2–1), b_0 is equal to the mean square value ψ_0 of I_N (and also of I_c and I_s).

Introducing the notation

$$v = Rb_0^{-1/2} = R/\text{rms } I_N$$
$$a = Ab_0^{-1/2} = Q/\text{rms } I_N, \tag{4.9}$$

which is the same as that of equations (3.10–15) of Reference A except that there P denotes the amplitude of the sine wave and plays the same role as Q does here, enables us to write (4.8) as

$$N_R = \left[\frac{b_2}{2\pi b_0}\right]^{1/2} vI_0(av)e^{-(v^2+a^2)/2} = \left[\frac{b_2}{2\pi b_0}\right]^{1/2} p(v). \tag{4.10}$$

The function $p(v)$ is plotted as a function of v for various values of a in Fig. 6, Section 3.10, of Reference A.

It is interesting to note that

$$(b_2/b_0)^{1/2}/\pi = \text{Expected number of zeros per second of } I_c \text{ (or of } I_s) \tag{4.11}$$

17

This relation, which is true even if the noise band is not symmetrical about f_q, follows from equation (3.3–11) of Reference A.

When $Q \gg$ rms I_N and f_q is not at the center of the noise band it is easier to obtain the asymptotic form of N_R from the approximation (3.6),

$$R \sim Q + I_c,$$

instead of the double integral (4.7). When $Q \gg$ rms I_N and R is in the neighborhood of Q (as it is most of the time in this case), N_R is approximately equal to the expected number of times I_c increases through the value $I_{c1} = R - Q$ in one second. Thus, regarding I_c as a random noise current we have from expression (3.3–14) of Reference A

$$N_R \sim e^{-I_{c1}^2/(2b_0)} \times [1/2 \text{ the expected number of zeros per second of } I_c]$$

and when we use equation (4.11) we obtain

$$\dot{N}_R \sim \frac{1}{2\pi} (b_2/b_0)^{1/2} e^{-(R-Q)^2/(2b_0)} = \frac{1}{2\pi} (b_2/b_0)^{1/2} e^{-(v-a)^2/2} \qquad (4.12)$$

TABLE 2

$w(f) = w_0 = b_0/\beta$ OVER A BAND OF WIDTH β

	b_2	N_R
1. Band extends from $f_q - \beta/2$ to $f_q + \beta/2$	$\pi^2\beta^2 b_0/3$	$(\pi/6)^{1/2}\beta p(v) = 0.724\,\beta p(v)$
2. Same as 1 and in addition $Q = 0$	``	$(\pi/6)^{1/2}\beta v e^{-v^2/2}$
3. Same as 1 and in addition $Q \gg$ rms I_N	``	$\sim \dfrac{\beta}{2\sqrt{3}} e^{-(v-a)^2/2}$
4. Band extends from f_q to $f_q + \beta$ and $Q \gg$ rms I_N	$4\pi^2\beta^2 b_0/3$	$\sim \dfrac{\beta}{\sqrt{3}} e^{-(v-a)^2/2}$

Table 2 lists the forms assumed by (4.10) and (4.12) when the power spectrum $w(f)$ of the noise current I_N is constant over a frequency band of width β. The quantity b_0 in the expressions for b_2 represents the mean square value of I_N.

In the general case where the band of noise is not centered on f_q and where R is not large enough to make (4.12) valid we are obliged to return to the double integral (4.7). Some simplification is possible, but not as much as could be desired. Introducing the notation

$$\alpha = RQ/b_0, \qquad \gamma = b_1 Q(Bb_0)^{-1/2}$$

$$x = (b_0 R' + b_1 Q\sin\theta)(Bb_0)^{-1/2}$$

18

enables us to write (4.7) as

$$N_R = (2\pi)^{-3/2}(R/b_0)(B/b_0)^{1/2}e^{-(R^2+Q^2)/(2b_0)}$$

$$\int_{-\pi}^{\pi} d\theta \int_{\gamma \sin \theta}^{\infty} (x - \gamma \sin \theta)e^{\alpha \cos \theta - x^2/2}\, dx \qquad (4.13)$$

Part of the integrand may be integrated with respect to x and the remaining portion integrated by parts with respect to θ. The double integral in the second line of (4.13) then becomes

$$\int_{-\pi}^{\pi} e^{\alpha \cos \theta - (\gamma \sin \theta)^2/2}\, d\theta + \int_{-\pi}^{\pi}\left[\int_{\gamma \sin \theta}^{\infty} e^{-x^2/2}\, dx\right]d[\gamma \alpha^{-1} e^{\alpha \cos \theta}]$$

$$= \int_{-\pi}^{\pi} (1 + \gamma^2 \alpha^{-1} \cos \theta)e^{\alpha \cos \theta - (\gamma \sin \theta)^2/2}\, d\theta \qquad (4.14)$$

$$= \gamma \alpha^{-1} e^{-(\gamma^2 + \alpha^2 \gamma^{-2})/2}\int_{-\pi}^{\pi}(\gamma \cos \theta + \alpha/\gamma)e^{(\gamma \cos \theta + \alpha/\gamma)^2/2}\, d\theta$$

$$= 2\pi \sum_{n=0}^{\infty} \frac{(\tfrac{1}{2})_n}{n!}\left(-\frac{\gamma^2}{\alpha}\right)^n [I_n(\alpha) + \gamma^2 \alpha^{-1} I_{n+1}(\alpha)].$$

The series is obtained by expanding $\exp[-(\gamma \sin \theta)^2/2]$ in the second equation in powers of $\sin \theta$ and integrating termwise.

5. Probability Density of $\dfrac{d\theta}{dt}$

As was pointed out in Section 3 the time derivative θ' of the phase angle θ associated with the envelope is closely related to the instantaneous frequency. The probability density $p(\theta')$ of θ' may be expressed in terms of modified Bessel functions as shown by equation (5.4). Curves for $p(\theta')$ are given when the sine wave frequency f_q lies at the middle of a symmetrical band of noise. Although the expressions for $p(\theta')$ are rather complicated, those for the averages $\bar{\theta}'$ and $\overline{|\theta'|}$ given by equations (5.7) and (5.16) are relatively simple.

The probability density $p(\theta')$ may be obtained by integrating the expression (4.5) for $p(R, R', \theta, \theta')$ with respect to R, R', θ. The integration with respect to R', the limits being $-\infty$ and $+\infty$, gives the probability density for R, θ, θ':

$$p(R, \theta, \theta') = \frac{R^2}{4\pi^2}\left(\frac{2\pi}{b_0 B}\right)^{1/2}\exp[-aR^2 + 2bR \cos \theta + c \sin^2 \theta$$

$$-b_2 Q^2/(2B)] \qquad (5.1)$$

19

where

$$B = b_0 b_2 - b_1^2 \qquad\qquad b = Q(b_2 - b_1\theta')/(2B)$$

$$a = (b_2 - 2b_1\theta' + b_0\theta'^2)/(2B) \qquad c = Q^2 b_1^2/(2Bb_0) = b_1^2\rho/B \qquad (5.2)$$

$$\rho = Q^2/(2b_0) \qquad\qquad \gamma = b^2/a$$

and b_0, b_1, b_2 are given in Appendix II.

Integrating with respect to R gives the probability density for θ, θ'. Expanding $\exp(2bR\cos\theta)$ in powers of R and integrating termwise,

$$p(\theta, \theta') = \frac{1}{16\pi a}\left(\frac{2}{ab_0 B}\right)^{1/2} e^{c\,\sin^2\theta - b_2 b_0\rho/B}$$

$$\sum_{n=0}^{\infty} \frac{n+1}{\Gamma\left(\dfrac{n}{2}+1\right)} \left(\frac{b\cos\theta}{a^{1/2}}\right)^n \qquad (5.3)$$

When we integrate θ from $-\pi$ to π to obtain $p(\theta')$ the terms for which n is odd disappear and we have to deal with the series, writing γ for b^2/a,

$$\sum_{m=0}^{\infty} \frac{2m+1}{m!}(\gamma\cos^2\theta)^m = (2\gamma\cos^2\theta + 1)\,\exp(\gamma\cos^2\theta)$$

Thus, the probability density of θ' is

$$p(\theta') = \frac{1}{16\pi a}\left(\frac{2}{ab_0 B}\right)^{1/2} \int_{-\pi}^{\pi} (2\gamma\cos^2\theta + 1)\,e^{c\,\sin^2\theta + \gamma\cos^2\theta - b_2 b_0\rho/B}\,d\theta$$

$$= \frac{1}{8a}\left(\frac{2}{ab_0 B}\right)^{1/2}\left[(\gamma+1)I_0\left(\frac{\gamma-c}{2}\right) + \gamma I_1\left(\frac{\gamma-c}{2}\right)\right] \qquad (5.4)$$

$$\exp\left[\frac{c+\gamma}{2} - \frac{b_2 b_0 \rho}{B}\right]$$

From (5.2)

$$\gamma - c = \rho\,\frac{b_2 - 2b_1\theta'}{b_2 - 2b_1\theta' + b_0\theta'^2}$$

$$\frac{c+\gamma}{2} - \frac{b_2 b_0\rho}{B} = -\frac{\rho}{2}\,\frac{b_2 - 2b_1\theta' + 2b_0\theta'^2}{b_2 - 2b_1\theta' + b_0\theta'^2} \qquad (5.5)$$

It will be noted that for large values of $|\theta'|$ the probability density of θ' varies as $|\theta'|^{-3}$. Although this makes the mean square value of θ' infinite, the average values $\overline{\theta'}$ and $\overline{|\theta'|}$ of θ' and $|\theta'|$ still exist. In order to obtain $\overline{\theta'}$ it is convenient to return to (4.5) and write

$$\overline{\theta'} = \int_{-\pi}^{\pi} d\theta \int_0^{\infty} dR \int_{-\infty}^{+\infty} dR' \int_{-\infty}^{\infty} d\theta'\,\theta'\,p(R, R', \theta, \theta') \qquad (5.6)$$

20

125

The integration with respect to θ' may be performed by setting $R\theta'$ equal to x and using

$$\int_{-\infty}^{+\infty} x\, e^{-\alpha x^2 + 2\beta x}\, dx = (\beta/\alpha)(\pi/\alpha)^{1/2}\, e^{\beta^2/\alpha}$$

The integral in R' reduces to a similar integral except that the factor x in the integrand is absent. Performing these two integrations and using the definition of B leads to

$$\bar{\theta}' = \frac{1}{2\pi}\frac{b_1}{b_0^2}\int_{-\pi}^{\pi} d\theta \int_0^{\infty} dR\, (R - Q\cos\theta)$$

$$\exp\left[-\frac{1}{2b_0}(R^2 - 2QR\cos\theta + Q^2)\right]$$

We may integrate at once with respect to R. When this is done $\cos\theta$ disappears and the integration with respect to θ becomes easy. Thus

$$\bar{\theta}' = (b_1/b_0)\exp\left[-Q^2/(2b_0)\right] = (b_1/b_0)e^{-\rho} \tag{5.7}$$

When the noise power spectrum is equal to w_0 in the band extending from $f_0 - \beta/2$ to $f_0 + \beta/2$ and is zero outside the band, $b_1 = 2\pi(f_0 - f_q)b_0$. Hence, from (3.11),

ave. instantaneous frequency $= f_q + \bar{\theta}'/(2\pi)$

$$= f_0 + (f_q - f_0)(1 - e^{-\rho}) \tag{5.8}$$

In the remainder of this section we assume the power spectrum of the noise current to be symmetrical about the sine wave frequency f_q. In this case b_1 and c are zero, B is equal to $b_0 b_2$ and (5.4) becomes

$$p(\theta') = \tfrac{1}{2}(b_0/b_2)^{1/2}(1 + z^2)^{-3/2}e^{-\rho + y/2}$$

$$[(y + 1)I_0(y/2) + yI_1(y/2)] \tag{5.9}$$

$$= \tfrac{1}{2}(b_0/b_2)^{1/2}(1 + z^2)^{-3/2}e^{-\rho}{}_1F_1\left(\frac{3}{2}; 1; y\right)$$

where ${}_1F_1$ denotes a confluent hypergeometric function[12] and

$$z^2 = b_0\theta'^2/b_2, \qquad y = (\gamma)_{b_1=0} = \rho/(1 + z^2) \tag{5.10}$$

When the noise power spectrum is constant in the band extending from $f_q - \beta/2$ to $f_q + \beta/2$ (see Table 2, Section 4)

$$(b_2/b_0)^{1/2} = 3^{-1/2}\beta\pi, \qquad z = 3^{1/2}\theta'/(\beta\pi) \tag{5.11}$$

[12] The relation used above follows from equation (66) (with misprint corrected) of W. R. Bennett's paper cited in connection with equation (1.2).

21

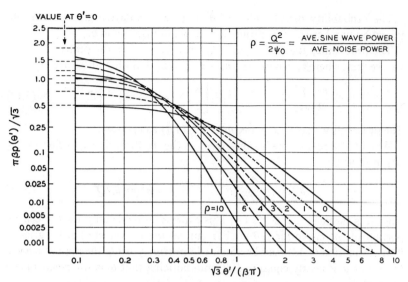

Fig. 5—Probability density of time derivative of phase angle.

$p(\theta')\,d\theta'$ = probability that the value of $d\theta/dt$ at an instant selected at random lies between θ' and $\theta' + d\theta'$. The power spectrum of I_N is constant in band of width β centered on f_q and is zero outside this band.

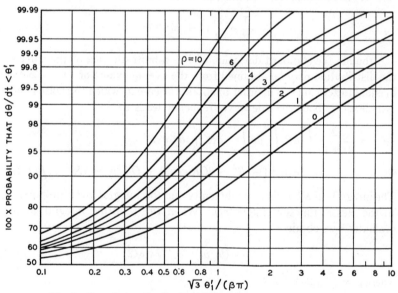

Fig. 6—Cumulative distribution of time derivative of phase angle.

Notation explained in Fig. 5.

22

127

The probability density $p(\theta')$ of θ' and its cumulative distribution, obtained by numberical integration, are shown in Figs. 5 and 6.

The probability that θ' exceeds a given θ_1' is equal to the probability that z exceeds z_1, where z_1 denotes $(b_0/b_2)^{1/2}\theta_1'$, and both probabilities are equal to

$$\frac{e^{-\rho}}{2}\int_{z_1}^{\infty}(1+z^2)^{-3/2}\,_1F_1\left[\frac{3}{2};1;\rho(1+z^2)^{-1}\right]dz \qquad (5.12)$$

The probability that $\theta' > \theta_1'$ becomes $e^{-\rho}/(4z_1^2)$ as $\theta_1' \to \infty$.

When $Q \gg$ rms I_N the leading term in the asymptotic expansion of the $_1F_1$ in (5.9) gives

$$p(\theta') \sim \frac{1}{\sigma\sqrt{2\pi}}e^{-\theta'^2/(2\sigma^2)}, \qquad \sigma^2 = b_2/Q^2 \qquad (5.13)$$

when it is assumed that $z^2 \ll 1$. This expression holds only for the central portion of the curve for $p(\theta')$. Far out on the curve, $p(\theta')$ still varies as θ'^{-3}. Equation (5.13) may be obtained directly by using the approximation (3.6) that θ' is nearly equal to I_s'/Q and noticing that b_2 is the mean square value of I_s'.

If the sine wave is absent, ρ is zero and

$$p(\theta') = \tfrac{1}{2}(b_0/b_2)^{1/2}(1+z^2)^{-3/2} \qquad (5.14)$$

which is consistent with the results given between equations (3.4–10) and (3.4–11) of Reference A. In this case (5.12) becomes

$$\frac{1}{2} - \frac{z_1}{2}(1+z_1^2)^{-1/2} \qquad (5.15)$$

Although the standard deviation of θ' is infinite an idea of the spread of the distribution may be obtained from the average value of $|\theta'|$. Setting b_1 equal to zero in (4.5) in order to obtain the case in which the noise band is symmetrical about the sine-wave frequency leads to

$$\overline{|\theta'|} = \frac{2}{4\pi^2 b_0 b_2}\int_0^{\infty}dR\int_{-\pi}^{\pi}d\theta\int_{-\infty}^{+\infty}dR'\int_0^{\infty}d\theta'\,\theta'\,R^2$$

$$\exp\tfrac{1}{2}\left[-(R^2 - 2QR\cos\theta + Q^2)/b_0 - (R'^2 + R^2\theta'^2)/b_2\right]$$

The integrals in R', θ' cause no difficulty and the integral in θ is proportional to the Bessel function $I_0(QR/b_0)$. When the resulting integral in R is evaluated[13] we obtain

$$\overline{|\theta'|} = (b_2/b_0)^{1/2}e^{-\rho/2}I_0(\rho/2) \qquad (5.16)$$

where $\rho = Q^2/(2b_0)$.

[13] See, for example, G. N. Watson, "Theory of Bessel Functions," Cambridge (1944), p. 394, equation (5).

When ρ is zero equation (5.16) agrees with a result given in Section 3.4 of reference A, namely, for an ideal band pass filter

$$\frac{\text{ave} \,|\tau - \tau_1|}{\tau_1} = \frac{f_b - f_a}{\sqrt{3}(f_b + f_a)}$$

where τ is the interval between two successive zeros and τ_1 is its average value. τ is equal to $t_1 - t_0$ of our equation (3.10) from which it follows that

$$(\tau - \tau_1)/\tau_1 \approx -\theta'/q \tag{5.17}$$

6. Expected Number of Crossings of θ and $\dfrac{d\theta}{dt}$ per Second

After a brief study of the expected number of times per second the phase angle θ increases through 0 and through π (where it is assumed that $-\pi < \theta \leq \pi$) expressions are obtained for the expected number $N_{\theta'}$ of times per second the time derivative of θ increases through the value θ'.

The point T shown in Fig. 4 of Section 3 wanders around, as time goes by, in the plane of the figure. How many times may we expect it to cross some preassigned section of the line OQ in one second? To answer this problem we note that, from expression (2.1), the probability that θ increases through zero during the interval $t, t + dt$ with the envelope lying between R and $R + dR$ is

$$dt \, dR \int_0^\infty \theta' p(R, 0, \theta') \, d\theta' \tag{6.1}$$

where the probability density in the integrand is obtained by setting θ equal to zero in equation (5.1). The expected number of such crossings per second is

$$(2\pi)^{-3/2} \, (b_0 \, B)^{-1/2} \, R^2 \, dR e^{-b_2 (R-Q)^2/(2B)}$$

$$\int_0^\infty d\theta' \, \theta' \exp \left[-b_0 \, R^2 \, \theta'^2/(2B) + b_1 R(R - Q)\theta'/B \right] \tag{6.2}$$

which may be evaluated in terms of error functions or the function $\varphi_{-1}(x)$ defined by equation (1.8). For the special case in which the power spectrum of the noise current I_N is symmetrical about the sine wave frequency, b_1 is zero and (6.2) yields

$$(2\pi)^{-3/2} b_0^{-1} b_2^{1/2} e^{-(R-Q)^2/(2b_0)} \, dR \tag{6.3}$$

From equation (6.1) onwards we have tacitly assumed that the range of θ is given by $-\pi < \theta \leq \pi$ because setting θ equal to any multiple of 2π in our equations leads to the same result as setting θ equal to zero. This is due to θ occurring only in $\cos \theta$ and $\sin \theta$. When θ increases through the value π,

24

129

as it does when the point T crosses, in the downward direction, the extension of the line OQ lying to the left of the point O in Fig. 4, we imagine the value of θ to change discontinuously to the value $-\pi$.

The expected number of times per second θ increases through π may be obtained from (6.2) and, in the symmetrical case, from (6.3) by changing the sign of Q since this produces the same effect as changing θ from 0 to π in $p(R, \theta, \theta')$.

The expected number of crossings per second when R lies between two assigned values may be obtained by integrating the above equations. For example, the number of times per second θ increases through zero with R between Q and R_1 is, from (6.3) for the symmetrical case,

$$(4\pi)^{-1}(b_2/b_0)^{1/2} \operatorname{erf} [(2b_0)^{-1/2} | R_1 - Q |] \qquad (6.4)$$

where we have used the absolute value sign to indicate that R_1 may be either less than or greater than Q and

$$\operatorname{erf} x = 2\pi^{-1/2} \int_0^x e^{-t^2} dt \qquad (6.5)$$

Expressions for b_0 and b_2 are given by equations (A2–1) of Appendix II. The mean square value of I_N is b_0, and when the power spectrum of I_N is constant over a band of width β, $b_2 = \pi^2\beta^2 b_0/3$.

In much the same way it may be shown that the expected number of times per second θ increases through π with R between 0 and R_1 is

$$(4\pi)^{-1}(b_2/b_0)^{1/2} \{\operatorname{erf} [(2b_0)^{-1/2}(R_1 + Q)] - \operatorname{erf} [(2b_0)^{-1/2}Q]\} \qquad (6.6)$$

A check on these equations may be obtained by noting that the expected number of zeros per second of I_s, given by equation (4.11), is equal to twice the number of times θ increases through zero plus twice the number of times θ increases through π. Setting R_1 equal to zero in (6.4), to infinity in both (6.4) and (6.6), and adding the three quantities obtained gives half of (4.11), as it should.

Now we shall consider the crossings of θ'. The equations in the first part of the analysis are quite similar to those encountered in Section 3.8 of Reference A where the maxima of R, for noise alone, are discussed. We start by introducing the variables $x_1, x_2, \cdots x_6$ where

$$x_1 = I_c = R\cos \theta - Q, \qquad x_4 = I_s = R\sin \theta \qquad (6.7)$$

and the remaining x's are defined in terms of the derivatives of I_c and I_s and are given by the equations just below (3.8–4) of Reference A.

Here we shall consider the noise band to be symmetrical about the sine

25

wave frequency f_q so that b_1 and b_3 are zero. Then from equations (3.8–3) and (3.8–4) of Reference A the probability density of $x_1, x_2, \cdots x_6$ is

$$\frac{1}{8\pi^3 b_2 D} \exp\left(-\frac{1}{2D}\left[b_4(x_1^2 + x_4^2) + 2b_2(x_1 x_3 + x_4 x_6)\right.\right.$$
$$\left.\left. + (D/b_2)(x_2^2 + x_5^2) + b_0(x_3^2 + x_6^2)\right]\right) \tag{6.8}$$

where $D = b_0 b_4 - b_2^2$ and the b_n's are given by equations (A2–1). Replacing the x's by their expressions in terms of R and θ, similar to those just above equation (3.8–5) of Reference A, shows that the probability density for $R, R', R'', \theta, \theta', \theta''$ is

$$p(R, R', R'', \theta, \theta', \theta'') = \frac{R^3}{8\pi^3 b_2 D} \exp\left(-\frac{1}{2D}\left[b_4(R^2 - 2RQ\cos\theta + Q^2)\right.\right.$$

$$+ (D/b_2)(R'^2 + R^2\theta'^2) + 2b_2(RR'' - R^2\theta'^2) \tag{6.9}$$

$$+ b_0(R''^2 - 2RR''\theta'^2 + 4R'^2\theta'^2 + 4RR'\theta'\theta'' + R^2\theta'^4 + R^2\theta''^2)$$

$$\left.\left. - 2b_2 Q(R''\cos\theta - R\theta'^2\cos\theta - 2R'\theta'\sin\theta - R\theta''\sin\theta)\right]\right)$$

It must be remembered that (6.9) applies only to the symmetrical case in which b_1 and b_3 are zero.

Integrating R' and R'' in (6.9) from $-\infty$ to ∞ gives the probability density of $R, \theta, \theta', \theta''$. The integration with respect to R'' is simplified by changing to the variable $R'' - R\theta'^2$. The result is

$$p(R, \theta, \theta', \theta'') = R^3(2\pi)^{-2}(b_0 b_2 D)^{-1/2}(1 + u)^{-1/2}$$

$$\exp\left(-\frac{1}{2b_0}\left[R^2 - 2RQ\cos\theta + Q^2 + b_0 R^2\theta'^2/b_2\right.\right. \tag{6.10}$$

$$\left.\left. + \frac{(Qb_2\sin\theta + b_0 R\theta'')^2}{(1 + u)D}\right]\right)$$

where $u = 4b_2 b_0 \theta'^2/D$. The expected number of times per second the time derivative of θ increases through the value θ' is

$$N_{\theta'} = \int_{-\pi}^{\pi} d\theta \int_0^\infty dR \int_0^\infty d\theta'' \, \theta'' p(R, \theta, \theta', \theta'')$$

$$= \pi^{-2}(b_2\delta/b_0)^{1/2} \int_{-\pi}^{\pi} d\theta \int_0^\infty r\,dr \int_0^\infty x\,dx \tag{6.11}$$

$$\exp\left[-\gamma r^2 + 2r\alpha\cos\theta - \alpha^2 - \delta(x + \alpha\sin\theta)^2\right]$$

26

where we have set

$$r = R(2b_0)^{-1/2} \qquad\qquad x = rb_0\theta''/b_2$$

$$\alpha = Q(2b_0)^{-1/2} = \rho^{1/2} \qquad \gamma = 1 + b_0\theta'^2/b_2 = 1 + z^2 \qquad (6.12)$$

$$\delta = \frac{b_2^2}{(1+u)D}$$

r being regarded as a constant when the variable of integration is changed from θ'' to x.

The double integral in θ and x occurring in (6.11) is of the same form as (A3–1) of Appendix III and hence may be transformed into (A3–3). Here $a = r\,\alpha, c = -\delta\alpha^2, c + b^2 = 0$. The diameter of the path of integration C may be chosen so large that the order of integration may be interchanged and the integration with respect to r performed. The result is again an integral of the form (A3–3) in which $a^2 = 0$. When this is reduced to (A3–6) it becomes

$$N_{\theta'} = e^{-\rho}(2\pi\gamma)^{-1}b_2^{1/2}(b_0\delta)^{-1/2}\,[e^{-\delta\rho/2}I_0(\delta\rho/2)$$
$$+ (1+\gamma\delta)(1+\gamma\delta/2)^{-1}e^{\rho/\gamma}Ie\,\{\gamma\delta(2+\gamma\delta)^{-1},\,\rho/\gamma+\delta\rho/2\}] \qquad (6.13)$$

where we have used $Ie(-k, x) = Ie(k, x)$ which follows from the definition (A1–1) given in Appendix I.

When there is no sine wave present, ρ is zero and (6.13) becomes

$$N_{\theta'} = \frac{1}{2\pi\gamma}\left(\frac{b_2}{b_0\delta}\right)^{1/2} = \frac{\sqrt{\dfrac{b_4}{b_2} - \dfrac{b_2}{b_0} + 4\theta'^2}}{2\pi\left(1 + \dfrac{b_0}{b_2}\theta'^2\right)} \qquad (6.14)$$

This gives a partial check on some of the above analysis since (6.14) may be obtained immediately by setting α equal to zero in (6.11). Another check may be obtained by letting $\rho \to \infty$ and using $Ie(k, \infty) = (1 - k^2)^{-1/2}$. (6.13) becomes

$$N_{\theta'} \sim (2\pi)^{-1}(b_4/b_2)^{1/2}e^{-\rho z^2} \qquad (6.15)$$

which agrees with the result obtained from $\theta' \sim I_s'/Q$.

For the case in which the power spectrum $w(f)$ of the noise is equal to the constant value w_0 over the frequency band extending from $f_q - \beta/2$ to $f_q + \beta/2$,

$$b_0 = \beta w_0, \quad b_2 = \pi^2\beta^3 w_0/3 = \pi^2\beta^2 b_0/3, \quad b_4 = \pi^4\beta^5 w_0/5 = \pi^4\beta^4 b_0/5 \qquad (6.16)$$

27

These lead to

$$z = (b_0/b_2)^{1/2}\theta' = 3^{1/2}\theta'/(\pi\beta) \qquad D/b_2^2 = b_4 b_0/b_2^2 - 1 = 9/5 - 1 = 4/5$$

$$u = 4\,b_2^2 z^2/D = 5z^2 \qquad \delta = \frac{5}{4(1 + 5z^2)} \qquad (6.17)$$

$$\gamma = 1 + z^2$$

and the coefficient in (6.13) may be simplified by means of

$$\frac{1}{2\pi\gamma}\left(\frac{b_2}{b_0\delta}\right)^{1/2} = \frac{\beta}{1 + z^2}\left(\frac{1 + 5z^2}{15}\right)^{1/2} \qquad (6.18)$$

From (6.14) we see that (6.18) is equal to $N_{\theta'}$ when noise alone is present (and is of constant strength in the band of width β). The curves of $N_{\theta'}/\beta$ versus z shown in Fig. 7 were obtained by setting (6.17) and (6.18) in (6.13). $N_{\theta'}/\beta$ approaches $e^{-\rho}/(z\sqrt{3})$ as $z \to \infty$.

When the wandering point T of Fig. 4 passes close to the point O, θ changes rapidly by approximately π and produces a pulse in θ'. In discussions of frequency modulation θ' is sometimes regarded as a noise voltage which is applied to a low pass filter. Although the closer T comes to O the higher the pulse, the area under the pulse will be of the order of π and the response of the low pass filter may be calculated approximately.

That the pulses in θ' arise in the manner assumed above may be checked as follows. We choose a point relatively far out on the curve for $\rho = 5$ in Fig. 7, say $z = \sqrt{3}\theta'/(\beta\pi) = 1.6$ or $\theta' = 2.9\beta$. The number of pulses per second having peaks higher than 2.9β is roughly $N_{\theta'} = .009\beta$, and half of these have peaks greater than $\theta' = 3.8\beta$ which is obtained from Fig. 7 for $N_{\theta'} = .0045\beta$. From Fig. 6 we see that θ' exceeds 2.9β about .0018 of the time. Thus the average width at the height 2.9β of the class of pulses whose peaks exceed this value is $.0018/(.009\beta) = .2/\beta$ seconds. This figure is to be checked by the width obtained from the assumption that the typical pulse arises when T moves along a straight line with speed v and passes within a distance b of O. We take $\tan \theta = vt/b = \alpha t$ so that $\theta' = \alpha/(1 + \alpha^2 t^2)$. From this expression for θ' it follows that a pulse of peak height 3.8β (the median height) has a width of $.3/\beta$ seconds at $\theta' = 2.9\beta$. This agreement seems to be fairly good in view of the roughness of our work. A similar comparison may be made for $\rho = 0$ by using the limiting forms of (5.15) and (6.18). Here it is possible to compute the average width instead of estimating it from the median peak value. Exact agreement is obtained, both methods leading to an average width of $\pi/(4\theta')$ seconds at height θ'.

28

133

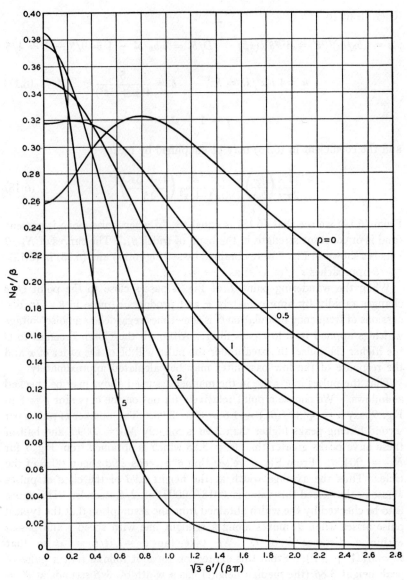

Fig. 7—Crossings of time derivative of phase angle.

$N_{\theta'}$ = expected number of times per second $d\theta/dt$ increases through the value θ'. ρ, β, and the power spectrum of I_N are the same as in Fig. 5.

29

7. Correlation Function for $\dfrac{d\theta}{dt}$

In this section we shall compute the correlation function $\Omega(\tau)$ of $\theta'(t)$. We are primarily interested in $\Omega(\tau)$ because it is, according to a fundamental result due to Wiener, the Fourier transform[14] of the power spectrum $W(f)$ of $\theta'(t)$.

We shall first consider the case in which the sine wave power is very large compared with the noise power so that, from (3.6), θ is approximately I_s/Q and θ' approximately I'_s/Q. Then using (A2–3) and (A2–1)

$$\Omega(\tau) = \overline{\theta'(t)\theta'(t+\tau)} \approx Q^{-2}\overline{I'_s(t)I'_s(t+\tau)}$$
$$= -g''Q^{-2} = 4\pi^2 Q^{-2}\int_0^\infty w(f)(f-f_q)^2 \cos 2\pi(f-f_q)\tau\, df \tag{7.1}$$

When $w(f)$ is effectively zero outside a relatively narrow band in the neighborhood of f_q, as it is in the cases with which we shall deal, (7.1) leads to the relation (divide the interval $(0, \infty)$ into $(0, f_q)$ and (f_q, ∞), introduce new variables of integration $f_1 = f_q - f, f_2 = f - f_q$ in the respective intervals, replace the upper limit f_q of the first integral by ∞, combine the integrals, and compare with (2.1–6) of Reference A)

Power spectrum of $\theta'(t) = W(f)$

$$\approx 4\pi^2 f^2 Q^{-2}[w(f_q + f) + w(f_q - f)] \tag{7.2}$$

This form is closely related to results customarily used in frequency modulation studies. It should be remembered that in (7.2) it is assumed that $0 < f \ll f_q$ and rms $I_N \ll Q$.

Additional terms in the approximation for $\Omega(\tau)$ may be obtained by expanding

$$\theta = \arctan \frac{I_s}{Q + I_c}$$

in descending powers of Q, multiplying two such series (one for time t and the other for time $t + \tau$) together, and averaging over t. If I_{c1}, I_{s1} and I_{c2}, I_{s2} denote the values of I_c, I_s at times t and $t + \tau$ respectively, the average values of the products of the I's may be obtained by expanding the characteristic function (obtainable from equation (7.5) given below by setting $z_5 = z_6 = z_7 = z_8 = 0$) of the four random variables $I_{c1}, I_{s1}, I_{c2}, I_{s2}$. This method is explained in Section 4.10 of Reference A. When $w(f)$ is symmetrical about f_q it is found that

[14] The form which we shall use is given by equation (2.1–5) of Reference A.

30

135

$$\overline{\theta_1\theta_2} = \frac{g}{Q^2} + \frac{g^2}{Q^4} + \frac{8g^3}{3Q^6} + \cdots$$

$$\Omega(\tau) = \overline{\theta_1'\theta_2'} = -\frac{d^2}{d\tau^2}\overline{\theta_1\theta_2} \tag{7.3}$$

$$= -\frac{g''}{Q^2} - \frac{2}{Q^4}(gg'' + g'^2) - \frac{8}{Q^6}(g^2g'' + 2g'^2g) + \cdots$$

From the exact expression for $\Omega(\tau)$ obtained below it is seen that the last equation in (7.3) is really asymptotic in character and the series does not converge. We infer that this is also true for the first equation of (7.3).

We shall now obtain the exact expression for the correlation function $\Omega(\tau)$ of $\theta'(t)$ when f_q is at the center of a symmetrical band of noise. At first sight it would appear that the easiest procedure is to calculate the correlation function for $\theta(t)$ and then obtain $\Omega(\tau)$ by differentiating twice. However, difficulties present themselves in getting θ outside the range $-\pi, \pi$ since θ enters the expressions only as the argument of trigonometrical functions. Because I could not see any way to overcome this difficulty I was forced to deal with θ' directly. Unfortunately this increases the complexity since now the distribution of the time derivatives of I_c and I_s also must be considered.

We have

$$\tan\theta = \frac{I_s}{Q + I_c}, \qquad \sec^2\theta = 1 + \left(\frac{I_s}{Q + I_c}\right)^2$$

$$\theta' = \frac{(Q + I_c)I_s' - I_sI_c'}{\sec^2\theta(Q + I_c)^2} = \frac{(Q + I_c)I_s' - I_sI_c'}{(Q + I_c)^2 + I_s^2}$$

and the value of $\overline{\theta'(t)\theta'(t + \tau)}$ is the eight-fold integral

$$\Omega(\tau) = \int_{-\infty}^{+\infty} dI_{c1} \cdots \int_{-\infty}^{+\infty} dI_{s2}' \, p(I_{c1}, \cdots, I_{s2}')$$

$$\frac{(Q + I_{c1})I_{s1}' - I_{s1}I_{c1}'}{(Q + I_{c1})^2 + I_{s1}^2} \times \frac{(Q + I_{c2})I_{s2}' - I_{s2}I_{c2}'}{(Q + I_{c2})^2 + I_{s2}^2} \tag{7.4}$$

where $p(I_{c1}, \cdots, I_{s2}')$ is an eight-dimensional normal probability density. As before, the subscripts 1 and 2 refer to times t and $t + \tau$, respectively. The most direct way of evaluating the integral (7.4) is to insert the expression for $p(I_{c1}, \cdots, I_{s2}')$ and then proceed with the integration. Indeed, this method was used the first time the integral (7.4) was evaluated. Later it was found that the algebra could be simplified by representing $p(I_{c1}, \cdots, I_{s2}')$ as the Fourier transform of its characteristic function. The second procedure will be followed here.

31

The characteristic function for $I_{c1}, I_{c2}, I_{s1}, I_{s2}, I'_{c1}, I'_{c2}, I'_{s1}, I'_{s2}$ is, from (A2–2) and (A2–3) of Appendix II and Section 2.9 of Reference A,

$$\text{ave. } \exp i[z_1 I_{c1} + z_2 I_{c2} + z_3 I_{s1} + z_4 I_{s2} + z_5 I'_{c1} + z_6 I'_{c2} + z_7 I'_{s1} + z_8 I'_{s2}]$$

$$= \exp\left(-\tfrac{1}{2}\right)\,[b_0(z_1^2 + z_2^2 + z_3^2 + z_4^2) + b_2(z_5^2 + z_6^2 + z_7^2 + z_8^2)$$

$$+2b_1(z_1 z_7 + z_2 z_8 - z_3 z_5 - z_4 z_6) \tag{7.5}$$

$$+2g(z_1 z_2 + z_3 z_4) + 2g'(z_1 z_6 - z_2 z_5 + z_3 z_8 - z_4 z_7)$$

$$-2g''(z_5 z_6 + z_7 z_8) + 2h(z_1 z_4 - z_2 z_3)$$

$$+2h'(z_1 z_8 + z_2 z_7 - z_3 z_6 - z_4 z_5) - 2h''(z_5 z_8 - z_6 z_7)].$$

Since we have included b_1, h, h', h'' this holds when f_q is not necessarily at the center of the noise band. However, henceforth we return to our assumption that f_q is placed at the center of a symmetrical noise band and take b_1, h, h', h'' to be zero.

The probability density of $I_{c1}, \cdots I'_{s2}$ which is to be placed in (7.4) is the eight-fold integral

$$p(I_{c1}, \cdots I'_{s2}) = (2\pi)^{-8} \int_{-\infty}^{+\infty} dz_1 \cdots \int_{-\infty}^{+\infty} dz_8$$

$$\exp[-iz_1 I_{c1} - \cdots - iz_8 I'_{s2}] \times [\text{ch.f.}] \tag{7.6}$$

where "ch.f." denotes the characteristic function obtained by setting b_1, h, h', h'' equal to zero on the right hand side of (7.5)

The integral (7.4) for $\Omega(\tau)$ may be written as

$$\Omega(\tau) = J_1 - J_2 - J_3 + J_4 \tag{7.7}$$

where J_1 is the 16-fold integral

$$J_1 = \int_{-\infty}^{+\infty} dI_{c1} \cdots \int_{-\infty}^{+\infty} dI'_{s2} (2\pi)^{-8} \int_{-\infty}^{+\infty} dz_1 \cdots \int_{-\infty}^{+\infty} dz_8$$

$$\exp[-iz_1 I_{c1} - \cdots - iz_8 I'_{s2}] \tag{7.8}$$

$$\frac{(Q + I_{c1})(Q + I_{c2})I'_{s1} I'_{s2}}{[(Q + I_{c1})^2 + I_{s1}^2][(Q + I_{c2})^2 + I_{s2}^2]} \times [\text{ch.f.}]$$

and J_2, J_3, J_4 are obtained from J_1 by replacing the product $(Q + I_{c1})$ $(Q + I_{c2})I'_{s1}I'_{s2}$ by $(Q + I_{c1})I'_{s1}I_{s2}I'_{c2}$, $I_{s1}I'_{c1}(Q + I_{c2})I'_{s2}$, $I_{s1}I'_{c1}I_{s2}I'_{c2}$ respectively.

The integration with respect to I_{c1} and I_{s1} in (7.8) may be performed at once. We replace $Q + I_{c1}$ and I_{s1} by x and y, respectively, and use

$$\int_{-\infty}^{+\infty} dx \int_{-\infty}^{+\infty} dy\, \frac{x}{x^2 + y^2}\, e^{-izx - i\zeta y} = \frac{-2\pi i z}{z^2 + \zeta^2}. \tag{7.9}$$

32

137

The integration with respect to I_{c2} and I_{s2} may be performed in a similar manner. In this way we obtain a 12-fold integral.

The integrations with respect to the I'''s may be performed by using

$$\frac{1}{2\pi} \int_{-\infty}^{+\infty} dI \int_{-\infty}^{+\infty} e^{-izI} f(z)\, dz = f(0)$$

$$\frac{1}{2\pi} \int_{-\infty}^{+\infty} I\, dI \int_{-\infty}^{+\infty} e^{-izI} f(z)\, dz = -i \left[\frac{df(z)}{dz}\right]_{z=0} \tag{7.10}$$

The result is the four-fold integral

$$J_1 = (2\pi)^{-2} \int_{-\infty}^{+\infty} dz_1 \cdots \int_{-\infty}^{+\infty} dz_4 \frac{z_1 z_2 (g'' - g'^2 z_3 z_4)}{(z_1^2 + z_3^2)(z_2^2 + z_4^2)} \tag{7.11}$$

$$\exp \left[-(b_0/2)(z_1^2 + z_2^2 + z_3^2 + z_4^2) - g(z_1 z_2 + z_3 z_4) + iQ(z_1 + z_2)\right].$$

In the same way J_2, J_3, J_4 may be reduced to the integrals obtained from (7.11) by replacing $z_1 z_2 (g'' - g'^2 z_3 z_4)$ by $-g'^2 z_1^2 z_4^2$, $-g'^2 z_2^2 z_3^2$ and $z_3 z_4 (g'' - g'^2 z_1 z_2)$, respectively. When the J's are combined in accordance with (7.7) we obtain an integral which may be obtained from (7.11) by replacing $z_1 z_2 (g'' - g'^2 z_3 z_4)$ by

$$g''(z_1 z_2 + z_3 z_4) + g'^2 (z_1 z_4 - z_2 z_3)^2 \tag{7.12}$$

The terms $z_1^2 + z_3^2$ and $z_2^2 + z_4^2$ in the denominator may be represented as infinite integrals. Interchanging the order of integration and expressing (7.12) in terms of partial derivatives of an exponential function leads to the six-fold integral

$$\Omega(\tau) = (4\pi)^{-2} \int_0^\infty du \int_0^\infty dv \left[-g''\frac{\partial}{\partial g} + g'^2 \frac{\partial^2}{\partial \alpha^2}\right]_{\alpha=0} \int_{-\infty}^{+\infty} dz_1 \cdots \int_{-\infty}^{+\infty} dz_4$$

$$\exp \left[-(b_0 + u)(z_1^2 + z_3^2)/2 - (b_0 + v)(z_2^2 + z_4^2)/2 \right. \tag{7.13}$$

$$\left. - g(z_1 z_2 + z_3 z_4) - \alpha(z_1 z_4 - z_2 z_3) + iQ(z_1 + z_2)\right]$$

where the subscript $\alpha = 0$ indicates that α is to be set equal to zero after the differentiations are performed.

When the four-fold integral in the z's is evaluated (7.13) becomes

$$\Omega(\tau) = \int_0^\infty du \int_0^\infty dv \left[-g'' \frac{\partial}{\partial g} + g'^2 \frac{\partial^2}{\partial \alpha^2}\right]_{\alpha=0}$$

$$\frac{1}{4D} \exp \left[-Q^2(2b_0 - 2g + u + v)/(2D)\right] \tag{7.14}$$

$$= \int_0^\infty du \int_0^\infty dv [(g'^2 - gg'')(2 - 2F + Q^2/g) - g'^2 Q^2/g] e^{-F}/(4D_0^2)$$

33

where

$$D = (b_0 + u)(b_0 + v) - g^2 - \alpha^2, \qquad F = Q^2(2b_0 - 2g + u + v)/(2D_0)$$

and D_0 denotes the value of D obtained by setting $\alpha = 0$. When differentiating with respect to α it is helpful to note that

$$\frac{\partial^2 f(D)}{\partial \alpha^2} = f''(D)\left(\frac{\partial D}{\partial \alpha}\right)^2 + f'(D)\frac{\partial^2 D}{\partial \alpha^2}$$

and that only $f'(D) = df/dD$ need be obtained since $\partial D/\partial \alpha$ vanishes when $\alpha = 0$.

In order to reduce the double integral to a single integral we make the change of variables

$$r = Q^2(b_0 + u - g)/(2D_0) \equiv \frac{Q^2(b_0 + u - g)}{2[(b_0 + u)(b_0 + v) - g^2]}$$

$$s = Q^2(b_0 + v - g)/(2D_0), \qquad F = r + s \tag{7.15}$$

$$\partial(r, s)/\partial(u, v) = -rs/D_0, \qquad 4srD_0 = Q^2[Q^2 - 2g(r + s)]$$

The limits of integration for r and s are obtained by noting that the points $(0, 0)$, $(\infty, 0)$, (∞, ∞), $(0, \infty)$ in the (u, v) plane go into $(Q^2/(2b_0 + 2g)$, $Q^2/(2b_0 + 2g))$, $(Q^2/(2b_0), 0)$, $(0, 0)$ $(0, Q^2/(2b_0))$, respectively, in the (r, s) plane. It may be verified that the region of integration in the (r, s) plane is the interior of the quadrilateral obtained by joining the above points by straight lines. Equation (7.14) may now be written as

$$\Omega(\tau) = \int\!\!\int \left\{ \frac{(g'^2 - gg'')(2 - 2r - 2s + Q^2/g) - g'^2 Q^2/g}{Q^2[Q^2 - 2g(r + s)]} \right\} e^{-r-s} \, dr \, ds$$

$$= \frac{g'^2 - gg''}{2g^2} y_1 - \frac{g'^2}{2g^2} y_2 \tag{7.16}$$

where y_1 and y_2 are the dimensionless quantities

$$y_1 = \int\!\!\int \frac{2g^2(2 - 2r - 2s + Q^2/g)}{Q^2[Q^2 - 2g(r + s)]} e^{-r-s} \, dr \, ds$$

$$y_2 = \int\!\!\int \frac{2g e^{-r-s} \, dr \, ds}{Q^2 - 2g(r + s)}$$

It is seen that

$$y_1 = 2gQ^{-2}\left\{ y_2 + \int\!\!\int e^{-r-s} \, dr \, ds \right\}. \tag{7.17}$$

Since the integrands are functions of $r + s$ alone we are led to apply the transformation

$$\int\!\!\int_A f(r + s) \, dr \, ds = \int_0^\alpha u f(u) \, du + \int_\alpha^{2\beta} \frac{\alpha(2\beta - u)}{2\beta - \alpha} f(u) \, du \tag{7.18}$$

34

139

where A is the area enclosed by the quadrilateral whose vertices are at the points (r, s) given by $(0, 0)$, $(0, \alpha)$, (β, β), $(\alpha, 0)$ and it is assumed that β and α are positive. u is a new variable and is not the one introduced in (7.13).

Setting $\alpha = Q^2/(2b_0)$ and $\beta = Q^2/(2b_0 + 2g)$, using (7.18), and introducing the notation

$$\rho = Q^2/(2b_0), \qquad k = g/b_0$$

$$\xi = Q^2/(2g) = \rho/k, \qquad \lambda = \frac{Q^2}{b_0 + g} = \frac{2\rho}{1 + k} \qquad (7.19)$$

permits us to write

$$\iint\limits_A e^{-r-s}\, dr\, ds = \int_0^\rho u e^{-u}\, du + \int_\rho^\lambda \frac{\rho(\lambda - u)}{\lambda - \rho}\, e^{-u}\, du$$

$$= 1 - \frac{\lambda e^{-\rho}}{\lambda - \rho} + \frac{\rho e^{-\lambda}}{\lambda - \rho} \qquad (7.20)$$

and (7.17) yields

$$y_2 = \frac{\rho}{k}\, y_1 - 1 + \frac{2}{1-k}\, e^{-\rho} - \frac{1+k}{1-k}\, e^{-2\rho/(1+k)} \qquad (7.21)$$

where we have expressed λ in terms of ρ and k.

The double integral defining y_2 may be treated in the same way as (7.20):

$$y_2 = \iint\limits_A \frac{e^{-r-s}\, dr\, ds}{\xi - r - s} = \int_0^\rho \frac{u e^{-u}\, du}{\xi - u} + \int_\rho^\lambda \frac{\rho(\lambda - u) e^{-u}\, du}{(\lambda - \rho)(\xi - u)}$$

Writing $u = \xi - (\xi - u)$ and $\lambda - u = \lambda - \xi + (\xi - u)$ in the two numerators leads to

$$y_2 = \xi \int_0^\rho \frac{e^{-u}\, du}{\xi - u} - \int_0^\rho e^{-u}\, du$$

$$- \xi \int_\rho^\lambda \frac{e^{-u}\, du}{\xi - u} + \frac{\rho}{\lambda - \rho} \int_\rho^\lambda e^{-u}\, du \qquad (7.22)$$

where we have used $\rho(\lambda - \xi)/(\lambda - \rho) = -\xi$ to simplify the coefficient of the third integral. When the second and fourth integrals are evaluated, their contribution to y_2 is found to be equal to the terms independent of y_1 on the right of (7.21). Hence, comparison of equations (7.21) and (7.22) shows that

$$y_1 = \int_0^\rho \frac{e^{-u}\, du}{\xi - u} - \int_\rho^\lambda \frac{e^{-u}\, du}{\xi - u} \qquad (7.23)$$

35

140

The integrals in (7.23) may be evaluated in terms of the exponential integral $Ei(x)$ defined by, for x real,

$$Ei(x) = \int_{-\infty}^{x} e^t \, dt/t = C + \tfrac{1}{2} \log_e x^2 + \sum_{n=1}^{\infty} \frac{x^n}{n!\,n}$$
$$\sim e^x \sum_{n=0} n!/x^{n+1}$$

where $C = .577 \cdots$ is Euler's constant and Cauchy's principal va'ue of the integral is to be taken when $x > 0$. We set $t = \xi - u$ and obtain

$$y_1 = e^{-\rho/k} \left\{ Ei[\rho/k] - 2Ei[\rho(1-k)/k] + Ei\left[\frac{\rho(1-k)}{k(1+k)}\right] \right\}$$

where we have again expressed ξ and λ in terms of ρ and k.

A power series for y_1 which converges when $-1/3 \le k < 1$ may be obtained by expanding the denominators of the integrands in (7.23) in powers of u/ξ and integrating termwise:

$$
\begin{aligned}
y_1 = {}& \xi^{-1}[1 - 2e^{-\rho} + e^{-\lambda}] \\
& + 1!\xi^{-2}[1 - 2(1 + \rho/1!)e^{-\rho} + (1 + \lambda/1!)e^{-\lambda}] \\
& + 2!\xi^{-3}[1 - 2(1 + \rho/1! + \rho^2/2!)e^{-\rho} + (1 + \lambda/1! + \lambda^2/2!)e^{-\lambda}] \\
& + \cdots
\end{aligned}
\tag{7.24}
$$

The following special values may be obtained from the equation given above. When $\rho = 0$

$$y_1 = -\log_e (1 - k^2)$$
$$y_2 = 0 \tag{7.25}$$

This result may also be obtained by evaluating the integral obtained when we set $Q = 0$, $z_1 = r_1 \cos \theta_1$, $z_3 = r_1 \sin \theta_1$, $z_2 = r_2 \cos \theta_2$, $z_4 = r_2 \sin \theta_2$ in (7.11) and (7.12).

Near $k = 1$,

$$y_1 \approx e^{-\rho}[Ei(\rho) - C - \log_e \rho(1 - k^2)]$$
$$y_2 \approx \rho y_1 - 1 + (1 + \rho)e^{-\rho} \tag{7.26}$$

Near $k = 0$,

$$y_1 \approx k(1 - e^{-\rho})^2/\rho, \qquad y_2 \approx y_1 \tag{7.27}$$

except when $\rho = 0$ in which case y_1 is approximately k^2.
When ρ is large

$$y_1 \sim \frac{k}{\rho} + \frac{1!\,k^2}{\rho^2} + \frac{2!\,k^3}{\rho^3} + \frac{3!\,k^4}{\rho^4} + \cdots$$

$$y_2 \sim -1 + \frac{\rho}{k} y_1 \sim \frac{1!\,k}{\rho} + \frac{2!\,k^2}{\rho^2} + \cdots \tag{7.28}$$

except near $k = 1$ where both y_1 and y_2 have logarithmic infinities. The asymptotic expansion (7.3) for $\Omega(\tau)$, which was obtained by the first method

36

141

of this section, may be checked by inserting (7.28) in the expression (7.16) for $\Omega(\tau)$ in terms of y_1 and y_2.

Values of y_1 and y_2 tabulated as functions of k for various values of ρ are given in Table 3. Negative values of k have not been considered since they

TABLE 3

VALUES OF y_1 AND y_2 USED IN COMPUTATION OF CORRELATION FUNCTION OF $d\theta/dt$

$$\Omega(\tau) = \overline{\theta'(t)\theta'(t+\tau)} = [g'^2(y_1 - y_2) - gg''y_1]/(2g^2)$$

$$g \equiv g(\tau) = \int_0^\infty w(f) \cos 2\pi(f - f_q)\tau \, df, \quad k = g(\tau)/g(0)$$

	Values of y_1					Values of y_2			
k	ρ					ρ			
	0	.5	1	2	5	.5	1	2	5
0	0	0	0	0	0	0	0	0	0
.1	.01005	.03526	.04224	.03854	.02000	.03171	.04147	.03936	.02051
.2	.04082	.08043	.09003	.07979	.04105	.06550	.08654	.08275	.04283
.3	.09431	.1379	.1452	.1246	.06292	.1022	.1363	.1315	.06702
.4	.1744	.2110	.2102	.1740	.08586	.1432	.1926	.1870	.09384
.5	.2877	.3056	.2886	.2296	.1101	.1914	.2579	.2515	.1238
.6	.4463	.4278	.3860	.2942	.1358	.2481	.3368	.3289	.1576
.7	.6733	.5953	.5129	.3721	.1636	.3220	.4379	.4269	.1975
.8	1.0216	.8416	.6914	.4729	.1941	.4275	.5803	.5602	.2461
.84	1.2228	.9798	.7888	.5242	.2075	.4866	.6593	.6318	.2693
.88	1.4890	1.1590	.9127	.5866	.2219	.5641	.7619	.7226	.2964
.90	1.6607	1.2742	.9898	.6241	.2296	.6138	.8260	.7752	.3114
.92	1.8734	1.4144	1.0834	.6686	.2378	.6753	.9058	.8486	.3294
.94	2.1507	1.5948	1.2024	.7217	.2466	.7550	1.0093	.9333	.3498
.96	2.5459	1.8486	1.3668	.7939	.2566	.8711	1.1558	1.0546	.3752
.97	2.8285	2.0251	1.4815	.8414	.2623	.9474	1.2605	1.1366	.3849
.98	3.2289	2.2762	1.6405	.9073	.2690	1.0704	1.4081	1.2548	.4119
.99	3.9170	2.7080	1.9066	1.0127	.2778	1.2773	1.6610	1.4505	.4429
.995	4.6072	3.1341	2.1721	1.1125	.2846	1.4838	1.9175	1.6416	.4705
.997	5.1175	3.4445	2.3622	1.1866	.2889	1.6367	2.1048	1.7859	.4893

are not required for the case in which I_N has a normal law power spectrum, the case discussed in the next section.

8. POWER SPECTRUM OF $\dfrac{d\theta}{dt}$ WHEN I_N HAS NORMAL LAW POWER SPECTRUM

The problem of computing the power spectrum $W(f)$ of $\theta'(t)$ appears to be a difficult one.* In order to obtain an answer without an excessive amount of work we have had to do two things which are rather restrictive. First, we confine our attention to the case in which the power spectrum $w(f)$ of

*Since the above was written the general f. m. problem has been studied by D. Middleton. He generalizes our (7.11) and (7.12), introduces polar coordinates, expands the integrand in powers of g, and integrates termwise. $W(f)$ then follows somewhat as in a.m. theory.

37

142

I_N is of the normal law type (our method could be applied to other types but g' and g'' would be more complicated functions of τ and Table 3 would have to be extended to negative values of k, if they should occur). Second, we resort to numerical integration to obtain a portion of $W(f)$. Because of the second item our results are either tabulated or are given as curves, shown in Figs. 8 and 9, except when $Q = 0$ (noise only) in which case the power spectrum of θ' is given by the series (8.7).

The power spectrum of I_N is assumed to be

$$w(f) = \frac{\psi_0}{\sigma\sqrt{2\pi}} e^{-(f-f_q)^2/(2\sigma^2)} \tag{8.1}$$

The mean square value of I_N is equal to that of a noise current whose power spectrum has the constant value of $\psi_0/(\sigma\sqrt{2\pi})$ over a band of width $f_b - f_a = \sigma\sqrt{2\pi} = \sigma 2.507$. The value of $w(f)$ is one quarter of its mid-band value at the points $f - f_q = \pm\sigma\sqrt{2\log_e 4} = \pm\sigma 1.665$ (the 6 db points) and the distance between these points is 3.330σ. Integration of (8.1) shows that the mean square value of I_N is ψ_0 in accordance with our customary notation. The mid-band value of $w(f)$ is $\psi_0/(\sigma\sqrt{2\pi})$.

Assuming $f_q \gg \sigma$ and evaluating the integrals (A2–1) of Appendix II defining b_0 and g gives

$$b_0 = \psi_0, \qquad g = \psi_0 e^{-2(\pi\sigma\tau)^2} = \psi_0 e^{-u^2/2}$$

$$g'/g = -uu' = -2\pi\sigma u, \qquad g''/g = -(2\pi\sigma)^2(1 - u^2)$$

$$\frac{g''g - g'^2}{g^2} = -(2\pi\sigma)^2, \qquad k = g/b_0 = e^{-u^2/2} \tag{8.2}$$

where we have set

$$u = 2\pi\sigma\tau, \qquad u' = 2\pi\sigma \tag{8.3}$$

and the primes on g and u denote differentiation with respect to τ. The correlation function is accordingly, from (7.16).

$$\Omega(\tau) = 2\pi^2\sigma^2(y_1 - u^2 y_2) \tag{8.4}$$

If $\theta'(t)$ be regarded as a noise current its power spectrum is

$$W(f) = 4\int_0^\infty \Omega(\tau) \cos 2\pi f\tau \, d\tau \tag{8.5}$$

When noise alone is present, ρ is zero and (7.25) yields

$$\Omega(\tau) = -2\pi^2\sigma^2 \log_e (1 - k^2) = -2\pi^2\sigma^2 \log_e (1 - e^{-u^2}) \tag{8.6}$$

38

143

In this case the power spectrum is, from (8.3), (8.5), and (8.6),

$$W_N(f) = -4\pi\sigma \int_0^\infty \cos(uf/\sigma) \log(1 - e^{-u^2})\, du$$

$$= 2\sigma\pi^{3/2} \sum_{n=1}^\infty n^{-3/2} e^{-f^2/(4n\sigma^2)}, \tag{8.7}$$

the series being obtained by expanding the logarithm and integrating term-wise. When this equation was used for computation it was found convenient to apply the Euler summation formula to sum the terms in the series beyond the $(N-1)$st. Writing b for $f^2/(4\sigma^2)$, the series in (8.7) becomes

$$1^{-3/2}e^{-b/1} + 2^{-3/2}e^{-b/2} + \cdots (N-1)^{-3/2}e^{-b/(N-1)}$$

$$+ (\pi/b)^{1/2} \operatorname{erf}[(b/N)^{1/2}] + N^{-3/2}e^{-b/N}\left[\frac{1}{2} - \frac{1}{12N}\left(-\frac{3}{2} + \frac{b}{N}\right)\right. \tag{8.8}$$

$$\left. + \frac{1}{720N^3}\left(-\frac{105}{8} + \frac{105}{4}\frac{b}{N} - \frac{21}{2}\frac{b^2}{N^2} + \frac{b^3}{N^3}\right) + \cdots\right]$$

When b is zero the sum[15] of the series is $2.61237\cdots$. The values for $\rho = 0$ in Table 4 were computed by taking $N = 12$ in (8.8). As $b \to \infty$ the dominant term in (8.8) is seen to be the one containing erf (choose N so that $b = N^{3/2}$). Hence as $f \to \infty$

$$W_N(f) \sim 4\pi^2\sigma^2/f. \tag{8.9}$$

When both noise and the sine wave are present it is convenient to split the power spectrum into three parts. The first part, $W_1(f)$, is proportional to $W_N(f)$, the power spectrum with noise alone. The second part $W_2(f)$ is proportional to the form $W(f)$ assumes when rms $I_N \ll Q$ and the third part $W_3(f)$ is of the nature of a correction term. This procedure is suggested when we subtract the leading terms in the expressions (7.26) and (7.27) (corresponding to $k = 1$ and $k = 0$, respectively) from y_1. Likewise we subtract the leading term in y_2, (7.27), at $k = 0$ but do not bother to do so at the end $k = 1$ because $u^2 y_2$ approaches zero there. We therefore write

$$y_1 - u^2 y_2 = [y_1 + e^{-\rho}\log(1 - k^2) - k(1 - e^{-\rho})^2/\rho - u^2 y_2$$

$$+ u^2 k(1 - e^{-\rho})^2/\rho] - e^{-\rho}\log(1 - k^2) + (1 - u^2)k(1 - e^{-\rho})^2/\rho \tag{8.10}$$

$$= Z(u) - e^{-\rho}\log(1 - k^2) - \frac{g''(2\pi\sigma)^{-2}}{b_0\rho}(1 - e^{-\rho})^2$$

[15] "Theory and Application of Infinite Series," Knopp, (1928), page 561.

39

144

where $Z(u)$ denotes the function enclosed by the brackets in the first equation and the expressions for g''/g and k in (8.2) have been used in the replacement of $(1 - u^2)k$.

TABLE 4
VALUES OF $W_3(f)/(4\pi^2\sigma)$

$\dfrac{6f}{\sigma\pi}$	$\rho = 0$	0.5	1.0	2.0	5.0
0	0	$-.03517$	$-.03891$	$-.02444$	$-.001948$
1	0	$-.03003$	$-.03196$	$-.01830$	$-.001814$
2	0	$-.01717$	$-.01486$	$-.003304$	.004052
3	0	$-.002436$	.004014	.01252	.008225
4	0	.008757	.01730	.02244	.01027
6	0	.01478	.02157	.02167	.007665
8	0	.01018	.01366	.01237	.003505
10	0	.005768	.007378	.006201	.001437
12	0	.004027	.004463	.003552	.0006439

VALUES OF $W(f)/(4\pi^2\sigma)$

0	.7369	.4118	.2322	.07529	.003017
1	.7098	.4294	.2672	.1134	.02342
2	.6439	.4516	.3231	.1784	.05828
3	.5542	.4225	.3225	.1947	.06852
4	.4623	.3496	.2654	.1580	.01590
6	.3195	.2178	.1508	.07554	.01540
8	.2390	.1553	.1019	.04506	.005325
10	.1908	.1215	.07768	.03206	.002726
12	.1595	.1003	.06306	.02511	.001719

Inserting (8.10) in the expression (8.4) for $\Omega(\tau)$ and taking the Fourier transform (8.5) leads to

$$W(f) = W_1(f) + W_2(f) + W_3(f)$$

$$W_1(f) = e^{-\rho}W_N(f)$$

$$W_2(f) = -\frac{2(1 - e^{-\rho})^2}{b_0\rho}\int_0^\infty g'' \cos 2\pi f\tau \, d\tau$$

$$= \frac{(1 - e^{-\rho})^2}{\rho}(2\pi f)^2 \frac{e^{-f^2/(2\sigma^2)}}{\sigma\sqrt{2\pi}}$$

$$W_3(f) = 4\pi\sigma \int_0^\infty Z(u) \cos(uf/\sigma) \, du$$

(8.11)

40

In these equations $W_N(f)$ is obtained from (8.7), and $W_2(f)$ by two-fold integration by parts to reduce g'' to g then evaluating the integral obtained

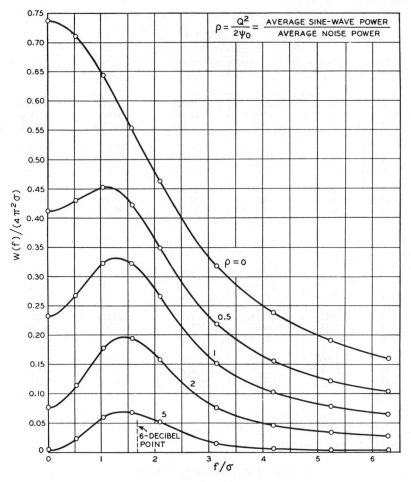

Fig. 8—Power spectrum of $d\theta/dt$.

Power spectrum of I_N is assumed to be

$$\psi_0(\sigma\sqrt{2\pi})^{-1} \exp\left[-(f - f_q)^2/(2\sigma^2)\right].$$

In this expression f is a frequency near f_q. The f in $W(f)$ and in the abscissa is a much lower frequency. $W(f)$ = power spectrum of $\theta' = d\theta/dt$, θ' being regarded as a random noise current. Dimensions of $W(f)\,df$ same as $(d\theta/dt)^2$ or $(\text{radians})^2/\text{sec.}^2$.

by substituting the expression (8.2) for g. That $W(f)$ approaches $W_2(f)$ as $\rho \to \infty$ follows when expression (8.11) for $W_2(f)$ is compared with the limiting form (8.13) given below.

41

Instead of dealing with $W(f)$ it is more convenient to deal with $(4\pi^2\sigma)^{-1}$ $W(f)$ which is the sum of the three components

$$(4\pi^2\sigma)^{-1}W_1(f) = \frac{e^{-\rho}}{2\sqrt{\pi}} \sum_{n=1}^{\infty} n^{-3/2} e^{-f^2/(4n\sigma^2)}$$

$$(4\pi^2\sigma)^{-1}W_2(f) = \frac{(1 - e^{-\rho})^2}{\rho\sqrt{2\pi}} \left(\frac{f}{\sigma}\right)^2 e^{-f^2/(2\sigma^2)} \qquad (8.12)$$

$$(4\pi^2\sigma)^{-1}W_3(f) = \frac{1}{\pi} \int_0^{\infty} Z(u) \cos (uf/\sigma)\, du$$

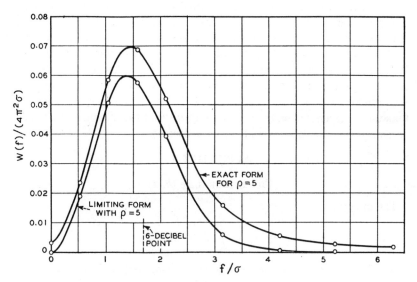

Fig. 9—Approach of $W(f)$ to limiting form.

As $\rho \to \infty$, $W(f) \to 4\pi^2\sigma\ (\rho\sqrt{2\pi})^{-1}\ (f/\sigma)^2 \exp\ [-f^2/(2\sigma^2)]$.

The integral involving $Z(u)$ has been computed by Simpson's rule, y_1 and y_2 being obtained from Table 3, with the results shown in the first section of Table 4. The value of $W_2(f)$ may be computed directly, and $W_1(f)$ may be obtained from $W_N(f)$. The values of these two functions together with those of $W_3\ (f)$ enable us to compute the values of $(4\pi^2\sigma)^{-1}W(f)$ given in Table 4 and plotted in Fig. 8.

Since, as is shown by (8.9), $W_N(f)$ varies as $1/f$ for large values of f, the areas under the curves of Fig. 8 become infinite. This agrees with the fact that the mean square value of θ' is infinite.

The values of $(4\pi^2\sigma)^{-1}W(0)$ for ρ equal to 0, .5, 1, 2, and 5 are .7369, 4118, 2322, .07529, and .003017 respectively. When these values are plotted on

42

147

semi-log paper they tend to lie on a straight line whose slope suggests that $W(0)$ decreases as $e^{-\rho}$ when ρ becomes large.

The limiting form assumed by $W(f)$ as $\rho \to \infty$ is given by equation (7.2). When the normal law expression (8.1) assumed in this section for the power spectrum of I_N is put in (7.2) we find that

$$W(f) \to \frac{4\pi^2 \sigma}{\rho\sqrt{2\pi}} \left(\frac{f}{\sigma}\right)^2 e^{-f^2/(2\sigma^2)} \tag{8.13}$$

Fig. 9 shows that for $\rho = 5$ the limiting form (8.13) agrees quite well with the exact form computed above.

Both (7.2) and (8.13) show that, for small values of f, the power spectrum of θ' varies as f^2 when $\rho >> 1$. This is in accord with Crosby's* result that the voltage spectrum of the random noise in the output of a frequency modulation receiver is triangular when the carrier to noise ratio is large. When this ratio becomes small he finds that the spectrum becomes rectangular. Fig. 8 shows this effect in that the areas under the curves between the ordinates at $f = 0$ and $f = \lambda\sigma$ (where λ is some number, generally less than unity, depending on the ratio of the widths of the i.f. and audio bands) become rectangles, approximately, as ρ decreases.

APPENDIX I

The Integral $Ie\,(k, x)$

The integral[16]

$$Ie(k,x) = \int_0^x e^{-u} I_0(ku)\, du, \tag{A1-1}$$

where $I_0(ku)$ denotes the Bessel function of imaginary argument and order zero, occurs in Sections 2 and 6. The following special cases are of interest.

$$Ie(0, x) = 1 - e^{-x}$$

$$Ie(1, x) = xe^{-x}[I_0(x) + I_1(x)] \tag{A1-2}$$

$$Ie(k, \infty) = \frac{1}{\sqrt{1 - k^2}}$$

The second of these relations is due to Bennett.[17]

* M. G. Crosby, "Frequency Modulation Noise Characteristics," Proc. I. R. E. Vol. 25 (1937), 472–514. See also J. R. Carson and T. C. Fry, "Variable Electric Circuit Theory with Application to the Theory of Frequency Modulation," B.S.T.J. Vol. 16 (1937), 513–540.

[16] The notation was chosen to agree with that used by Bateman and Archibald (Guide to Tables of Bessel Functions appearing in "Math. Tables and Aids to Comp.", Vol. 1 (1944) pp. 205–308) to discuss integrals used by Schwarz (page 248).

[17] It is given in equation (62) of the reference cited in connection with our equation (1.2) in Section 1.

43

The values in the table given below were computed by Simpson's rule for numerical integration. The work was checked at several points by using

$$Ie(k, x) = \sum_{n=0}^{\infty} (k/2)^{2n} \frac{(2n)!}{n!\,n!} A_n$$

where

$$A_n = 1 - \left[1 + x + \frac{x^2}{2!} \cdots + \frac{x^{2n}}{(2n)!}\right] e^{-x}$$

When x is so large that $Ie(k, x)$ is nearly equal to $Ie(k, \infty)$ we have

$$Ie(k, x) \sim (1 - k^2)^{-1/2} - [2k(1 - k)]^{-1/2}(2/\sqrt{\pi}) \int_{t_1}^{\infty} e^{-t^2}\, dt$$

where $t_1 = \sqrt{x(1 - k)}$. However, this was not found to be especially useful in checking the values given in the table.

TABLE OF $Ie(k, x) = \int_0^x e^{-u} I_0(ku)\, du$

x	k						
	0	.2	.4	.6	.8	.9	1.0
0	0	0	0	0	0		0
.2	.1813	.1813	.1814	.1815	.1816		.1818
.4	.3297	.3298	.3303	.3311	.3322		.3337
.6	.4512	.4517	.4530	.4554	.4586		.4629
.8	.5507	.5516	.5545	.5593	.5661		.5749
1.0	.6321	.6337	.6386	.6468	.6584		.6736
.2	.6988	.7012	.7086	.7209	.7386		.7620
.4	.7534	.7567	.7669	.7841	.8089		.8422
.6	.7981	.8025	.8157	.8383	.8712		.9157
.8	.8347	.8401	.8566	.8850	.9267		.9839
2.0	.8647	.8712	.8910	.9255	.9766		1.0476
.2	.8892	.8968	.9201	.9607	1.0217		1.1075
.4	.9093	.9179	.9446	.9916	1.0627		1.1642
.6	.9257	.9354	.9655	1.0186	1.1001		1.2183
.8	.9392	.9499	.9831	1.0424	1.1345		1.2699
3.0	.9502	.9618	.9982	1.0635	1.1661		1.3195
.2	.9592	.9718	1.0110	1.0822	1.1953		1.3672
.4	.9666	.9800	1.0220	1.0988	1.2223		1.4132
.6	.9727	.9868	1.0314	1.1136	1.2475		1.4578
.8	.9776	.9925	1.0394	1.1268	1.2708		1.5010
4.0	.9817	.9971	1.0463	1.1386	1.2926		1.5430
.2	.9830	1.0010	1.0522	1.1492	1.3130		1.5839
.4	.9877	1.0043	1.0574	1.1587	1.3320		1.6237
.6	.9899	1.0070	1.0619	1.1672	1.3499		1.6625
.8	.9918	1.0092	1.0657	1.1749	1.3666		1.7005
5.0	.9933	1.0111	1.0690	1.1818	1.3823		1.7376
5.4	.9955	1.0140	1.0743	1.1937	1.4110		1.8095

44

149

x	k						
	0	.2	.4	.6	.8	.9	1.0
5.8	.9970	1.0160	1.0783	1.2034	1.4364		1.8786
6.2	.9980	1.0174	1.0814	1.2114	1.4590		1.9452
6.6	.9986	1.0183	1.0837	1.2180	1.4792		2.0097
7.0	.9991	1.0190	1.0854	1.2234	1.4972		2.0722
7.4	.9994	1.0195	1.0867	1.2278	1.5134		2.1328
7.8	.9996	1.0198	1.0876	1.2375	1.5279		2.1917
8.2	.9997	1.0201	1.0885	1.2346	1.5409		2.2491
8.6	.9998	1.0202	1.0891	1.2371	1.5526		2.3050
9.0	.9999	1.0203	1.0896	1.2393	1.5631		2.3597
10.0	1.0000	1.0205	1.0902	1.2431	1.5852	1.9207	2.4910
11.0	1.0000	1.0206	1.0907	1.2456	1.6024	1.9668	2.6157
12.0	1.0000	1.0206	1.0909	1.2471	1.6158	2.0066	2.7347
13.0	1.0000	1.0206	1.0910	1.2482	1.6263	2.0411	2.8487
14.0	1.0000	1.0206	1.0910	1.2488	1.6346	2.0711	2.9584
15.0	1.0000	1.0206	1.0911	1.2492	1.6412	2.0973	3.0641
∞	1.0000	1.0206	1.0911	1.2500	1.6667	2.2942	∞

x	k			
	.86	.90	.96	1.0
15.0	1.8773	2.0973	2.5810	3.0641
16.0	1.8899	2.1201	2.6371	3.1663
17.0	1.9006	2.1403	2.6894	3.2653
18.0	1.9095	2.1579	2.7381	3.3614
19.0	1.9171	2.1737	2.7837	3.4548
20.0	1.9235	2.1870	2.8263	3.5457
∞	1.9597	2.2942	3.5714	∞

APPENDIX II

SECOND MOMENTS ASSOCIATED WITH I_c AND I_s

The in-phase and quadrature components of the noise current I_N

$$I_c(t) = \sum_{n=1}^{M} c_n \cos\left[(\omega_n - q)t - \varphi_n\right]$$

$$I_s(t) = \sum_{n=1}^{M} c_n \sin\left[(\omega_n - q)t - \varphi_n\right]$$

(3.3)

are closely related to the envelope R and phase angle θ of the total current, this relationship being being shown by the equations (3.4) and (3.5). $I_c(t)$ and $I_s(t)$ and their time derivatives may be regarded as random variables. In much of our work we have to deal with the probability distribution of these random variables. By virtue of the representation (3.3) and the central limit theorem[18] this distribution is normal in the several variables. The coefficients in the quadratic form occurring in the exponent are deter-

[18] Section 2.10 of Reference A.

45

150

mined by the second moments of the variables.[19] Here we state these moments. Some of the moments have already been given in Sections 3.7 and 3.8 of Reference A. For the sake of completeness we shall also give them here. The new results given below are derived in much the same way as those given in Reference A.

Let

$$b_n = (2\pi)^n \int_0^\infty w(f)(f - f_q)^n \, df$$

$$b_0 = \int_0^\infty w(f) \, df = \psi_0$$

$$g = \int_0^\infty w(f) \cos 2\pi(f - f_q)\tau \, df \qquad \text{(A2-1)}$$

$$h = \int_0^\infty w(f) \sin 2\pi(f - f_q)\tau \, df$$

and let g', g'', h', h'' denote the first and second derivatives of g and h with respect to τ. For example,

$$g' = -2\pi \int_0^\infty w(f)(f - f_q) \sin 2\pi(f - f_q)\tau \, df$$

Incidentally, in many of our cases $w(f)$ is assumed to be symmetrical about f_q. This introduces considerable simplification because b_1, b_3, b_5, $\cdots$, h, h', h'', reduce to zero.

The following table gives values of b_n's and g for two cases of frequent occurrence

	Ideal band pass filter centered on f_q	Normal law filter centered on f_q, $f_q \gg \sigma$
$w(f)$	w_0 for $f_a < f < f_b$ and zero elsewhere	$\dfrac{\psi_0}{\sigma\sqrt{2\pi}} e^{-(f-f_q)^2/2\sigma^2}$
b_0	$w_0(f_b - f_a)$	ψ_0
b_2	$\pi^2 w_0(f_b - f_a)^3/3$	$4\pi^2\sigma^2\psi_0$
b_4	$\pi^4 w_0(f_b - f_a)^5/5$	$48\pi^4\sigma^4\psi_0$
g	$(\pi\tau)^{-1}w_0 \sin \pi(f_b - f_a)\tau$	$\psi_0 e^{-2(\pi\sigma\tau)^2}$

If we write I_c, I_c', I_c'' for $I_c(t)$, $I_c'(t)$, $I_c''(t)$, where the primes denote differ-

[19] Section 2.9 of Reference A.

46

151

entiation with respect to t, and do the same for $I_s(t)$ and its derivatives we have, from Section 3.8 of Reference A,

$$\overline{I_c^2} = \overline{I_s^2} = b_0, \qquad \overline{I_c I_s} = 0$$

$$\overline{I_c I_s'} = -\overline{I_c' I_s} = b_1, \qquad \overline{I_s I_c'} = \overline{I_s I_s'} = 0$$

$$\overline{I_c'^2} = \overline{I_s'^2} = -\overline{I_c I_c''} = -\overline{I_s I_s''} = b_2, \qquad \overline{I_c' I_s'} = \overline{I_c I_s''} = \overline{I_s I_c''} = 0$$

$$\overline{I_c' I_s''} = -\overline{I_s' I_c''} = b_3, \qquad \overline{I_c' I_c''} = \overline{I_s' I_s''} = 0$$

$$\overline{I_c''^2} = \overline{I_s''^2} = b_4, \qquad \overline{I_c'' I_s''} = 0 \qquad (A2\text{-}2)$$

When we deal with moments in which the arguments of the two variables are separated by an interval τ as in (see the last of equations (3.7–11) of Reference A)

$$\overline{I_c(t) I_s(t + \tau)} = h,$$

it is convenient to denote the argument t by the subscript 1 and the argument $t + \tau$ by 2. Then our example becomes

$$\overline{I_{c1} I_{s2}} = h$$

We shall need the following moments of this type.

$$\overline{I_{c1} I_{c2}} = \overline{I_{s1} I_{s2}} = g, \qquad \overline{I_{c1} I_{s2}} = -\overline{I_{c2} I_{s1}} = h$$

$$\overline{I_{c1} I_{c2}'} = \overline{I_{s1} I_{s2}'} = -\overline{I_{c1}' I_{c2}} = -\overline{I_{s1}' I_{s2}} = g'$$

$$\overline{I_{c1} I_s'} = \overline{I_{c2} I_{s1}'} = -\overline{I_{c1}' I_{s2}} = -\overline{I_{c2}' I_{s1}} = h' \qquad (A2\text{-}3)$$

$$\overline{I_{c1}' I_{c2}'} = \overline{I_{s1}' I_{s2}'} = -g'', \qquad \overline{I_{c1}' I_{s2}'} = -\overline{I_{c2}' I_{s1}'} = -h''$$

It should be remembered that in these equations the primes on the I's denote differentiation with respect to t while the primes on g and h denote differentiation with respect to τ.

APPENDIX III

EVALUATION OF A MULTIPLE INTEGRAL

Several multiple integrals encountered during the preparation of this paper were initially evaluated by the following procedure. The integral was first converted into a multiple series by expanding a portion of the integrand and integrating termwise. It was found possible to sum these series when one of the factorials in the denominator was represented as a contour integral. This reduced the multiple integral to a contour integral and sometimes the latter could be evaluated.

47

We shall illustrate this procedure by examining the integral

$$I = \int_{-\pi}^{\pi} d\theta \int_0^{\infty} dx\, x \exp\left[-x^2 + 2a \cos\theta + 2bx \sin\theta + c \sin^2\theta \right] \quad \text{(A3-1)}$$

Expanding that part of the exponential which contains the trigonometrical terms and integrating termwise gives

$$I = \sum_{m=0}^{\infty} \sum_{n=0}^{\infty} \sum_{\ell=0}^{\infty} \frac{a^{2n} b^{2m} c^{\ell} \pi \Gamma(\ell + m + \frac{1}{2})}{n!\, \ell!\, (\ell + m + n)!\, \Gamma(m + \frac{1}{2})}$$

where we have used

$$2^{2n} \Gamma(n + \tfrac{1}{2}) n! = \sqrt{\pi}\,(2n)!$$

We next make the substitution

$$\frac{1}{(\ell + m + n)!} = \frac{1}{2\pi i} \int_C \frac{e^t\, dt}{t^{\ell+m+n+1}} \quad \text{(A3-2)}$$

where the path of integration C is a circle chosen large enough to ensure the convergence of the series obtained when the order of summation and integration is changed. The summations may now be performed:

$$I = \frac{1}{2i} \int_C dt\, e^{t+a^2/t} \sum_{m=0}^{\infty} b^{2m} t^{-m-1} (1 - ct^{-1})^{-m-1/2}$$

$$= \frac{1}{2i} \int_C \frac{t^{-1/2} (t - c)^{1/2}}{t - c - b^2} e^{t+a^2/t}\, dt \quad \text{(A3-3)}$$

C encloses the pole at $c + b^2$ and the branch point at c as well as the origin.

When a^2 is zero the integral may be reduced still further. Let c be complex and b such that the point $c + b^2$ does not lie on the line joining 0 to c. Deform C until it consists of an isolated loop about $c + b^2$ and a loop about 0 and c, the latter consisting of small circles about 0 and c joined by two straight portions running along the line joining 0 to c. The contributions of the small circles about 0 and c vanish in the limit. Along the portion starting at 0 and running to c, $\arg (t - c) = -\pi + \arg c$, and along the portion starting at c and running to 0, $\arg (t - c) = \pi + \arg c$. On both portions $\arg t = \arg c$. Bearing this in mind and setting $t = c \sin^2\theta$ on the two portions gives

$$I_{a=0} = \pi b(c + b^2)^{-1/2} e^{c+b^2} + 2c \int_0^{\pi/2} \frac{\cos^2\theta\, e^{c \sin^2\theta}}{b^2 + c \cos^2\theta}\, d\theta \quad \text{(A3-4)}$$

The integral may be expressed in terms of the function

$$Ie(k, x) = \int_0^{x} e^{-u} I_0(ku)\, du$$

48

by noting that

$$\int_0^\pi \frac{e^{-\alpha-\beta\cos v}}{\alpha + \beta\cos v}\, dv = \int_0^\pi dv \left[\frac{1}{\alpha + \beta\cos v} - \int_0^1 e^{-t(\alpha+\beta\cos v)}\, dt \right]$$

$$= \pi(\alpha^2 - \beta^2)^{-1/2} - \pi\int_0^1 e^{-\alpha t}\, I_0(\beta t)\, dt \qquad \text{(A3-5)}$$

$$= \pi(\alpha^2 - \beta^2)^{-1/2} - (\pi/\alpha)Ie(\beta/\alpha,\, \alpha)$$

Thus

$$I_{a=0} = \pi e^{c/2} I_0(c/2) + (\pi b^2/\alpha)e^{b^2+c} Ie\left(\frac{c}{2\alpha},\, \alpha\right) \qquad \text{(A3-6)}$$

where

$$\alpha = b^2 + c/2 \qquad \text{(A3-7)}$$

49

154

PAPER NO. 5

Reprinted from *Proc. IRE*, Vol. 36, No. 9, pp. 1081–1092, Sept. 1948

Theory of Frequency-Modulation Noise*

F. L. H. M. STUMPERS†

Summary—The energy spectrum of frequency-modulation noise is computed for different ratios of signal to noise. Numerical values are given for some simple filter amplitude characteristics. The theory is based on the Fourier concept of noise and treated in three steps: no signal, signal without modulation, and modulated signal. The result is given in the form of a series, and it is shown that this development is convergent. The suppression of the modulation by noise is also discussed.

I. Introduction

SINCE THE PAPER by Armstrong[1] drew attention to the possibilities of frequency modulation with regard to the reduction of noise, a considerable amount of work has been published in this field. So far as is known, however, the theoretical treatment of noise and signal has been confined to the case in which the noise energy is small compared to the signal energy. In this paper we will try to give a rigorous treatment valid for all signal-to-noise energy ratios. The theory is developed by methods which Fränz[2,3] and Rice[4] applied to similar problems and which are based on the Fourier spectrum of the noise. An interesting idea of Mann[5] has been used for the counting of the number of zeros.

Usually the instantaneous frequency of a frequency-modulated signal is defined as the derivative of the phase with respect to the time. (For this definition, see van der Pol.[6]) In this section an alternative definition is given, which is more suitable for our further computations. It will be shown that, for a normal frequency-modulated signal, it gives the same result as the usual definition. Using this starting point, we further deduce a mathematical expression for the energy spectrum, and give a first example of its application. We shall confine ourselves to signals consisting of high-frequency components in such a way that all important components lie within a relatively narrow band $\omega_0 \pm \Delta\omega$ where $\Delta\omega \ll \omega_0$.

For a sinusoidal signal $\cos \omega_0 t$, the angular frequency is equal to the number of zeros in a time interval of π

seconds. Now we choose a time interval τ, large compared to π/ω_0 but small compared to $\pi/2\Delta\omega$:

$$\omega_0\tau \gg \pi, \qquad 2\Delta\omega\tau \ll \pi. \tag{1}$$

The *instantaneous frequency* is defined at the time t as the ratio of the number of zeros between $t-\tau/2$ and $t+\tau/2$ to τ/π, or as the mean density of the zeros averaged over the time interval τ/π.

As an example, let us take the signal

$$\cos \left\{ \omega_0 t + f(t) \right\}.$$

If this function has consecutive zeros at $t=t_1$ and $t=t_2$, then

$$\omega_0(t_1 - t_2) + f(t_1) - f(t_2) = \pi.$$

If we assume that $f'(t)$ changes slowly compared to $\cos \omega_0 t$, then we can replace $f(t_1)-f(t_2)$ by $(t_1-t_2) f'(t_1)$, and thus obtain:

$$t_1 - t_2 = \pi / \left\{ \omega_0 + f'(t_1) \right\}.$$

τ being defined in such a way that $f'(t)$ is practically constant during a time interval τ, the number of zeros within the time τ is $\tau \left\{ \omega_0 + f'(t) \right\}/\pi$. The definition of the instantaneous frequency above thus gives the result:

$$\omega_i(t) = \omega_0 + f'(t). \tag{2}$$

This, as we have stated, is the same result as is obtained on the basis of the usual definition.

The counting of the number of zeros of a function $v(t)$ within an interval τ is best done with the help of δ-functions such as are used in the operational calculus. The δ-function is defined by:

$$\delta(x) = 0, \qquad x \neq 0,$$
$$\delta(x) = \infty, \qquad x = 0,$$
$$\int_{-\infty}^{+\infty} \delta(x)dx = 1.$$

It is plausible to consider the integral

$$\int_{t_0-\tau/2}^{t_0+\tau/2} \delta\left\{ v(t) \right\} v'(t)dt.$$

If $v(t)$ has a simple zero in a certain interval, the absolute value of the integral over that interval is 1. In the subsequent interval, which also is assumed to contain only one simple zero, the sign of the result will be different. The reason is that, when we introduce $v(t)$ as a new variable in the integral, this variable of integration runs from a negative to a positive value in one interval, and in the opposite direction in the next.

Therefore we modify our procedure so as to count only those zeros passed through with a positive slope.

---

* Decimal classification: R148.2. Original manuscript received by the Institute, August 8, 1947; revised manuscript received, March 11, 1948.

† Natuurkundig Laboratorium der N. V. Philips Gloeilampenfabrieken, Eindhoven, the Netherlands.

[1] E. H. Armstrong, "A method of reducing disturbances in radio-signalling by a method of frequency-modulation," Proc. I.R.E., vol. 24, pp. 689–740; May, 1936.

[2] K. Fränz, "Beiträge zur Berechnung des Verhältnisses von Signalspannung zu Rauschspannung am Ausgang von Empfängern," Elek. Nach. Tech., vol. 17, pp. 215–230; 1940. Also, vol. 19, pp. 285–287; 1942.

[3] K. Fränz and T. Vellat, "Der Einfluss von Trägern auf das Rauschen hinter Amplitudenbegrenzern und linearen Gleichrichtern," Elek. Nach. Tech., vol. 20, pp. 183–189; 1943.

[4] S. O. Rice, "Mathematical analysis of random noise," Bell Sys. Tech. Jour., vol. 33, pp. 282–332; July, 1944. Also vol. 34, pp. 46–156; January, 1945.

[5] P. A. Mann, "Der Zeitablauf von Rauschspannungen," Elek. Nach. Tech., vol. 20, pp. 232–237; 1943.

[6] Balth. van der Pol, "The fundamental principles of frequency-modulation," Jour. I.E.E., part III, vol. 93, pp. 153–158; 1946.

The instantaneous frequency is then

$$\omega_i(t_0) = \frac{2\pi}{\tau} \int_{t_0-\tau/2}^{t_0+\tau/2} \delta(v) v' U(v') dt \qquad (3)$$

in which

$$U(x) = 0, \qquad x < 0,$$
$$U(x) = \tfrac{1}{2}, \qquad x = 0,$$
$$U(x) = 1, \qquad x > 0.$$

The result is a function of τ, but as long as τ is subjected to the inequalities (1), the variation with τ will be unimportant.

The use of δ functions in the integrals can be avoided when we use the Stieltjes integral. In this way (3) is written:

$$\omega_i(t_0) = \frac{2\pi}{\tau} \int_{t_0-\tau/2}^{t_0+\tau/2} U(v') dU(v). \qquad (3a)$$

In a frequency-modulation receiver, a device is used which, when a signal is applied to it, gives an output voltage proportional to the instantaneous frequency of that signal. This device is the frequency detector, or discriminator. In general, the instantaneous frequency will not be constant. We can make a registration of it during a certain time. A Fourier analysis of that registration will give a spectrum of components at different frequencies. It is in this spectrum of the detected signal that we are now interested.

When we choose a time interval of 2π seconds for the Fourier analysis we have the formulas:

$$\omega_i(t) = \sum f_m e^{imt}$$
$$f_m = (2\pi)^{-1} \int_0^{2\pi} \omega_i(t) e^{-imt} dt. \qquad (4)$$

Let us consider a zero t_0 of $v(t)$, where the function has a positive slope whereas t_0 is not too near to 0 or 2π. Its contribution to the integral (4) is then

$$\tau^{-1} \int_{t_0-\tau/2}^{t_0+\tau/2} e^{-imt} dt = 2(m\tau)^{-1} \sin (m\tau/2) e^{-imt_0}.$$

In this result, $2(m\tau)^{-1} \sin (m\tau/2)$ approximates 1 for small values of τ. We have chosen $\tau \ll \pi/2\Delta\omega$. Therefore, for all frequencies smaller than $\Delta\omega$ we may replace $2(m\tau)^{-1} \sin (m\tau/2)$ by 1. Accordingly, for these frequencies we may use

$$f_m = \int_0^{2\pi} \delta(v) v' U(v') e^{-imt} dt. \qquad (5)$$

Using the Laplace transform, we get

$$\delta(v) = (2\pi)^{-1} \int_{-\infty-ic}^{+\infty-ic} e^{iuv} du, \quad c > 0 \qquad (6a)$$

$$v' U(v') = -(2\pi)^{-1} \int_{-\infty-ic}^{\infty-ic} e^{iuv'} u^{-2} du, \quad c > 0. \qquad (6b)$$

The path of integration can also be taken from $-\infty$ to ∞ along the real axis with a small indentation below the origin.

$$f_m = -(4\pi^2)^{-1} \int_{-\pi}^{\pi} e^{-imt} dt \int_{-\infty}^{+\infty} du_1 \int_{-\infty}^{+\infty} du_2 u_2^{-2} e^{iu_1 v + iu_2 v'}$$

This formula gives the spectral composition of the instantaneous frequency. If the frequency detector gives a potential difference of 1 volt over a resistance of 1 ohm for a frequency deviation of 1 radian per second, the same formula applies to the output of the frequency detector. We shall calculate the distribution of the energy, dissipated in that resistance, over the spectrum. The result is the energy spectrum of the output. In practical cases there will be a proportionality factor, which is omitted in our calculations. The energy corresponding to a certain frequency m is given by $2f_m f_m^*$ if $m \neq 0$, or by $f_0 f_0^*$ for the dc term. f_m^* is the complex conjugate of f_m. Hence,

$$2f_m f_m^* = (8\pi^4)^{-1} \int_{-\pi}^{\pi} e^{-imt_1} dt_1 \int_{-\pi}^{\pi} e^{imt_2} dt_2$$

$$\iiiint_{-\infty}^{+\infty} u_2^{-2} u_4^{-2} e^{iu_1 v(t_1) + iu_2 v'(t_1) + iu_3 v(t_2) + iu_4 v'(t_2)}$$

$$du_1 du_2 du_3 du_4. \qquad (7)$$

As a first example, we shall apply this formula to a frequency-modulated signal $\cos (\omega_0 t + m_0 \sin pt)$. $(m_0 = \Delta\omega/p)$. In this case,

$$e^{iu_1 v(t_1) + iu_2 v'(t_1)} = \exp \{ iu_1 \cos (\omega_0 t + m_0 \sin pt_1)$$
$$- iu_2(\omega_0 + \Delta\omega \cos pt_1) \sin (\omega_0 t_1 + m_0 \sin pt_1) \}.$$

We develop this form into a series:

$$\sum_{k=0}^{\infty} \frac{2^{-k}}{k!} \{ (iu_1 - u_{2a}) e^{i\omega_0 t_1 + i m_0 \sin p t_1}$$
$$+ (iu_1 + u_{2a}) e^{-i\omega_0 t_1 - i m_0 \sin p t_1} \}^k$$

in which $u_{2a} = u_2(\omega_0 + \Delta\omega \cos pt_1)$.

Since, in (7), we have to integrate the last result with e^{-imt_1}, where $m \ll \omega_0$ we are only interested in those terms in the binomial development that have no $i \omega_0 t_1$ in the exponent. Thus the series reduces to

$$\sum_{k=0}^{\infty} \frac{2^{-2k}}{k! k!} (u_1^2 + u_{2a}^2)^k.$$

This is the well-known development of the Bessel function of order zero. A relation between Bessel functions gives

$$J_0 \{ u_1^2 + u_2^2(\omega_0 + \Delta\omega \cos pt)^2 \}^{1/2}$$
$$= \sum_{-\infty}^{\infty} (-1)^m J_{2m}(u_1) J_{2m} \{ u_2(\omega_0 + \Delta\omega \cos pt_1) \}.$$

Now we perform the integration with respect to u_1 and u_2:

$$\int_{-\infty}^{+\infty} J_{2m}(u_1)du_1 = 2$$

$$\int_{-\infty}^{\infty} u_2^{-2} J_{2m}\{u_2(\omega_0 + \Delta\omega \cos pt_1)\}du_2$$
$$= (\omega_0 + \Delta\omega \cos pt_1)\{(2m-1)^{-1} - (2m+1)^{-1}\}.$$

(The last integral reduces to an easier type by one partial integration.) For $m=0$ the singularity at the origin is avoided by the small indentation. The result of the integration with respect to u_1 and u_2 is, therefore,

$$(\omega_0 + \Delta\omega \cos pt_1)\{-4 + 4(-1 + \tfrac{1}{3}) + 4(\tfrac{1}{3} - \tfrac{1}{5}) + \cdots \}$$
$$= -2\pi(\omega_0 + \Delta\omega \cos pt_1).$$

In the same way, the integrations with respect to u_3 and u_4 yield

$$-2\pi(\omega_0 + \Delta\omega \cos pt_2),$$

and, in total,

$$2f_m f_m^* = (2\pi^2)^{-1} \int_{-\pi}^{\pi} \int_{-\pi}^{\pi} e^{-imt_1} e^{+imt_2} (\omega_0 + \Delta\omega \cos pt_1)$$
$$(\omega_0 + \Delta\omega \cos pt_2)dt_1 dt_2.$$

Only for $m=0$ and $m=p$ do we get a result different from zero. The dc energy is

$$f_0 f_0^* = \omega_0^2.$$

For the frequency p, the energy is

$$2f_p f_p^* = \frac{\Delta\omega^2}{2}.$$

These results are in complete agreement with the customary definition of instantaneous frequency. We have chosen this simple problem because the way in which it is solved will again be used in the more complicated problems further on. Its aim is also to give the reader confidence in the following computations, where the result is less obvious.

II. FREQUENCY-MODULATION NOISE WITHOUT SIGNAL

In this section we will first recall some properties of a noise spectrum, and then apply (7) to a noise band. As is usual in noise problems, an averaging procedure will be necessary.

By means of a filter we select a certain band of frequencies from a normal noise source, and apply these components as an input signal to an ideal frequency detector. As a first example, we shall take a filter with a rectangular amplitude-versus-frequency characteristic. This filter is not realizable but, as the phase characteristic is not important for these computations, it can be approximated. Later on we shall consider a filter with a gaussian amplitude-versus-frequency characteristic.

If we register the noise from a normal noise source during a time interval $(-T, T)$, we can make a Fourier analysis:

$$v(t) = \sum a_n \cos n\pi t/T - b_n \sin n\pi t/T. \tag{8}$$

When such a Fourier ana'ysis is made a great many times consecutively, each time over an interval of the same length, the Fourier components will show a gaussian probability distribution:

$$W(a_n)da_n = (\pi C)^{-1/2} e^{-a_n^2/C} da_n \tag{9a}$$
$$W(b_n)db_n = (\pi C)^{-1/2} e^{-b_n^2/C} db_n. \tag{9b}$$

The value of a_n in one particular Fourier analysis is independent of the value of the other coefficients. This subject is treated extensively by Fränz and Rice.

The number of noise lines in a band of $2\Delta\omega$ radians will be $2\Delta\omega T/\pi$. As it makes the formulas simpler, we shall choose $T = 2\pi$. If we made another choice, the formulas (4) to (7) which are also based on a time interval 2π, would need an appropriate modification. The choice of 2π is, however, quite arbitrary, and, whenever we find it advisable to increase the number of lines in a part of the spectrum, we shall do so.

The average energy per component is

$$\overline{\tfrac{1}{2}(a_n^2 + b_n^2)} = (2\pi C)^{-1} \int\int_{-\infty}^{+\infty} (a_n^2 + b_n^2) e^{-(a_n^2+b_n^2)/C} da_n db_n$$
$$= C/2.$$

The effective voltage corresponding to a noise band extending from $\omega_0 - \Delta\omega$ to $\omega_0 + \Delta\omega$ is

$$v_{n0} = (\Delta\omega C)^{1/2}.$$

As the number of lines in the band increases proportionally to the length of the considered time interval, the average amplitude has to be reduced so as to keep the average power constant.

When we introduce the $v(t)$ of (8) in (7), the result will be a function of the $4\Delta\omega$ variables a_n and b_n. As is usual in noise computations, the average of the result of (7) over all a's and b's is used to obtain the effective energy spectrum after detection. Thus,

$$\overline{2f_m f_m^*} = 2 \int\int\int \cdots \int da_{N_1} \cdots db_{N_2}$$
$$W(a_{N_1} \cdots b_{N_2}) f_m f_m(a_{N_1} \cdots b_{N_2})$$
$$N_1 = \omega_0 - \Delta\omega; \qquad N_2 = \omega_0 + \Delta\omega.$$

The integration does not lead to great difficulties (see Appendix I). After introduction of a new variable $s = n - \omega_0$, we obtain

$$\overline{f_0 f_0^*} = \omega_0^2 + \Delta\omega^2/3. \tag{13a}$$

The dc corresponding to the central frequency is usually suppressed by balanced detection. For the frequency m we obtain the energy in the form of an integral:

$$\overline{2f_m f_m{}^*}$$

$$= (2\pi)^{-1} \sum_{k=1}^{\infty} k^{-1} N^{-2k} \int_{-\pi}^{\pi} e^{-imv} \left\{ \left(\sum e^{isv} \right)^{2k-1} \left(\sum s^2 e^{isv} \right) \right.$$

$$\left. - \left(\sum e^{isv} \right)^{2k-2} \left(\sum s e^{isv} \right)^2 \right\} dv \quad (N = 2\Delta\omega). \quad (13b)$$

The summation over s has to be taken over all integers satisfying

$$-\Delta\omega \leqq s \leqq \Delta\omega.$$

To get the energy in a part of the spectrum, the results for all frequencies m in this part are totaled. Now we are free to increase the number of lines by enlargement of the intervals $2T$, which we have so far chosen $2T = 2\pi$. In this way a continuous energy distribution $E_0(u)$ will be approximated, and the sum

$$N^{-1} \sum e^{isv}$$

can be replaced by the integral

$$\int_{-1/2}^{1/2} e^{iuv} du.$$

Instead of s/N we have introduced the continuous variable u. In the same way,

$$N^{-1} \sum_{-\Delta\omega}^{\Delta\omega} s^2 e^{isv} = N^2 \int_{-1/2}^{1/2} u^2 e^{iuv} du.$$

Equation (13b) can now be written:

$$E_0(u) = \sum_{k=1}^{\infty} 4 (k)^{-1} \Delta\omega^2 h_{2k}(u) \quad (14)$$

where $h_{2k}(u)$ is given by the relation

$$\int_{-\infty}^{+\infty} h_{2k}(u) e^{iuv} du$$

$$= \left\{ \int_{-1/2}^{1/2} e^{iuv} du \right\}^{2k-1} \left\{ \int_{-1/2}^{1/2} u^2 e^{iuv} du \right\}$$

$$- \left\{ \int_{-1/2}^{1/2} e^{iuv} du \right\}^{2k-2} \left\{ \int_{-1/2}^{1/2} u e^{iuv} \right\}^2. \quad (15)$$

The values of $h_{2k}(u)$ can be derived directly from this integral, but a shorter computation will be treated in the next section. There we shall also see that, for k large,

$$h_{2k}(u) \approx (12)^{-1}(5k - 3)^{-1/2}\pi^{-1/2}(15)^{1/2}e^{-15u^2/(5k-3)}. \quad (16)$$

The series in (14) is convergent, since for large k the general term behaves as $k^{-3/2}$. We have calculated the values for $E_0(u)$ as shown in Table I.

Let a filter with a symmetrical, but otherwise arbitrary, amplitude characteristic be used, the calculations being slightly modified. If now the characteristic is given by $f(\omega)e^{i\phi(\omega)}$, the input signal will be:

$$v(t) = \sum f(n) \left\{ a_n \cos(nt + \phi_n) - b_n \sin(nt + \phi_n) \right\} \quad (8a)$$

$f(\omega)$ be normalized in such a way that its maximum

TABLE I

Noise energy and noise voltage as a function of the frequency $(u = \omega/2\Delta\omega)$. Energy per unit bandwidth. No carrier wave present. Rectangular filter amplitude characteristic.

$2u = \omega/\Delta\omega$	$E_0(u)$	$v_0(u) = \{E_0(u)\}^{1/2}$
0	$1.2241 \, \Delta\omega^2$	$1.1064 \, \Delta\omega$
0.1	1.1274	1.0634
0.2	1.0381	1.0189
0.3	0.9557	0.9776
0.4	0.8799	0.9380
0.5	0.8101	0.9001
0.6	0.7462	0.8638
0.7	0.6881	0.8295
0.8	0.6354	0.7971
0.9	0.5877	0.7666
1.0	0.5445	0.7379

value is 1, and the bandwidth $2\Delta\omega$ of the filter be defined by

$$\int_{-\infty}^{+\infty} f^2(\omega) d\omega = 2\Delta\omega.$$

Then (13a) does not change.

As before, we introduce a new variable $u = \omega/2\Delta\omega$. The function $F(u)$ is so defined that $F(u) = f^2(\omega - \omega_0)$; then

$$\int_{-\infty}^{+\infty} F(u) du = 1. \quad (18)$$

Instead of (14), we obtain (see Appendix I)

$$E_0(u) = \sum_{k=1}^{\infty} 4(k)^{-1}(\Delta\omega)^2 \quad H_{2k}(u)$$

in which $H_{2k}(u)$ is now given by

$$\int_{-\infty}^{\infty} H_{2k}(u) e^{iuv} du$$

$$= \left\{ \int F(u) e^{iuv} du \right\}^{2k-1} \left\{ \int F(u) u^2 e^{iuv} du \right\}$$

$$- \left\{ \int F(u) e^{iuv} du \right\}^{2k-2} \left\{ \int u F(u) e^{iuv} du \right\}^2. \quad (15a)$$

When $F(u)$ is given, all further functions can be found successively by direct integration. Here, too, the operational calculus may furnish a shorter method of calculation, as is shown in the next section.

As an example, take a gaussian amplitude characteristic. The normalized squared amplitude characteristic is given by $F(u) = e^{-\pi w^2}$. Then, as is shown in the next section,

TABLE II

Noise energy and noise voltage (per unit bandwidth) as a function of the frequency $(u = \omega/2\Delta\omega)$. No carrier wave. Gaussian amplitude characteristic.

$2u = \omega/\Delta\omega$	$E_0(u)$	$V_0(u) = \{E_0(u)\}^{1/2}$
0	$1.17594 \, \Delta\omega^2$	$1.08441 \, \Delta\omega$
0.2	1.15719	1.07573
0.4	1.10377	1.05061
0.6	1.02373	1.01180
0.8	0.92702	0.96282
1	0.82527	0.90844

$$H_{2k}(u) = 2^{-3/2}k^{-1/2}\pi^{-1}e^{-\pi u^2/2k}. \qquad (18)$$

Table II shows the result. For filters with a nonsymmetrical amplitude characteristic, the computation leads to longer formulas, as shown in Appendix I. For this case we have not computed a numerical example.

III. Some Distribution Functions

In our computations of the energy spectra some functions occur regularly, and we shall treat them together in this section. At first let us consider the problem of finding the product distribution when the two functions f_1 and f_2 are given.

$$\int_{-\infty}^{+\infty} f_1(u)e^{iuv}du \int_{-\infty}^{+\infty} f_2(w)e^{iwv}dw = \int_{-\infty}^{+\infty} f_3(x)e^{ixv}dx. \quad (19)$$

This can be done directly by considering the product as a double integral and by the introduction of $x = u + w$ as a new variable in this integral. The Laplace transforms, when known, are of much help. Let $f_{p,1}(p)$ be the image of $f_1(u)$:

$$f_{p,1}(p) \doteq f_1(u),$$

which shorthand notation stands for:

$$f_{p,1}(p) = p\int_{-\infty}^{+\infty} f_1(u)e^{-pu}du.$$

Then, upon introducing $p = -iv$ in (19), we get at once

$$f_{p,3} = p^{-1}f_{p,1}f_{p,2}.$$

Thus, for the rectangular distribution, the product functions are found by

$$\left\{\int_{-1/2}^{1/2} e^{iuv}du\right\}^k = \int_{-\infty}^{+\infty} f_k(u)e^{iuv}du \qquad (20a)$$

$$f_1(u) = 1, \qquad -\tfrac{1}{2} \leq u \leq \tfrac{1}{2}$$
$$= 0, \qquad -\infty < u < -\tfrac{1}{2}, \quad \tfrac{1}{2} < u < \infty.$$
$$f_1(u) \doteq e^{p/2} - e^{-p/2} = 2\sinh p/2$$
$$f_k(u) \doteq p^{1-k}(e^{p/2} - e^{-p/2})^k = p^{1-k}(2\sinh p/2)^k.$$

Therefore,

$$f_k(u) = \sum_{r=0}^{k} {}_r\binom{k}{r}\frac{(u+k/2-r)^{k-1}}{(k-1)!} U(u+k/2-r). \quad (20b)$$

For the definition of U, see (3).

A function is computed from its Laplace transform by means of the inversion integral (Bromwich); for instance,

$$f_k(u) = (2\pi i)^{-1}\int_{c-i\infty}^{c+i\infty} p^{-1}f_{p,k}(p)e^{pu}dp.$$

For large k the integrand of this integral (here to be taken along the imaginary axis) has its maximum for $p = 0$, and can be approximated by its development for small p.

$$p^{1-k}(e^{p/2} - e^{-p/2})^k \approx pe^{p^2k/24}.$$

The result is an approximation of $f_k(u)$ for large k:

$$f_k(u) \approx 6^{1/2}k^{-1/2}\pi^{-1/2}e^{-6u^2/k}. \qquad (21)$$

We arrive at another type of distribution function by differentiation of (20a):

$$\left\{\int_{-1/2}^{1/2} e^{iuv}du\right\}^{k-1}\left\{\int_{-1/2}^{1/2} ue^{iuv}du\right\} = \int_{-\infty}^{+\infty} c_k(u)e^{iuv}du$$

$$c_k(u) = uf_k(u)/k.$$

A third type, of which we have already met examples, is

$$\left\{\int_{-1/2}^{1/2} e^{iuv}du\right\}^{k-1}\left\{\int_{-1/2}^{1/2} u^2e^{iuv}du\right\} = \int_{-\infty}^{+\infty} a_k(u)e^{iuv}du$$

$$a_1(u) = u^2, \qquad -\tfrac{1}{2} \leq u \leq \tfrac{1}{2}$$
$$a_1(u) = 0, \qquad -\infty < u < -\tfrac{1}{2}, \quad \tfrac{1}{2} < u < \infty.$$
$$a_1(u) \doteq 4^{-1}p^{-2}(p^2 - 4p + 8)e^{p/2}$$
$$\qquad\qquad - 4^{-1}p^{-2}(p^2 + 4p + 8)e^{-p/2}$$
$$a_k(u) \doteq 4^{-1}p^{-1-k}\{(p^2 - 4p + 8)e^{p/2}$$
$$\qquad\qquad - (p^2 + 4p + 8)e^{-p/2}\}(e^{p/2} - e^{-p/2})^{k-1}.$$

This gives, for instance,

$$a_2(u) = \tfrac{1}{3}u^3 + \tfrac{1}{2}u^2 + \tfrac{1}{4}u + \tfrac{1}{12}, \qquad -1 \leq u \leq 0$$
$$= -\tfrac{1}{3}u^3 + \tfrac{1}{2}u^2 - \tfrac{1}{4}u + \tfrac{1}{12}, \qquad 0 \leq u \leq 1$$
$$= 0, \qquad -\infty < u < -1, \quad 1 < u < \infty.$$

Approximation of the inversion integral leads to the result:

$$a_k(u) \doteq a_{p,k}(p) \approx \frac{p}{12}e^{(5k+4)p^2/120}$$

$$a_k(u) \approx (12)^{-1}(30)^{1/2}(5k+4)^{-1/2}\pi^{-1/2}e^{-30u^2/(5k+4)},$$
$$\qquad\qquad\qquad\qquad\qquad k \text{ large.} \quad (22)$$

Another function worth consideration is

$$\left\{\int_{-1/2}^{1/2} ue^{iuv}du\right\}^2 = \int_{-\infty}^{+\infty} b_2(u)e^{iuv}du$$

$$\left\{\int_{-1/2}^{1/2} e^{iuv}du\right\}^{k-2}\left\{\int_{-1/2}^{1/2} ue^{iuv}du\right\}^2$$

$$= \int_{-\infty}^{+\infty} b_k(u)e^{iuv}du \qquad k \geq 2.$$

$$b_1(u) = 0.$$

The Laplace transform gives in this case:

$$b_k(u) \doteq 4^{-1}p^{-1-k}\{(2 - p)e^{p/2}$$
$$\qquad\qquad - (2 + p)e^{-p/2}\}^2(e^{p/2} - e^{-p/2})^{k-2}.$$

Again the originals are easily found; for instance,

$$b_2(u) = \tfrac{1}{6}u^3 - \tfrac{1}{4}u - \tfrac{1}{12}, \qquad -1 \leq u \leq 0$$
$$= -\tfrac{1}{6}u^3 + \tfrac{1}{4}u - \tfrac{1}{12}, \qquad 0 \leq u \leq 1$$
$$= 0, \qquad -\infty < u < -1, \quad 1 < u < \infty.$$

For large k we get the approximation:

$$b_k(u) \approx -(72)^{-1}(30)^{3/2}(\pi)^{-1/2}(5k-4)^{-3/2}$$
$$\{1 - 60u^2/(5k-4)\}e^{-30u^2/(5k-4)}. \quad (23)$$

The function $h_k(u)$ which we have used in the second section is defined by $h_k(u) = a_k(u) - b_k(u)$. Therefore,

$$h_k(u) \doteqdot p^{-1-k}\{e^p - (p^2+2) + e^{-p}\}(e^{p/2} - e^{-p/2})^{(k-2)}.$$

For $h_2(u)$, the result is

$$h_2(u) = \tfrac{1}{6}(u+1)^3, \qquad -1 \leq u \leq 0$$
$$= \tfrac{1}{6}(1-u)^3, \qquad 0 \leq u \leq 1$$
$$= 0, \qquad -\infty < u < -1, \quad 1 < u < \infty.$$

For large k,

$$h_k(u) \doteqdot h_{p,k}(p) \approx \tfrac{1}{12}pe^{(5k-6)p^2/120}$$
$$h_k(u) \approx (12)^{-1}(30)^{1/2}(5k-6)^{-1/2}\pi^{-1/2}e^{-30u^2/(5k-6)}. \quad (24)$$

The approximate formulas are already fairly good for low values of k; for instance,

$$h_4(u) = 0.0667 \text{ (exact)} \qquad 0.0688 \text{ (approx.)}$$
$$h_6(u) = 0.0512 \text{ (exact)} \qquad 0.0526 \text{ (approx.)}$$

By differentiating the equation (20a) twice, we arrive at a relation between the functions:

$$kh_k(u) = -k^2 b_k(u) + u^2 f_k(u).$$

The functions $f_k(u)$ in particular have been treated frequently since De Moivre.[7] All these functions have a place in the theory of averages. One may compare Maurer's[8] paper, where some asymptotic formulas are derived in a more precise way. So far the rectangular distribution has been our starting point, but the computation can be made for another type as well. The gaussian frequency distribution is attractive because it gives simple results. Moreover, we have already pointed out that a gaussian amplitude-versus-frequency characteristic may be better approximated by real conditions than a rectangular one.

Corresponding to the original distribution $F_1(u) = e^{-\pi u^2}$, we get, in the same way as before,

$$\left\{\int_{-\infty}^{+\infty} e^{-\pi u^2 + iuv} du\right\}^k = \int_{-\infty}^{+\infty} F_k(u)e^{iuv} du$$
$$F_k(u) = k^{-1/2}e^{-\pi u^2/k}. \quad (21a)$$

Analogous to $a_k(u)$, we get

$$A_k(u) = \tfrac{1}{2}k^{-5/2}e^{-\pi u^2/k}\{2u^2 + k(k-1)/\pi\} \quad (22a)$$
$$k = 1, 2, 3, \cdots,$$

[7] A. De Moivre, "Mensura sortis," 1711; "Miscellanea analytica," 1730.

[8] L. Maurer, "Ueber die Mittelwerthe der Funktionen einer reellen Variabelen," *Math. Ann.*, vol. 47, pp. 263–280; 1896.

and, instead of $b_k(u)$, we get

$$B_k(u) = \tfrac{1}{2}k^{-5/2}e^{-\pi u^2/k}(2u^2 - k/\pi) \quad (23a)$$
$$k = 2, 3, \cdots, \qquad B_1(u) = 0.$$

Here,

$$\int_{-\infty}^{+\infty} A_k(u)e^{iuv} du$$
$$= \left\{\int_{-\infty}^{+\infty} F_1(u)e^{iuv} du\right\}^{k-1}\left\{\int_{-\infty}^{+\infty} u^2 F_1(u)e^{iuv} du\right\}.$$

In the same way, $h_k(u)$ is replaced by

$$H_k(u) = \frac{1}{2\pi} k^{-1/2}e^{-\pi u^2/k} \quad (24a)$$
$$k = 2, 3, \cdots, \qquad H_1(u) = A_1(u).$$

All these functions have simple Laplace transforms, and are therefore easily found by this method.

IV. Frequency-Modulation Noise in the Presence of a Nonmodulated Carrier Wave

When an unmodulated carrier-wave is present, together with a rectangular noise spectrum symmetrically around it, the input signal is given by:

$$v(t) = \cos\omega_0 t + \sum_{\omega_0 - \Delta\omega}^{\omega_0 + \Delta\omega}(a_n\cos nt - b_n\sin nt) \quad (25)$$

This function is substituted in (7) and the average is taken in the same way as in (12). Some comments on the integration are given in Appendix II.

The dc energy is now

$$\overline{f_0 f_0}^* = \omega_0^2 + \tfrac{1}{3}(\Delta\omega)^2 e^{-1/NC} \quad (29)$$

$1/NC$ is the quotient of signal energy and noise energy at the input of the frequency detector ($N = 2\Delta\omega$). After introduction of the continuous variable u in the same way as in Section 2, the energy spectrum is given by:

$$E_1(u)/4(\Delta\omega)^2 = \sum_{r=1}^{\infty} r^{-1}e^{-2/NC} {}_1F_1^2(-r+1, 1, 1/NC)h_{2r}(u)$$
$$+\sum_{k=1}^{\infty} \sum_{r=0}^{1/2(k-1)} \binom{k-r}{r} \frac{e^{-2/NC}}{(k-r)^2(k-2r)!(NC)^{k-2r}}$$
$${}_1F_1^2(-r+1, k-2r+1, 1/NC)\{kh_k(u)+(k-2r)^2 b_k(u)\}. \quad (30)$$

In this expression ${}_1F_1$ is the confluent hypergeometric function, and $h_k(u)$ and $b_k(u)$ are the functions defined in Section 3. For the calculation we begin with the term for $k=1$ and add the terms for the higher values of k until they are sufficiently small. The convergence of the development is shown in Appendix II. The first term in the development of $E_1(u)$ is

$$4(\Delta\omega)^2 NC(1 - e^{-1/NC})^2 h_1(u)$$
$$= 4(\Delta\omega)^2 NC(1 - e^{-1/NC})^2 u^2, \qquad 0 \leq u \leq \tfrac{1}{2}.$$

As we are interested in the energy spectrum, only non-negative values of u are important. For small values of NC the first term gives a good approximation of the energy. Then the effective noise voltage is $2\Delta\omega u(NC)^{1/2}$. This gives the well-known triangular noise spectrum which is already given by the simplified analysis.

The second term in the development of $E_1(u)$ is

$$4\Delta\omega^2[e^{-2/NC}h_2(u) + N^2C^2\{1 - (1 + 1/NC)e^{-1/NC}\}^2$$
$$\cdot\{h_2(u) + 2b_2(u)\}].$$

The third term is

$$4\Delta\omega^2[2N^3C^3\{1 - (1 + 1/NC + 1/2N^2C^2)e^{-1/NC}\}^2$$
$$\cdot\{h_3(u) + 3b_3(u)\} + (2NC)^{-1}e^{-2/NC}\{3h_3(u) + b_3(u)\}].$$

For small values of NC the terms containing N^kC^k form an asymptotic expansion (asymptotic for $NC\to 0$). We were led to this development when trying to get a more precise estimate from the same starting point as the simplified analysis.[9] However, for the calculation of the output noise for larger ratios of input noise energy to signal energy, one has to take into account the full

terms of the development (30). For very large values of NC it is seen that the terms of the development (14) are predominant, thus affirming the result of Section 2.

With the help of (30) we have calculated the energy spectrum by adding up the terms up to $k = 10$, or $2r = 10$, and making a graphical estimate for the remainder. In the following table the effective noise voltage $\{E_1(u)\}^{1/2}$ is given for $u = 0, 0.1, 0.2, 0.3, 0.4,$ and 0.5 (corresponding to frequencies $0, 0.2\Delta\omega, 0.4\Delta\omega, 0.6\Delta\omega, 0.8\Delta\omega,$ and $\Delta\omega$). It is seen that for $NC = 0.1$ the deviation from the triangular spectrum is still very small. For this value of NC, the output noise still grows linearly with the input

TABLE III

Effective noise voltage as a function of frequency and input noise-to-signal energy ratio. Rectangular amplitude characteristic.

Input noise-to-signal ratio NC	$\{E_1(u)\}^{1/2}/\Delta\omega$					
	$u=0$	$u=0.1$	$u=0.2$	$u=0.3$	$u=0.4$	$u=0.5$
0.01	0	0.0200	0.0400	0.0600	0.0800	0.1000
0.1	0	0.06485	0.1294	0.1935	0.2574	0.3208
0.2	0.04032	0.1014	0.1900	0.2802	0.3702	0.4594
0.5	0.2763	0.2988	0.3679	0.4564	0.5658	0.6750
1	0.5275	0.5191	0.5500	0.5933	0.6739	0.7616
2	0.7071	0.6664	0.6500	0.6582	0.6882	0.7377
5	0.8836	0.8143	0.7593	0.7207	0.6991	0.6930
10	1.0260	0.9512	0.8868	0.8336	0.7920	0.7630

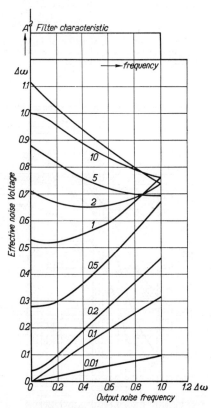

Fig. 1—Spectrum of effective noise voltage after detection. Parameter is input noise-to-signal energy ratio (NC). Rectangular amplitude-versus-frequency characteristic of the filter. Remark the triangular spectrum for $NC = 0.01$ and 0.1. The rms voltage of the noise in a small band of B cps is $(B/2\Delta\omega)^{1/2}$ times the value given by the curve.

[9] F. L. H. M. Stumpers, "Eenige onderzoekingen over trillingen et frequentiemodulatie," (in Dutch), diss. *Delft*, pp. 38–46; 1946.

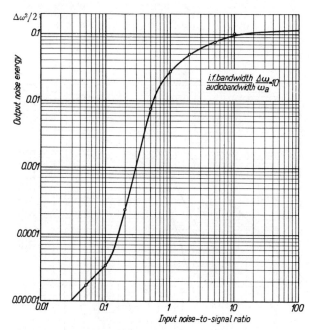

Fig. 2—Output noise energy as a function of input noise-to-signal energy ratio. If bandwidth is 10 times af bandwidth. Rectangular amplitude characteristic. Owing to the slow convergence of the series, the values for $NC = 5$ and 10 are less accurate.

noise. (Strict linearity would give 0.3162 instead of 0.3208). There is already a marked deviation from the triangular spectrum for $NC = 0.2$. All results are shown graphically in Fig. 1.

For radio reception only the audible noise is important. The ratio of the audio-frequency bandwidth to the intermediate-frequency bandwidth can vary between 0.1 and 1. The noise energy is computed by integration of the noise energy $E_1(u)$ between appropriate boundaries. For a ratio of 0.1 one has to take into account the noise between 0 and $0.1\Delta\omega$ (or $0 \leqq u \leqq 0.05$). Figs. 2 and 3 show the energy of the output noise as a function of the ratio of input noise to signal. Typical is the strong increase of the noise above $NC = 0.1$ in the curves for $\Delta\omega/\omega_a = 5$, or 10, as compared to the curve for $\Delta\omega/\omega_a = 1$. This effect was found experimentally by Guy and Morris.[10] The influence of pre-emphasis can be calculated by multiplying the energy distribution after detection by $(1 + R^2C^2\omega^2)^{-1}$. This we have done for an audio-fre-

quency bandwidth of 15,000 cps and an RC time of 75.10^{-6} seconds. The result is shown in Figs. 4 and 5.

TABLE IV

Effective noise voltage (per unit frequency) as a function of frequency and input noise-to-signal ratio. Gaussian amplitude characteristic.

NC	$\{E_1(u)\}^{1/2}/\Delta\omega$					
	$u=0$	$u=0.1$	$u=0.2$	$u=0.3$	$u=0.4$	$u=0.5$
0.01	0	0.01969	0.03756	0.05209	0.06222	0.06752
0.1	0.004684	0.06369	0.1214	0.1686	0.2009	0.2200
0.2	0.06681	0.1129	0.1865	0.2513	0.2983	0.3253
0.5	0.3486	0.3706	0.4223	0.4809	0.5278	0.5554
1	0.5932	0.6054	0.6363	0.6724	0.7008	0.7137
2	0.7999	0.8050	0.8176	0.8311	0.8380	0.8334
5	0.9579	0.9577	0.9563	0.9513	0.9399	0.9203
10	1.0183	1.0160	1.0090	0.9963	0.9772	0.9578

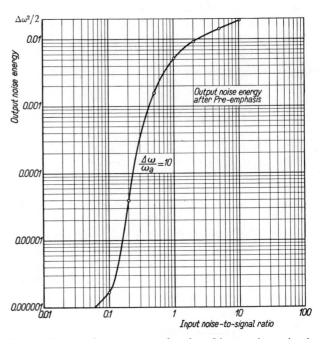

Fig. 4—Output noise energy as a function of input noise-to-signal ratio, when pre-emphasis is applied with an RC time constant of 75.10^{-6} seconds. Analogous to Fig. 2.

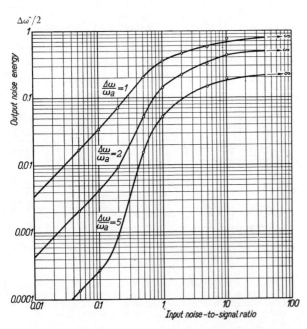

Fig. 3—Output noise energy as a function of input noise-to-signal energy ratio. If bandwidth is 5, 2, or 1 times af bandwidth. Rectangular filter characteristic.

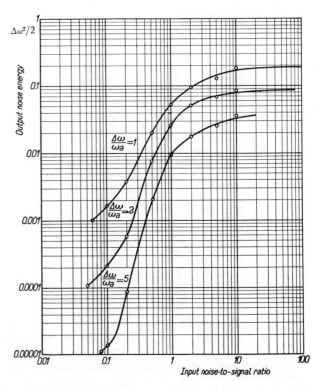

Fig. 5—Output noise energy as a function of input noise-to-signal energy ratio. Pre-emphasis applied. RC time constant, 75.10^{-6} seconds. Analogous to Fig. 3.

[10] R. F. Guy and R. M. Morris, "N.B.C. frequency modulation field test," *RCA Rev.*, vol. 5, pp. 190–225; October, 1940.

So far, the filter amplitude characteristic has been idealized to a rectangular form. As in Section 2, we shall consider now a gaussian amplitude characteristic, which provides a better approximation of actual conditions. All calculations are similar to those already given, and we have only to replace the functions $h_k(u)$ by $H_k(u)$, $a_k(u)$ by $A_k(u)$, $b_k(u)$ by $B_k(u)$, etc., in the final result. Compare (21a) to (24a). The effective noise voltage is given as a function of the frequency and the input noise-to-signal ratio in Table IV.

The results are shown in Fig. 6. Comparison with Fig. 1 makes it clear that the general behavior does not change, although there are minor deviations. In Figs. 7 and 8 the energy of the noise is drawn as a function of the input noise-to-signal ratio in the same way as in Figs. 2 and 3, but now for a gaussian amplitude characteristic.

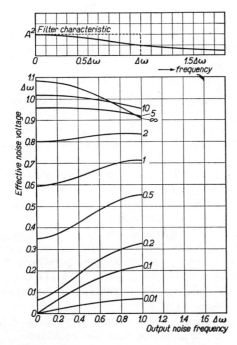

Fig. 6—Spectrum of effective noise voltage (per unit frequency bandwidth) after detection. Parameter is input noise-to-signal ratio (NC). Gaussian amplitude-versus-frequency characteristic of the filter. See also Fig. 1.

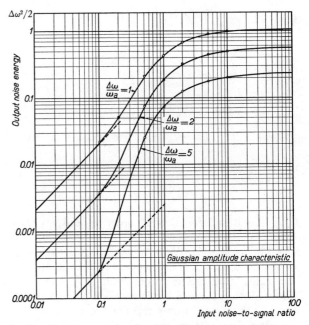

Fig. 8—Analogous to Fig. 7, but if bandwidth 5, 2, or 1 times the af bandwidth.

V. Noise in the Presence of a Frequency-Modulated Signal. Suppression of the Modulation by Noise

When a frequency-modulated signal is amplified in a receiver, there may be some distortion of the modulation due to insufficient bandwidth or to a nonlinear phase characteristic. In the following calculations we shall leave this effect out of account and assume that the signal passes the filter undistorted.

With a rectangular amplitude characteristic of the filter, the input-signal is given by

$$v(t) = \cos(\omega_0 t + m_1 \sin pt)$$
$$+ \sum_{\omega_0 - \Delta\omega}^{\omega_0 + \Delta\omega} (a_n \cos nt - b_n \sin nt) \qquad (31)$$
$$m_1 = \Delta\omega_1/p.$$

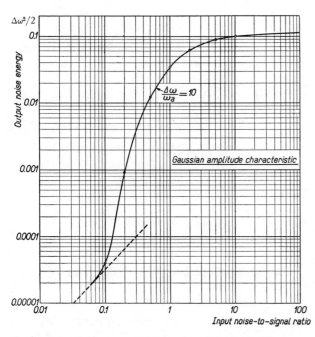

Fig. 7—Output noise energy of a receiver with an ideal frequency detector as a function of input-noise-to-signal energy ratio (NC). Gaussian amplitude characteristic of the filter. If bandwidth (energetically defined) 10 times af bandwidth.

It is necessary to substitute this function $v(t)$ in (7) and to take the average, as in (12). Some remarks on the integration are given in Appendix III. Use is made of the following abbreviations:

$$\alpha = \frac{1}{(k-r)^2(k-2r)!}\binom{k-r}{r}\frac{e^{-2/NC}}{(NC)^{k-2r}}$$

$$J_m\{(k-2r)m_1\} = J_m$$

$${}_1F_1(-r+1, k-2r+1, 1/NC) = X$$

$${}_1F_1(-r, k-2r+1, 1/NC) = Y$$

$$u - mp/2\Delta\omega = u_m.$$

The noise energy is then given by

$$E_2(u)/4\Delta\omega^2 = \sum_{r=1}^{\infty} r^{-1}e^{2/NC}{}_1F_1(-r+1, 1, 1/NC)h_{2r}(u)$$

$$+ \sum_{k=1}^{\infty}\sum_{r=0}^{1/2(k-1)}\sum_{m=-\infty}^{+\infty}[\alpha X^2 J_m{}^2\{kh_k(u_m)$$

$$+ (k-2r)^2 b_k(u_m)\}$$

$$+ 2\alpha k^{-1}X\{kX - 2(k-r)Y\}mp(2\Delta\omega)^{-1}J_m{}^2 u_m f_k(u_m)$$

$$+ \alpha(k-2r)^{-2}\{kX - 2(k-r)Y\}^2m^2p^2(2\Delta\omega)^{-2}J_m{}^2 f_k(u_m)]$$

$$+ \sum_{r=0}^{\infty}\tfrac{1}{2}(NC)^{-2}e^{-2/NC}{}_1F_1{}^2(-r+1, 2, 1/NC)$$

$$(\Delta\omega_1)^2(2\Delta\omega)^{-2}\{f_{2r}(u_{-1}) + f_{2r}(u_{+1})\}. \tag{33}$$

J_m is the Bessel function of order m and argument $(k-2r)m_1$. The functions $h_k(u)$, $b_k(u)$, $f_k(u)$ are discussed in Section III. We have not yet used $f_0(u) = \delta(u)$ (this is the same δ-function as used in Section I).

In calculating the spectrum from (33), one has to start with the terms of the lowest order. Here the last term of (33) gives the only term of order zero. It gives a result different from zero only if $u = p/2\Delta\omega$; that is, only for the frequency p. The energy for that frequency is

$$\tfrac{1}{2}(NC)^{-2}e^{-2/NC}{}_1F_1{}^2(1, 2, 1/NC)(\Delta\omega_1)^2$$

$$= \tfrac{1}{2}(\Delta\omega_1)^2(1 - e^{-1/NC})^2.$$

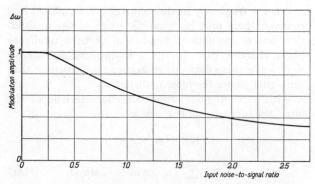

Fig. 9—Suppression of the modulation by noise. Ordinate: amplitude of the modulation. Abscissa: noise-to-signal energy ratio.

Whereas, in general, $E_2(u)$ gives the energy per unit bandwidth, here the energy is concentrated in a single line (this is the meaning of the δ-function). It is the energy of the modulation. If no noise is present, the amplitude of the modulation is $\Delta\omega_1$. In the presence of noise this amplitude is modified to $\Delta\omega_1(1-e^{-1/NC})$ where $1/NC$ is the ratio of signal energy to noise energy (if). (See Fig. 9). Thus (33) takes into account the suppression of the modulation by noise.

The first-order term in (33) gives:

$$4(\Delta\omega)^2NC(1 - e^{-1/NC})u^2 + 4(\Delta\omega_1)^2(NC)^{-1}e^{-2/NC}.$$

For small values of NC, this is a fair approximation for the output noise. Then the sweep of the modulation has no effect on the noise energy. We see from this term, however, that, when the noise energy is not small compared to the signal energy, the sweep of the modulation affects the noise after detection. This effect was found experimentally by Guy and Morris,[10] and is fully described by (33). When, instead of the result for a rectangular filter, one wishes to know the result for another symmetrical filter, one has only to substitute the appropriate functions for $h_k(u)$, $b_k(u)$, $f_k(u)$. For a gaussian amplitude characteristic, these functions have been discussed in Section III. As the amount of work involved in numerical calculations of the noise by means of (33) is considerable, a numerical example is omitted.

All of the above calculations refer to the noise energies inherent in the system of frequency modulation. They will give an increasingly better approximation of the practical results as the frequency detector more nearly approaches the ideal.

Appendix I[11]

Starting from (12), we integrate first with respect to a_n. This integral has the form

$$(\pi C)^{-1/2}\int_{-\infty}^{+\infty}\exp\{-a_n{}^2/C + ia_n(u_1\cos nt_1 - nu_2\sin nt_1$$

$$+ u_3\cos nt_2 - nu_4\sin nt_2)\}da_n$$

$$\exp\{-(C/4)(u_1\cos nt_1 - nu_2\sin nt_1$$

$$+ u_3\cos nt_2 - nu_4\sin nt_2)^2\}.$$

In the same way, the integration over b_n gives

$$\exp\{-C/4(u_1\sin nt_1 + nu_2\cos nt_1$$

$$+ u_3\sin nt_2 + nu_4\cos nt_2)^2\}.$$

Multiplying all probability integrals, we get

$$\exp[-(C/4)\sum\{u_1{}^2 + n^2u_2{}^2 + u_3{}^2 + n^2u_4{}^2$$

$$+ 2(u_1u_3 + n^2u_2u_4)\cos nv + 2n(u_1u_4 - u_2u_3)\sin nv\}] \tag{16}$$

[11] In the Appendixes a more specified outline of the calculations is given, but for space considerations much ordinary algebra has been left to the reader.

where the summation has to be taken over all n satisfying

$$\omega_0 - \Delta\omega \leqq n \leqq \omega_0 + \Delta\omega$$

and

$$v = t_1 - t_2.$$

Let us introduce new variables $\beta_2 = u_2\omega_0$, $\beta_4 = u_4\omega_0$, $s = n - \omega_0$, and make a series development of that part of the exponential form which contains $\cos(\omega_0 v + sv)$ and $\sin(\omega_0 v + sv)$:

$$\sum_{k=0}^{\infty} (-C/4)^k (1/k!) \Big[\sum_{s=-\Delta\omega}^{\Delta\omega} \{ 2\alpha_s \cos(\omega_0 v + sv)$$

$$+ 2\gamma_s \sin(\omega_0 v + sv) \} \Big]^k.$$

Here

$$\alpha_s = u_1 u_3 + \beta_2 \beta_4 (1 + s/\omega_0)^2;$$

$$\gamma_s = (u_1 \beta_4 - \beta_2 u_3)(1 + s/\omega_0).$$

This form can also be written:

$$\sum_{k=0}^{\infty} (-C/4)^k (1/k!) \Big[\sum_{s=-\Delta\omega}^{\Delta\omega} \{ (\alpha_s - i\gamma_s) e^{i\omega_0 v + isv}$$

$$+ (\alpha_s + i\gamma_s) e^{-i\omega_0 v - isv} \} \Big]^k.$$

As in Section I, we are only interested in such values of m in (7) and (12) which are small compared to ω_0. Therefore, as in the example treated in Section I, we use the binomial formula and retain only those terms which contain no $\omega_0 v$. The result is

$$\sum (C/4)^{2k} (k!k!)^{-1} \{ \sum (\alpha_s - i\gamma_s) e^{isv} \}^k$$

$$\cdot \{ \sum (\alpha_s + i\gamma_s) e^{-isv} \}^k. \quad (17a)$$

Now we develop the integrand with respect to s/ω_0 and stop at $(s/\omega_0)^2$. The result can be integrated straightforwardly. The following types of integrals occur ($\gamma = NC/4$, $N = 2\Delta\omega$):

$$\iint_{-\infty}^{+\infty} e^{-\gamma(u_1^2 + u_2^2)} (u_1^2 + u_2^2)^k u_2^{-2} du_1 du_2 = 0, \text{ if } k \neq 0$$

$$= -2\pi, \text{ if } k = 0$$

$$\iint_{-\infty}^{+\infty} e^{-\gamma(u_1^2 + u_2^2)} (u_1^2 + u_2^2)^{k-1} u_1^2 u_2^{-2} du_1 du_2$$

$$= -(k-1)! \pi \gamma^{-k}, \quad k \geqq 1$$

$$\iint_{-\infty}^{+\infty} e^{-\gamma(u_1^2 + u_2^2)} (u_1^2 + u_2^2)^{k-1} du_1 du_2$$

$$= (k-1)! \pi \gamma^{-k}, \quad k \geqq 1$$

$$\iint_{-\infty}^{+\infty} e^{-\gamma(u_1^2 + u_2^2)} (u_1^2 + u_2^2)^{k-1} u_1^4 u_2^{-2} du_1 du_2$$

$$= -3/2(k-1)! \pi \gamma^{-k}, \quad k \geqq 1.$$

This leads directly to (13a) and (13b). The introduction of (8a) (symmetrical amplitude characteristic) modifies (16) into

$$\exp\Big[-C/4 \sum f^2(n) \{ u_1^2 + n u_2^2 + u_3^2 + n^2 u_4^2$$

$$+ 2(u_1 u_3 + n^2 u_2 u_4) \cos nv + 2n(u_1 u_4 - u_2 u_3) \sin nv \} \Big]. \quad (16a)$$

The phase characteristic does not influence the calculations. On account of the symmetry in the characteristic, we only have to introduce an extra factor $f^2(s)$ in both sums of (17a). After changing to the new variable u, this leads directly to (15a). In case the amplitude characteristic is not symmetrical, the change in (17a) is greater. Instead of (17a), we now get:

$$\sum (C/4)^{2k} (k!k!)^{-1} \{ \sum f^2(s)(\alpha_s - i\gamma_s) e^{isv} \}^k$$

$$\cdot \{ \sum f^2(s)(\alpha_s + i\gamma_s) e^{-isv} \}^k.$$

After introduction of the new variable u, the analogue of form (15a) is, then,

$$\int_{-\infty}^{+\infty} H_{2k}(u) e^{iuv} du = \frac{1}{4} \Big\{ \int F(u) e^{iuv} du \int u^2 F(-u) e^{iuv} du$$

$$+ \int F(-u) e^{iuv} du \int u^2 F(u) e^{iuv} du$$

$$- 2 \int u F(u) e^{iuv} du \int u F(-u) e^{iuv} du \Big\}$$

$$\cdot \Big\{ \int F(u) e^{iuv} du \int F(-u) e^{iuv} du \Big\}^{k-1}.$$

In this expression all integrals are from $-\infty$ to ∞.

Appendix II

Equation (25) is substituted in (7) and the average is taken as in (12). The integration over a_n and b_n goes exactly in the same way as in Appendix I. As in (16), the result is a function of $t_2 - t_1 = v$. When we introduce new variables t_1 and v, instead of t_1 and t_2, the integration over t_1 gives the result:

$$1/(4\pi^3) \int dv e^{-inv} J_0 \{ u_1^2 + u_2^2 \omega_0^2 + u_3^2 + u_4^2 \omega_0^2$$

$$+ 2(u_1 u_3 + u_2 u_4 \omega_0^2) \cos nv$$

$$+ 2(u_1 u_4 - u_2 u_3)\omega_0 \sin nv \}^{1/2}. \quad (26)$$

The Bessel function can also be written

$$\sum_{-\infty}^{+\infty} (-1)^q J_q(u_1^2 + u_2^2 \omega_0^2)^{1/2} J_q(u_3^2 + u_4^2 \omega_0^2)^{1/2} e^{iq(nv + \phi)} \quad (27)$$

where ϕ is defined by

$$\cos\phi = \frac{u_1 u_3 + u_2 u_4 \omega_0^2}{(u_1^2 + u_2^2 \omega_0^2)^{1/2} (u_3^2 + u_4^2 \omega_0^2)^{1/2}}$$

and

$$\sin\phi = \frac{(u_2 u_3 - u_1 u_4)\omega_0}{(u_1^2 + u_2^2 \omega_0^2)^{1/2} (u_3^2 + u_4^2 \omega_0^2)^{1/2}}.$$

Introduce new variables as in Appendix I, and expand into a series that part of the exponent which contains $\cos(\omega_0 v + sv)$ and $\sin(\omega_0 v + sv)$. This gives the same result as in (17). Binomial development of the terms of this sum results in the double sum:

$$\sum_{k=1}^{\infty} \sum_{r=0}^{k} (C/4)^k \{(k-r)!r!\}^{-1} \left\{ \sum (\alpha_s - i\gamma_s) e^{isv} \right\}^{k-r}$$

$$\cdot \left\{ \sum (\alpha_s + i\gamma_s) e^{-isv} \right\}^r \cdot e^{i(k-2r)\omega_0 v}. \quad (28)$$

Now we have to integrate the product of (27) and (28) and

$$(4\pi^3)^{-1} e^{-imv} \exp \left\{ -(C/4) \sum (u_1^2 + \beta_{2a}^2 + u_3^2 + \beta_{4a}^2) \right\}$$

in which

$$\beta_{2a} = \beta_2(1 + s/\omega_0); \qquad \beta_{4a} = \beta_4(1 + s/\omega_0).$$

As m is small compared to ω_0, we have to choose $q+k-2r=0$. $q=2r-k$. Again we expand into a series with respect to s/ω_0, and stop at $(s/\omega_0)^2$. We introduce the continuous variable u and we use the functions introduced in Section III. The result is

$$\sum_{k=1}^{\infty} \sum_{r=0}^{k} \left\{ 2u^2 f_k(u) - 8(k-r)r b_k(u) \right\} \{(k-r)!r!\}^{-1} \gamma^k.$$

$$\cdot \iint_0^\infty dx\, dy\, J_{k-2r}(x) J_{k-2r}(y) x^{k-1} y^{k-1} e^{-\gamma(x^2+y^2)}.$$

Here $\gamma = NC/4$. These integrals are of the type called by Watson[12] "Weber's first exponential integral." Their computation leads to equation (30).

To show the convergence of the development, we return to a single sum. The part of the formula containing $b_k(u)$ is modified into

$$\sum_2^\infty -8b_k(u)(k!)^{-1}(2\gamma)^k(2\pi)^{-1}$$

$$\cdot \int_0^{2\pi} d\psi \int_0^\infty dx \int_0^\infty dy\, J_0 \{(x^2 + y^2 - 2xy \cos \psi)^{1/2}\}$$

$$\cdot (\cos \psi)^{k-2} e^{-\gamma(x^2+y^2)} x^{k-1} y^{k-1}.$$

This series is even convergent when J_0 is replaced by 1 when the rest of the integrand is positive, and by -1 when the rest is negative. For k large, the general term of the series behaves as b_k, or is smaller, and converges to zero at least with $k^{-3/2}$. The part of the formula containing $f_k(u)$ is still faster convergent.

[12] G. N. Watson, "A Treatise on the Theory of Besselfunctions," second edition, p. 393; Cambridge, 1944.

APPENDIX III

After the introduction of (31) in (7), the averaging has to be done as in (12). The first steps in the computation are the averaging over all a_n, b_n, and the removal of all terms containing $\cos \omega_0 t$. Further, we introduce new variables as in Appendix II. If, now,

$$G(\beta_2, \beta_4, \psi) = u_1^2 + \beta_2^2 + u_3^2 + \beta_4^2$$
$$+ 2(u_1^2 + \beta_2^2)^{1/2}(u_3^2 + \beta_4^2)^{1/2} \cos \psi,$$

the result of these first steps can be written in the form:

$$\overline{2f_m f_m^*} = (8\pi^4)^{-1}\omega_0^2 \int dt_1 \int dv\, e^{-inv}$$

$$\cdot \iiiint \exp -(C/4G) \sum (\beta_{2a}, \beta_{4a}, \psi_a)$$

$$\cdot J_0 \big[\{G(\beta_{2b}, \beta_{4b}, \psi_b)\}^{1/2} \big] \beta_2^{-2} \beta_4^{-2} du_1 d\beta_2 du_3 d\beta_4. \quad (32)$$

In this formula we have used the following abbreviations:

$$\beta_{2a} = \beta_2(1 + s/\omega_0);$$

$$\beta_{2b} = \beta_2 \left(1 + \frac{\Delta\omega_1}{\omega_0} \cos pt_1 \right);$$

$$\beta_{4a} = \beta_4(1 + s/\omega_0);$$

$$\beta_{4b} = \beta_4 \left(1 + \frac{\Delta\omega_1}{\omega_0} \cos p(t_1 - v) \right);$$

$$\psi_a = \omega_0 v + sv - \phi_a;$$

$$\psi_b = \omega_0 v - m_1 \sin p(t - v) + m_1 \sin pt - \phi_b;$$

$$m_1 = \Delta\omega_1/p$$

$$\cos \phi(\beta_2, \beta_4) = \frac{u_1 u_3 + \beta_2 \beta_4}{(u_1^2 + \beta_2^2)^{1/2}(u_3^2 + \beta_4^2)^{1/2}};$$

$$\sin \phi(\beta_2, \beta_4) = \frac{u_1 \beta_4 - u_3 \beta_2}{(u_1^2 + \beta_2^2)^{1/2}(u_3^2 + \beta_4^2)^{1/2}};$$

$$\phi_a = \phi(\beta_{2a}, \beta_{4a}); \qquad \phi_b = \phi(\beta_{2b}, \beta_{4b}).$$

As before, we take the terms in the exponent containing $\omega_0 v$ and expand into a series. We also use the series of (27) for J_0, and take the terms together in such a way that $\omega_0 v$ disappears from the result ($q=2r-k$).

This time we are interested in terms up to $(s/\omega_0)^2$, $(s\Delta\omega_1/\omega_0^2)$, and $(\Delta\omega_1^2/\omega_0^2)$.

The integrals are of the same type as in Appendix II. The result of the integration is given in (33).

PAPER NO. 6

Reprinted from *RCA Rev.*, Vol. 7, No. 4, pp. 522–560, Dec. 1946

FREQUENCY MODULATION DISTORTION CAUSED BY COMMON- AND ADJACENT- CHANNEL INTERFERENCE*

By

Murlan S. Corrington

Home Instruments Department, RCA Victor Division,
Camden, N. J.

Summary—*During frequency-modulated radio broadcasting the signal is liable to be badly distorted whenever multipath transmission occurs or when any other interfering signal is present on the same or an adjacent channel. During hot weather, or before a storm, long-distance reception has been observed from frequency modulation broadcast stations on the 42–50 megacycle band. When such a distant station was in the same channel as a desired station, it sometimes happened that for short intervals the undesired station became stronger than the desired one. When this happened there was a small amount of noise and the programs suddenly changed. This interchange often lasted for several seconds but sometimes was limited to a word or two or a few notes of music.*

Formulas are given for computing the amplitudes of the harmonics and cross-modulation frequencies produced by the interference. These enable the calculation of the effect of a de-emphasis network following the discriminator, of a low-pass audio filter, and of nonlinear phase shift in the amplifiers.

Introduction

FOR several years it has been evident that frequency-modulated radio broadcasting offers certain advantages in noise reduction when compared with the usual amplitude-modulation systems. Many papers describe and discuss frequency modulation systems and their noise-suppressing properties.[1-8] Extensive field tests showed[9] that

---

* Decimal Classification: R148.2 × R430.

[1] Edwin H. Armstrong, "A Method of Reducing Disturbances in Radio Signaling by a System of Frequency Modulation," *Proc. I.R.E.*, Vol. 24, No. 5, pp. 689–740; May, 1936.

[2] Murray G. Crosby, "Frequency Modulation Noise Characteristics," *Proc. I.R.E.*, Vol. 25, No. 4, pp. 472–514; April, 1937.

[3] H. Roder, "Noise in Frequency Modulation," *Electronics*, Vol. 10, No. 5, pp. 22–25, 60, 62, 64; May, 1937.

[4] E. H. Plump, "Störverminderung durch Frequenzmodulation," *Hochfrequenztechnik und Elektroakustik*, Vol. 52, pp. 73–80; September, 1938.

[5] Stanford Goldman, "F-M Noise and Interference," *Electronics*, Vol. 14, No. 8, pp. 37–42; August, 1941.

[6] Harold A. Wheeler, "Common-Channel Interference Between Two Frequency-Modulated Signals," *Proc. I.R.E.*, Vol. 30, No. 1, pp. 34-50; January, 1942.

when frequency modulation was used there was less interference produced by two stations operating at the same frequency than for the corresponding case of amplitude modulation, and that less power was required to cover a given area. It was also found that when the ratio of the carrier voltage to the noise voltage is high, the signal-to-noise ratio improvement due to frequency modulation is considerable. As the interfering noise voltage is increased with respect to the desired carrier-wave voltage, the improved noise suppression is obtained as long as the desired signal is several times as strong as the noise.

When a definite carrier-to-noise voltage ratio is reached (a ratio of 2 or 3 for wide-band frequency modulation) the amount of distortion in the audio output increases rapidly. When the noise voltage exceeds the signal voltage during all parts of the audio cycle, the noise eliminates the desired signal. This means that when frequency modulation is used the signal is either good or bad; there is only a small range for the ratio of carrier voltage to noise voltage that gives a noisy, but tolerable, signal.

Multipath transmission occurs when two or more interfering signals come from the same transmitter, but one is delayed with respect to the others because of a longer transmission path. Considerable distortion has been observed when multipath transmission occurs in frequency-modulated broadcasting and fairly complete discussions of this problem are available.[10-13] If the second wave comes from a different station than the desired wave, the result is common- or adjacent-channel interference according to whether the two carrier frequencies are nearly the same or are separated by the width of one channel.

There is not much information available on the amount of interference to be expected in the new frequency modulation band. The effects to be described were observed on the old 42–50 megacycle band and on the 30–42 megacycle police bands. The frequency of occurrence and the magnitude of these effects will not be known for the new 88–108 megacycle band until a reasonable number of transmitters with normal power and antenna gains are in operation. If such interference does occur, the analysis given here will be applicable.

[7] Herbert J. Reich, "Interference Suppression in A-M and F-M," *Communications*, Vol. 22, No. 8, pp. 7, 16, 19, 20; August, 1942.

[8] Robert N. Johnson, "Interference in F-M Receivers," *Electronics*, Vol. 18, No. 9, pp. 129–131; September, 1945.

[9] I. R. Weir, "Field Tests of Frequency- and Amplitude-Modulation With Ultra-High-Frequency Waves," *Gen. Elec. Rev.*, Vol. 42, Nos. 5 and 6, pp. 188–191, May, 1939; pp. 270–273, June, 1939.

[10] Murray G. Crosby, "Observations of Frequency-Modulation Propagation on 26 Megacycles," *Proc. I.R.E.*, Vol. 29, No. 7, pp. 398–403; July, 1941.

[11] A. D. Mayo and Charles W. Sumner, "F.M. Distortion in Mountainous Terrain," *Q.S.T.*, Vol. 28, No. 3, pp. 34–36; March, 1944.

[12] Murlan S. Corrington, "Frequency-Modulation Distortion Caused by Multipath Transmission," *Proc. I.R.E.*, Vol. 33, No. 12, pp. 878–891; Dec., 1945.

[13] S. T. Meyers, "Nonlinearity in frequency-modulation radio systems due to multipath propagation," *Proc. I.R.E.*, Vol. 34, No. 5, pp. 256–265; May, 1946.

Sometimes during hot weather, or before a storm, long-distance transmission has been observed from frequency modulation broadcast stations. During the summer of 1944, station WSM-FM in Nashville, Tenn. was heard often in Camden, New Jersey. During July it was very strong and free of noise for nine evenings in succession and it was heard several other evenings. Occasionally, long-distance reception from stations in all directions was observed. On July 7, 1944 nearly all the mid-western stations and several from other directions could be received in Camden, for about 2½ hours with a standard commercial receiver and indoors antenna. The following list of stations received was compiled that evening:

Call	*Station*	*Megacycles*
WWZR	Zenith Radio Corp., Chicago	45.1
WGNB	WGN, Inc., Chicago	45.9
WBBM-FM	Columbia Broadcasting System, Chicago	46.7
WDLM	Moody Bible Institute, Chicago	47.5
WSBF	South Bend Tribune, South Bend, Ind.	47.1
WMLL	Evansville on the Air, Evansville, Indiana	44.5
WENA	Evening News Assn., Detroit	44.5
WMFM	The Journal Company, Milwaukee, Wisconsin	45.4
WSM-FM	National Life & Accident Ins. Co., Nashville, Tenn.	44.7
WMIT	Gordon Gray, Winston-Salem, N. C.	44.1
WMTW	Yankee Network, Mt. Washington, N. H.	43.9
W2XMN	Edwin H. Armstrong, New York	43.1
WHNF	Marcus Loew Booking Agency, New York	46.3
WBAM	Bamberger Broadcasting Service, New York	47.1
WABC-FM	Columbia Broadcasting System, New York	46.7
WABF	Metropolitan Television, Inc., New York	47.5
WIP-FM	Pennsylvania Broadcasting Co., Philadelphia	44.9

Some interesting common-channel phenomena were observed. Stations WENA, Detroit, and WMLL, Evansville, were of nearly equal strength. First one, and then the other was received; they changed about every fifteen seconds. There would be a slight amount of noise and the programs would suddenly be interchanged. This continued for about one-half hour. Sometimes the carrier-wave voltage levels dropped below the level at which the limiter in the receiver operated and both programs could be heard simultaneously.

Stations WSBF, South Bend, and WBAM, New York, were also in a common channel. WSBF was stronger and was clear most of the time; WBAM would come in with sudden bursts of a word or two or a bit of music as station WSBF faded rapidly. These bursts occurred at intervals of about ten seconds.

Some of the state police frequency-modulation systems have reported serious skip interference on numerous occasions. In Missouri, on the talk-back frequency of 39.78 megacycles, the interfering signals are usually those of the New Jersey State Police and the North Carolina Highway Patrol Cars, although cars of the Ohio State Patrol and those of Rhode Island occasionally cause interference. The signal strengths of the undesired stations are greatest during May, June, and July and range from weak to strong. The strong signals are of sufficient intensity to swamp out the local cars and may be received for an hour or two or for the whole day, from about two hours after sunrise to an hour or so after sunset.

The Florida State Patrol have reported considerable interference on frequency modulation from stations in California, New Jersey, Connecticut, and Massachusetts, and they have made car-to-car contacts with Pittsfield, Massachusetts. The Michigan State Police reported that signals from the Alabama State Patrol stations were received by their patrol cars with signal levels at the input to the receiver as high as 300 microvolts, and these stations in Alabama have taken control of their receivers throughout Michigan for hours at a time.

The Indiana State Police have had their cars blocked out by stations in Virginia and Oklahoma for all cars more than three miles from the transmitter. During the hunt for escaped German war prisoners near Carlisle, Indiana, on June 10th, the interference was so bad they had considerable difficulty maintaining contact with their cars. On June 22nd, during a man-hunt and road blockade following a bank holdup at San Pierce, Indiana, cars were completely blocked out at various times by cars in Virginia and Massachusetts. Further disruption of service was caused many afternoons by the second harmonic of short-wave broadcast stations in Massachusetts and New York.

Recent observations by the Federal Communications Commission show that such bursts or sudden increases in strength of signals received beyond the line of sight occur regularly.[14-15] The long-distance transmission that occurs during such bursts can be interpreted as reflections from media of height comparable to the E layer, but lying at each side of the great-circle plane. It is assumed that when meteors pass through the upper atmosphere, the air is ionized and this causes the bursts.

If a local station is on the same channel as a distant one which is being received in bursts, interference may be expected to occur for intervals as

---

[14] "Measurement of V-H-F Bursts," *Electronics*, Vol. 18, No. 1, p. 105; January, 1945.

[15] K. A. Norton and E. W. Allen, Jr., "Very-High-Frequency and Ultra-High-Frequency Signal Ranges as Limited by Noise and Co-Channel Interference," *Proc. I.R.E.*, Vol. 33, No. 1, p. 58; January, 1945.

long as several seconds. This might even cause the program to change suddenly from one station to the other during these short intervals.

Analysis of Fundamental Case

The most elementary case of frequency modulation interference is that produced when two unmodulated radio-frequency carriers, having nearly the same frequency, are added together. This gives the usual heterodyne envelope as the two voltages beat together. In addition there is a variation in the phase of the resultant which is equivalent to frequency modulation. If the difference in frequency of the two carriers is now varied sinusoidally by changing the frequency of one, keeping the two amplitudes constant, the result is common-channel interference or adjacent-channel interference, depending upon the way the one frequency is varied. It is thus evident that, if the most elementary case is properly analyzed, the frequency modulation interference is merely a generalization of the results.

Heterodyne Envelope

As shown in Appendix I, if two radio-frequency carriers $e_1 \sin \omega t$ and $e_2 \sin (\omega + 2\pi\mu)t$ are added, the heterodyne envelope is given by

$$\text{Envelope} = e_1\sqrt{1 + x^2 + 2x \cos 2\pi\mu t} \qquad (1)$$

where
e_1 = amplitude of first carrier
e_2 = amplitude of second carrier
$x = e_2/e_1$
ω = angular frequency of first carrier, radians per second
μ = difference in frequency, cycles per second

This is the voltage that will be obtained if the resultant signal is sent through a linear rectifier and filtered. Figure 1 shows the variation of the envelope over one beat-note cycle as the ratio of the amplitudes of the two signals, x, is changed. For small values of x the envelope is approximately,

$$\text{Envelope} = e_1(1 + x \cos 2\pi\mu t) \qquad x \ll 1 \qquad (2)$$

As the ratio x is increased gradually, the higher harmonics increase in amplitude; so the peaks become broader and the hole in the carrier becomes deeper and narrower. In the limit, as $x \to 1$, the envelope becomes a series of rectified cosine waves, or:

$$\text{Envelope} = 2e_1 \mid \cos \pi\mu t \mid \qquad x = 1 \qquad (3)$$

Average Value of Envelope. If the resultant heterodyne voltage is sent through a linear rectifier, the direct-current voltage across the rectifier output increases gradually as x is increased. Figure 2 shows that this voltage increases 27.3 per cent when x changes from zero to one. As shown in Appendix I, this voltage is given by:

$$\text{Average voltage} = \frac{2(1+x)e_1}{\pi} \; E\left\{\frac{2\sqrt{x}}{1+x}\right\} \tag{4}$$

where $E\left\{\dfrac{2\sqrt{x}}{1+x}\right\}$ is a complete elliptic integral of the second kind with modulus $\dfrac{2\sqrt{x}}{1+x}$.

Root-Mean-Square Value of Envelope. If a square-law rectifier instead of a linear rectifier is used, the root-mean-square value of the rectified envelope can be read with an average-reading direct-current voltmeter. The root-mean-square voltage will increase more rapidly with x than the average

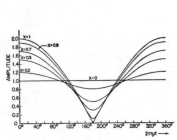

Fig. 1—The heterodyne envelope.

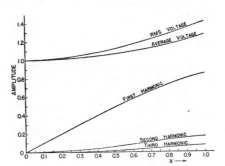

Fig. 2—Harmonic content of the heterodyne envelope.

voltage, as shown by Figure 2. The voltage is given by:

$$\text{Root-mean-square voltage} = e_1\sqrt{1+x^2} \tag{5}$$

and it increases 41.4 per cent when x increases from zero to one.

Fourier-Series Analysis of Envelope. If the heterodyne envelope is rectified with a linear rectifier, and the radio frequency is filtered out, the resultant audio signal (shown by Figure 1) can be expanded in a Fourier series. The coefficients of this series are given in Appendix I and the zero-frequency component is the same as the average value which is shown by Figure 2. The fundamental component increases almost linearly with increasing x to a maximum value of $\frac{2}{3}$ of the corresponding direct current voltage. The second harmonic increases slowly until it equals 20 per cent of the fundamental when $x = 1$, and the third harmonic has a maximum value of 8.6 per cent of the fundamental.

Phase-Angle Variations

The two signals $e_1 \sin \omega t$ and $e_2 \sin (\omega + 2\pi\mu)t$ are in phase when $t = 0$. Since the frequency of the second signal is higher than the frequency of the first signal, this means that a vector representing e_2 will rotate with respect to one representing e_1. If e_1 is a vector rotating at ω radians per second, then e_2 will rotate at $\omega + 2\pi\mu$ radians per second.

Figure 3 shows the variation of the phase angle φ which the resultant, R, of e_1 and e_2 makes at any given instant with the vector e_1. When $t = 0$, the two vectors are in phase and $\varphi = 0$. At a later time $2\pi\mu t = 90$ degrees, so e_2 and e_1 are at right angles and $\tan \varphi = e_2/e_1 = x$. When $2\pi\mu t = 180$ degrees, φ is again zero. This process gives the variations in φ shown by Figure 4. The maximum value of φ is equal to $\sin^{-1} x$, as shown

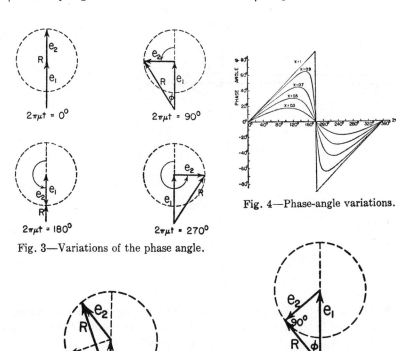

Fig. 4—Phase-angle variations.

Fig. 3—Variations of the phase angle.

Fig. 6—Variations of φ.   Fig. 5—Maximum value of φ.

by Figure 5. As x approaches one, the angle φ varies more and more rapidly near $2\pi\mu t = 180$ degrees. When $e_2 = e_1$ or $x = 1$, φ increases linearly from zero to 90 degrees as e_2 turns through 180 degrees.

As shown by Figure 6, φ is then an inscribed angle, and since an inscribed angle is measured by one-half its intercepted arc, φ increases linearly when e_2 turns uniformly. As e_2 approaches cancellation of e_1, R is an infinitesimal vector and $\varphi \rightarrow + 90$ degrees. As e_2 swings past cancellation, the direction of R suddenly reverses so $\varphi = -90$ degrees; i. e., there is an instantaneous change of φ equal to 180 degrees. Beyond that point φ increases linearly toward 0 degrees, as shown by Figure 4.

Instantaneous Frequency

The output from a linear discriminator is proportional to the instantaneous frequency, where the instantaneous frequency is defined by:[16]

$$f = \frac{1}{2\pi} \frac{d}{dt} \text{ (argument of sine function)}. \tag{6}$$

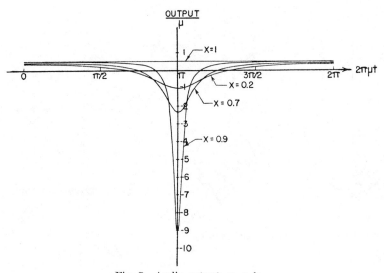

Fig. 7—Audio output, $x < 1$.

For a balanced linear discriminator, tuned to frequency ω, the output is proportional to the deviation in frequency from the center frequency ω. As shown in Appendix I, the output is given by

$$\text{Output} \propto \frac{\mu}{\dfrac{\cos 2\pi\mu t + 1/x}{\cos 2\pi\mu t + x} + 1} \tag{7}$$

Obviously this output is proportional to the slope of the curves of Figure 4, since it represents the first derivative with respect to time.

The curves of Figure 7 show the wave form in the audio output from

---

[16] J. R. Carson, "Notes on the Theory of Modulation," *Proc. I.R.E.*, Vol. 10, No. 2, p. 57; February, 1922.

a frequency modulation receiver, with perfect limiting and linear phase shift in the tuned circuits. As x approaches one, the output becomes more and more like an impulse, until at $x = 1$, the output has the constant value one-half except when $2\pi\mu t = \pi$; here the output becomes infinite. The area between the line one-half unit above the time axis and the curve for the instantaneous frequency over one cycle is constant for all values of x and equals $-\pi\mu$. This means that as $x \to 1$ the output is constant except at $2\pi\mu t = \pi$ and at that point is an impulse equal to $\pi\mu$ times a unit-impulse function.

When x becomes greater than one, the polarity of the impulse changes, but the shape is the same, as shown by Figure 8.

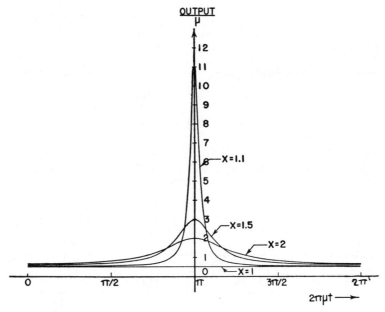

Fig. 8—Audio output, $x > 1$.

Average Value of Instantaneous Frequency. If the discriminator is tuned to the frequency ω, the average audio output is zero when $x < 1$. As shown in Appendix I, the average output is proportional to μ when $x > 1$. The curves of Figure 8 show this shift in average value when e_2 becomes stronger than e_1 and takes control.

Root-Mean-Square Value of Instantaneous Frequency. If the audio output from the discriminator is measured with an root-mean-square meter, the readings will vary as shown by Figure 9. The output increases uniformly from zero when $x = 0$ until it rapidly approaches infinity when $x = 1$. When $x > 1$ the output decreases uniformly to one as x becomes large.

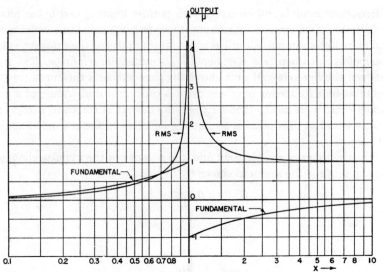

Fig. 9—Root-mean-square voltage output.

As shown in Appendix I:

$$\text{Root-mean-square output} \propto \frac{x\mu}{\sqrt{2(1-x^2)}} \qquad \text{when } x<1 \qquad (8)$$

$$= \mu \sqrt{\frac{2x^2-1}{2(x^2-1)}} \qquad \text{when } x>1. \qquad (9)$$

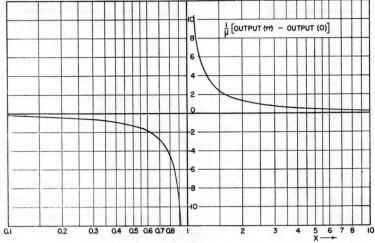

Fig. 10—Peak-to-peak audio output.

Peak-to-Peak Value of Instantaneous Frequency. The output when $2\pi\mu t = \pi$ minus the output at $2\pi\mu t = 0$ gives the peak-to-peak value of the instantaneous frequency. This is given by:

$$\text{Output } (\pi) - \text{Output } (0) = \frac{2x\mu}{x^2 - 1} \qquad (10)$$

The curves of Figure 10 show how the peak-to-peak output varies as x increases. When $x = 1$, the peak-to-peak output becomes infinite, and it decreases uniformly beyond this point.

Harmonic Analysis of Instantaneous Frequency. If the harmonic content of the audio output is calculated by means of a Fourier-series analysis, the result can be expressed as:

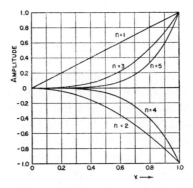

Fig. 11—Harmonic content of audio output.

$$\text{Output} \propto - \mu \sum_{n=1}^{\infty} (- x)^n \cos n(2\pi\mu t). \qquad (11)$$

This means that the nth harmonic amplitude is proportional to μx^n. Figure 11 shows the increase of the harmonic amplitudes with increasing x for the first five harmonics. For small values of x, the higher harmonics are much smaller than the fundamental; but as x approaches one, the higher harmonics increase rapidly, until at $x = 1$ all harmonics are equal.

Effect of Limited Band Width. If the audio output from the discriminator is sent through a low-pass filter, having approximately linear phase-shift, the resultant wave form will depend upon how many harmonics are passed by the filter. In Figure 12, the case of $x = 0.9$ is shown for various low-pass filters. The case $n = 1$ means that only one harmonic, the fundamental, is passed by the filter. If two harmonics are passed, $n = 2$, the center begins to dip more because both harmonics are in phase at that point. The cases for $n = 3$ and $n = 5$ are also shown. The effect, there-

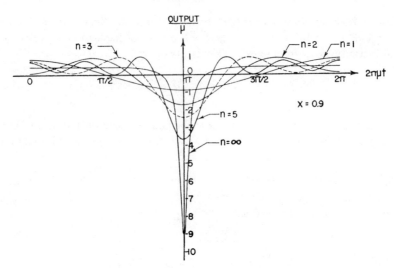

Fig. 12—Effect of low-pass filter.

fore, of limited band width is to reduce the output at $2\pi\mu t = \pi$ and to cause the resulting wave to oscillate about the curve that would be obtained with unlimited band width. For the case when $n = 5$, the peak output is reduced from 9.0 to 3.69, or the output becomes 41 per cent of that for unlimited band width. The curves of Figure 13 show the effect of limited band width. The variable on the axis of abscissas shows the number of harmonics passed by the low-pass filter, and the other axis shows the percent of peak amplitude compared to that for unlimited band width. Thus, if $x = 0.9$ and 10 harmonics are passed by the filter, the peak output

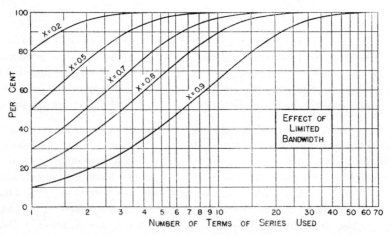

Fig. 13—Effect of limited band width.

will be approximately 65 per cent of what it would be if all harmonics were passed. If $x = 0.5$, it is evident that five or six harmonics will give nearly undistorted output.

As shown by Appendix I, this ratio of the peak output to the corresponding peak for unlimited band width is equal to $1 - x^n$ where n is the number of harmonics passed.

COMMON- AND ADJACENT-CHANNEL INTERFERENCE

The simplest case of frequency modulation interference (that of two

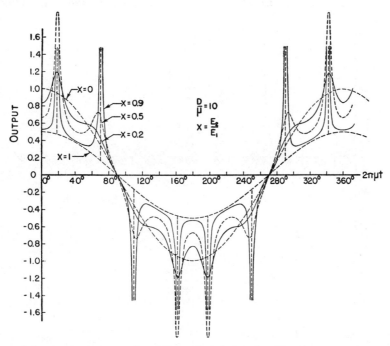

Fig. 14—Variation of distortion as interfering signal becomes stronger.

unmodulated carriers of slightly different frequency) has already been discussed. If now the amplitudes of the two waves are kept constant, but the frequency of one carrier is changed, the problem becomes one of common- or adjacent-channel interference depending upon what range of frequencies the swings of the modulated carrier cover. If the deviations of the one wave are about a mean frequency which coincides with the frequency of the second carrier, the result is common-channel interference. If the mean frequencies are separated by the width of one channel, the result will be adjacent-channel interference.

Common-Channel Interference, Interfering Signal Unmodulated. If a frequency-modulated signal and an unmodulated carrier produce the beat-note interference, the output from a frequency-modulation receiver with limiter will be as shown by Figure 14. This shows the wave form for the various ratios of the interfering signal voltage x. When $x = 0$ (i.e., no interference) the output is an undistorted cosine wave, as shown by the dotted line. As the interference increases, the peaks and dips increase in size, until finally, in the limit, they become very narrow pulses superimposed on a cosine wave of one-half the amplitude obtained with no interference.

As x becomes greater than one, the interfering signal takes control and

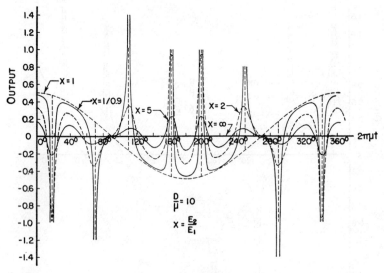

Fig. 15—Variation of distortion as interfering signal becomes stronger.

the modulation of the desired signal is suppressed. Figure 15 shows how the peaks and dips in output decrease when x increases from one to infinity. The envelope of the carrier amplitude corresponding to Figures 14 and 15 is shown by Figure 16. There is one cancellation or hole in the carrier amplitude corresponding to each peak or dip in the output, since the rapid phase change which occurs at cancellation produces the large frequency deviation. If the limiter is not able to maintain a constant voltage input to the discriminator, the amplitude variations of the carrier will cause a reduction in the peaks in the output.

Envelope of Beat-note Pattern. As shown in Appendix II, the beat-note produced in the output of a receiver with a perfect limiter during common-channel interference is a series of peaks and dips which are limited by the

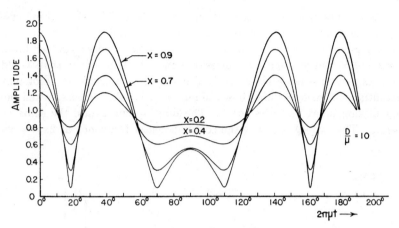

Fig. 16—Heterodyne envelope.

two curves $\dfrac{D}{1+x}\cos 2\pi\mu t$ and $\dfrac{D}{1-x}\cos 2\pi\mu t.$

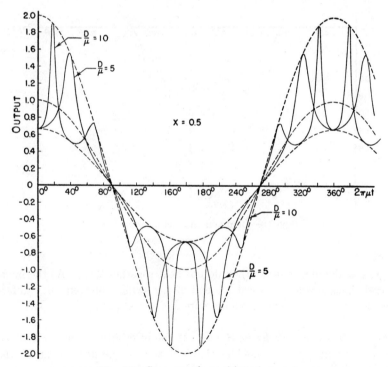

Fig. 17—Common-channel interference.

181

This will be true for all modulation indexes D/μ. The effect of increasing the modulation index is to produce more peaks and dips in output with no change in the limits. Figure 17 shows these two limits as dotted lines, for $x = 0.5$, and modulation indexes of 5 and 10. The output that would be obtained with no interference is also shown as a cosine wave of unit amplitude. Figure 18 shows how the number of peaks increases when the modulation index increases to 30. The limiting curves are the same as before. The two signals have the common center frequency at $2\pi\mu t = 90$

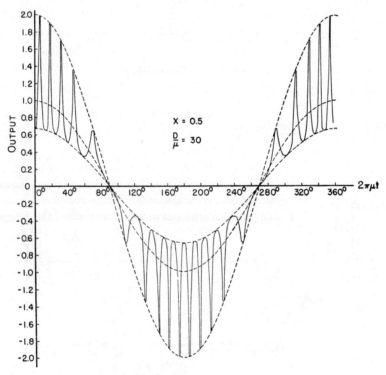

Fig. 18—Common-channel interference.

degrees and 270 degrees so they do not beat together there. As the modulated signal deviates toward the end of the swing, the frequency difference is large, and the peaks come more and more rapidly.

Effect of Detuning Interfering Signal. If the interfering signal is detuned by an amount equal to one-half the deviation of the desired signal, the effect is to move the frequency at which zero beat occurs to that point.

Figure 19 shows how the beat-note then becomes unsymmetrical. At one end of the swing the two signals have nearly the same frequency and the beats come slowly. At the other end of the swing there is a considerable frequency difference and the beats are very much more rapid. The peaks and dips are limited by the two curves:

$$\text{Envelope} = \frac{D}{1+x} \cos 2\pi\mu t + \frac{\alpha}{2\pi} \frac{x}{x+1} \tag{12}$$

$$\text{and} \qquad \frac{D}{1-x} \cos 2\pi\mu t + \frac{\alpha}{2\pi} \frac{x}{x-1} \tag{13}$$

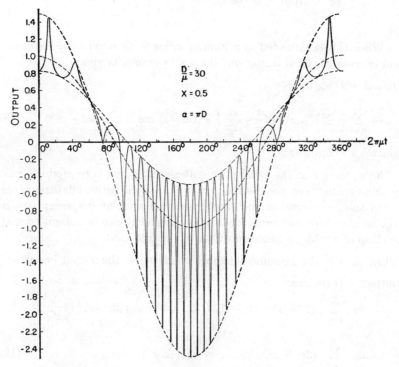

Fig. 19—Common-channel interference, interfering signal detuned.

Fourier-Series Analysis of Distorted Output. If the desired frequency-modulated signal is:

$$e_1 = E_1 \sin \left(\omega t + \frac{D}{\mu} \sin 2\pi\mu t \right) \tag{14}$$

and the interference is an unmodulated r-f carrier of angular frequency

$\omega + \alpha$, and phase angle θ, or:

$$e_2 = E_2 \sin \{(\omega + \alpha)t + \theta\} \qquad (15)$$

then, as shown in Appendix II, the envelope of the resultant carrier is given by:

$$\text{Envelope} = E_1 \sqrt{1 + x^2 + 2x \cos \beta} \qquad (16)$$

where:

$$\beta = \frac{D}{\mu} \sin 2\pi\mu t - \alpha t - \theta$$

and the audio output is given by:

$$\text{Output} = D \cos 2\pi\mu t - \frac{D \cos 2\pi\mu t - \alpha/2\pi}{\dfrac{\cos \beta + 1/x}{\cos \beta + x} + 1} \qquad (17)$$

When this is expanded in a Fourier series to determine the harmonic and cross-modulation distortion, the audio output is given by:

Output $= D \cos 2\pi\mu t$

$$+ \sum_{n=1}^{\infty} \sum_{r=-\infty}^{\infty} (-x)^n \left\{ \frac{\mu r}{n} - \frac{\alpha}{2\pi} \right\} J_r(nD/\mu) \cos (r\varepsilon - n\alpha t - n\theta) \qquad (18)$$

where $\varepsilon = 2\pi\mu t$, and $x < 1$.

This shows that the effect of the interfering signal is to produce cross modulation between the desired signal modulated with audio frequency μ and the interfering unmodulated carrier of angular frequency $\omega + \alpha$. The amplitude of each cross-modulation frequency can be computed with the help of a table of Bessel functions of the first kind.

When $\alpha = 0$, (i.e., common-channel interference) the output becomes:

Output $= D \cos 2\pi\mu t$

$$+ 2\mu \sum_{r=1}^{\infty} (2r - 1) C(2r - 1, D/\mu; x, \theta) \cos \{(2r - 1) (2\pi\mu t)\}$$

$$+ 2\mu \sum_{r=1}^{\infty} (2r) S(2r, D/\mu; x, \theta) \sin \{(2r) (2\pi\mu t)\} \qquad (19)$$

where the C- and S-functions are defined as follows:

$$C(m, n; x, \theta) = \sum_{s=1}^{\infty} \frac{(-x)^s}{s} J_m(sn) \cos s\theta \qquad (20)$$

$$S(m, n; x, \theta) = \sum_{s=1}^{\infty} \frac{(-x)^s}{s} J_m(sn) \sin s\theta \qquad (21)$$

$$x^2 \leqq 1$$

To find the amplitudes of the various harmonics produced during common-channel interference, compute the value of the desired $C-$ or $S-$ function from equations 20 and 21, and multiply by the proper factor, which is shown by the above equation 19 for the audio output. A special table of Bessel functions has been prepared for this purpose.[17]

The effect of a de-emphasis network following the discriminator, and of a low-pass audio filter, can be determined by computing the amplitude of each harmonic that falls within the working range, correcting each one for amplitude and phase changes in the audio amplifier and filters, and then recombining them by superposition.

If the signal-noise ratio is defined as the desired audio output with no interfering carrier present, divided by the peak noise (i.e., the maximum departure from the desired audio output when no interference is present), then, as shown by Figures 17 and 18, the signal-noise ratio is independent of the modulation index, but depends only on the ratio of the two voltages, x. This assumes a perfect limiter, adequate band width in the amplifiers and discriminator, and linear-phase-shift circuits.

If a de-emphasis network and a low-pass audio filter are used, many of the harmonics will be attentuated or removed, and the nonlinear phase shift will prevent the remaining harmonics from coming into phase all at the same time. The peaks of noise are therefore reduced considerably. When the modulation index, D/μ, is large, the noise beat-note peaks come very rapidly. This means that the harmonics will be of high order and they will be reduced or removed by the audio selectivity. This accounts for the observed noise reduction with wide-band frequency modulation and shows that it is very important to use a de-emphasis network and low-pass filter.

Common-Channel Interference, Both Signals Modulated. The preceding cases have described the interference produced by an unmodulated carrier on the same channel as the desired signal, and the effect of detuning the interfering carrier. This section is a discussion of the case when both the desired and undesired signals are modulated sinusoidally, and of the resultant distortion, which is even more complicated.

In order to illustrate this form of interference, assume the following conditions:

$D_1/\mu_1 = 10$, $D_2/\mu_2 = 5$, $D_1 = 4D_2$, $\mu_1 = 2\mu_2$, $x = E_2/E_1$
For example, $D_1 = 60$ kc, $\mu_1 = 6$ kc, $D_2 = 15$ kc, $\mu_2 = 3$ kc, $x = 0.5$ and 0.9 could be one set of numerical values.

---

[17] Murlan S. Corrington and William Miehle, "Tables of Bessel Functions $J_n(x)$ for Large Arguments," *Jour. Math. Phys.*, Vol. 24, No. 1, pp. 30–50; Feb., 1945.

The beat-note envelope produced in this case is shown by Figure 20. The characteristic peaks and holes in the resultant carrier amplitude are present, but some of them are modified in shape because the two audio frequencies are present simultaneously.

Near 40 degrees and again near 130 degrees the two voltages start to go out-of-phase, but the two vectors then begin to reverse themselves and only a small decrease in amplitude occurs.

If this signal is sent through a receiver with a perfect limiter and linear discriminator, the resultant audio output will be as shown by Figure 21. Two cycles of the desired signal are shown as a dotted curve. This corresponds to one cycle of the undesired signal. As x increases toward one, the beat-note interference increases in amplitude until in the limit as $x \to 1$, the pulses become very narrow and long. If x becomes greater than one,

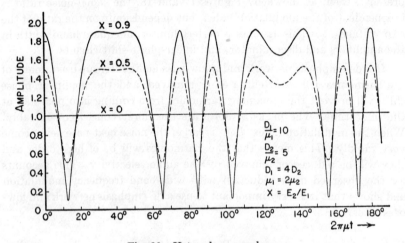

Fig. 20—Heterodyne envelope.

the polarity of the pulses is reversed (as shown by Figure 22) and this undesired signal gains control. When x becomes very large, only the undesired signal is received, as shown by the dotted cosine wave of unit amplitude.

The equations for the envelope and the beat-note interference are derived in Appendix III. If D_1 and D_2 are the two deviations and μ_1 and μ_2 are the corresponding audio frequencies, the envelope of the carrier is:

$$\text{Envelope} = E_1 \sqrt{1 + x^2 + 2x \cos \{D_1/\mu_1 \sin 2\pi\mu_1 t - D_2/\mu_2 \sin 2\pi\mu_2 t\}}$$

(22)

and the audio output from a receiver with limiter and balanced discriminator is:

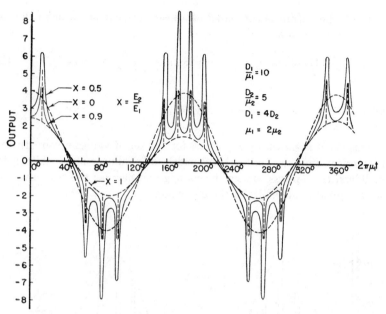

Fig. 21—Beat-note interference, $x < 1$.

$$\text{Output} \propto D_1 \cos 2\pi\mu_1 t - \frac{D_1 \cos 2\pi\mu_1 t - D_2 \cos 2\pi\mu_2 t}{\dfrac{\cos \theta + 1/x}{\cos \theta + x} + 1} \tag{23}$$

where

$$\theta = \frac{D_1}{\mu_1} \sin 2\pi\mu_1 t - \frac{D_2}{\mu_2} \sin 2\pi\mu_2 t. \tag{24}$$

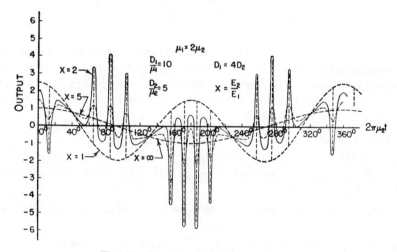

Fig. 22—Beat-note interference, $x > 1$.

The audio output is composed of a beat-note pattern which is limited by the two envelopes:

$$\text{Envelope} = \frac{D_1}{1+x} \cos 2\pi\mu_1 t + \frac{D_2 x}{1+x} \cos 2\pi\mu_2 t \qquad (25)$$

and

$$\frac{D_1}{1-x} \cos 2\pi\mu_1 t + \frac{D_2 x}{x-1} \cos 2\pi\mu_2 t \qquad (26)$$

This effect is shown by Figure 23 for the set of values given. In case of imperfect limiting, limited band width, or nonlinear phase shift in the amplifiers, these peaks will not be so long and narrow; the two envelopes shown represent the limits of the distortion.

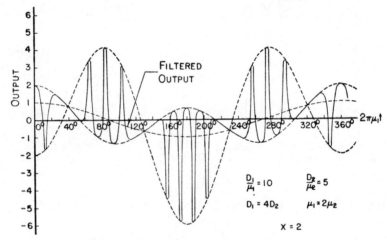

Fig. 23—Envelope of beat-note pattern.

The effect of low-pass filters or other audio selectivity can be determined from a study of the harmonic content of the distortion. As shown in Appendix III, the audio output can be expressed as a Fourier series which gives the cross modulation terms produced and their amplitudes.

Thus: Output $\propto D_1 \cos 2\pi\mu_1 t$

$$+ \sum_{r=-\infty}^{\infty} \sum_{s=-\infty}^{\infty} (r\mu_1 - s\mu_2)\, C(r, D_1/\mu_1; s, D_2/\mu_2: x, 0) \cos(r\alpha - s\beta) \qquad (27)$$

where $\alpha = 2\pi\mu_1 t$, $\beta = 2\pi\mu_2 t$ and the generalized C-function is defined as:

$$C(k, l; m, n: x, \theta) = \sum_{s=1}^{\infty} \frac{(-x)^s}{s} J_k(sl)\, J_m(sn) \cos s\theta. \qquad (28)$$

The amplitude of any desired combination tone can be determined by choosing the appropriate values of r and s and by computing the desired C-function. Since the C-function cannot exceed unity for a given combination tone, it is evident that if $D_1 \gg \mu_1$ or μ_2 the distortion will be reduced with increasing modulation index.

CONCLUSIONS

Frequency-modulated radio broadcasting offers the advantage of improved noise reduction when compared with the usual amplitude-modulation systems. There is less interference between stations operating on the same frequency than for the corresponding case of amplitude modulation, and less power is required to cover a given area.

A difficulty arose occasionally in the 42–50 megacycle frequency modulation band because long-distance transmission could be observed from frequency modulation broadcast stations during hot weather or before a storm. It sometimes happened that such an interfering station became stronger than a desired station in the same channel for short intervals. When this happened there was a small amount of noise and the programs suddenly were interchanged. This change often lasted for several seconds but sometimes was limited to a word or two or a few notes of music. If the proposed new frequency modulation stations are all completed, this interference may occur again. When the interfering station has nearly the same carrier frequency as the desired station this effect is called common-channel interference. If the two carrier frequencies are separated by the width of one channel the result is called adjacent-channel interference.

The simplest case of frequency modulation interference occurs when two modulated carriers, having nearly the same frequency, beat together to produce a resultant signal. As the two voltages alternately reinforce and cancel each other, the result is a heterodyne envelope consisting of a series of broad peaks and sharp dips. Each time the two interfering voltages cancel each other to produce a hole in the envelope, there is a rapid phase shift of the resultant voltage. Since the audio output from a frequency-modulation receiver is proportional to the rate of change of the phase of this resultant, the rapid phase shift produces a distorted audio output, which becomes more and more like an impulse as the interfering carrier voltage becomes nearly equal to the desired carrier voltage.

When the two amplitudes of the interfering voltages are kept constant but the frequency of one is changed, the result is common- or adjacent-channel interference depending upon what range of frequencies the swings of the modulated carrier cover. The beat-note produced by this interference consists of a series of sharp peaks and dips of noise and is super-

189

imposed on the desired audio output. When the modulation index is increased, these peaks occur more and more rapidly, and the harmonics produced are redistributed to higher and higher orders. If the receiver has sufficient band width, a perfect limiter, and a wide-band audio system, the signal-noise ratio does not depend on the modulation index, but is determined solely by the ratio of the desired signal voltage to the undesired signal voltage.

Formulas are given for computing the amplitudes of the harmonics and cross-modulation frequencies produced by the interference. The effect of a de-emphasis network following the discriminator, of a low-pass audio filter, and of nonlinear phase-shift can be determined by computing the amplitude of each harmonic that falls within the working range. Each such harmonic is then corrected for amplitude and phase changes in the audio amplifier and filters, and they are then recombined by superposition to obtain the filtered audio output.

When the modulation index is large, the beat-notes of the noise come very rapidly, and since this means that the harmonics are then of high order, most of the distortion will be removed by the audio selectivity. This accounts for the observed noise reduction with wide-band frequency modulation and shows that it is very important that a de-emphasis network and low-pass filter be used to obtain maximum performance. In order to obtain the maximum signal-noise ratio, it is necessary to use some means for removing the variations in the amplitude of the resultant signal so that the discriminator responds to the variations in the instantaneous frequency, but is not affected by amplitude variations.

* * *

APPENDIX I.

ANALYSIS OF FUNDAMENTAL CASE

Let the two interfering signals be $e_1 \sin \omega t$ and $e_2 \sin (\omega + 2\pi\mu)t$
The resultant voltage is then:

$$e_1 \sin \omega t + e_2 \sin (\omega + 2\pi\mu)t$$
$$= e_1 \sqrt{1 + x^2 + 2x \cos 2\pi\mu t}\, \sin (\omega t + \varphi) \tag{29}$$

where $e_2/e_1 = x$ and $\tan \varphi = \dfrac{x \sin 2\pi\mu t}{1 + x \cos 2\pi\mu t}$.

The instantaneous frequency becomes:

$$f = \frac{1}{2\pi} \frac{d}{dt} (\omega t + \varphi) = \frac{\omega}{2\pi} + \frac{1}{2\pi} \frac{d}{dt} \tan^{-1} \frac{x \sin 2\pi\mu t}{1 + x \cos 2\pi\mu t}$$

$$= \frac{\omega}{2\pi} + \mu \frac{x \cos 2\pi\mu t + x^2}{1 + x^2 + 2x \cos 2\pi\mu t}$$

$$= \frac{\omega}{2\pi} + \frac{\mu}{\dfrac{\cos 2\pi\mu t + 1/x}{\cos 2\pi\mu t + x} + 1} \tag{30}$$

This is valid for all values of x.

For a balanced linear discriminator, the audio output is proportional to:

$$\text{Output} \propto \frac{\mu}{\dfrac{\cos 2\pi\mu t + 1/x}{\cos 2\pi\mu t + x} + 1} \tag{31}$$

When $x \ll 1$, this is, approximately,

$$\text{Output} \propto \mu x \cos 2\pi\mu t \tag{32}$$

The instantaneous frequency can be written:

$$f = \frac{\omega}{2\pi} + \mu - \mu \frac{\dfrac{1}{x} \cos 2\pi\mu t + \dfrac{1}{x^2}}{1 + \dfrac{1}{x^2} + \dfrac{2}{x} \cos 2\pi\mu t} \tag{33}$$

This means that as x goes from less than one to greater than one (i.e., if x is changed to $1/x$) there is an apparent change in frequency equal to μ and a reversal in polarity of the modulation. This means that e_2 becomes stronger than e_1 and takes control.

Average Voltage of Rectified Envelope

The average voltage of the carrier envelope is:

$$\text{Average voltage} = \frac{1}{\pi} \int_0^\pi e_1 \sqrt{1 + x^2 + 2x \cos \theta} \, d\theta$$

$$= \frac{2(1 + x) e_1}{\pi} \int_0^{\pi/2} \sqrt{1 - \frac{4x}{(1 + x)^2} \sin^2 \alpha} \, d\alpha$$

$$= \frac{2(1 + x)e_1}{\pi} E\left\{\frac{2\sqrt{x}}{1 + x}\right\} \tag{34}$$

where $E\left\{\dfrac{2\sqrt{x}}{1+x}\right\}$ is a complete elliptic integral of the second kind with

modulus $\dfrac{2\sqrt{x}}{1+x}$.

Root-Mean-Square Voltage of Rectified Envelope

The Root-Mean-Square voltage of the rectified carrier envelope is:

$$\text{Root-Mean-Square voltage} = e_1 \sqrt{\frac{1}{\pi} \int_0^\pi (1 + x^2 + 2x \cos \theta)\, d\theta}$$

$$= e_1 \sqrt{\frac{1}{\pi}\left[(1+x^2)\theta + 2x \sin \theta\right]_0^\pi}$$

$$= e_1 \sqrt{1+x^2} \tag{35}$$

Fourier-Series Analysis of Envelope

The envelope of the carrier is given by:

$$\text{Envelope} = e_1 \sqrt{1 + x^2 + 2x \cos 2\pi\mu t} \qquad \text{where } x \leqq 1. \tag{36}$$

Consider the expression:

$$\sqrt{1 + x^2 + 2x \cos \beta} = (1 + xe^{i\beta})^{\frac{1}{2}}(1 + xe^{-i\beta})^{\frac{1}{2}}$$

$$= \left\{1 + \frac{1}{2} xe^{i\beta} - \frac{1(1)}{2(4)} x^2 e^{2i\beta} + \frac{1(1)\ (3)}{2(4)\ (6)} x^3 e^{3i\beta} - \frac{1(1)\ (3)\ (5)}{2(4)\ (6)\ (8)} x^4 e^{4i\beta} + \dots\right\}$$

$$\times \left\{1 + \frac{1}{2} xe^{-i\beta} - \frac{1(1)}{2(4)} x^2 e^{-2i\beta} + \frac{1(1)\ (3)}{2(4)\ (6)} x^3 e^{-3i\beta} - \frac{1(1)\ (3)\ (5)}{2(4)\ (6)\ (8)} x^4 e^{-4i\beta} + \dots\right\}$$

$$\tag{37}$$

by the usual binomial series expansion.[18]

Multiply these factors together, term by term, then:

$$\sqrt{1 + x^2 + 2x \cos \beta} = a_0 + a_1 \cos \beta + a_2 \cos 2\beta + \dots \tag{38}$$

where:

$$a_0 = 1 + \frac{x^2}{4} + \frac{x^4}{64} + \frac{x^6}{256} + \frac{25x^8}{16384} + \dots \tag{39}$$

$$a_1 = x \left\{1 - \frac{x^2}{8} - \frac{x^4}{64} - \frac{5x^6}{1024} - \frac{35x^8}{16384} - \dots\right\} \tag{40}$$

---

[18] Edwin P. Adams, SMITHSONIAN MATHEMATICAL FORMULAE AND TABLES OF ELLIPTIC FUNCTIONS, Smithsonian Institution, Washington, D. C., 1939, p. 117.

192

$$a_2 = -\frac{x^2}{4}\left\{1 - \frac{x^2}{4} - \frac{5x^4}{128} - \frac{7x^6}{512} - \frac{105x^8}{16384} - \cdots\right\} \tag{41}$$

$$a_3 = \frac{x^3}{8}\left\{1 - \frac{5x^2}{16} - \frac{7x^4}{128} - \frac{21x^6}{1024} - \frac{165x^8}{16384} - \cdots\right\} \tag{42}$$

$$a_4 = -\frac{5x^4}{64}\left\{1 - \frac{7x^2}{20} - \frac{21x^4}{320} - \frac{33x^6}{1280} - \frac{429x^8}{32768} - \cdots\right\} \tag{43}$$

$$a_n = 2(-1)^n\left\{\frac{1(3)\cdots(2n-1)}{n!}\right\}\frac{x^n}{2^n}\left\{\frac{-1}{2n-1} + \frac{1}{1(n+1)}\frac{x^2}{2^2}\right.$$

$$\left. + \sum_{k=2}^{\infty}\frac{1(3)\cdots(2k-3)}{k!\,2^{2k}}\frac{(2n+1)(2n+3)\cdots(2n+2k-3)}{(n+1)(n+2)\cdots(n+k)}x^{2k}\right\} \tag{44}$$

This expression for a_n was previously obtained by Vigoureux[19] and Moullin[20]

In the limit as $x \to 1$:

$$\text{Envelope} = \sqrt{2}\,e_1\sqrt{1 + \cos 2\pi\mu t} = 2e_1\,|\cos \pi\mu t|$$

$$= \frac{4e_1}{\pi}\left\{1 + \frac{2}{3}\cos\theta - \frac{2}{15}\cos 2\theta\right.$$

$$\left. + \frac{2}{35}\cos 3\theta - \frac{2}{63}\cos 4\theta - \cdots\right\}$$

$$= \frac{4e_1}{\pi}\left\{1 - 2\sum_{n=1}^{\infty}\frac{(-1)^n\cos n\theta}{(2n)^2 - 1}\right\}$$

where $\theta = 2\pi\mu t$. \hfill (45)

Calculation of Average Value of Instantaneous Frequency

Consider the integral:

$$I = \frac{\mu}{\pi}\int_0^{\pi}\frac{x^2 + x\cos\varepsilon}{1 + x^2 + 2x\cos\varepsilon}\,d\varepsilon \tag{46}$$

Make the transformation: $\cos\varepsilon = \dfrac{1 - t^2}{1 + t^2}$, $d\varepsilon = \dfrac{2dt}{1 + t^2}$

[19] F. M. Colebrook, "A Note on the Frequency Analysis of the Heterodyne Envelope. Its Relation to Problems of Interference." *Wireless Engineer & Experimental Wireless*, Vol. 9, p. 200, April, 1932.

[20] E. B. Moullin, "The Detection by a Straight Line Rectifier of Modulated and Heterodyne Signals," *Wireless Engineer & Experimental Wireless*, Vol. 9, pp. 378–383; July, 1932.

Then.

$$I = \frac{2x\,\mu}{\pi} \int_0^\infty \frac{(1+x) - (1-x)t^2}{(1+x)^2 + (1-x)^2 t^2} \frac{dt}{1+t^2}$$

$$= \frac{(x^2 - 1)\mu}{\pi} \int_0^\infty \frac{dt}{(1+x)^2 + (1-x)^2 t^2} + \frac{\mu}{\pi} \int_0^\infty \frac{dt}{1+t^2}$$

$$= \frac{(x^2 - 1)\mu}{\pi} \left\{ \frac{\pi}{2(1-x^2)} \right\} + \frac{\mu}{\pi}\left(\frac{\pi}{2}\right) = 0 \text{ when } x < 1$$

$$= \mu \text{ when } x > 1. \qquad (47)$$

The average value of the instantaneous frequency therefore equals zero when $x < 1$ and is proportional to μ when $x > 1$.

Calculation of the Root-Mean-Square Value of Instantaneous Frequency

Consider the integral:

$$I = \frac{1}{\pi} \int_0^\pi \left\{ \frac{x^2 + x\cos\varepsilon}{1 + x^2 + 2x\cos\varepsilon} \right\}^2 d\varepsilon \qquad (48)$$

Make the transformation: $\cos\varepsilon = \dfrac{1 - t^2}{1 + t^2},\ d\varepsilon = \dfrac{2\,dt}{1 + t^2}.$

Then:

$$I = \frac{2x^2}{\pi} \int_0^\infty \left\{ \frac{(1+x) - (1-x)t^2}{(1+x)^2 + (1-x)^2 t^2} \right\}^2 \frac{dt}{1+t^2}$$

$$= \frac{-2x(1+x)^2}{\pi} \int_0^\infty \frac{dt}{\{(1+x)^2 + (1-x)^2 t^2\}^2}$$

$$+ \frac{(3x-1)(1+x)}{2\pi} \int_0^\infty \frac{dt}{\{(1+x)^2 + (1-x)^2 t^2\}} + \frac{1}{2\pi} \int_0^\infty \frac{dt}{1+t^2} \qquad (49)$$

Consider the integral:

$$\int_0^\infty \frac{dt}{\{a^2 + b^2 t^2\}^2} = \frac{1}{2a^2}\left[\frac{t}{a^2 + b^2 t^2}\right]_0^\infty + \frac{1}{2a^2}\int_0^\infty \frac{dt}{a^2 + b^2 t^2}$$

$$= \frac{1}{2a^2}\int_0^\infty \frac{dt}{a^2 + b^2 t^2} = \frac{1}{2a^3 b}\left[\tan^{-1}\frac{b}{a}\,t\right]_0^\infty = \frac{\pi}{4a^3 b}\text{when } x < 1 \quad (50)$$

$$= -\frac{\pi}{4a^3 b}\text{ when } x > 1. \quad (51)$$

Therefore I becomes:

$$I = \frac{-\,2x(1+x)^2}{\pi}\left\{\frac{\pi}{4(1+x)^3\,(1-x)}\right\}$$

$$+\ \frac{(3x-1)\,(1+x)}{2\pi}\left\{\frac{\pi}{2(1+x)\,(1-x)}\right\}+\frac{1}{2\pi}\left\{\frac{\pi}{2}\right\}$$

$$=\frac{-x}{2(1-x^2)}+\frac{3x-1}{4(1-x)}+\frac{1}{4}=\frac{x^2}{2(1-x^2)}\ \text{when}\ x<1, \qquad (52)$$

$$=\frac{1-2x^2}{2(1-x^2)}\ \text{when}\ x>1. \qquad (53)$$

The root-mean-square voltage is proportional to:

$$\mu\sqrt{I}=\frac{x\mu}{\sqrt{2(1-x^2)}}\ \text{when}\ x<1 \qquad (54$$

$$=\sqrt{\frac{2x^2-1}{2(x^2-1)}}\ \mu\ \text{when}\ x>1. \qquad (55)$$

Calculation of the Area of One Cycle of the Instantaneous Frequency

The area bounded by one cycle of the variation of the instantaneous frequency and a line one-half unit above the time axis, as shown by Figure 7, will now be computed. From equation 2 the instantaneous frequency is given by:

$$f=\frac{\omega}{2\pi}+\frac{1}{2\pi}\frac{d}{dt}\,\tan^{-1}\frac{x\sin 2\pi\mu t}{1+x\cos 2\pi\mu t} \qquad (56)$$

$$\text{Area}=2\int_0^\pi\left\{\frac{1}{2\pi}\frac{d}{dt}\,\tan^{-1}\frac{x\sin 2\pi\mu t}{1+x\cos\ 2\pi\mu t}-\frac{1}{2}\right\}d(2\pi\mu t)$$

$$=2\mu\int_0^\pi\left\{\frac{d}{d\theta}\tan^{-1}\frac{x\sin\theta}{1+x\cos\theta}-\frac{1}{2}\right\}d\theta$$

$$=2\mu\left[\tan^{-1}\frac{x\sin\theta}{1+x\cos\theta}-\frac{\theta}{2}\right]_0^\pi=-\pi\mu. \qquad (57)$$

for all values of x.

Fourier-Series Analysis of Instantaneous Frequency

The audio output is proportional to:

$$\mu\frac{x\cos 2\pi\mu t+x^2}{1+x^2+2x\cos 2\pi\mu t}=\frac{1}{2\pi}\frac{d}{dt}\tan^{-1}\frac{x\sin 2\pi\mu t}{1+x\cos 2\pi\mu t} \qquad (58)$$

195

Let $2\pi\mu t = \beta$ and $\tan\alpha = \dfrac{x \sin\beta}{1 + x \cos\beta}$

Then $k \sin\alpha = x \sin\beta$ (59)

$k \cos\alpha = 1 + x \cos\beta$ (60)

where: $k = \sqrt{1 + x^2 + 2x \cos\beta}$ (61)

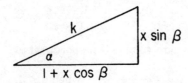

Fig. 24—Determination of k.

Multiply equation 59 by i and add equation 60.

Then: $1 + x \cos\beta + ix \sin\beta = k(\cos\alpha + i \sin\alpha)$ (62)

or: $1 + x e^{i\beta} = k e^{i\alpha}$.   Take logarithms of both sides. Then:

$\log(1 + x e^{i\beta}) = \log k + i\alpha.$

Since:

$$\log(1 + x) = x - \frac{x^2}{2} + \frac{x^3}{3} - \frac{x^4}{4} + \ldots \qquad -1 < x \leqq 1$$

$$\log(1 + x e^{i\beta}) = xe^{i\beta} - \frac{x^2}{2} e^{2i\beta} + \frac{x^3}{3} e^{3i\beta} - \ldots$$

so:

$$\log k + i\alpha = x(\cos\beta + i \sin\beta) - \frac{x^2}{2}(\cos 2\beta + i \sin 2\beta)$$

$$+ \frac{x^3}{3}(\cos 3\beta + i \sin 3\beta) - \ldots \tag{63}$$

Equate imaginary terms:

$$\alpha = x \sin\beta - \frac{x^2}{2} \sin 2\beta + \frac{x^3}{3} \sin 3\beta - \ldots \tag{64}$$

Differentiate:

$$\frac{1}{2\pi}\frac{d\alpha}{dt} = \mu\,(x \cos\beta - x^2 \cos 2\beta + x^3 \cos 3\beta - \ldots) \tag{65}$$

196

The audio output is therefore proportional to:

$$\text{Output} \propto - \mu \sum_{n=1}^{\infty} (-x)^n \cos n\beta \qquad -1 < x \leq 1 \tag{66}$$

When $\beta = 0$,
$$\text{Output} \propto - \mu \sum_{n=1}^{\infty} (-x)^n = \frac{\mu x}{x+1} \tag{67}$$

When $\beta = \pi$, $\text{Output} \propto - \mu \sum_{n=1}^{\infty} x^n = \frac{\mu x}{x-1} \tag{68}$

Effect of Limited Band Width

To show the effect of a limited band width, consider the geometrical progression:

$$S = - \mu \sum_{n=1}^{n} x^n \tag{69}$$

By ordinary long division:

$$\frac{p^n - 1}{p - 1} = 1 + p + p^2 + \ldots + p^{n-1} \tag{70}$$

so,

$$S = - \mu x \frac{x^n - 1}{x - 1} \tag{71}$$

The ratio of the partial sum to the output at $\beta = \pi$ equals $1 - x^n$.

APPENDIX II.

COMMON- AND ADJACENT-CHANNEL INTERFERENCE

In order to show the effect of common- and adjacent-channel interference, let the desired frequency-modulated signal be:

$$e_1 = E_1 \sin \left(\omega t + \frac{D}{\mu} \sin 2\pi\mu t \right) \tag{72}$$

and let the interference be an unmodulated radio-frequency carrier at angular frequency $\omega + \alpha$, and phase angle θ, or

$$e_2 = E_2 \sin \{ (\omega + \alpha)t + \theta \} \tag{73}$$

197

Then:

$$e_1 + e_2 = E_1 \sqrt{1 + x^2 + 2x \cos \beta} \; \sin \left\{ \omega t + \frac{D}{\mu} \sin 2\pi\mu t - \varphi \right\} \quad (74)$$

where $\quad x = E_2/E_1, \; \beta = \dfrac{D}{\mu} \sin 2\pi\mu t - \alpha t - \theta$

and $\quad \tan \varphi = \dfrac{x \sin \beta}{1 + x \cos \beta}$

The instantaneous frequency becomes:

$$\frac{\omega}{2\pi} + D \cos 2\pi\mu t - \frac{1}{2\pi} \frac{d}{dt} \tan^{-1} \frac{x \sin \beta}{1 + x \cos \beta}$$

$$= \frac{\omega}{2\pi} + D \cos 2\pi\mu t - \frac{x \cos \beta + x^2}{1 + x^2 + 2x \cos \beta} \left\{ D \cos 2\pi\mu t - \frac{\alpha}{2\pi} \right\}$$

$$= \frac{\omega}{2\pi} + D \cos 2\pi\mu t - \frac{D \cos 2\pi\mu t - \alpha/2\pi}{\dfrac{\cos \beta + 1/x}{\cos \beta + x} + 1} \quad (75)$$

Envelope of Beatnote Pattern

The beatnote produced in the output of a receiver with a perfect limiter is given by:

$$\text{Output} = D \cos 2\pi\mu t - \frac{D \cos 2\pi\mu t - \alpha/2\pi}{\dfrac{\cos \beta + 1/x}{\cos \beta + x} + 1} \quad (76)$$

where: $\quad \beta = \dfrac{D}{\mu} \sin 2\pi\mu t - \alpha t - \theta$. The two envelopes of the maxima and minima of the beat-note pattern are obtained by setting $\beta = 2n\pi$ or $(2n + 1)\pi$ where n is an integer. This gives the two envelopes:

$$\text{Envelope} = \frac{D}{1 + x} \cos 2\pi\mu t + \frac{\alpha}{2\pi} \frac{x}{x + 1} \quad (77)$$

and:

$$\frac{D}{1 - x} \cos 2\pi\mu t + \frac{\alpha}{2\pi} \frac{x}{x - 1} \quad (78)$$

Fourier-Series Analysis of Instantaneous Frequency

In accordance with the analysis of Appendix I, equation 65:

$$\frac{d}{dt} \tan^{-1} \frac{x \sin \beta}{1 + x \cos \beta} = -\sum_{n=1}^{\infty} (-x)^n \cos n\beta \, \frac{d\beta}{dt} \tag{79}$$

When: $\beta = \dfrac{D}{\mu} \sin 2\pi\mu t - \alpha t - \theta$ and $\varepsilon = 2\pi\mu t$

the instantaneous frequency is:

$$f = \frac{\omega}{2\pi} + D \cos \varepsilon + \sum_{n=1}^{\infty} (-x)^n \cos n \left\{ \frac{D}{\mu} \sin \varepsilon - \alpha t - \theta \right\} \left\{ D \cos \varepsilon - \frac{\alpha}{2\pi} \right\} \tag{80}$$

Let: $\gamma = D/\mu$, then

$$f = \frac{\omega}{2\pi} + D \cos \varepsilon$$

$$+ \frac{1}{4} \sum_{n=1}^{\infty} (-x)^n \left\{ D(e^{i\varepsilon} + e^{-i\varepsilon}) - \frac{\alpha}{\pi} \right\} \left\{ e^{in\gamma \sin \varepsilon - in\alpha t - in\theta} + e^{-in\gamma \sin \varepsilon + in\alpha t + in\theta} \right\}$$

$$= \frac{\omega}{2\pi} + D \cos \varepsilon$$

$$+ \frac{1}{4} \sum_{n=1}^{\infty} (-x)^n \left\{ D \, e^{i(\varepsilon - n\alpha t - n\theta)} \, e^{in\gamma \sin \varepsilon} + D \, e^{i(\varepsilon + n\alpha t + n\theta)} \, e^{-in\gamma \sin \varepsilon} \right.$$

$$+ D \, e^{-i(\varepsilon + n\alpha t + n\theta)} \, e^{in\gamma \sin \varepsilon} + D \, e^{-i(\varepsilon - n\alpha t - n\theta)} \, e^{-in\gamma \sin \varepsilon}$$

$$\left. - \frac{\alpha}{\pi} \, e^{-i(n\alpha t + n\theta)} \, e^{in\gamma \sin \varepsilon} - \frac{\alpha}{\pi} \, e^{i(n\alpha t + n\theta)} \, e^{-in\gamma \sin \varepsilon} \right\} \tag{81}$$

Using the identities:

$$e^{ix \sin \varepsilon} = \sum_{k=-\infty}^{\infty} J_k(x) \, e^{ik\varepsilon} \tag{82}$$

and $J_k(-x) = J_{-k}(x)$ where $J_k(x)$ is a Bessel function of the first kind of order k and argument x, the instantaneous frequency becomes:

$$f = \frac{\omega}{2\pi} + D \cos \varepsilon$$

$$+ \frac{1}{4} \sum_{n=1}^{\infty} (-x)^n \left\{ D \, e^{i(\varepsilon - n\alpha t - n\theta)} \sum_{k=-\infty}^{\infty} J_k(n\gamma) \, e^{ik\varepsilon} \right.$$

$$+ D\, e^{i(\epsilon + n\alpha t + n\theta)} \sum_{k=-\infty}^{\infty} J_{-k}(n\gamma)\, e^{ik\epsilon}$$

$$+ D\, e^{-i(\epsilon + n\alpha t + n\theta)} \sum_{k=-\infty}^{\infty} J_{k}(n\gamma)\, e^{ik\epsilon}$$

$$+ D\, e^{-i(\epsilon - n\alpha t - n\theta)} \sum_{k=-\infty}^{\infty} J_{-k}(n\gamma)\, e^{ik\epsilon}$$

$$- \frac{\alpha}{\pi}\, e^{-i(n\alpha t + n\theta)} \sum_{k=-\infty}^{\infty} J_{k}(n\gamma)\, e^{ik\epsilon}$$

$$- \frac{\alpha}{\pi}\, e^{i(n\alpha t + n\theta)} \sum_{k=-\infty}^{\infty} J_{-k}(n\gamma)\, e^{ik\epsilon} \Bigg\}$$

$$= \frac{\omega}{2\pi} + D \cos \epsilon$$

$$+ \frac{1}{4} \sum_{n=1}^{\infty} (-x)^n \left\{ D \sum_{k=-\infty}^{\infty} J_{k}(n\gamma)\, e^{i\{(k+1)\epsilon - n\alpha t - n\theta\}} \right.$$

$$+ D \sum_{k=-\infty}^{\infty} J_{-k}(n\gamma)\, e^{i\{(k+1)\epsilon + n\alpha t + n\theta\}}$$

$$+ D \sum_{k=-\infty}^{\infty} J_{k}(n\gamma)\, e^{i\{(b-1)\epsilon - n\alpha t - n\theta\}}$$

$$+ D \sum_{k=-\infty}^{\infty} J_{-k}(n\gamma)\, e^{i\{(k-1)\epsilon + n\alpha t + n\theta\}}$$

$$- \frac{\alpha}{\pi} \sum_{k=-\infty}^{\infty} J_{k}(n\gamma)\, e^{i\{k\epsilon - n\alpha t - n\theta\}}$$

$$\left. - \frac{\alpha}{\pi} \sum_{k=-\infty}^{\infty} J_{-k}(n\gamma)\, e^{i\{k\epsilon + n\alpha t + n\theta\}} \right\} \tag{83}$$

Make the substitutions:

$k + 1 = r$ in term 1	$k - 1 = -r$ in term 4
$k + 1 = -r$ in term 2	$k = r$ in term 5
$k - 1 = r$ in term 3	$k = -r$ in term 6

Then:

$$f = \frac{\omega}{2\pi} + D \cos \epsilon$$

$$+ \frac{1}{4} \sum_{n=1}^{\infty} (-x)^n \left\{ D \sum_{r=-\infty}^{\infty} \left[J_{r-1}(n\gamma) + J_{r+1}(n\gamma) \right] e^{i(r\varepsilon - n\alpha t - n\theta)} \right.$$

$$+ D \sum_{r=-\infty}^{\infty} \left[J_{r-1}(n\gamma) + J_{r+1}(n\gamma) \right] e^{-i(r\varepsilon - n\alpha t - n\theta)}$$

$$\left. - \frac{\alpha}{\pi} \sum_{r=-\infty}^{\infty} J_r(n\gamma) \left[e^{i(r\varepsilon - n\alpha t - n\theta)} + e^{-i(r\varepsilon - n\alpha t - n\theta)} \right] \right\} \qquad (84)$$

Apply the relation:

$$J_{r-1}(n\gamma) + J_{r+1}(n\gamma) = \frac{2r}{n\gamma} J_r(n\gamma) \qquad (85)$$

Then:

$$f = \frac{\omega}{2\pi} + D \cos \varepsilon$$

$$+ \sum_{n=1}^{\infty} (-x)^n \left\{ D \sum_{r=-\infty}^{\infty} \frac{r}{n\gamma} J_r(n\gamma) \cos (r\varepsilon - n\alpha t - n\theta) \right.$$

$$\left. - \frac{\alpha}{2\pi} \sum_{r=-\infty}^{\infty} J_r(n\gamma) \cos (r\varepsilon - n\alpha t - n\theta) \right\} \qquad (86)$$

$$= \frac{\omega}{2\pi} + D \cos 2\pi\mu t$$

$$+ \sum_{n=1}^{\infty} \sum_{r=-\infty}^{\infty} (-x)^n \left\{ \frac{\mu r}{n} - \frac{\alpha}{2\pi} \right\} J_r\left(\frac{nD}{\mu} \right) \cos (2\pi r\mu t - n\alpha t - n\theta) \qquad (87)$$

where: $x < 1$.

This shows that the effect of the interfering signal is to produce cross modulation between the desired audio signal of frequency μ and the difference angular frequency α. The amplitude of each frequency can be computed from this equation. When $\alpha = 0$, i.e., common-channel interference, this reduces to:

$$f = \frac{\omega}{2\pi} + D \cos 2\pi\mu t$$

$$+ \sum_{n=1}^{\infty} \sum_{r=-\infty}^{\infty} (-x)^n \frac{\mu r}{n} J_r\left(\frac{nD}{\mu} \right) \left\{ \cos 2\pi r\mu t \cos n\theta + \sin 2\pi r\mu t \sin n\theta \right\}$$

$$= \frac{\omega}{2\pi} + D \cos 2\pi\mu t$$

$$+ 2\mu \sum_{r=1}^{\infty} (2r - 1)\ C\left(2r - 1, \frac{D}{\mu}; x, \theta\right) \cos\left\{\ (2r - 1)\ (2\pi\mu t)\right\}$$

$$+ 2\mu \sum_{r=1}^{\infty} (2r)\ S\left(2r, \frac{D}{\mu}; x, \theta\right) \sin\left\{(2r)\ (2\pi\mu t)\right\} \tag{88}$$

where the C- and S- functions are defined as follows:

$$C(m, n; x, \theta) = \sum_{s=1}^{\infty} \frac{(-x)^s}{s}\ J_n(ms)\ \cos s\theta \tag{89}$$

$$S(m, n; x, \theta) = \sum_{s=1}^{\infty} \frac{(-x)^s}{s}\ J_n(ms)\ \sin s\theta \tag{90}$$

$$x^2 \leqq 1.$$

APPENDIX III.

COMMON-CHANNEL INTERFERENCE, BOTH SIGNALS MODULATED

When two frequency-modulated signals with a common carrier frequency produce interference, the effect is similar to that which occurs when one wave is not modulated. The exact relations can be obtained in the following way:

Let: $\qquad e_1 = E_1 \sin\left(\omega t + \dfrac{D_1}{\mu_1} \sin 2\pi\mu_1 t\right) \tag{91}$

and: $\qquad e_2 = E_2 \sin\left(\omega t + \dfrac{D_2}{\mu_2} \sin 2\pi\mu_2 t\right) \tag{92}$

be the two interfering waves. Then:

$$e_1 + e_2 = \sqrt{E_1{}^2 + E_2{}^2 + 2E_1 E_2 \cos \psi}\ \ \sin\left(\omega t + \frac{D_1}{\mu_1} \sin 2\pi\mu_1 t - \varphi\right) \tag{93}$$

where: $\qquad \tan \varphi = \dfrac{x \sin \psi}{1 + x \cos \psi}$

and: $\qquad \psi = \dfrac{D_1}{\mu_1} \sin 2\pi\mu_1 t - \dfrac{D_2}{\mu_2} \sin 2\pi\mu_2 t$

The instantaneous frequency becomes:

$$f = \frac{\omega}{2\pi} + D_1 \cos 2\pi\mu_1 t - \frac{D_1 \cos 2\pi\mu_1 t - D_2 \cos 2\pi\mu_2 t}{\dfrac{\cos \psi + 1/x}{\cos \psi + x} + 1} \qquad (94)$$

Envelope of Beat-note Pattern

The beat-note produced in the output of a receiver with a perfect limiter and balanced discriminator is given by:

$$\text{Output} = D_1 \cos 2\pi\mu_1 t - \frac{D_1 \cos 2\pi\mu_1 t - D_2 \cos 2\pi\mu_2 t}{\dfrac{\cos \psi + 1/x}{\cos \psi + x} + 1} \qquad (95)$$

The two envelopes of the maxima and minima of the beat-note pattern are obtained by setting $\psi = 2n\pi$ or $\psi = (2n + 1)\pi$, where n is an integer. This gives the result:

$$\text{Envelope} = \frac{D_1}{1 + x} \cos 2\pi\mu_1 t + \frac{D_2 x}{x + 1} \cos 2\pi\mu_2 t \qquad (96)$$

and:

$$\frac{D_1}{1 - x} \cos 2\pi\mu_1 t + \frac{D_2 x}{x - 1} \cos 2\pi\mu_2 t \qquad (97)$$

Fourier-Series Analysis of Instantaneous Frequency

The distortion present in the instantaneous frequency is given by

$$-\frac{d}{dt} \tan^{-1} \frac{x \sin \psi}{1 + x \cos \psi} = \sum_{n=1}^{\infty} (-x)^n \cos n\psi \frac{d\psi}{dt} \qquad (98)$$

Consider the expression:

$$\cos n\psi \frac{d\psi}{dt} = 2\pi \cos n\psi \, (D_1 \cos 2\pi\mu_1 t - D_2 \cos 2\pi\mu_2 t) \qquad (99)$$

Make the substitutions:

$$\alpha = 2\pi\mu_1 t \qquad\qquad \gamma = n \frac{D_1}{\mu_1}$$

$$\beta = 2\pi\mu_2 t \qquad\qquad \delta = n \frac{D_2}{\mu_2}$$

Then the first term becomes:

$$D_1 \cos \alpha \cos \{\gamma \sin \alpha - \delta \sin \beta\}$$

$$= \frac{D_1}{4} \{e^{i\alpha} + e^{-i\alpha}\} \{e^{i\gamma \sin \alpha} e^{-i\delta \sin \beta} + e^{-i\gamma \sin \alpha} e^{i\delta \sin \beta}\}$$

$$= \frac{D_1}{4} \{e^{i\alpha} \sum_{r=-\infty}^{\infty} J_r(\gamma) \, e^{ir\alpha} \sum_{s=-\infty}^{\infty} J_{-s}(\delta) \, e^{is\beta}$$

$$+ e^{-i\alpha} \sum_{r=-\infty}^{\infty} J_r(\gamma)\, e^{ir\alpha} \sum_{s=-\infty}^{\infty} J_{-s}(\delta)\, e^{is\beta}$$

$$+ e^{i\alpha} \sum_{r=-\infty}^{\infty} J_{-r}(\gamma)\, e^{ir\alpha} \sum_{s=-\infty}^{\infty} J_s(\delta)\, e^{is\beta}$$

$$+ e^{-i\alpha} \sum_{r=-\infty}^{\infty} J_{-r}(\gamma)\, e^{ir\alpha} \sum_{s=-\infty}^{\infty} J_s(\delta)\, e^{is\beta} \Big\}$$

$$= \frac{D_1}{4} \Big\{ \sum_{r=-\infty}^{\infty} \Big[J_r(\gamma)\, e^{i(r+1)\alpha} + J_r(\gamma)\, e^{i(r-1)\alpha} \Big] \sum_{s=-\infty}^{\infty} J_{-s}(\delta)\, e^{is\beta}$$

$$+ \sum_{r=-\infty}^{\infty} \Big[J_{-r}(\gamma)\, e^{i(r+1)\alpha} + J_{-r}(\gamma)\, e^{i(r-1)\alpha} \Big] \sum_{s=-\infty}^{\infty} J_s(\delta)\, e^{is\beta} \Big\} \qquad (100)$$

Make the following substitutions:

$r + 1 = k$ in the first expression in the first bracket
$r - 1 = k$ in the second expression in the first bracket
$r + 1 = -k$ in the first expression in the second bracket
$r - 1 = -k$ in the second expression in the second bracket

and apply the identity:

$$J_{k-1}(\gamma) + J_{k+1}(\gamma) = \frac{2k}{\gamma} J_k(\gamma). \qquad (101)$$

This gives:

$$D_1 \cos\alpha \cos\{\gamma \sin\alpha - \delta \sin\beta\}$$

$$= \frac{D_1}{4} \Big\{ \sum_{k=-\infty}^{\infty} \Big[J_{k-1}(\gamma) + J_{k+1}(\gamma) \Big] e^{ik\alpha} \sum_{s=-\infty}^{\infty} J_s(\delta)\, e^{-is\beta}$$

$$+ \sum_{k=-\infty}^{\infty} \Big[J_{k-1}(\gamma) + J_{k+1}(\gamma) \Big] e^{-ik\alpha} \sum_{s=-\infty}^{\infty} J_s(\delta)\, e^{is\beta} \Big\}$$

$$= \frac{D_1}{2} \Big\{ \sum_{k=-\infty}^{\infty} \frac{k}{\gamma} J_k(\gamma)\, e^{ik\alpha} \sum_{s=-\infty}^{\infty} J_s(\delta)\, e^{-is\beta}$$

$$+ \sum_{k=-\infty}^{\infty} \frac{k}{\gamma} J_k(\gamma)\, e^{-ik\alpha} \sum_{s=-\infty}^{\infty} J_s(\delta)\, e^{is\beta} \Big\}$$

$$= D_1 \sum_{r=-\infty}^{\infty} \sum_{s=-\infty}^{\infty} \frac{r}{\gamma} J_r(\gamma)\, J_s(\delta)\, \cos(r\alpha - s\beta) \qquad (102)$$

By the same process:

$$D_2 \cos\beta \cos\{\gamma \sin\alpha - \delta \sin\beta\}$$

$$= D_2 \sum_{r=-\infty}^{\infty} \sum_{s=-\infty}^{\infty} \frac{s}{\delta} J_r(\gamma)\, J_s(\delta)\, \cos(r\alpha - s\beta) \qquad (103)$$

These two results give the expression:

$$\cos n\psi \frac{d\psi}{dt} = 2\pi \sum_{r=-\infty}^{\infty} \sum_{s=-\infty}^{\infty} \left\{ \frac{rD_1}{\gamma} - \frac{sD_2}{\gamma} \right\} J_r(\gamma) J_s(\delta) \cos(r\alpha - s\beta)$$

$$= \frac{2\pi}{n} \sum_{r=-\infty}^{\infty} \sum_{s=-\infty}^{\infty} (r\mu_1 - s\mu_2) J_r(\gamma) J_s(\delta) \cos(r\alpha - s\beta) \tag{104}$$

The audio output from a balanced discriminator thus becomes:

Output $\propto D_1 \cos 2\pi\mu_1 t$

$$\sum_{n=1}^{\infty} \sum_{r=-\infty}^{\infty} \sum_{s=-\infty}^{\infty} \frac{(-x)^n}{n} (r\mu_1 - s\mu_2) J_r\left(\frac{nD_1}{\mu_1}\right) J_s\left(\frac{nD_2}{\mu_2}\right) \cos(r\alpha - s\beta)$$

$$= D_1 \cos 2\pi\mu_1 t + \sum_{r=-\infty}^{\infty} \sum_{s=-\infty}^{\infty} (r\mu_1 - s\mu_2) C(r, \frac{D_1}{\mu_1}; s, \frac{D_2}{\mu_2}: x, 0)$$

$$\cos(r\alpha - s\beta) \tag{105}$$

where the generalized C-function is defined as follows:

$$C(k, l; m, n: x, \theta) = \sum_{s=1}^{\infty} \frac{(-x)^s}{s} J_k(s l) J_m(sn) \cos s\theta, \tag{106}$$

$$\alpha = 2\pi\mu_1 t \qquad \text{and} \qquad \beta = 2\pi\mu_2 t.$$

PAPER NO. 7

Reprinted from *Time Series Analysis*, ed. M. Rosenblatt, pp. 395–422 (Wiley 1963)

CHAPTER 25

Noise in FM Receivers

S. O. Rice, Bell Telephone Laboratories,
Murray Hill, New Jersey

ABSTRACT

This chapter is concerned with the noise in the output of an FM receiver when the input contains both signal and Gaussian noise. For large values of ρ, defined as the carrier-to-noise power ratio at the input, the output noise is small. As ρ is reduced, individual snaps or clicks are heard in the output. As ρ is reduced still further, these clicks merge into a crackling or sputtering noise. Here expressions are given for the expected number of clicks per second. These results are used to obtain approximate expressions for the output signal-to-noise ratio which hold in the breaking region.

1. INTRODUCTION

This chapter is concerned with the noise in the output of an idealized FM receiver when the input has the form

$$Q \cos \left[2\pi f_c t + \phi(t)\right] + I_N(t) = R(t) \cos \left[2\pi f_c t + \phi(t) + \theta(t)\right]. \qquad (1)$$

The amplitude Q and the carrier frequency f_c are given constants, and $I_N(t)$ is a narrow-band Gaussian noise current whose power spectrum is $w(f)$. Throughout the chapter $w(f)$ is taken to be symmetrical about f_c.

The current $I_N(t)$ represents the noise accepted by the receiver. Its power spectrum is different from 0 only over the radio-frequency band required to transmit the FM signal. In actual receivers this is also the intermediate frequency band. We refer to the band specified by $w(f)$ as the "input band."

The signal is $\phi'(t)$, the time derivative of the phase angle $\phi(t)$. It is measured in radians per second. The output of the frequency detector or discriminator is assumed to be the derivative $\phi'(t) + \theta'(t)$. The output of the receiver is obtained by applying $\phi'(t) + \theta'(t)$ to a low-pass filter which passes the signal $\phi'(t)$ but removes the high-frequency components of the noise $\theta'(t)$. We assume that this output filter is ideal and cuts off at frequency f_a (a for "audio").

The average carrier power is $Q^2/2$, the average input noise power is $\overline{I_N^2(t)}$ and their ratio is

$$\rho = \frac{Q^2}{2\overline{I_N^2(t)}}. \tag{2}$$

When $\rho \to \infty$ and certain simplifying assumptions are made, the ratio S_o/N_o of the average signal power to the average noise power at the receiver output (subscript o for "output") is customarily taken to be

$$\frac{S_o}{N_o} = 3\rho \left(\frac{\beta}{2f_a}\right)^3, \tag{3}$$

a result due essentially to Crosby [1]. Here the top frequency in the signal is supposed to be equal to the cutoff frequency f_a of the low-pass output filter, and β is the input bandwidth in cycles per second.

It is found in practice that as the input noise is increased, so that ρ decreases from ∞, the FM receiver "breaks." At first, individual clicks are heard in the output. As ρ decreases still further, the clicks rapidly merge into a "crackling" or "sputtering" sound. Near the (not precisely defined) breaking point (3) begins to fail by predicting values of S_o/N_o larger than the actual ones.

The purpose of the present chapter is to discuss the behavior of the receiver in the region around the breaking point. Particular attention is paid to the relation between the breaking point and the expected number of clicks per second in the output.*

Earlier papers by Stumpers [2], Middleton [3], and the writer [4] give expression for the power spectrum $W(f)$ of $\theta'(t)$ from which the output noise power N_o may be computed. However, the complexity of the expressions make it difficult to carry out the computations with the desired accuracy in the regions of greatest interest.

Here it is shown that around the breaking point S_o/N_o is given approximately by (26) for a rectangular $w(f)$ or by (27) for a Gaussian $w(f)$. Both approximations are extensions of (3). Although (26) and (27) are approximations, they are based on an exact result, namely, that when the carrier is unmodulated (no signal present) the expected number of clicks per second is

$$r(1 - \text{erf } \sqrt{\rho}), \tag{4}$$

where r is the radius of gyration, in cycles per second, of $w(f)$ about its axis of symmetry at f_c. Values of r are given in Table 2.1 for rectangular and Gaussian $w(f)$. Expression (4) follows from results given in Section 6 of [4].

Modulating the carrier increases the number of clicks per second and also

---

* A similar approach has been developed independently by N. M. Blachman. This was brought out at the Brussels Information Theory Symposium (September 1962) during the discussion of a paper by G. Battail which dealt with FM reception. Mr. Blachman was led to the basic idea while studying the zero crossings of Gaussian noise.

affects S_o/N_o.   However, it does not change the position of the breaking point much.   Figure 3 shows that the breaking point occurs at values of $10 \log \rho$ around 10 or 11 db and is slightly higher when the signal is present.

It appears that both S_o/N_o and the number of clicks per second should be considered in judging the effect of Gaussian noise on FM receivers.

2. REVIEW OF RESULTS, SOME KNOWN AND SOME NEW

The first portion of this review is concerned with the case in which the signal is absent.   This case has received most attention in the past because of its simplicity.   The second portion deals with the case in which the signal is present.   The arguments given in this section are intended merely to make the results seem plausible.   The detailed analysis is reserved for later sections.

Unmodulated Carrier

When there is no signal, the carrier is unmodulated and (1) becomes

$$Q \cos \omega_c t + I_N = R \cos (\omega_c t + \theta), \qquad \omega_c = 2\pi f_c$$
$$= (Q + I_c) \cos \omega_c t - I_s \sin \omega_c t, \qquad (5)$$

where I_c and I_s are the in-phase and quadrature components of I_N with respect to the carrier frequency f_c:

$$I_N = I_c \cos \omega_c t - I_s \sin \omega_c t. \qquad (6)$$

Both I_c and I_s have the power spectrum $2w(f_c + f)$.   This follows from the assumption that the power spectrum $w(f)$ of $I_N(t)$ is symmetrical about f_c.   The arguments (t) of the various functions in (5) and (6) have been omitted—as they are in subsequent sections when it is convenient to do so.

EXPRESSION (3) FOR S_o/N_o.   It is helpful to review the reasoning that leads to expression (3) for the output signal-to-noise ratio S_o/N_o.   Figure 1 shows the phase relations between the various currents in (5).   The point P wanders around the point Q as the amplitudes and phases of $I_c(t)$ and $I_s(t)$ change in a random manner.   When the carrier-to-noise power ratio ρ is large, I_c and I_s are much smaller than Q most of the time and, to within a multiple of 2π,

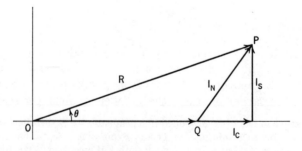

Figure 1.   A graphical interpretation of Equation (5).

$\theta \approx I_s/Q.$ Differentiation gives

$$\theta' \approx \frac{I_s'}{Q}. \tag{7}$$

Since the power spectrum of I_s' is $(2\pi f)^2$ times the power spectrum $2w(f_c + f)$ of I_s, it follows that

$$W(f) \sim \frac{(2\pi f)^2 2w(f_c + f)}{Q^2}, \tag{8}$$

where $W(f)$ is the power spectrum of θ'. In fact, the right-hand side of (8) is the first term in an asymptotic expansion for $W(f)$ which holds for $\rho \to \infty$.

Since θ' is measured in radians per second, $W(f)\,df$ has the dimensions of (radians per second)^2. The average noise power in the receiver output is

$$N_o = \int_0^{f_a} W(f)\,df \quad (\text{radians/sec})^2 \tag{9}$$

where f_a is the cutoff frequency of the output filter.

Consider a rectangular input band of width β cps centered at f_c:

$$w(f) = \begin{cases} w_0, & f_c - 2^{-1}\beta < f < f_c + 2^{-1}\beta \\ 0, & \text{elsewhere} \end{cases} \tag{10}$$

$$\overline{I_N^2} = \beta w_0, \qquad \rho = \frac{Q^2}{2\beta w_0}$$

From (8) and (9) with $f_a < \beta/2$ and ρ large,

$$N_0 \sim \frac{8\pi^2 w_0 f_a^3}{3Q^2} = \frac{4\pi^2 f_a^3}{3\rho\beta}. \tag{11}$$

If $f_s < f_a$, the sinusoidal signal $\phi' = A \sin \omega_s t$, $\omega_s = 2\pi f_s$, passes through the ideal receiver unchanged when no noise is present. Hence the average output signal power is $S_o = A^2/2$ (radians/sec)^2. In forming the S_o/N_o ratio (3), A is taken to be $2\pi(\beta/2)$ so that the carrier swings back and forth across the entire input band and

$$S_o = \tfrac{1}{2}(\pi^2\beta^2). \tag{12}$$

Dividing (12) by (11) gives (3) for S_o/N_o.

EXPECTED NUMBER OF CLICKS PER SECOND. Our approximation for S_0/N_0 in the breaking region makes use of a special case of Equation 6.6 of [4]: When $\theta(t)$ is given by (5), the expected number of times per second $\theta(t)$ increases through an odd multiple of π is

$$N_+ = \frac{r}{2}(1 - \text{erf }\sqrt{\rho}),$$

$$1 - \text{erf }\sqrt{\rho} = \frac{2}{\sqrt{\pi}}\int_{\sqrt{\rho}}^{\infty} e^{-t^2}\,dt \sim e^{-\rho}(\rho\pi)^{-\frac{1}{2}} \tag{13}$$

Here

$$r = (2\pi)^{-1}(b_2/b_0)^{1/2}, \qquad \rho = \frac{Q^2}{2b_0}$$

$$b_0 = \overline{I_N^2} = \overline{I_c^2} = \overline{I_s^2} = \int_0^\infty w(f)\, df = \int_0^\infty 2w(f_c + f)\, df \qquad (14)$$

$$b_2 = (2\pi)^2 \int_0^\infty (f - f_c)^2 w(f)\, df.$$

The parameter r may be regarded as either the "radius of gyration" of $w(f)$ about its axis of symmetry at $f = f_c$ or as the "representative frequency" of $I_s(t)$ [or $I_c(t)$]. The second interpretation follows from the fact that $2r$ is the expected number of zeros of $I_s(t)/\text{sec}$. Two special cases of interest are shown in Table 1.

TABLE 1

VALUES OF r

$w(f)$	b_0	b_2	r
Rectangle of width β cps centered on f_c, $w(f) = w_0$ for $f_c - 2^{-1}\beta < f < f_c + 2^{-1}\beta$ and $w(f) = 0$ elsewhere	$w_0\beta$	$\dfrac{\pi^2 w_0 \beta^3}{3}$	$\dfrac{\beta}{\sqrt{12}}$
Normal law filter $w(f) = \dfrac{b_0}{\sigma \sqrt{2\pi}} \exp\left[-\dfrac{(f - f_c)^2}{2\sigma^2} \right]$ Equivalent rectangular bandwidth is $\sigma \sqrt{2\pi}$ cps.	b_0	$4\pi^2 \sigma^2 b_0$	σ

Because of the symmetry of $w(f)$ about f_c, the expected number, N_-, of times per second $\theta(t)$ decreases through an odd multiple of π is equal to N_+. For conciseness we refer to N_+ and N_-, respectively, as the number of upward and downward clicks per second. These names are suggested by the following considerations.

When $\rho \gg 1$, the wandering point P in Figure 1 spends most of its time near Q. However, it occasionally sweeps around the origin and θ increases or decreases by 2π. Figure 2 shows in a rough way how these excursions produce impulses in θ'. The impulses have different heights depending on how close P comes to the origin O, but all have areas nearly equal to $\pm 2\pi$ radians. When the impulses shown in Figure 2b are applied to the low-pass output filter, corresponding but wider impulses are excited in the output and are heard as clicks. Clicks are produced only when θ changes by $\pm 2\pi$. When point P in Figure 1 leaves the region Q, cuts across the segment OQ close to O, and then returns to Q, θ changes rapidly by nearly $\pm \pi$ during the sweep past O. However, the resulting pulse in θ' has little low-frequency content since the integral of θ' taken over such an excursion is nearly zero. Hence the output of the

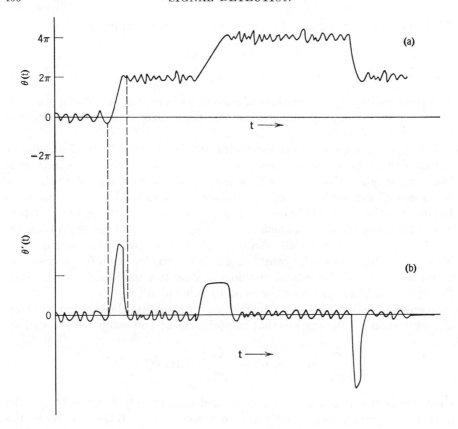

Figure 2. Sketch showing impulses in $\theta'(t)$ produced by changes of $\pm 2\pi$ in $\theta(t)$. The time scale has been expanded in the vicinity of the impulses in order to show differences in shape.

low-pass filter is much smaller than that for an excursion in which O changes by $\pm 2\pi$.

When a television signal is transmitted by FM, the upward clicks appear as white spots on the screen and the downward clocks as black spots, or vice versa, depending on the polarity of the connections.

It should be observed that $N_\pm$ represent number of clicks per second whereas N_o stands for the average noise power in the output.

APPROXIMATION FOR $W(0)$ WHEN $\rho \gg 1$. The asymptotic expression (8) for the power spectrum $W(f)$ of θ' gives the value 0 for $W(0)$. It was pointed out by Crosby [1] and shown in detail by Stumpers [2], Middleton [3], and the writer [4] that although $W(0)$ is small when $\rho \gg 1$, it is not zero. The value of $W(0)$ is of interest because of its importance in determining the behavior of $W(f)$ for small f. This in turn determines the values of ρ for which S_o/N_o begins to deviate appreciably from (3).

It has been conjectured [4] that $W(0)$ decreases as $e^{-\rho}$ as $p \to \infty$. In Section

3 it is shown for unmodulated carrier that we have the approximation

$$W(0) \approx 8\pi^2 r(1 - \text{erf } \sqrt{\rho})$$

$$\sim 8\pi^2 r(\rho\pi)^{-\frac{1}{2}} e^{-\rho} \tag{15}$$

as ρ becomes large. The method of derivation indicates that the approximation is actually asymptotic in the sense that the ratio of $W(0)$ to the right-hand side of (15) approaches unity as $\rho \to \infty$.

The approximation (15) is in line with a physical picture of the discriminator output θ' suggested to the writer by H. E. Curtis. He regards θ' as the sum of two components. One is a small Gaussian noise current and the other is a succession of randomly occurring impulses or clicks. This is illustrated in Figure 2b (the impulse durations are exaggerated). The power spectrum of the Gaussian noise component is given by the asymptotic expression (8) for $W(f)$ and is zero at $f = 0$. Since impulses of the sort shown in Figure 2b have spectra that are nearly constant (and not zero) for small frequencies, the power spectrum of the second component does not vanish at $f = 0$. It is this component that produces the nonzero value of $W(0)$.

In order to make (15) seem plausible, we approximate the random sequence of positive and negative pulses that comprise the second component of $\theta'(t)$ by

$$\sum_{k=-\infty}^{\infty} 2\pi\delta(t - t_k) + \sum_{l=-\infty}^{\infty} (-2\pi)\,\delta(t - t_l). \tag{16}$$

Here $\delta(x)$ is the unit impulse function, and t_k, t_l represent, respectively, the instants at which positive and negative pulses occur. When ρ is large, the pulses tend to occur at random and each sum in (16) can be regarded as a shot-effect current. Such a current, say $I = \Sigma_k F(t - t_k)$, has the d-c component

$$I_{\text{dc}} = \nu \int_{-\infty}^{\infty} F(t)\,dt, \tag{17}$$

where ν is the expected number of arrivals per second. It is also known that the power spectrum of $I - I_{\text{dc}}$ is $2\nu|s(f)|^2$, where

$$s(f) = \int_{-\infty}^{\infty} e^{-2\pi ift} F(t)\,dt. \tag{18}$$

When these results are applied to the first sum in (16), we have $F(t) = 2\pi\delta(t)$, $\nu = N_+$, $I_{\text{dc}} = 2\pi N_+$, $s(f) = 2\pi$ and the power spectrum $8\pi^2 N_+$. Similarly, the second sum has the d-c component $-2\pi N_-$ and the power spectrum $8\pi^2 N_-$. Since the sums represent independent currents, the expression (16) has the d-c value

$$2\pi(N_+ - N_-) \tag{19}$$

and the power spectrum

$$8\pi^2(N_+ + N_-). \tag{20}$$

Because the delta functions in (16) only approximate the pulses, we expect the actual power spectrum to agree with (20) only for frequencies small compared to the reciprocal of the pulse duration. For the higher frequencies, the actual power spectrum is less than (20).

The expression (15) for $W(0)$ follows from (20), since in the present case $N_+ = N_-$ and both are given by (13). It should be noted that $W(0)$ has the dimensions of (radians per second)^2/cycles per second, which works out to be $4\pi^2$ cps or 1/(time) as required by $W(0) \approx 16\pi^2 N_+$.

Approximation for $W(f)$ When $\rho \gg 1$. Picturing $\theta'(t)$ as the sum of a Gaussian noise component with power spectrum (8) plus a sequence of random pulses leads to the conjecture that $W(f)$ is the sum of the two corresponding power spectra:

$$W(f) \approx 8\pi^2 r(1 - \operatorname{erf} \sqrt{\rho}) + (2\pi f)^2 \, 2w(f_c + f)Q^{-2}. \tag{21}$$

This is supposed to hold when ρ is large and the carrier is unmodulated.

For the rectangular and normal law input spectra of Table 1, (21) takes the respective forms

$$\frac{W(f)}{4\pi^2 r} \approx 2(1 - \operatorname{erf} \sqrt{\rho}) + \left(\frac{f}{\beta}\right)^2 \frac{\sqrt{12}}{\rho}, \qquad r = \frac{\beta}{\sqrt{12}} \tag{22}$$

$$\frac{W(f)}{4\pi^2 \sigma} \approx 2(1 - \operatorname{erf} \sqrt{\rho}) + \left(\frac{f}{\sigma}\right)^2 \frac{e^{-f^2/(2\sigma^2)}}{\rho \sqrt{2\pi}}; \tag{23}$$

(22) holds only for $0 \leqslant f \leqslant \beta/2$; $W(f)$ is approximately 0 for $f > \beta/2$.

An idea of how much the conjecture (21) is in error may be obtained from Table 2, which shows the approximate values computed from (23) and the

TABLE 2
COMPARISON OF EXACT AND APPROXIMATE VALUES OF $W(f)/4\pi^2\sigma$

| f/σ | $\rho = 1$ | | $\rho = 2$ | | $\rho = 5$ | |
	Exact	Approx.	Exact	Approx.	Exact	Approx.
0	0.2322	0.315	0.07529	0.0911	0.003017	0.00314
$\pi/6$	0.2672	0.410	0.1134	0.139	0.02342	0.0222
$2\pi/6$	0.3231	0.567	0.1784	0.217	0.05828	0.0537

corresponding exact values taken from Table 4 of [4]. It is seen that the agreement is best in the important region in which ρ is large and f is small. Better agreement for $\rho = 5$ may be obtained by noting that expression (8) is only the first of the terms of an asymptotic series for $W(f)$ that correspond to the terms in the expansion of $\tan^{-1} I_s(Q + I_c)^{-1}$ in descending powers of Q; (8) corresponds to the leading term $I_s Q^{-1}$. When the second term of the series for $W(f)/4\pi^2\sigma$ is evaluated (by calculating the power spectrum of $I_s I_c Q^{-2}$ or by using Equation 7.3 of [4]) and added to (23), the approximate values 0.0222, 0.0537 given in Table 2 increase to 0.0237, 0.0584. Although these values were computed on a slide rule, they agree quite well with the exact values.

S_o/N_o IN THE BREAKING REGION. The expression (3) for S_o/N_o holds when ρ is large. The value of the average output noise power N_o used in (3) is obtained by integrating the second term in the expression (22) for $W(f)$. A better approximation to S_o/N_o is obtained by including the first term as well. Thus integrating (22) and (23) from $f = 0$ to $f = f_a < \beta/2$ gives

$$N_o \approx \frac{8\pi^2 \beta f_a}{(12)^{\frac{1}{2}}} (1 - \mathrm{erf} \sqrt{\rho}) + \frac{4\pi^2 f_a^3}{3\rho\beta} \tag{24}$$

$$N_o \approx 8\pi^2 \sigma f_a (1 - \mathrm{erf} \sqrt{\rho}) + \frac{4\pi^2 f_a^3}{3\rho\sigma(2\pi)^{\frac{1}{2}}} \left(1 - \frac{3f_a^2}{10\sigma^2} + \cdots \right) \tag{25}$$

for the rectangular and normal law input spectra of Table 1.

Taking the output signal power to be $S_o = \pi^2 \beta^2/2$ in the rectangular case, as in (12), and $S_o = \pi^2 B^2/2$, $B = \sigma(2\pi)^{\frac{1}{2}}$ in the normal law case leads to the respective ratios

$$\frac{S_o}{N_o} \approx \frac{3\rho\beta^3 (2f_a)^{-3}}{\rho \sqrt{3} (1 - \mathrm{erf} \sqrt{\rho})(\beta/f_a)^2 + 1} \tag{26}$$

$$\frac{S_o}{N_o} \approx \frac{3\rho B^3 (2f_a)^{-3}}{\rho(18/\pi)^{\frac{1}{2}}(1 - \mathrm{erf} \sqrt{\rho})(B/f_a)^2 + (1 - 0.3\sigma^{-2}f_a^2 + \cdots)}. \tag{27}$$

It will be recalled that $B = \sigma(2\pi)^{\frac{1}{2}}$ is the "equivalent" rectangular bandwidth of the normal law input filter.

As $\rho \to \infty$, the denominator of (26) tends to unity and S_o/N_o tends to the customary value (3). Curve A of Figure 3 is a plot of (26) when the deviation ratio $\beta/2f_a$ is equal to 5. It shows that S_o/N_o deviates appreciably from a linear function of ρ when ρ becomes less than 10. The plot is not continued below $\rho = 2$ because Table 2 indicates that the approximation (24) for N_o does not hold for smaller values. Furthermore, the decrease in the output signal amplitude due to the presence of the noise begins to be appreciable for smaller values of ρ (this is discussed briefly in the Appendix).

Curve A also shows the expected number of clicks per second computed from $N_+ + N_- = r(1 - \mathrm{erf} \sqrt{\rho})$ for $\beta = 150$ kc so that $r = \beta/(12)^{\frac{1}{2}} \approx 43,300$ cps. The click rate is about 1/sec near $\rho = 10$, where the break in the S_o/N_o curve occurs. This rate is proportional to the input bandwidth β so that if $\beta/2$ were 750 kc instead of 75 kc the numbers given in Figure 3 would have to be multiplied by 10.

Signal Present

The foregoing discussion was concerned with the unmodulated carrier case. When the signal is present (5) is replaced by the general expression (1),

$$Q \cos (\omega_c t + \phi) + I_N = R \cos (\omega_c t + \phi + \theta), \tag{28}$$

where ϕ' is the signal and θ' is the noise in the discriminator output. As in the case of no signal, it is assumed that a click is produced every time θ changes

214

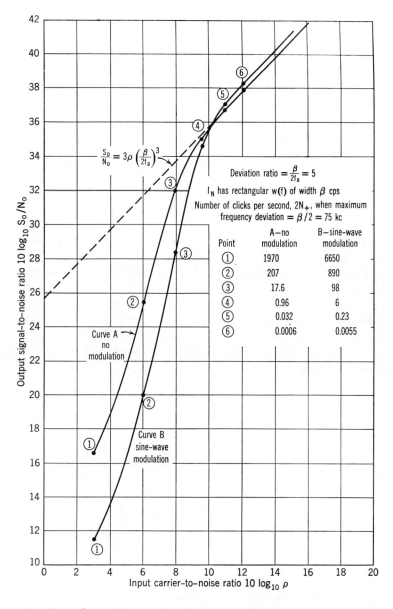

Figure 3.   Dependence of output signal-to-noise ratio on input carrier-to-noise ratio.

by $\pm 2\pi$. The notation N_+, N_- is used again to denote the expected number of times per second θ increases or decreases through an odd multiple of π.

EXPECTED NUMBER OF CLICKS PER SECOND. The presence of the signal changes the simple error function expression (13) for N_+ and N_- into the more complicated integrals derived in Sections 4 and 5. When ϕ' fluctuates symmetrically about zero, N_+ is equal to N_- and both tend to increase as rms ϕ'

increases.   In the general case

$$N_+ - N_- = (2\pi)^{-1}e^{-\rho} \lim_{T\to\infty} \frac{\phi(0) - \phi(T)}{T} \tag{29}$$

where the limit is supposed to exist.

Expressions for N_+ when the signal is the sine wave $\phi' = A \sin \omega_s t$ are given in Section 5.   For ρ large we have the approximation (83),

$$N_+ \sim \frac{A}{2\pi} \frac{e^{-\rho}}{\pi} + \frac{re^{-\rho}}{(4\pi\rho)^{1/2}} e^{-a\rho} I_0(a\rho),$$

$$a = \frac{(A/2\pi)^2}{2r^2} = \left[\frac{\mathrm{rms}(\phi'/2\pi)}{r}\right]^2. \tag{30}$$

where $I_0(a\rho)$ is a Bessel function of imaginary argument.   This holds when $I_N(t)$ has any symmetrical power spectrum $w(f)$ and r is defined by (14). It reduces to the asymptotic form of (13) when $A = 0$.   The parameter a is somewhat like $\rho = Q^2/2b_0$ with the maximum frequency deviation $A/2\pi$ playing the role of Q and the representative frequency r of the noise envelope playing the role of rms $I_N(t)$.   The value of N_+ does not depend on the signal frequency ω_s.   Also $N_+ = N_-$.

From (30) it is seen that modulating the carrier tends to increase the number of clicks per second in the ratio

$$\frac{N_+ \text{ for } A \neq 0}{N_+ \text{ for } A = 0} \approx \frac{A}{2\pi r}\left(\frac{4\rho}{\pi}\right)^{1/2} + e^{-a\rho}I_0(a\rho). \tag{31}$$

The case when the signal ϕ' is a Gaussian noise with average 0 and variance $\overline{\phi'^2}$ is examined in Section 5.   It turns out that the approximation corresponding to (30) is

$$N_+ = N_- \sim re^{-\rho}\left(\frac{1 + 2a\rho}{4\pi\rho}\right)^{1/2}$$

$$a = \frac{\overline{\phi'^2}}{(2\pi r)^2} \tag{32}$$

where $\rho \gg 1$.

Several curves showing how N_+/β varies with ρ are plotted in Figure 4 for the case of the sine-wave signal $\phi' = A \sin \omega_s t$.   The input spectrum is the usual rectangular $w(f)$ of width β cps.   The dashed lines are computed from the approximation (30) with $r = \beta/(12)^{1/2}$ and the solid lines by numerical integration of the exact integrals given in Section 5. $A = 0$ corresponds to no signal and $A/2\pi = \beta/2$ to a maximum frequency deviation which makes the carrier swing back and forth across the entire input band.   This is the same as the signal used in the S_o/N_o ratios (3) and (26).   The exact curves show that for maximum frequency swing the ratio on the left side of (31) increases from about 1.5 at $\rho = 0$ to 4.4 at $\rho = 4$.

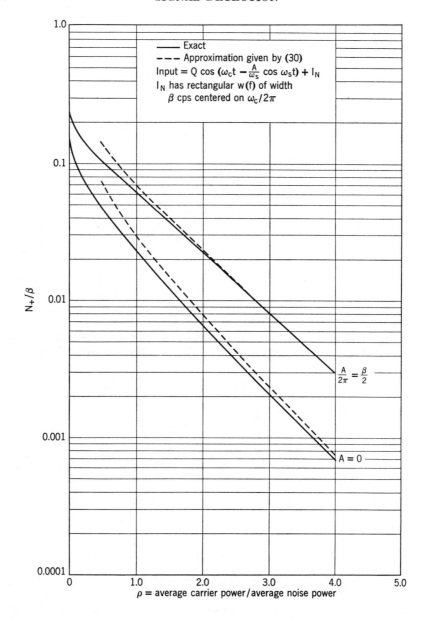

Figure 4. The expected number N_+ of upward clicks per second with and without sine-wave modulation.

APPROXIMATION FOR $W(f)$. The heuristic argument leading to (19) and (20) indicates that when a signal is present $\theta'(t)$ has the d-c component

$$\overline{\theta'} = 2\pi(N_+ - N_-) = e^{-\rho} \lim_{T \to \infty} \frac{\phi(0) - \phi(T)}{T}, \qquad (33)$$

where we have used (29). Furthermore, the power spectrum $W(f)$ of $\theta'(t) - \overline{\theta'}$ at $f = 0$ is given by

$$W(0) \approx 8\pi^2(N_+ + N_-) \tag{34}$$

when $\rho \gg 1$. This is proved in Section 6 for several different kinds of signals.

The conjecture (21) generalizes to

$$W(f) \approx 8\pi^2(N_+ + N_-) + (2\pi f)^2 Q^{-2} W_y(f). \tag{35}$$

Here $W_y(f)$ is the power spectrum of

$$y(t) = I_s(t) \cos \phi(t) - I_c(t) \sin \phi(t), \tag{36}$$

in which $I_s(t)$ and $I_c(t)$ are to be regarded as independent Gaussian noise currents with power spectra $2w(f_c + f)$. In Figure 5, y is shown as the component of the noise current I_N which is perpendicular to the signal vector $Qe^{i\phi}$. Consequently, when $\rho \gg 1$,

$$\theta' \approx \frac{y'}{Q} \tag{37}$$

is the analogue of (7) and $(2\pi f)^2 Q^{-2} W_y(f)$ is the analogue of (8).

It would be desirable to test the accuracy of (35) by constructing a table similar to Table 2. However, no exact values of $W(f)$ seem to be available for the cases of interest.

For the sine-wave signal $\phi' = A \sin \omega_s t$ it is shown in Section 7 that

$$W_y(f) = 2 \sum_{n = -\infty}^{\infty} w(f_c + f + nf_s) J_n^2(A\omega_s^{-1}), \tag{38}$$

where $J_n(z)$ is a Bessel function of order n. An equivalent result has been given by N. M. Blachman [5]. When $w(f)$ is the usual rectangle of height w_0 and width β cps, the range of n in (38) is given by $-\beta/2 < f + nf_s < \beta/2$, $\omega_s = 2\pi f_s$. For example, if $\beta/2f_s = 5$ and $0 < f < f_s$, n runs from -5 to 4 and (38) becomes

$$W_y(f) = [J_0^2 + J_5^2 + 2(J_1^2 + J_2^2 + J_3^2 + J_4^2)]2w_0, \tag{39}$$

where the argument of the Bessel functions is the frequency deviation ratio $A/2\pi f_s$. Since $\Sigma J_n^2 = 1$, when the summation extends from $n = -\infty$ to $+\infty$, $W_y(f)$, as given by (39), is at most equal to $2w_0$, the value for no signal.

S_o/N_o COMPUTED FROM $W(f)$ OF (35). Consider the case of a rectangular $w(f)$ and a sine-wave signal $\phi' = \pi\beta \sin 2\pi f_a t$ which swings back and forth across the entire input band at the maximum audio frequency $f_s = f_a$. Assume that $\beta/2f_a$ is an integer so that, as in example (39), $W_y(f)$ is constant over the band 0 to f_a and is given by

$$W_y(f) = 2cw_0 \tag{40}$$

in the range $0 < f < f_a$. The constant c is $\leqslant 1$.

Setting $2N_+$ for $N_+ + N_-$ and $2cw_0$ for $W_y(f)$ in the approximation (35) for $W(f)$ and then integrating from 0 to f_a gives

$$N_0 \approx 16\pi^2 N_+ f_a + 4\pi^2 3^{-1} Q^{-2} 2cw_0 f_a^3 \tag{41}$$

for the average output noise power.

The average signal power S_o in the output is taken to be $(\pi\beta)^2/2$ as in (26). Strictly speaking, this should be multiplied by $(1 - e^{-\rho})^2$, since the noise reduces the output signal power by this factor, as mentioned in the appendix. However, this refinement is omitted because the factor is nearly unity in the region of large ρ where the approximation (41) for N_0 is supposed to hold.

Dividing S_o by (41) and using $\rho = Q^2/2\beta w_0$ leads to

$$\frac{S_0}{N_0} \approx \frac{3\rho\beta^3 (2f_a)^{-3}}{12\rho(N_+/\beta)(\beta/f_a)^2 + c}, \tag{42}$$

where ρ is large.   From (30) with $r = \beta/(12)^{1/2}$, $A/2\pi = \beta/2$, and $a = \frac{3}{2}$,

$$\frac{N_+}{\beta} \approx \frac{e^{-\rho}}{2\pi}\left[1 + \left(\frac{\pi}{12\rho}\right)^{1/2} e^{-3\rho/2} I_0\left(\frac{3\rho}{2}\right)\right]. \tag{43}$$

Values of N_+/β for $\rho < 4$ may be read off the upper solid curve in Figure 4.

In (42) the signal and the noise are supposed to be present at the same time. On the other hand, the N_0 in expression (26) for S_o/N_o is computed on the assumption that the signal is absent (i.e., the carrier is unmodulated).   Both expressions are plotted in Figure 3 for a deviation ratio of $\beta/2f_a = 5$ (Curve A for unmodulated carrier, Curve B for modulated carrier).   The expected number of clicks per second, $2N_+$, listed in Figure 3, are computed for $\beta/2 = 75$ kc and, for Curve B, a signal amplitude of $A = \pi\beta$.

As mentioned in the preceding paragraph, Curve B in Figure 3 was computed by setting $\beta/2f_a = 5$ in (42).   The value of c for this deviation ratio is the sum of the terms within the brackets on the right side of (39).   Setting the argument of the Bessel functions equal to 5 and performing the addition gives

$$c = 0.89,$$

which is less than unity, as expected.

Figure 3 shows that as ρ decreases from infinity Curve B starts to deviate appreciably from linearity at about $10 \log \rho = 11$ db.   The corresponding point on Cuve A is at about 10 db.   The number of clicks per second for a given value of ρ in this region ranges from roughly 10 to 0.1 and is about six times greater for Curve B than for Curve A.   Experimentally, it is found that occasional clicks are heard in the output at a carrier-to-noise ratio of 13 db. This appears to be a little higher than the theory indicates.   The discrepancy, if any, may be due to our idealization of the receiver or, possibly, the actual input noise may not be strictly Gaussian.

3. ASYMPTOTIC EXPRESSION FOR $W(0)$—NO SIGNAL

This section is concerned with the asymptotic behavior of $W(0)$, the value of the power spectrum $W(f)$ of $\theta'(t)$ at $f = 0$.   Only the unmodulated carrier case, that is, when $\theta(t)$ is given by

$$Q \cos \omega_c t + I_N = R \cos (\omega_c t + \theta), \tag{44}$$

is considered in this section. It is shown that as $\rho = Q^2/(2\overline{I_N^2})$ becomes large

$$W(0) \sim 16\pi^2 N_+ = 8\pi^2 r(1 - \operatorname{erf} \sqrt{\rho}). \tag{45}$$

This result was mentioned earlier in connection with (15) of Section 2. The representative frequency r is defined by (14).

When a signal is present, the results are more complicated. Therefore, their discussion is delayed until Section 6.

For large values of ρ a plot of $\theta(t)$ would resemble Figure 2a and would show jumps of $\pm 2\pi$ at irregular intervals. Let n_+, n_- denote the respective number of upward and downward jumps of $\theta(t)$ during a long interval $(0, T)$. Then

$$W(0) = \lim_{T \to \infty} \frac{2}{T} \left\langle \left| \int_0^T \theta'(t) \, dt \right|^2 \right\rangle, \tag{46}$$

where the value of the integral is

$$\theta(T) - \theta(0) = (n_+ - n_-)2\pi + F(T) - F(0), \tag{47}$$

in which $F(0)$, $F(T)$ are $0(1)$. $F(0)$ and $F(T)$ are the small initial and final deviations from levels of the sort shown in Figure 2a. Here $\langle \ \rangle$ denotes the "ensemble average" taken over the ensemble of noise currents $I_N(t)$.

Equation (46) follows from the expression

$$w_x(f) = \lim_{T \to \infty} \frac{2}{T} \left\langle \left| \int_0^T e^{-2\pi i f t} x(t) \, dt \right|^2 \right\rangle \tag{48}$$

for the power spectrum of the random function $x(t)$. This holds when $x(t)$ has no periodic component at the particular frequency f being considered (or no d-c component if $f = 0$). The random function $\theta'(t)$ obtained from (44) has no d-c component because the power spectrum of I_N is symmetrical about the carrier frequency f_c. Hence it is legitimate to set $f = 0$ in (48) to get (46).

The expected values of n_+ and n_- are both equal to $TN_+ = TN_-$, where, from equation 6.6 of [4],

$$N_+ = N_- = \frac{r}{2}(1 - \operatorname{erf} \sqrt{\rho}). \tag{49}$$

We now take ρ to be so large and the successive jumps so far apart that they occur "individually and collectively at random." In other words, we assume that upward jumps are from a Poisson process with an average rate of N_+/sec. The same assumption holds for the downward jumps, and the two Poisson processes are independent. This assumption implies a sufficiently rapid decrease in the correlation function $\Omega(\tau)$ of $\theta'(t)$ as $\tau \to \infty$, a condition that is apparently satisfied in most practical cases.

Since the variance of the Poisson distribution is equal to its mean,

$$\langle (n_+ - TN_+)^2 \rangle = TN_+, \tag{50}$$

and a similar expression holds for the downward jumps. We rewrite (47) as

$$\theta(T) - \theta(0) = (n_+ - TN_+)2\pi - (n_- - TN_-)2\pi + F(T) - F(0) \tag{51}$$

square both sides, take the ensemble average, and substitute in (46). Since the upward and downward jumps occur independently, the average value of $(n_+ - TN_+)(n_- - TN_-)$ is zero. When it is further assumed that $F(T), F(0)$ have zero means and are independent of n_+, n_-, and each other, (46) becomes

$$W(0) \sim \lim_{T \to \infty} \frac{8\pi^2}{T} \left[\langle (n_+ - TN_+)^2 \rangle + \langle (n_- - TN_-)^2 \rangle + \left\langle \frac{F^2(T)}{4\pi^2} \right\rangle + \left\langle \frac{F^2(0)}{4\pi^2} \right\rangle \right]. \tag{52}$$

When (50) is used, this reduces to

$$W(0) \sim 8\pi^2(N_+ + N_-), \tag{53}$$

which is equivalent to (45), since $N_+ = N_-$ for the present case of no signal.

4. VALUES OF N_+ AND N_- WHEN CARRIER IS MODULATED

As described earlier, the wave entering the receiver is

$$Q \cos(\omega_c t + \phi) + I_N = R \cos(\omega_c t + \phi + \theta), \tag{54}$$

where ϕ' is the signal and θ' the noise at the discriminator output. Equation (54) is the real part of

$$(Qe^{i\phi} + I_c + iI_s)e^{i\omega_c t} = Re^{i\omega_c t + i\phi + i\theta} \tag{55}$$

where I_c and I_s are the components of I_N with respect to the carrier frequency f_c, as shown in (6). By dividing both sides of (55) by $\exp(i\omega_c t + i\phi)$ and setting $x + iy$ equal to $(I_c + iI_s)e^{-i\phi}$, we have

$$Q + x + iy = Re^{i\theta},$$
$$x = I_c \cos\phi + I_s \sin\phi, \tag{56}$$
$$y = I_s \cos\phi - I_c \sin\phi.$$

Figure 5 shows the various phase relations. Two resolutions of the noise vector are given. The first is the usual one in which the components are I_c, I_s, and in the second the components x, y are parallel and perpendicular, respectively, to the signal vector $Qe^{i\phi}$. Since we are interested primarily in the behavior of θ, it is convenient to redraw the second resolution as in Figure 5b.

It is seen that θ increases through an odd multiple of π whenever y decreases through $y = 0$ and, at the same time, $x < -Q$. In order to calculate the expected number N_+ of times this happens in one second, we make use of the following result. Let x and y be functions of time given by

$$x = F(a_1, \cdots, a_N; t)$$
$$y = G(a_1, \cdots, a_N; t), \tag{57}$$

where the parameters $a_1, \cdots, a_N$ are random variables. The joint probability that (1) y will decrease through $y = y_1$ in the interval $t_1, t_1 + dt_1$ and (2) x will be less than x_1 at $t = t_1$ is

$$dt_1 \int_{-\infty}^{x_1} d\xi \int_{-\infty}^{0} |\zeta| p(\xi, y_1, \zeta; t_1)\, d\zeta, \tag{58}$$

where $p(\xi, \eta, \zeta; t_1)$ is the joint probability density of the random variables

$$\xi = F(a_1, \cdots, a_N; t_1)$$
$$\eta = G(a_1, \cdots, a_N; t_1)$$
$$\zeta = \left[\frac{\partial}{\partial t} G(a_1, \cdots, a_N; t)\right]_{t=t_1}. \tag{59}$$

Furthermore, the expected number of such crossings per second is the limit, as $T \to \infty$, of $1/T$ times the integral of (58) taken from $t_1 = 0$ to $t_1 = T$, provided the limit exists. The expression (58) is much the same as those used in the study of the zeros of random functions.

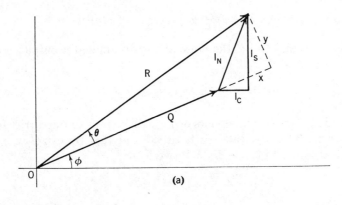

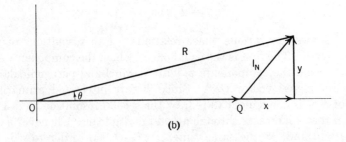

Figure 5. Phase relations between the signal and noise components.

In our application $a_1, \cdots, a_N$ are the random parameters in the representation of I_c, I_s; and F, G are given by the expressions (56) for x, y. Since I_c and I_s are Gaussian, so are ξ, η, ζ. We have

$$\xi = I_c \cos \phi + I_s \sin \phi$$

$$\eta = I_s \cos \phi - I_c \sin \phi \tag{60}$$

$$\zeta = I_s' \cos \phi - I_c' \sin \phi - \phi' \xi,$$

with the understanding that primes denote differentiation with respect to t and that all quantities on the right are evaluated at $t = t_1$.

It may be shown that the required second moments (see, for example, equation A2-2 of [4]) are given by

$$\overline{I_c^2} = \overline{I_s^2} = b_0, \qquad \overline{I_c'^2} = \overline{I_s'^2} = b_2,$$

$$\overline{I_c I_s} = \overline{I_c I_s'} = -\overline{I_c' I_s} = \overline{I_c I_c'} = \overline{I_s I_s'} = \overline{I_c' I_s'} = 0, \tag{61}$$

where b_0, b_2 are defined by (14) and the symmetry of $w(f)$ about f_c has been used. Hence

$$\overline{\xi^2} = \overline{\eta^2} = b_0, \qquad \overline{\zeta^2} = b_2 + b_0 \phi'^2,$$

$$\overline{\xi \eta} = \overline{\eta \zeta} = 0, \qquad \overline{\xi \zeta} = -b_0 \phi', \tag{62}$$

$$p(\xi, \eta, \zeta; t_1) = (2\pi)^{-3/2} b_0^{-1} b_2^{-1/2} \exp \left[-\frac{\xi^2 + \eta^2}{2b_0} - \frac{(\zeta + \xi \phi')^2}{2b_2} \right].$$

Setting $y_1 = 0$, $x_1 = -Q = -(2\rho b_0)^{1/2}$ in (58), introducing the parameter

$$u = \phi' \left(\frac{b_0}{b_2} \right)^{1/2} = \frac{\phi'}{(2\pi r)}$$

$$= \frac{(\text{``instantaneous'' signal frequency in cps at } t = t_1)}{[\text{representative frequency of } I_c(t)]}, \tag{63}$$

making the change of variable $\xi = -\xi_a (b_0)^{1/2}$, $\zeta = -\zeta_a (b_2)^{1/2}$, and then dropping the subscript a from ξ_a, ζ_a gives

$$H_+(t_1) \, dt_1 = \frac{r \, dt_1}{(2\pi)^{1/2}} \int_{(2\rho)^{1/2}}^{\infty} d\xi \int_0^{\infty} d\zeta \, \zeta \exp \left[-\frac{\xi^2}{2} - \frac{(\zeta + u\xi)^2}{2} \right] \tag{64}$$

for the chance that a $\theta(t)$ picked at random from our universe of $\theta(t)$'s will increase through an odd multiple of π between t_1 and $t_1 + dt_1$. In line with (60), u is supposed to be evaluated at t_1.

The double integral in (64) may be reduced to the sum of two single integrals by first writing the exponential function as $\exp z$. The partial derivatives of $\exp z$ with respect to ξ and ζ give two simultaneous equations which may be solved for $\zeta \exp z$:

$$\zeta e^z = u \frac{\partial e^z}{\partial \xi} - (1 + u^2) \frac{\partial e^z}{\partial \zeta}.$$

This leads to

$$H_+(t_1) = \frac{r(1+u^2)}{(2\pi)^{1/2}} \int_{(2\rho)^{1/2}}^{\infty} d\xi \exp\left[-\frac{(1+u^2)\xi^2}{2}\right]$$
$$- \frac{rue^{-\rho}}{(2\pi)^{1/2}} \int_0^{\infty} d\zeta \exp\left\{-\frac{[\zeta + u(2\rho)^{1/2}]^2}{2}\right\} \quad (65)$$

$$= 2^{-1}r\{(1+u^2)^{1/2}[1 - \operatorname{erf}(\rho + \rho u^2)^{1/2}] - ue^{-\rho}[1 - \operatorname{erf}(u\sqrt{\rho})]\}.$$

When dealing with the last expression, it is sometimes convenient to use

$$1 - \operatorname{erf} z = \frac{2}{\sqrt{\pi}} \int_z^{\infty} e^{-t^2} dt \sim \frac{e^{-z^2}}{z\sqrt{\pi}}\left(1 - \frac{1}{2z^2} + \frac{1\cdot 3}{4z^4} - \frac{1\cdot 3\cdot 5}{8z^8} + \cdots\right).$$
$$\quad (66)$$

Several remarks may now be made.

1. When $Q \to 0$, ρ tends to zero but the "reference" provided by the signal still remains; (65) becomes

$$H_+(t_1) = 2^{-1}r[(1+u^2)^{1/2} - u].$$

2. The chance that $\theta(t)$ will decrease through an odd multiple of π between t_1 and $t_1 + dt_1$ is $H_-(t_1)\,dt_1$ where $H_-(t_1)$ is obtained by changing the sign of u in $H_+(t_1)$. This follows from the symmetry of $w(f)$ about f_c. From (65)

$$H_-(t_1) = H_+(t_1) + (2\pi)^{-1}\phi'e^{-\rho}. \quad (67)$$

3. The expected number of times per second $\theta(t)$ increases or decreases through odd multiples of π is given by

$$N_\pm = \lim_{T \to \infty} \frac{1}{T} \int_0^T H_\pm(t_1)\,dt_1, \quad (68)$$

where the limit is assumed to exist. From (67)

$$N_+ - N_- = \frac{e^{-\rho}}{2\pi} \lim_{T \to \infty} \frac{[\phi(0) - \phi(T)]}{T}. \quad (69)$$

4. If the signal is such that $p(u)\,du$ is the fraction of time $\phi'/2\pi r$ spends between u and $u + du$, the time average (68) can be replaced by

$$N_\pm = \int_{-\infty}^{\infty} H_\pm(t_1)\,p(u)\,du, \quad (70)$$

where $H_\pm(t_1)$ depends on t_1 only through u.

5. When $\phi' = 2\pi f_0 = \text{constant}$, the value of u is f_0/r. The quantity $H_+(t_1)$ does not depend on t_1 and

$$N_+ = H_+(t_1) = 2^{-1}\{(r^2 + f_0^2)^{1/2}[1 - \operatorname{erf}(\rho + \rho f_0^2 r^{-2})^{1/2}]$$
$$- f_0 e^{-\rho}[1 - \operatorname{erf}(f_0 r^{-1}\sqrt{\rho})]\} \quad (71)$$

$$N_- = H_-(t_1) = N_+ + f_0 e^{-\rho}.$$

It is seen that when f_0 is positive N_- is greater than N_+. Setting $f_0 = 0$ gives the special case of unmodulated carrier:

$$N_+ = N_- = 2^{-1}r(1 - \text{erf } \sqrt{\rho}). \tag{72}$$

The result (71) may also be obtained from expression 6.2 of [4].

5. VALUES OF N_+ FOR TWO TYPES OF SIGNALS

In this section the formulas of Section 4 are applied to determine N_+ for the special cases of (1) the sine-wave signal $\phi' = A \sin \omega_s t$ and (2) a Gaussian noise signal. In both cases symmetry gives $N_+ = N_-$.

Sine-wave Signal

When $\phi' = A \sin \omega_s t$, it is convenient to write

$$u = \frac{\phi'}{2\pi r} = \frac{A \sin \omega_s t}{2\pi r} = \alpha \sin x,$$

$$\alpha = \frac{A}{2\pi r}, \quad x = \omega_s t, \tag{73}$$

$$N_+ = \frac{1}{2\pi} \int_0^{2\pi} H_+\left(\frac{x}{\omega_s}\right) dx = L_1 + L_2$$

where the separation into L_1 and L_2 is suggested by the two parts of (65):

$$L_1 = \frac{r}{\pi} \int_0^{\pi/2} (1 + \alpha^2 \sin^2 x)^{\frac{1}{2}} [1 - \text{erf } (\rho + \rho\alpha^2 \sin^2 x)^{\frac{1}{2}}] \, dx$$

$$L_2 = -\frac{r\alpha e^{-\rho}}{4\pi} \int_0^{2\pi} \sin x[1 - \text{erf } (\alpha \sqrt{\rho} \sin x)] \, dx. \tag{74}$$

The integral for L_2 may be evaluated by using $\sin x \, dx = d(-\cos x)$ to integrate by parts. It is found that

$$L_2 = \frac{re^{-\rho}}{(4\pi\rho)^{\frac{1}{2}}} \{2ze^{-z}[I_0(z) + I_1(z)]\}_{z=a\rho} \tag{75}$$

where $a = \alpha^2/2$ and $I_n(z)$ denotes the Bessel function of imaginary argument

$$I_n(z) = \frac{1}{2\pi} \int_0^{2\pi} \cos nye^{z \cos y} \, dy, \quad n \text{ integer}. \tag{76}$$

The integral for L_1 is more difficult to handle analytically, but it is well suited for numerical integration. It may also be expressed as

$$L_1 = \frac{re^{-\rho-a\rho}}{(4\pi\rho)^{\frac{1}{2}}} \left\{ I_0(a\rho) - 1 + 2\left(\frac{c-\rho}{\pi}\right)^{\frac{1}{2}} E\left[\left(\frac{2a}{1+2a}\right)^{\frac{1}{2}}\right]\left[1 - \operatorname{erf}\left(\frac{c}{2}\right)^{\frac{1}{2}}\right]e^{c/2} \right.$$

$$-\frac{c}{2}\sum_{n=1}^{\infty}\frac{(\frac{1}{2})_{2n-1}}{n!n!}\left(\frac{a}{2+2a}\right)^{2n}\left[1 + \frac{c}{1\cdot 3} + \frac{c^2}{1\cdot 3\cdot 5} + \cdots\right.$$

$$\left.\left.+ \frac{c^{2n-2}}{1\cdot 3\cdots(4n-3)}\right]\right\}, \quad (77)$$

where $c = 2\rho(1+a)$, $E(k)$ is the complete elliptic integral of the second kind with modulus k, and $(\frac{1}{2})_{2n-1} = \frac{1}{2}\cdot\frac{3}{2}\cdot\frac{5}{2}\cdots(2n-\frac{3}{2})$. When ρ is large, L_1 is given asymptotically by

$$L_1 \sim \frac{re^{-\rho}}{(4\pi\rho)^{\frac{1}{2}}}\left\{\left(1 - \frac{1}{2\rho} + \frac{3}{4\rho^2}\right)e^{-z}I_0(z) + \frac{ze^{-z}}{2\rho^2}[I_0(z) - I_1(z)]\right\}_{z=a\rho}. \quad (78)$$

Both (77) and (78) are based on

$$L_1 = \frac{r}{\sqrt{4\pi}}\left[\frac{e^{-\rho-a\rho}I_0(a\rho)}{\sqrt{\rho}} - \frac{\sqrt{a}}{2}\int_{a\rho}^{\infty}e^{-(1+a^{-1})z}I_0(z)z^{-\frac{3}{2}}\,dz\right] \quad (79)$$

which is obtained by (1) replacing the integrand in (74) by the corresponding first integral in (65) (with $u = \alpha\sin x$), (2) interchanging the order of integration, evaluating the integrals in x by setting $y = 2x$ and using (76), (3) setting $z = a\xi^2/2$, and (4) eliminating $I_1(z)$ by an integration by parts based upon $I_1(z)\,dz = dI_0(z)$.

To obtain the series (77) for L_1 replace $I_0(z)$ in (79) by its power series and integrate term by term. The first term in the expansion of $I_0(z)$, namely unity, gives

$$\int_{a\rho}^{\infty}e^{-\gamma z}z^{-\frac{3}{2}}\,dz = \frac{2}{\sqrt{a\rho}}e^{-a\rho\gamma} - 2\sqrt{\gamma}\int_{a\rho\gamma}^{\infty}v^{-\frac{1}{2}}e^{-v}\,dv$$

$$= \frac{2}{\sqrt{a\rho}}e^{-c/2} - 2\sqrt{\gamma}\int_{c/2}^{\infty}v^{-\frac{1}{2}}e^{-v}\,dv \quad (80)$$

where $\gamma = 1 + a^{-1}$ and $c = 2a\rho\gamma = 2\rho(1+a)$. The remaining terms in the expansion of $I_0(z)$ give rise to integrals of the form

$$\int_{a\rho}^{\infty}e^{-\gamma z}z^{2n-(3/2)}\,dz = (\tfrac{1}{2})_{2n-1}\gamma^{-2n+(1/2)}\left\{\int_{c/2}^{\infty}v^{-\frac{1}{2}}e^{-v}\,dv\right.$$

$$\left.+ (2c)^{\frac{1}{2}}e^{-c/2}\left[1 + \frac{c}{1\cdot 3} + \frac{c^2}{1\cdot 3\cdot 5} + \cdots + \frac{c^{2n-2}}{1\cdot 3\cdots(4n-3)}\right]\right\}. \quad (81)$$

The result (81) is obtained by setting $f(v) = v^{2n-(3/2)}$ and $m = 2n-2$ in

$$\int e^{-v}f(v)\,dv = -[f(v) + f'(v) + \cdots + f^{(m)}(v)]e^{-v} + \int e^{-v}f^{(m+1)}(v)\,dv. \quad (82)$$

When (80) and (81) are used in (79), the coefficient of

$$\int_{c/2}^{\infty} v^{-\frac{1}{2}} e^{-v} \, dv = \sqrt{\pi} \left[1 - \text{erf} \left(\frac{c}{2} \right)^{\frac{1}{2}} \right]$$

is a series proportional to

$$\sqrt{\gamma} \, F(-\tfrac{1}{4}, \tfrac{1}{4}; 1; \gamma^{-2}) = \frac{2}{\pi} (\gamma + 1)^{\frac{1}{2}} E(k)$$

$$E(k) = \int_0^{\pi/2} (1 - k^2 \sin^2 \theta)^{\frac{1}{2}} \, d\theta$$

$$k = [2/(1 + \gamma)]^{\frac{1}{2}} = [2a/(1 + 2a)]^{\frac{1}{2}}.$$

The series (77) for L_1 is obtained by combining these results.

The asymptotic expression (78) for L_1 may be obtained by setting $z = av$ in (79). The resulting integral is proportional to

$$\int_{\rho}^{\infty} e^{-v} f(v) \, dv, \qquad f(v) = [e^{-z} I_0(z) z^{-\frac{3}{2}}]_{z=av}.$$

When this is integrated by parts by using (82), an asymptotic series good for large ρ is obtained. Taking only the first two terms, $f(v) + f'(v)$, leads to (78).

When ρ is large, the largest term in L_1 is the first term in (79), and the asymptotic behavior of the Bessel functions in (75) shows that $L_2 \sim 2^{-1} \pi^{-2} A e^{-\rho}$. Addition then gives

$$N_+ \sim \frac{A}{2\pi} \frac{e^{-\rho}}{\pi} + \frac{r e^{-\rho}}{(4\pi\rho)^{\frac{1}{2}}} e^{-a\rho} I_0(a\rho), \tag{83}$$

which is given in Section 2 as (30).

It should be mentioned that in computing the upper exact curve of Figure 4 L_1 was evaluated by using Simpson's rule on (74). For the accuracy required, this was more convenient than summing the series (77).

Gaussian Signal

Let $\phi'(t)$ be a Gaussian noise signal. Since the ensemble of values $\phi'(t_1)$ is Gaussian, $u = \phi'(t_1)/2\pi r$ is distributed normally about zero and its variance is

$$a = \frac{\overline{\phi'^2}}{(2\pi r)^2} \tag{84}$$

where $\overline{\phi'^2}$ denotes the mean square value of the signal. Both ϕ' and $2\pi r$ are measured in radians per second and a is dimensionless. The parameter a defined by (84) has the same significance as the $a = \alpha^2/2 = (A^2/2)/(2\pi r)^2$ used in the preceding section because the mean square value of $\phi' = A \sin \omega_s t$ is $A^2/2$.

Equations (70) and (65) give

$$N_+ = \int_{-\infty}^{\infty} H_+(t_1)(2\pi a)^{-\frac{1}{2}} e^{-u^2/2a} \, du$$

$$= \frac{r}{(8\pi a)^{\frac{1}{2}}} \int_{-\infty}^{\infty} e^{-u^2/2a}\{(1 + u^2)^{\frac{1}{2}}[1 - \mathrm{erf}\,(\rho + \rho u^2)^{\frac{1}{2}}]$$

$$+ ue^{-\rho}\,\mathrm{erf}\,(u\,\sqrt{\rho})\} \, du$$

$$= \frac{r}{\sqrt{\pi}} \int_{\sqrt{\rho}}^{\infty} e^{-v^2}(1 + 2av^2)^{\frac{1}{2}} \, dv, \tag{85}$$

where the last expression was obtained by several partial integrations. It may be verified by differentiating the last two lines with respect to $\sqrt{\rho}$ and then integrating by parts to remove $\mathrm{erf}\,(u\,\sqrt{\rho})$ from the integrand.

When ρ is large, an integration by parts shows that

$$N_+ \sim re^{-\rho} \left(\frac{1 + 2a\rho}{4\pi\rho}\right)^{\frac{1}{2}}. \tag{86}$$

6. ASYMPTOTIC EXPRESSION FOR $W(0)$—SIGNAL PRESENT

In Section 3 it was shown that when the carrier is unmodulated $W(0)$ is given asymptotically by (45) as $\rho \to \infty$. When a signal is present, the work is complicated by the fact that $\theta'(t)$ may have a d-c component. Furthermore, the ensemble average $\langle n_+ \rangle$ of the number of upward jumps in the long interval $(0, T)$ may not be exactly TN_+ but may oscillate around TN_+ as T increases.

In this section $\theta(t)$ is defined by the equation

$$Q \cos (\omega_c t + \phi) + I_N = R \cos (\omega_c t + \phi + \theta),$$

but we still have

$$\int_0^T \theta'(t) \, dt = (n_+ - n_-)2\pi + F(T) - F(0), \tag{87}$$

where n_+, n_- are the numbers of upward and downward jumps in $(0, T)$ and $F(T)$ and $F(0)$ are $0(1)$ just as in (47).

First we consider the case in which the signal $\phi'(t)$ is a Gaussian noise. Then we go on to the case in which $\phi'(t)$ is some specified function of time, for example $\phi' = A \sin \omega_s t$.

Gaussian Signal

In this case the difficulties mentioned in the introductory paragraph do not occur. Considerations of symmetry show that $\theta'(t)$ has no d-c component, hence $W(0)$ is given by (46). However, now the ensemble average on the right extends over the ensemble of the Gaussian signals $\phi'(t)$ as well as the ensemble of the Gaussian noise currents $I_N(t)$. Considerations of stationarity and symmetry show that the expected values of n_+ and n_- are both equal to TN_+ where N_+ is given by (85). The argument given in Section 3 still holds and it follows that

$$W(0) \sim 16\pi^2 N_+ \tag{88}$$

228

as $\rho \to \infty$. To repeat, N_+ in (88) is given exactly by the integral (85) and approximately by the expression (86).

Signal Given as a Definite Function of Time

When $\phi'(t)$ is a given function of time, the d-c component of $\theta'(t)$ is

$$\bar{\theta'} = \lim_{T \to \infty} \frac{1}{T} \int_0^T \theta'(t)\, dt$$

$$= (N_+ - N_-)2\pi = \lim_{T \to \infty} \frac{[\phi(0) - \phi(T)]e^{-\rho}}{T}, \tag{89}$$

where (87) has been divided by T and the limit $N_\pm$ of $n_\pm / T$ is supposed to exist. The last expression in (89) may be obtained from (69). It may also be obtained by integrating the known result

$$\langle \theta'(t_1) \rangle = -\phi'(t_1)e^{-\rho}, \tag{90}$$

in which the left-hand side is the value of $\theta'(t)$ at time t_1 averaged over the noise ensemble; this equation is discussed briefly in the appendix.

We now show that when rather general conditions are satisfied (as they are for $\phi' = A \sin \omega_s t$ and for $\phi' = $ constant)

$$W(0) \sim 8\pi^2 (N_+ + N_-) \tag{91}$$

as $\rho \to \infty$. Here $W(f)$ is the power spectrum of $\theta'(t) - \bar{\theta'}$ and N_+, N_- are the expected number of upward and downward clicks per second.

The proof of (91) is along the lines developed in Section 3 for the no-signal case. In place of (46) we have

$$W(0) = \lim_{T \to \infty} \frac{2}{T} \left\langle \left[\int_0^T (\theta'(t) - \bar{\theta'})\, dt \right]^2 \right\rangle \tag{92}$$

where, from (87) and (89), the value of the integral is

$$(n_+ - n_- - TN_+ + TN_-)2\pi + F(T) - F(0). \tag{93}$$

From Section 4 it follows that when a particular $\theta(t)$ is selected at random from the universe of $\theta(t)$'s the chance that it will jump upward (more precisely, pass upward through an odd multiple of π) during the short interval $(t_1, t_1 + dt_1)$ is $H_+(t_1)\, dt_1 + o(dt_1)$. We now assume that for large ρ the probability of a jump in $(t_1, t_1 + dt_1)$ becomes independent of jumps that have occurred. For example, it is assumed that the chance of upward jumps in $(t_1, t_1 + dt_1)$ and $(t_2, t_2 + dt_2)$ is simply the product $H_+(t_1)\, dt_1 H_+(t_2)\, dt_2$. It is known that such a process behaves much like a Poisson process in that the chance of exactly k upward jumps in $(0, T)$, that is, $n_+ = k$, is

$$(a_+^k / k!) \exp(-a_+),$$

where

$$a_+ = \langle n_+ \rangle = \int_0^T H_+(t_1)\, dt_1.$$

Furthermore, the variance of n_+ is equal to its mean value just as in the Poisson process:

$$\langle (n_+ - a_+)^2 \rangle = \langle n_+ \rangle. \tag{94}$$

The symbol a_+ is used to denote the ensemble average $\langle n_+ \rangle$ in order to simplify the appearance of some of the later expressions.

Writing (93) as

$$[(n_+ - a_+) - (n_- - a_-) + (a_+ - a_- - TN_+ + TN_-)]2\pi + F(T) - F(0) \tag{95}$$

carries it into the analogue to the right-hand side of (51). As we go from one member of the ensemble to another, $a_\pm$, $N_\pm$ remain fixed while $n_\pm$, $F(T)$, $F(0)$ behave like random variables. Squaring (95), averaging over the noise ensemble, and making the same assumptions regarding $F(T)$ and $F(0)$ as in Section 3 takes (92) into

$$W(0) \sim \lim_{T \to \infty} \frac{8\pi^2}{T} \left[\langle (n_+ - a_+)^2 \rangle + \langle (n_- - a_-)^2 \rangle \right.$$
$$\left. + (a_+ - a_- - TN_+ + TN_-)^2 + \left\langle \frac{F^2(T)}{4\pi^2} \right\rangle + \left\langle \frac{F^2(0)}{4\pi^2} \right\rangle \right].$$

The terms in $F(T)$ and $F(0)$ vanish in the limit. From (94) and the assumption that $N_\pm$ is the limit of $\langle n_\pm \rangle / T$,

$$W(0) \sim 8\pi^2 (N_+ + N_-) + 8\pi^2 \lim_{T \to \infty} \left[\frac{a_+ - a_-}{\sqrt{T}} - (N_+ - N_-)\sqrt{T} \right]^2. \tag{96}$$

The value of the limit in (96) may be determined for any particular $\phi'(t)$ with the help of expression (89) and

$$\frac{(a_+ - a_-)2\pi}{\sqrt{T}} = \frac{[\phi(0) - \phi(T)]e^{-\rho}}{\sqrt{T}}, \tag{97}$$

which follows from (67) and the definition of $a_\pm$.

Examples. (a) $\phi'(t) = constant = 2\pi f_0$. Integration gives $-2\pi f_0 T$ for $\phi(0) - \phi(T)$ and it follows from (89) and (97) that the limit term in (96) is zero. Hence (96) becomes (91) for $W(0)$. In the present example the expressions (71) for N_+ and N_- lead to

$$W(0) \sim 8\pi^2 \{ (r^2 + f_0^2)^{1/2} [1 - \mathrm{erf}\,(\rho + \rho f_0^2 r^{-2})^{1/2}] + f_0 e^{-\rho}\,\mathrm{erf}\,(\rho f_0^2 r^{-2})^{1/2} \}. \tag{98}$$

This reduces to the central result (45) of Section 3 when $f_0 = 0$.

(b) $\phi' = A \sin \omega_s t$. Integration gives $-A(1 - \cos \omega_s T)/\omega_s$ for $\phi(0) - \phi(T)$. From symmetry, or the last equation in (89), it is seen that $N_+ = N_-$. Furthermore, the limit of the right-hand side of (97) is zero. Expression (96) again reduces to expression (91) for $W(0)$. Equations for the appropriate value of N_+ are given in Section 5.

7. THE POWER SPECTRUM OF $y(t)$

The conjectured approximation (35) for the power spectrum $W(f)$ of θ' when a signal ϕ' is present contains the power spectrum $W_y(f)$ of

$$y = I_s \cos \phi - I_c \sin \phi, \tag{99}$$

where y is the component of the noise I_N which is perpendicular to the signal vector $Qe^{i\phi}$. Here we mention several expressions for $W_y(f)$. Although the results probably are not new, they are given for the sake of completeness.

The average value of $y(t_1)\, y(t_1 + \tau)$, taken over the ensemble of noise currents, is

$$\langle y_1 y_2 \rangle = \langle I_{s1} I_{s2} \rangle \cos \phi_1 \cos \phi_2 + \langle I_{c1} I_{c2} \rangle \sin \phi_1 \sin \phi_2$$

$$= \langle I_{s1} I_{s2} \rangle (\cos \phi_1 \cos \phi_2 + \sin \phi_1 \sin \phi_2) \tag{100}$$

where the subscripts 1 and 2 refer to the instants t_1 and $t_1 + \tau$ and use has been made of the independence of I_c and I_s. Averaging (100) with respect to the time t_1 gives the autocorrelation function of y. Since $\langle I_{s1} I_{s2} \rangle$ is the autocorrelation function of I_s and the time average of $\cos \phi_1 \cos \phi_2$ is the autocorrelation function of $\cos \phi$, taking the Fourier cosine transform of (100) and using the convolution theorem gives

$$W_y(f) = \int_0^\infty w(f_c + x)[w_{c\phi}(|f - x|) + w_{c\phi}(f + x) + w_{s\phi}(|f - x|) \\ + w_{s\phi}(f + x)]\, dx. \tag{101}$$

Here $w_{c\phi}(f)$, $w_{s\phi}(f)$ are the power spectra of $\cos \phi(t)$ and $\sin \phi(t)$, and $2w(f_c + f)$ is the power spectrum of I_c and I_s.

For the signal $\phi' = A \sin \omega_s t$ we take $\phi = -A\omega_s^{-1} \cos \omega_s t$. Then $\cos \phi$ and $\sin \phi$ are given by the real and imaginary parts of

$$\exp\left(-i\, A\omega_s^{-1} \cos \omega_s t\right) = J_0(A\omega_s^{-1}) + 2 \sum_{n=1}^\infty (-i)^n\, J_n(A\omega_s^{-1}) \cos \omega_s nt,$$

and the sum of the power spectra of $\cos \phi$ and $\sin \phi$ is

$$w_{c\phi}(f) + w_{s\phi}(f) = 2J_0^2(A\omega_s^{-1})\, \delta(f) + 2 \sum_{n=1}^\infty J_n^2(A\omega_s^{-1})\, \delta(f - nf_s), \tag{102}$$

where $\delta(f)$ denotes the unit impulse function. In (102) f is supposed to be $\geqslant 0$ and the integral of $\delta(f)$ from $f = 0$ to ∞ is $\frac{1}{2}$. Substituting (102) in (101) gives

$$W_y(f) = 2 \sum_{n=-\infty}^{n} w(f_c + f + nf_s)\, J_n^2(A\omega_s^{-1}). \tag{103}$$

This result is stated as (38) in Section 2.

When the signal ϕ' is a Gaussian noise with power spectrum $w_{\phi'}(f)$ $[w_{\phi'}(f)\, df$

is measured in (radians per second)^2], it may be shown that

$$W_y(f) = 4 \int_0^\infty R_{Is}(\tau) \cos 2\pi f\tau \exp\left[-F(\tau)\right] d\tau, \qquad (104)$$

where

$$R_{Is}(\tau) = \int_0^\infty 2w(f_c + f) \cos 2\pi f\tau \, df$$

$$F(\tau) = \int_0^\infty w_{\phi'}(f) \cdot (1 - \cos 2\pi f\tau)(2\pi f)^{-2} \, df.$$

APPENDIX. PROOF OF EQUATION (90)

Equation (90), namely,

$$\langle\theta'(t_1)\rangle = -\phi'(t_1)e^{-\rho}, \qquad (105)$$

holds when the first moment of $w(f)$ about f_c is zero. Here $w(f)$ is the power spectrum of $I_N(t)$ and f_c is the carrier frequency. In particular, it holds when $w(f)$ is symmetrical about f_c, as is assumed throughout this chapter. Results equivalent to (105) for the special cases in which ϕ' is a constant and a sine wave are given in Reference 4 and in Stumper's paper [2]. A more general result, which contains (90) as a special case, is given by Middleton[3].

Equation (105) is of interest because it shows that the discriminator output $\phi' + \theta'$ contains the component $(1 - e^{-\rho})\phi'$, that is, the presence of the noise reduces the output signal by the factor $1 - e^{-\rho}$.

The result (105) may be established by writing the relation between the phase angles [see (55)]

$$Qe^{i\phi} + I_c + iI_s = Re^{i\phi+i\theta}.$$

Then logarithmic differentiation gives

$$\frac{R'}{R} + i(\phi' + \theta') = \frac{i\phi'Qe^{i\phi} + I'_c + iI'_s}{Qe^{i\phi} + I_c + iI_s}.$$

The ensemble average of the left-hand side at time t_1 is obtained by averaging the right-hand side over all values of I_c, I_s, I'_c, I'_s. As indicated in (61), these four Gaussian variables are independent when $w(f)$ is symmetrical about f_c. Averaging over I'_c, I'_s removes $I'_c + iI'_s$ from the numerator, leaving

$$\langle R'R^{-1} + i(\phi' + \theta')\rangle = \frac{i\phi'Q}{2\pi b_0} \int_{-\infty}^\infty dI_c \int_{-\infty}^\infty dI_s \frac{\exp\left[i\phi - (2b_0)^{-1}(I_c^2 + I_s^2)\right]}{Qe^{i\phi} + I_c + iI_s}.$$

The integral converges at the zero of the denominator. It may be evaluated by setting $I_c = r\cos\alpha - Q\cos\phi$, $I_s = r\sin\alpha - Q\sin\phi$, $\beta = \alpha - \phi$, and using

$$\frac{Q}{2\pi b_0} \int_0^\infty dr \int_0^{2\pi} d\beta \cos\beta \exp\left[-(2b_0)^{-1}(r^2 - 2rQ\cos\beta + Q^2)\right] = 1 - e^{-\rho}.$$

The resulting expression for $\langle\phi' + \theta'\rangle$ is $(1 - e^{-\rho})\phi'$ and (105) follows immediately.

REFERENCES

1. Crosby, M. G.   Frequency modulation noise characteristics.   *Proc. IRE*, **25,** 472–514 (1937).
2. Stumpers, F. L. H. M.   Theory of frequency-modulation noise.   *Proc. IRE*, **36,** 1081– 1092 (1948).
3. Middleton, D.   The spectrum of frequency-modulated waves after reception in random noise I, II.   *Quart. Appl. Math.*, **7,** 129–174 (1949); **8,** 59–80 (1950).
4. Rice, S. O.   Statistical properties of a sine wave plus random noise.   *Bell System Tech. J.*, **27,** 109–157 (1948).   Good accounts of noise in FM systems are given in a number of textbooks.   Among the more mathematical treatments we mention, for example, Chapter 15 in *Statistical Communication Theory*, by D. Middleton, McGraw-Hill, 1960, and Chapter 13 (describing work by M. C. Wang) in *Threshold Signals*, by J. L. Lawson and G. E. Uhlenbeck, MIT Rad. Lab. Ser. Vol. 24, McGraw-Hill, 1950.   For additional references, see Middleton's book.
5. Blachman, N. M.   Theoretical study of the demodulation of a frequency-modulated carrier and random noise by an FM receiver.   Cruft Laboratory Tech. Report No. 31 (March 1948).

II

FM Circuit Theory

Reprinted from *BSTJ*, Vol. 16, No. 4, pp. 513–540, Oct. 1937

Variable Frequency Electric Circuit Theory with Application to the Theory of Frequency-Modulation

By JOHN R. CARSON AND THORNTON C. FRY

In this paper the fundamental formulas of variable frequency electric circuit theory are first developed. These are then applied to a study of the transmission, reception and detection of frequency modulated waves. A comparison with amplitude modulation is made and quantitative formulas are developed for comparing the noise-to-signal power ratio in the two modes of modulation.

FREQUENCY modulation was a much talked of subject twenty or more years ago. Most of the interest in it then centered around the idea that it might afford a means of compressing a signal into a narrower frequency band than is required for amplitude modulation. When it was shown that not only could this hope not be realized,* but that much wider bands might be required for frequency modulation, interest in the subject naturally waned. It was revived again when engineers began to explore the possibilities of radio transmission at very short wave lengths where there is little restriction on the width of the frequency band that may be utilized.

During the past eight years a number of papers have been published on frequency modulation, as reference to the attached bibliography will show. That by Professor E. H. Armstrong † deals with this subject in comprehensive fashion. In his paper the problem of discrimination against extraneous noise is discussed, and it is pointed out that important advantages result from a combination of wide frequency bands together with severe amplitude limitation of the received signal waves. His treatment is, however, essentially nonmathematical in character, and it is therefore believed that a mathematical study of this phase of the problem will not be unwelcome. This the present paper aims to supply by developing the basic mathematics of frequency modulation and applying it to the question of noise discrimination with or without amplitude limitation.

The outstanding conclusions reached in the present paper, as regards discrimination against noise by frequency modulation, may be briefly summarized as follows:

* See Bibliography, No. 1.
† See Bibliography, No. 12.

(1) To secure any advantage by frequency modulation as distinguished from amplitude modulation, the frequency band width must be much greater in the former than in the latter system.

(2) Frequency modulation in combination with severe amplitude limitation for the received wave results in substantial reduction of the noise-to-signal power ratio. Formulas are developed which make possible a quantitative estimate of the noise-to-signal power ratio in frequency modulation, with and without amplitude limitation, as compared with amplitude modulation.

It is a pleasure to express our thanks to several colleagues who have been helpful in various ways: to Dr. Ralph Bown who in a brief but very incisive memorandum, which was not intended to be a mathematical study, disclosed all the essential ideas of the quasi-stationary method of attack; to Mr. J. G. Chaffee,* who has been conducting experimental work on frequency modulation in these Laboratories for some years past, by means of which quantitative checks on the accuracy of some of the principal results have been possible; and to various associates, especially Mr. W. R. Bennett and Mrs. S. P. Mead, for detailed criticism of certain portions of the work.

I

In the well-known steady-state theory of alternating currents, the e.m.f. and the currents in all the branches of a network in which the e.m.f. is impressed involve the time t only through the common factor $e^{i\omega t}$ where $i = \sqrt{-1}$ and ω is the *constant* frequency. To this fact is attributable the remarkable simplicity of alternating current theory and calculation, and also the fact that the network is completely specified by its complex admittance $Y(i\omega)$. Thus, if the e.m.f. is $Ee^{i\omega t}$, the steady-state current is

$$I_{ss} = EY(i\omega)e^{i\omega t}. \tag{1}$$

In the present paper we shall deal with the case where the frequency is *variable*, and write the impressed e.m.f. as

$$E \exp\left(i \int_0^t \Omega(t)dt \right). \tag{2}$$

$\Omega(t)$ will be termed the *instantaneous* frequency. This agrees with the usual definition of frequency when Ω is a constant; it is the rate of change of the phase angle at time t; and in addition the interval T between adjacent zeros of sin $\int\Omega(t)dt$ or cos $\int\Omega(t)dt$ is approximately $\pi/\Omega(t)$ in cases of practical importance.

* See Bibliography, No. 11.

Instead of dealing with an arbitrary instantaneous frequency $\Omega(t)$ we shall suppose that

$$\Omega(t) = \omega + \mu(t), \tag{3}$$

where ω is a constant and $\mu(t)$ is the variable part of the instantaneous frequency. In practical applications $\mu(t)$ will be written as $\lambda s(t)$ where λ is a real parameter and the mean square value $\overline{s^2}$ of $s(t)$ is taken as equal to $1/2$. Other restrictions on $\mu(t)$ will be imposed in the course of the theory to be developed in this paper. Fortunately these restrictions do not interfere with the application of the theory to important problems.

The steady-state current as given by (1) varies with time in precisely the same way as the impressed e.m.f. When the frequency is variable this is no longer true. On the other hand, formula (1) suggests a "quasi-stationary" or "quasi-steady-state current" component, I_{qss}, defined by the formula

$$I_{qss} = E Y(i\Omega) \cdot \exp\left(i \int_0^t \Omega dt \right), \tag{4}$$

which corresponds exactly to (1) with the distinction that the admittance is now an explicit function of time. We are thus led to examine the significance of I_{qss} as defined above and the conditions under which it is a valid approximate representation of the actual response of the network to a variable frequency electromotive force, as given by (2).

We start with the fundamental formula of electric circuit theory.[1] Let an e.m.f. $F(t)$ be impressed at time $t = 0$, on a network of indicial admittance $A(t)$; then the current $I(t)$ in the network is given by

$$I(t) = \int_0^t F(t - \tau) \cdot A'(\tau) d\tau. \tag{5}$$

Here $A'(t) = d/dt \cdot A(t)$ and it is supposed that $A(0) = 0$. (This restriction does not limit our subsequent conclusions and is introduced merely to simplify the formulas. Furthermore $A(0)$ is actually zero in all physically realizable networks.)

Omitting the superfluous amplitude constant E we have

$$F(t) = \exp\left(i \int_0^t \Omega dt \right)$$

$$= \exp\left(i\omega t + i \int_0^t \mu dt \right), \tag{6}$$

[1] See J. R. Carson, "Electric Circuit Theory and Operational Calculus," p. 16.

$$F(t - \tau) = \exp\left[i(t - \tau)\omega + i \int_0^{t-\tau} \mu d\tau_1 \right]$$

$$= \exp\left[i(t - \tau)\omega + i \int_0^t \mu d\tau_1 - i \int_{t-\tau}^t \mu d\tau_1 \right]$$

$$= \exp\left[i\Omega(t) \right] \cdot \exp\left[- i\omega\tau - i \int_0^\tau \mu(t - \tau_1)d\tau_1 \right]. \quad (7)$$

Substituting this expression in (5) for $F(t - \tau)$ and writing

$$\exp\left(- i \int_0^\tau \mu(t - \tau_1)d\tau_1 \right) = M(t, \tau), \quad (8)$$

we have for the current in the network

$$I = e^{i \int \Omega dt} \cdot \int_0^t M(t, \tau)e^{-i\omega\tau}A'(\tau)d\tau. \quad (9)$$

We now split the integral into two parts, thus:

$$\int_0^t = \int_0^\infty - \int_t^\infty .$$

The second integral on the right represents an initial transient which dies away for sufficiently large values of time, t, while the infinite integral represents the total current, I, for sufficiently large values of t. We have therefore

$$I = e^{i \int \Omega dt} \cdot \int_0^\infty M(t, \tau)e^{-i\omega\tau}A'(\tau)d\tau + I_T \quad (10)$$

$$= Y(i\omega, t)e^{i \int \Omega dt} + I_T,$$

where

$$Y(i\omega, t) = \int_0^\infty M(t, \tau)e^{-i\omega\tau}A'(\tau)d\tau. \quad (11)$$

The transient current,[2] I_T, is then given by

$$I_T = e^{i \int \Omega dt} \int_t^\infty M(t, \tau)e^{-i\omega\tau}A'(\tau)d\tau. \quad (12)$$

The foregoing formulas correspond precisely with the formulas for a constant frequency impressed e.m.f.; these are

$$I_{ss} = e^{i\omega t} \int_0^\infty e^{-i\omega\tau}A'(\tau)d\tau, \quad (10a)$$

---

[2] Hereafter the transient term I_T of (10) will be consistently neglected and the symbol I will refer only to the quasi-stationary current.

$$Y(i\omega) = \int_0^\infty e^{-i\omega\tau}A'(\tau)d\tau, \tag{11a}$$

$$I_T = e^{i\omega t}\int_t^\infty e^{-i\omega\tau}A'(\tau)d\tau, \tag{12a}$$

to which the more general formulas reduce when $\mu = 0$ and consequently $M = 1$.

We have now to evaluate $Y(i\omega, t)$ as given by (11). We shall assume tentatively, at the outset, that $\mu = \lambda s(t)$ has the following properties:

$$\lambda s(t) \ll \omega \quad \text{for all values of } t,$$
$$-1 \leq s(t) \leq 1,$$
$$-1 \leq \int_0^t s\,dt \leq 1.$$

With these restrictions the instantaneous frequency lies within the limits $\omega \pm \lambda$.

Let us now replace $M(t, \tau)$ by the formal series expansion

$$M(t, \tau) = M(t, 0) + \frac{\tau}{1!}\left[\frac{\partial}{\partial\tau}M(t, \tau)\right]_{\tau=0}$$
$$+ \frac{\tau^2}{2!}\left[\frac{\partial^2}{\partial\tau^2}M(t, \tau)\right]_{\tau=0} + \cdots, \tag{13}$$

which converges in the vicinity of all values of t for which s has a complete set of derivatives. Then, if we write

$$\left[\frac{\partial^n}{\partial\tau^n}M(t, \tau)\right]_{\tau=0} = (-i)^n C_n(t) \tag{13a}$$

and substitute (13) in (11), we get

$$Y(i\omega, t) = \int_0^\infty e^{-i\omega\tau}A'(\tau)d\tau + \sum_1^\infty (-i)^n C_n \int_0^\infty \frac{\tau^n}{n!}e^{-i\omega\tau}A'(\tau)d\tau. \tag{14}$$

From (11a) it follows at once that

$$\int_0^\infty \frac{\tau^n}{n!}e^{-i\omega\tau}A'(\tau)d\tau = \frac{i^n}{n!}\frac{d^n}{d\omega^n}Y(i\omega), \tag{15}$$

so that

$$Y(i\omega, t) = Y(i\omega) + \sum_1^\infty \frac{1}{n!}C_n(t)\frac{d^n}{d\omega^n}Y(i\omega). \tag{16}$$

The coefficients C_n are easily evaluated from (8) and (13a); they are [3]

$$C_1 = \mu(t),$$

$$C_2 = \mu^2 - i\frac{d}{dt}\mu, \tag{17}$$

$$\cdot \quad \cdot \quad \cdot \quad \cdot \quad \cdot \quad \cdot \quad \cdot$$

$$C_{n+1} = \left(\mu - i\frac{d}{dt}\right)C_n.$$

Now consider the quasi-stationary admittance $Y(i\Omega)$. Writing $\Omega = \omega + \mu(t)$ and expanding as a power series, we have (assuming that the series is convergent)

$$Y(i\Omega) = Y(i\omega) + \sum_{1}^{\infty}\frac{\mu^n}{n!}\frac{d^n}{d\omega^n}Y(i\omega). \tag{18}$$

From (16), (17) and (18) we have at once

$$Y(i\omega, t) = Y(i\Omega) + \sum_{2}^{\infty}\frac{1}{n!}D_n(t)\frac{d^n}{d\omega^n}Y(i\omega), \tag{19}$$

where

$$D_2 = -i\frac{d}{dt}\mu(t),$$

$$D_3 = -i3\mu\frac{d\mu}{dt} - \frac{d^2}{dt^2}\mu, \tag{20}$$

$$\cdot \quad \cdot \quad \cdot \quad \cdot \quad \cdot \quad \cdot \quad \cdot$$

$$D_{m+1} = C_{m+1} - \mu^{m+1}.$$

Consequently, the total current, after initial transients have died away, is given by

$$I = I_{qss} + \Delta(t)$$

$$= \exp\left(i\int_0^t \Omega dt\right) \cdot \left[Y(i\Omega) - \frac{i}{2!}\frac{d\mu}{dt}\frac{d^2Y}{d\omega^2}\right.$$

$$\left. - \frac{1}{3!}\left(i3\mu\frac{d\mu}{dt} + \frac{d^2\mu}{dt^2}\right)\frac{d^3Y}{d\omega^3} + \cdots\right]. \tag{21}$$

We have thus succeeded in expressing the response of the network in terms of the quasi-stationary current

$$I_{qss} = Y(i\Omega) \cdot \exp\left(i\int \Omega dt\right) \tag{22}$$

[3] From these recursion formulas C_n can be derived in the compact form

$$C_n = \left(\mu - i\frac{d}{dt}\right)\left(\mu - i\frac{d}{dt}\right)\cdots\left(\mu - \frac{d}{dt}\right)\mu$$

$$= \left(\mu - i\frac{d}{dt}\right)^{n-1}\mu \quad symbolically.$$

and a correction series Δ, which depends on the derivatives of the steady-state admittance $Y(i\omega)$ with respect to frequency and the derivatives of the variable frequency $\mu(t)$ with respect to time.

If the parameter λ is sufficiently large and the derivatives of s are small enough so that C_n may be replaced by the two leading terms, we get

$$C_n = \mu^n - i\frac{(n-1)n}{2}\mu'\mu^{n-2}, \quad \mu' = \frac{d\mu}{dt}.$$

Then by (16) and (18)

$$
\begin{aligned}
Y(i\omega, t) &= Y(i\Omega) - \frac{i\mu'}{2}\sum_{2}^{\infty}\frac{\mu^{n-2}}{(n-2)!}\frac{d^n}{d\omega^n}\,Y(i\omega) \\
&= Y(i\Omega) - \frac{i\mu'}{2}\frac{\partial^2}{\partial\mu^2}\,Y(i\Omega) \\
&= Y(i\Omega) - \frac{i\mu'}{2}\frac{d^2}{d\Omega^2}\,Y(i\Omega) \\
&= Y(i\Omega) + \frac{i\mu'}{2}\,Y^{(2)}(i\Omega).
\end{aligned}
\tag{16a}
$$

The preceding formulas are so fundamental to variable frequency theory and the theory of frequency modulation that an alternative derivation seems worth while. We take the applied e.m.f. as

$$E\exp\left(i\omega_c t + i\theta + i\int_0^t \mu dt\right),\tag{23}$$

the phase angle θ being included for the sake of generality.

Now in any finite epoch $0 \le t \le T$, it is always possible to write

$$\exp\left(i\int_0^t \mu dt\right) = \int_{-\infty}^{\infty} F(i\omega)e^{i\omega t}d\omega,\tag{24}$$

thus expressing the function on the left as a Fourier integral. For present purposes it is quite unnecessary to evaluate the Fourier function $F(i\omega)$.

Substitution of (24) in (23) gives for the current

$$I = E\cdot\exp\left(i\omega_c t + i\theta\right)\cdot\int_{-\infty}^{\infty} F(i\omega)\,Y(i\omega_c + i\omega)e^{i\omega t}d\omega.\tag{25}$$

We suppose as before that, in the interval $0 \le t \le T$, $\mu(t)$ and its derivatives are continuous. We can then expand the admittance function Y in the form

$$Y(i\omega_c + i\omega) = Y(i\omega_c) + \frac{i\omega}{1!} Y^{(1)}(i\omega_c) + \frac{(i\omega)^2}{2!} Y^{(2)}(i\omega_c) + \cdots$$

$$= Y(i\omega_c) + \sum_1^\infty \frac{(i\omega)^n}{n!} Y^{(n)}(i\omega_c)$$

$$= Y(i\omega_c) + \sum_1^\infty \frac{\omega^n}{n!} \frac{d^n}{d\omega_c^n} Y(i\omega_c). \qquad (26)$$

Substitution of (26) in (25) gives

$$I = E \cdot \exp\,(i\omega_c t + i\theta) \sum_0^\infty \frac{1}{n!} \frac{d^n}{d\omega_c^n} Y(i\omega_c) \int_{-\infty}^\infty \omega^n F(i\omega) e^{i\omega t} d\omega. \quad (27)$$

But by the identity (24) and repeated differentiations with respect to t, we have

$$\int_{-\infty}^\infty \omega F(i\omega) e^{i\omega t} d\omega = \mu \exp\left(i \int_0^t \mu dt \right),$$

$$\int_{-\infty}^\infty \omega^2 F(i\omega) e^{i\omega t} d\omega = \left(\mu^2 - i\frac{d\mu}{dt} \right) \exp\left(i \int_0^t \mu dt \right),$$

$$\cdot \quad \cdot \quad \cdot \quad \cdot \quad \cdot \quad \cdot \quad \cdot \quad \cdot \quad \cdot \quad \cdot \quad \cdot \quad \cdot \quad \cdot \quad \cdot \quad \cdot \qquad (28)$$

$$\int_{-\infty}^\infty \omega^n F(i\omega) e^{i\omega t} d\omega = C_n \exp\left(i \int_0^t \mu dt \right).$$

Substitution of (28) in (27) gives

$$I = E \exp\left(i \int_0^t \Omega dt + i\theta \right) \cdot \left\{ Y(i\omega_c) + \sum_1^\infty \frac{1}{n!} C_n \frac{d^n}{d\omega_c^n} Y(i\omega_c) \right\}, \quad (29)$$

which agrees with (16).

Formula (25), as it stands, includes the initial transients at time $t = 0$ as well as any which occur at discontinuities in $\mu(t)$. Differentiation with respect to t under the integral sign, however, in effect eliminates these transients and (29) leaves only the quasi-stationary current (plus the correction series given in (19)).

The series appearing in formula (29) may not be convergent; in any case its computation is laborious. Furthermore, in its application to the theory of frequency modulation, terms beyond the first two represent distortion. For these reasons it is often preferable to proceed as follows:

Returning to formula (25), we write

$$Y(i\omega_c + i\omega) = \left(1 + \frac{\omega}{1!} \frac{d}{d\omega_c} + \cdots + \frac{\omega^n}{n!} \frac{d^n}{d\omega_c^n} \right) Y(i\omega_c) + R_n(\omega_c, \omega), \quad (30)$$

thus defining the *remainder R_n*. Then (29) becomes

$$I = E \exp\left(i \int_0^t \Omega dt + i\theta \right) \cdot \left[1 + \frac{C_1}{1!} \frac{d}{d\omega_c} + \cdots + \frac{C_n}{n!} \frac{d^n}{d\omega_c{}^n} \right] Y(i\omega_c)$$

$$+ E \exp\left(i\omega_c t + i\theta \right) \int_{-\infty}^{\infty} R_n(\omega_c, \omega) F(i\omega) e^{i\omega t} d\omega. \quad (31)$$

In practice it is usually desirable to take $n = 1$.

Now the infinite integral

$$D(t) = \int_{-\infty}^{\infty} R_n(\omega_c, \omega) F(i\omega) e^{i\omega t} d\omega \qquad (32)$$

must be kept small if the finite series in (31) is to be an accurate representation of the current I. While it is not in general computable, we see that, in order to keep it small, $R_n(\omega_c, \omega)$ must be small over the essential range of frequencies of $F(i\omega)$. In cases of practical importance we shall find (see Appendix 1) this range is from $\omega = -\lambda$ to $\omega = +\lambda$.

If the transducer introduces a large phase shift, the linear part of which is predominant in the neighborhood of $\omega = \omega_c$, it is preferable to express the received current I in terms of a "retarded" time. To do this, return to (25) and write

$$Y(i\omega_c + i\omega) = | Y(i\omega_c + i\omega) | e^{-i\phi}, \qquad (33)$$
$$\phi = \omega_c \tau + \omega \tau + \beta(\omega) + \theta_c,$$
$$\beta(0) = \beta'(0) = 0,$$

so that

$$I = E \exp\left(i\omega_c t' + i\theta' \right) \int_{-\infty}^{\infty} | Y(i\omega_c + i\omega) | e^{-i\beta(\omega)} F(i\omega) e^{i\omega t'} d\omega, \quad (34)$$

where $t' = t - \tau$ is the "retarded" time and $\theta' = \theta - \theta_c$. Formula (34) is identical with (25) but is expressed in the "retarded" time.

Now we can expand the function

$$| Y(i\omega_c + i\omega) | e^{-i\beta(\omega)}$$

in powers of ω; thus

$$\left(1 + \omega \frac{d}{d\omega_c} \right) | Y(i\omega_c) | + \sum_2^{\infty} r_n(\omega_c) \omega^n,$$

where

$$r_n(\omega_c) = \frac{1}{n!} \left\{ \frac{\partial^n}{\partial\omega_c{}^n} | Y(i\omega_c + i\omega) | e^{-i\beta(\omega)} \right\}_{\omega=0};$$

245

and by substitution in (34) we get

$$I = E \exp\left(i \int_0^{t'} \Omega(\tau)d\tau + i\theta' \right)$$
$$\times \left[\left(1 + \lambda s(t')\frac{d}{d\omega_c} \right) |Y(i\omega_c)| + \sum_2^\infty \frac{r_n}{n!} C_n(t') \right], \quad (35)$$

which corresponds precisely with (29) except that it is expressed in terms of the retarded time t'. If the transducer introduces a large phase delay, (35) may be much more rapidly convergent than (29) and should be employed in preference thereto.

Corresponding to (30) we may write

$$Y(i\omega_c + i\omega)e^{-i\beta(\omega)} = \left(1 + \omega\frac{d}{d\omega_c} \right) |Y(i\omega_c)| + R,$$

which defines the remainder. Then

$$I = E \exp\left(i \int_0^{t'} \Omega d\tau + i\theta' \right) \cdot \left[|Y(i\omega_c)| + \lambda s(t')\frac{d}{d\omega_c} |Y(i\omega_c)| \right]$$
$$+ E \exp (i\omega_c t' + i\theta')D(t'), \quad (36)$$

where

$$D(t') = \int_{-\infty}^{\infty} R(\omega_c, \omega) \cdot F(i\omega)e^{-i\omega t'}d\omega. \quad (37)$$

Formulas (36) and (37) correspond precisely with (31) and (32) and the same remarks apply.

II

The foregoing will now be applied to the Theory of Frequency Modulation. A pure frequency modulated wave may be defined as a high frequency wave of constant amplitude, the "instantaneous" frequency of which is varied in accordance with a low frequency signal wave. Thus

$$W = \exp i\left(\omega_c t + \lambda \int_0^t s(t)dt \right) \quad (38)$$

is a pure frequency modulated wave. Here ω_c is the constant carrier frequency and $s(t)$ is the low frequency signal which it is desired to transmit. λ is a real parameter which will be termed the modulation index. The "instantaneous" frequency is then defined as

$$\omega_c + \lambda s(t).$$

It is convenient to suppose that $s(t)$ varies between ± 1; in this case the instantaneous frequency varies between the limits

$$\omega_c \pm \lambda.$$

In all cases it will be postulated that $\lambda \ll \omega_c$.

With the method of producing the frequency modulated wave (38) we are not here concerned beyond stating that it may be gotten by varying the capacity or inductance of a high frequency oscillating circuit by and in accordance with the signal $s(t)$.

Corresponding to (38), the pure *amplitude* modulated wave (carrier suppressed) is of the form

$$s(t) \cdot e^{i\omega_c t}. \tag{39}$$

If the maximum essential frequency in the signal $s(t)$ is ω_a, the wave (39) occupies the frequency band lying between $\omega_c - \omega_a$ and $\omega_c + \omega_a$, so that the band width is $2\omega_a$. In the pure *frequency* modulated wave the "instantaneous" frequency band width is 2λ. In practical applications $\lambda \gg \omega_a$. We shall now examine in more detail the concept of "instantaneous" frequency and the conditions under which it has physical significance.

The instantaneous frequency is, as stated, $\omega_c + \lambda s(t)$; a steady-state analysis is of interest and importance. To this end we suppose $s(t) = \cos \omega t$ so that ω is the frequency of the signal. Then the wave (38) may be written

$$e^{i\omega_c t}\left\{ \cos\left(\frac{\lambda}{\omega}\sin \omega t\right) + i \sin\left(\frac{\lambda}{\omega}\sin \omega t\right) \right\},$$

and, from known expansions,

$$W = \sum_{n=-\infty}^{\infty} J_n(\lambda/\omega)e^{i(\omega_c+n\omega)t}, \tag{40}$$

where J_n is the Bessel function of the first kind. Thus the frequency modulated wave is made up of sinusoidal components of frequencies

$$\omega_c \pm n\omega, \qquad n = 0, 1, 2, \cdots, \infty.$$

If $\lambda/\omega \gg 1$ (the case in which we shall be interested in practice) the terms in the series (40) beyond $n = \lambda/\omega$ are negligible; this follows from known properties of the Bessel functions. In this case the frequencies lie in the range

$$\omega_c \pm n\omega = \omega_c \pm \lambda,$$

which agrees with the result arrived at from the idea of instantaneous frequency. On the other hand, suppose we make λ so small that $\lambda/\omega \ll 1$. Then (40) becomes to a first order

$$e^{i\omega_c t} + \frac{1}{2}\left(\frac{\lambda}{\omega}\right) e^{i(\omega_c+\omega)t} - \frac{1}{2}\left(\frac{\lambda}{\omega}\right) e^{i(\omega_c-\omega)t},$$

so that the frequencies ω_c, $\omega_c + \omega$, $\omega_c - \omega$ are present in the pure frequency modulated wave.

It is possible to generalize the foregoing and build up a formal steady-state theory by supposing that

$$s(t) = \sum_{m=1}^{M} A_m \cos (\omega_m t + \theta_m). \tag{41}$$

On this assumption, it can be shown that the frequency modulated wave (38) is expressible as

$$W = \exp (i\omega_c t) \, \Pi_m \sum_{n=-\infty}^{\infty} J_n(v_m) \exp [in(\omega_m t + \theta_m)], \tag{42}$$

$$v_m = \lambda A_m/\omega_m.$$

The corresponding current is then

$$\exp (i\omega_c t) \, \Pi_m \sum_{n=-\infty}^{\infty} J_n(v_m) Y(i\omega_c + n\omega_m) \exp [in(\omega_m t + \theta_m)]. \tag{43}$$

Formulas (42) and (43) are purely formal and far too complicated for profitable interpretation. Consequently this line of analysis will not be carried farther.[4]

If we compare the pure *frequency* modulated wave, as given by (38), with the pure *amplitude* modulated wave, as given by (39), it will be observed that, in the latter, the low frequency signal $s(t)$, which is ultimately wanted in the receiver, is *explicit* and methods for its detection and recovery are direct and simple. In the pure frequency modulated wave, on the other hand, the low frequency signal is *implicit;* indeed it may be thought of as concealed in minute phase or frequency variations in the high frequency carrier wave.

If we differentiate (38) with respect to time t, we get

$$dW/dt = [\omega_c + \lambda s(t)] \exp \left(i\omega_c t + i\lambda \int_0^t s\,dt \right). \tag{44}$$

[4] See Appendix 1.

The first term,

$$\omega_c \exp\left(i\omega_c t + i\lambda \int_0^t s \, dt \right), \tag{45}$$

is still a pure frequency modulated wave. The second term,

$$\lambda s(t) \cdot \exp\left(i\omega_c t + i\lambda \int_0^t s \, dt \right), \tag{46}$$

is a "hybrid" modulated wave, since it is modulated with respect to both *amplitude* and *frequency*. The important point to observe is that, by differentiation, we have "rendered explicit" the wanted low frequency signal. We infer from this that the detection of a pure frequency modulated wave involves in effect its differentiation. The process of rendering explicit the low frequency signal has been termed "frequency detection." Actually it converts the *pure frequency* modulated wave into a *hybrid* modulated wave.

Every frequency distorting transducer inherently introduces frequency detection or "hybridization" of the pure frequency-modulated wave, as may be seen from formula (16). The transmitted current is conveniently written in the form

$$I = Y(i\omega_c) \exp\left(i \int_0^t \Omega \, dt \right) \cdot \left\{ 1 + \frac{1}{\omega_1}\lambda s + \frac{1}{2!}\frac{1}{\omega_2{}^2}C_2 \right.$$
$$\left. + \frac{1}{3!}\frac{1}{\omega_3{}^3}C_3 + \cdots \right\}, \tag{47}$$

where

$$\frac{1}{\omega_n{}^n} = \frac{1}{Y(i\omega_c)}\frac{d^n}{d\omega_c{}^n}Y(i\omega_c). \tag{48}$$

(Note that ω_n has the dimensions of frequency. It may be and usually is complex.)

Every term in (47) except the first, is a hybrid modulated wave.

In passing it is interesting to compare the distortion, as given by (47), undergone by the pure *frequency*-modulated wave, with that suffered by the pure *amplitude*-modulated wave (39), in passing through the same transducer. The transmitted current corresponding to the amplitude-modulated wave (39) is

$$I = Y(i\omega_c)e^{i\omega_c t}\left\{ s(t) + \frac{1}{i\omega_1}\frac{ds}{dt} + \frac{1}{2!(i\omega_2)^2}\frac{d^2s}{dt^2} \right.$$
$$\left. + \frac{1}{3!(i\omega_3)^3}\frac{d^3s}{dt^3} + \cdots \right\}. \tag{49}$$

This equation corresponds to (47) for the pure frequency-modulated wave.

III

In this section we consider the recovery of the wanted low frequency signal $s(t)$ from the frequency-modulated wave. This involves two distinct processes: (1) rendering explicit the low frequency signal "implicit" in the high frequency wave; that is, "frequency detection" or "hybridization" of the high frequency wave; and (2) detection proper.

It is convenient and involves no loss of essential generality to suppose that the transducer proper is equalized in the neighborhood of the carrier frequency ω_c; that is,

$$\frac{d}{d\omega_c} Y(i\omega_c), \qquad \frac{d^2}{d\omega_c^2} Y(i\omega_c), \cdots \qquad (50)$$

are negligible.

Frequency detection is then effected by a terminal network. We therefore take as the over-all transfer admittance

$$y(i\omega) \cdot Y(i\omega). \qquad (51)$$

$y(i\omega)$ represents the terminal receiving network; it is under control and can be designed for the most efficient performance of its function. As we shall see, it should approximate as closely as possible a pure reactance.

Taking the over-all transfer admittance as (51), we have from (47),

$$I = y(i\omega_c) Y(i\omega_c) \cdot \exp\left(i \int_0^t \Omega dt \right)$$

$$\times \left\{ 1 + \frac{1}{\omega_1} \lambda s + \frac{1}{2!\omega_2^2} C_2 + \frac{1}{3!\omega_3^3} C_3 + \cdots \right\}, \quad (52)$$

where now

$$1/\omega_n{}^n = \frac{1}{y(i\omega_c)} \frac{d^n}{d\omega_c{}^n} y(i\omega_c). \qquad (53)$$

Inspection of (52) shows that the terms beyond the second simply represent distortion. The terminal network or frequency detector should be so designed as to make the series

$$1 + \frac{\lambda}{\omega_1} + \left(\frac{\lambda}{\omega_2}\right)^2 + \left(\frac{\lambda}{\omega_3}\right)^3 + \cdots$$

rapidly convergent from the start.[5] In fact the ideal frequency detector is a network whose admittance $y(i\omega)$ can be represented with

[5] See note at end of this section (p. 528) for specific example.

sufficient accuracy in the neighborhood of $\omega = \omega_c$ by the expression

$$y(i\omega) = y(i\omega_c)\left(1 + \frac{\omega - \omega_c}{\omega_1}\right). \qquad (53a)$$

This approximation should be valid over the frequency range from $\omega = \omega_c - \lambda$ to $\omega = \omega_c + \lambda$.

Supposing that this condition is satisfied, the wave, after passing over the transducer and through the terminal frequency detector, is (omitting the constant $y \cdot Y$)

$$I = \left(1 + \frac{\lambda}{\omega_1}s(t)\right) \cdot \exp\left(i\int_0^t \Omega dt\right). \qquad (54)$$

If y is a pure reactance, ω_1 is a pure real; due to unavoidable dissipation it will actually be complex. To take this into account we replace ω_1 in (54) by $\omega_1 e^{-i\alpha}$ where now ω_1 is real; (54) then becomes

$$I = \left\{1 + \frac{\lambda}{\omega_1}\cos\alpha \cdot s(t) + i\frac{\lambda}{\omega_1}\sin\alpha \cdot s(t)\right\}\exp\left(i\int_0^t \Omega dt\right). \qquad (55)$$

The amplitude A of this wave is then

$$A = \left\{\left(1 + \frac{\lambda}{\omega_1}\cos\alpha \cdot s(t)\right)^2 + \left(\frac{\lambda}{\omega_1}\sin\alpha \cdot s(t)\right)^2\right\}^{1/2}. \qquad (56)$$

Now let λ/ω_1 be *less than unity* and let the wave (55) be impressed on a straight-line rectifier. Then the rectified or detected output is

$$\left(1 + \frac{\lambda}{\omega_1}\cos\alpha \cdot s(t)\right)\left\{1 + \left(\frac{\lambda \sin\alpha \cdot s(t)}{\omega_1 + \lambda\cos\alpha \cdot s(t)}\right)^2\right\}^{1/2}, \qquad (57)$$

or, to a first order,

$$1 + \frac{\lambda}{\omega_1}\cos\alpha \cdot s(t) + \frac{1}{2}\frac{\lambda^2}{\omega_1^2}\sin^2\alpha \cdot s^2(t). \qquad (58)$$

The second term is the recovered signal and the third term is the first order non-linear distortion.

Inspection of the foregoing formulas shows at once that, for detection by straight rectification, the following conditions should be satisfied:

(1) λ/ω_1 *must* be less than unity.
(2) The terminal network should be as nearly as possible a pure reactance to make the phase angle α as nearly zero as possible.

(3) To minimize both linear and non-linear distortion it is necessary that the sequence

$$\frac{\lambda}{\omega_1}, \quad \left(\frac{\lambda}{\omega_2}\right)^2, \quad \left(\frac{\lambda}{\omega_3}\right)^3, \cdots$$

be rapidly convergent from the start.

The first term of (58) is simply direct current and has no significance as regards the recovered signal. When we come to consider the problem of noise in the next section, we shall find that its elimination is important. This can be effected by a scheme which may be termed *balanced rectification*. Briefly described the scheme consists in terminating the transducer in two frequency detectors y_1 and y_2 in parallel; these are so adjusted that $y_1(i\omega_c) = -y_2(i\omega_c)$ and $dy_1/d\omega_c = dy_2/d\omega_c$. ω_1 is therefore of opposite sign in the two frequency detectors. The rectified outputs of the two parallel circuits are then differentially combined in a common low frequency circuit. Corresponding to (58), the resultant detected output is

$$2\frac{\lambda}{\omega_1}\cos\alpha \cdot s(t). \tag{59}$$

This arrangement therefore eliminates first order non-linear distortion, as well as the constant term.

Rectification is the simplest and most direct mode of detection of frequency-modulated waves. However, in connection with the problem of noise reduction other methods of detection will be considered.

Note

As a specific example of the foregoing let the terminal frequency detector, specified by the admittance $y(i\omega)$, be an oscillation circuit consisting simply of an inductance L in series with a capacitance C. Then

$$y(i\omega) = i\sqrt{\frac{C}{L}}\frac{\omega/\omega_R}{1 - \omega^2/\omega_R{}^2},$$

where $\omega_R{}^2 = 1/LC$.

Then, if ω_c/ω_R is nearly equal to unity, that is, if

$$\omega_R = (1 + \delta)\omega_c,$$
$$|\delta| \ll 1,$$

we have approximately,

$$\frac{1}{\omega_n{}^n} \doteq \frac{n!}{(\omega_R - \omega_c)^n},$$

$$y(i\omega_c) \doteq \frac{i}{2}\frac{\sqrt{C/L}}{\omega_R - \omega_c}.$$

Formula (42) thus becomes

$$I = y(i\omega_c) \cdot Y(i\omega_c) \cdot \exp\left(i \int_0^t \Omega dt \right) \cdot \left\{ 1 + \frac{\lambda s}{\omega_R - \omega_c} + \frac{C_2}{(\omega_R - \omega_c)^2} + \frac{C_3}{(\omega_R - \omega_c)^3} + \cdots \right\}.$$

In order that the distortion shall be small it is necessary that

$$\lambda \ll |\omega_R - \omega_c|.$$

If the two networks y_1 and y_2 are oscillation circuits so adjusted that

$$C_1/L_1 = C_2/L_2,$$
$$\omega_{R_1} = (1 + \delta)\omega_c = 1/\sqrt{L_1 C_1},$$
$$\omega_{R_2} = (1 - \delta)\omega_c = 1/\sqrt{L_2 C_2},$$

then the combined rectified output of the two parallel circuits is proportional to

$$\frac{\lambda s}{\delta \cdot \omega_c} + \frac{C_3}{(\delta \cdot \omega_c)^3} + \frac{C_5}{(\delta \cdot \omega_c)^5} + \cdots.$$

Thus the constant term and the first order distortion are eliminated in the low frequency circuit.

IV

The most important advantage known at present of *frequency*-modulation, as compared with *amplitude*-modulation, lies in the possibility of substantial reduction in the low frequency noise-to-signal power ratio in the receiver. Such reduction requires a correspondingly large increase in the width of the high frequency transmission band. For this reason frequency-modulation appears to be inherently restricted to short wave transmission.

In the discussion of the theory of noise which follows, it is expressly assumed *that the high frequency noise is small compared with the high frequency signal wave.* Also ideal terminal networks, filters and detectors are postulated.

In view of the assumption of a low noise power level, the calculation of the low frequency noise power in the receiver proper can be made to depend on the calculation of the noise due to the typical high frequency noise element

$$A_n \exp\left(i\omega_c t + i\omega_n t + i\theta_n \right). \tag{60}$$

Corresponding to the noise element (60), the output of the ideal frequency detector is

$$\exp\left(i \int_0^t \Omega dt \right) \cdot \left\{ 1 + \frac{\lambda s}{\omega_1} + \left(1 + \frac{\omega_n}{\omega_1} \right) A_n \exp\left(i\omega_n t + i\theta_n \right. \right.$$
$$\left. \left. - i\lambda \int_0^t s dt \right) \right\}. \quad (61)$$

Since the expression

$$\exp\left(i\omega_n t + i\theta_n - i\lambda \int_0^t s dt \right)$$

occurs so frequently in the analysis which is to follow, it is convenient to adopt the notation

$$\Omega_n = \omega_n - \lambda s(t),$$
$$\int_0^t \Omega_n dt = \omega_n t - \lambda \int_0^t s dt. \quad (61a)$$

With this notation and on the assumption that $A_n \ll 1$ and ω_1 real, the amplitude of the wave (61) is

$$1 + \frac{\lambda s}{\omega_1} + \left(1 + \frac{\omega_n}{\omega_1} \right) A_n \cos\left(\int_0^t \Omega_n dt \right). \quad (62)$$

In this formula the argument of the cosine function should be strictly

$$\int_0^t \Omega_n dt + \theta_n.$$

The phase angle θ_n is random however and does not affect the final formulas; it may therefore be omitted at the outset. Consequently, if the wave (61) is passed through a straight line rectifier, the rectified or low frequency current is proportional to

$$\lambda s(t) + (\omega_1 + \omega_n) A_n \cos\left(\int_0^t \Omega_n dt \right). \quad (63)$$

The first term is the recovered signal and the second term the low frequency noise or interference corresponding to the high frequency element (60).

Now the wave (63), before reaching the receiver proper, is transmitted through a low-pass filter, which cuts off all frequencies above ω_a; ω_a is the highest essential frequency in the signal $s(t)$. Consequently, in order to find the noise actually reaching the receiver proper, it is necessary in one way or another to make a frequency analysis of the wave (63). This is done in Appendix 2, attached hereto, where however, instead of dealing with the special formula (63), a more general expression

$$\lambda s(t) + (\omega_1 + \omega_n + \mu s) A_n \cos \int_0^t \Omega_n dt, \tag{64}$$

is used for the low frequency current. This will be found to include, as special cases, several other important types of rectification, as well as amplitude limitation, which we shall wish to discuss later.[6] Then, subject to the limitation that the noise energy is uniformly distributed over the spectrum, it is shown in Appendix 2 that

$$P_S = \lambda^2 \overline{s^2}, \tag{65}$$

$$P_N = (\tfrac{1}{3}\omega_a{}^2 + \omega_1{}^2 + (1 + \nu)^2 \lambda^2 \overline{s^2}) \omega_a N^2, \tag{66}$$

$$\nu = \mu/\lambda, \tag{67}$$

N^2 = mean high frequency power level.

These formulas are quite important because they make the calculation of low frequency noise-to-signal power ratio very simple for all the modes of frequency detection and demodulation which we shall discuss.

Applying them to formula (63) we find for *straight line rectification*

$$P_N = (\tfrac{1}{3}\omega_a{}^2 + \omega_1{}^2 + \lambda^2 \overline{s^2}) \omega_a N^2, \tag{68}$$

$$P_S = \lambda^2 \overline{s^2}.$$

It is known that in practice $\omega_1{}^2 \gg \lambda^2 \overline{s^2}$ and $\lambda^2 \overline{s^2} \gg \omega_a{}^2$. Consequently in the factor $(\tfrac{1}{3}\omega_a{}^2 + \omega_1{}^2 + \lambda^2 \overline{s^2})$ the largest term is $\omega_1{}^2$. Therefore it is important, if possible, to eliminate this term. This can be effected by the scheme briefly discussed at the close of section III; parallel rectification and differential recombination. For this scheme the low frequency current is found to be proportional to

$$\lambda s + \omega_n A_n \cos \left(\int_0^t \Omega_n dt \right). \tag{69}$$

Consequently, for *parallel rectification* and *differential recombination*,

$$P_N = (\tfrac{1}{3}\omega_a{}^2 + \lambda^2 \overline{s^2}) \omega_a N^2. \tag{70}$$

[6] The formula is also general enough to include detection by a product modulator, which however is not discussed in the text as no advantage over linear rectification was found.

Here, in the factor $(\frac{1}{3}\omega_a{}^2 + \lambda^2\overline{s^2})$, the term $\lambda^2\overline{s^2}$ is predominant. The elimination of the term $\omega_1{}^2$ has resulted in a substantial reduction in the noise power.

Returning to the general formula (66) for P_N, it is clear, that, if in addition to eliminating the term $\omega_1{}^2$, the parameter $\nu = \mu/\lambda$ can be made equal to -1, the noise power will be reduced to its lowest limits:

$$P_N = \tfrac{1}{3}\omega_a{}^3 N^2.$$

This highly desirable result can be effected by *amplitude limitation*, the theory of which will now be discussed.

V

When amplitude limitation is employed in frequency-modulation, the incoming high frequency signal is drastically reduced in amplitude. If no interference is present this merely results in an equal reduction in the low frequency recovered signal which is *per se* undesirable. When, however, noise or interference is present, amplitude limitation prevents the interference from affecting the *amplitude* of the resultant high frequency wave; its effect then can appear only as *variations in the phase or instantaneous frequency* of the high frequency wave. To this fact is to be ascribed the potential superiority of *frequency-modulation* as regards the reduction of noise power. This superiority is only possible with wide band high frequency transmission; that is, the index of frequency-modulation λ must be large compared with the low frequency band width ω_a. Insofar as the present paper is concerned, the potential superiority of frequency-modulation with amplitude limitation is demonstrated only for the case where the high frequency noise is small compared with the high frequency signal wave.

If, to the frequency-modulated wave $\exp\left(i \int_0^t \Omega dt \right)$, there is added the typical noise element $A_n \exp\left(i\omega_c + i\omega_n t + \theta_n\right)$, the resultant wave may be written as

$$\exp\left(i \int_0^t \Omega dt \right) \cdot \left(1 + A_n \exp\left(i \int_0^t \Omega_n dt \right) \right). \tag{71}$$

Postulating that $A_n \ll 1$ and therefore neglecting terms in $A_n{}^2$, the real part of (71) is

$$\left(1 + A_n \cos\left(\int_0^t \Omega_n dt \right) \right) \cdot \cos\left(\int_0^t \Omega dt + A_n \sin\left(\int_0^t \Omega_n dt \right) \right). \tag{72}$$

If this wave is subjected to amplitude limitation, the amplitude variation is suppressed, leaving a pure frequency-modulated wave, *proportional* to the *real part* of

$$\exp\left[i\left(\int_0^t \Omega dt + A_n \sin\left(\int_0^t \Omega_n dt \right) \right) \right] \tag{73}$$

(but drastically reduced in amplitude).

After frequency detection the wave (73) is, within a constant,

$$\exp\left[\left(i\int_0^t \Omega dt + A_n \sin\left(\int_0^t \Omega_n dt \right) \right) \right]$$
$$\times \left[1 + \frac{1}{\omega_1}\frac{d}{dt}\left(\lambda \int_0^t s\, dt + A_n \sin\left(\int_0^t \Omega_n dt \right) \right) \right]. \tag{74}$$

Consequently, since

$$\int_0^t \Omega_n dt = \omega_n t + \theta_n - \lambda \int_0^t s\, dt, \tag{75}$$

the amplitude of the wave (74) is

$$1 + \frac{1}{\omega_1}\left\{ \lambda s + (\omega_n - \lambda s)A_n \cos\left(\int_0^t \Omega_n dt \right) \right\}. \tag{76}$$

This is the amplitude of the low frequency wave after rectification; it is obviously proportional to

$$\lambda s + (\omega_n - \lambda s)A_n \cos\left(\int_0^t \Omega_n dt \right), \tag{77}$$

which is a special case of (64) and may be used in calculating the relative signal and noise power with amplitude limitation. Hence we have, by aid of (65) and (66),

$$\begin{aligned} P_S &= \lambda^2 \overline{s^2}, \\ P_N &= \tfrac{1}{3}\omega_a^3 N^2. \end{aligned} \tag{78}$$

(These are, of course, relative values and take no account of the absolute reduction in power due to amplitude limitation.)

Comparing (78) with (68) it is seen that, for detection by straight line rectification, the ratio of the noise power *with* to that *without* amplitude limitation is

$$\frac{1}{1 + 3\omega_1^2/\omega_a^2 + 3\lambda^2\overline{s^2}/\omega_a^2}; \tag{79}$$

or taking $\overline{s^2} = 1/2$,

$$\frac{1}{1 + 3\omega_1^2/\omega_a^2 + 3\lambda^2/2\omega_a^2}.$$ (80)

Since in practice $\omega_1 \gg \omega_a$ and $\lambda \gg \omega_a$, amplitude limitation results in a very substantial reduction in low frequency noise power in the receiver proper. Reference to formula (70) shows that, as compared with parallel rectification and recombination, amplitude limitation reduces the noise power by the factor

$$\frac{1}{1 + 3\lambda^2/2\omega_a^2}.$$ (81)

It should be observed that *without* amplitude limitation little reduction in the noise-to-signal power ratio results from increasing the modulation index λ (and consequently the high frequency transmission band width). On the other hand, *with* amplitude limitation, the ratio ρ of noise-to-signal power is

$$\rho = P_N/P_S = \frac{2}{3}\left(\frac{\omega_a}{\lambda}\right)^2 \omega_a N^2.$$ (82)

The ratio ρ is then (within limits) inversely proportional to the square of the modulation index λ, so that a large value of λ is indicated. It should be noted that, within limits ($\lambda \ll \omega_c$), the power transmitted from the sending station is independent of the modulation index λ.

It might be inferred from formula (82) that the noise power ratio ρ can be reduced indefinitely by indefinitely increasing the modulation index λ. Actually there are practical limits to the size of λ. First, if λ is made sufficiently large, the variable frequency oscillator generating the frequency-modulated wave may become unstable or function imperfectly. Secondly, the frequency spread of the frequency modulated wave is 2λ (from $\omega_c - \lambda$ to $\omega_c + \lambda$) and, if this is made too large, interference with other stations will result. Finally, the stationary distortion of the recovered low frequency signal $s(t)$ increases rapidly with the size of λ.

To summarize the results of the foregoing analysis the potential advantages of frequency-modulation depend on two facts. (1) By increasing the modulation index λ it is possible to increase the recovered low frequency signal power at the receiving station without increasing the high frequency power transmitted from the sending station. (2) It is possible to employ amplitude limitation (inherently impossible with amplitude-modulation) whereby the effect of interference or noise is reduced to a phase or "instantaneous frequency" variation of the high frequency wave.

Appendix 1

Formula (40) *et sequa* establish the fact that the actual frequency of the wave (29) varies between the limits

$$\omega_c \pm \lambda$$

provided $s(t)$ is a pure sinusoid $\lambda \sin \omega t$ and $\lambda \gg \omega$. This agrees with the concept of instantaneous frequency.

When $s(t)$ is a complex function—say a Fourier series—the frequency range of W can be determined qualitatively under certain restrictions, as follows:

We write

$$W = \exp\left(i\omega_c t + i\lambda \int_0^t s\,dt \right) \tag{1a}$$

$$= e^{i\omega_c t} \int_{-\infty}^{\infty} F(i\omega) e^{i\omega t} d\omega. \tag{2a}$$

The Fourier formulation is supposed to be valid in the epoch $0 \leq t \leq T$ and T can be made as great as desired. Then

$$F(i\omega) = \pi \int_0^T \exp\left(i\lambda \int_0^t s\,dt - i\omega t \right) dt. \tag{3a}$$

We now suppose that, in the epoch $0 \leq t \leq T$,

$$\left| \lambda \int_0^T s\,dt \right| \tag{4a}$$

becomes very large compared with 2π. On this assumption, it follows from the Principle of Stationary Phase, that, for a fixed value of ω, the important contributions to the integral $(3a)$ occur for those values of the integration variable t for which

$$\frac{d}{dt}\left(\lambda \int_0^t s\,dt - \omega t \right) = 0,$$

or

$$\omega = \lambda s(t).$$

Consequently the important part of the spectrum $F(i\omega)$ corresponds to those values of ω in the range

$$\lambda s_{\min} \leq \omega \leq \lambda s_{\max}.$$

Therefore the frequency spread of W lies in the range from $\omega_c + \lambda s_{\min}$ to $\omega_c + \lambda s_{\max}$ or $\omega_c \pm \lambda$ if $s_{\max} = -s_{\min} = 1$.

Appendix 2

We take the frequency modulated wave as

$$\cos\left(\omega_c t + \lambda \int_0^t s\,dt\right),\tag{1b}$$

where ω_c is the carrier frequency and $s = s(t)$ is the low frequency signal. λ is a real parameter, which fixes the amplitude of the frequency spread.

Correspondingly, we take the typical noise element as

$$A_n \cos\left((\omega_c + \omega_n)t + \theta_n\right).\tag{2b}$$

For reasons stated in the text, we take the more general formula for the low frequency current as proportional to

$$\lambda s + (\omega_0 + \omega_n + \mu s)A_n \cos\left(\omega_n t + \theta_n - \lambda \int_0^t s\,dt\right),\tag{3b}$$

where ω_0, λ, μ are real parameters. The term λs is the recovered signal and the second term is the low frequency noise corresponding to the high frequency noise element (2b).

We suppose that the noise is uniformly distributed over the frequency spectrum, at least in the neighborhood of $\omega = \omega_c$, so that, corresponding to the noise element

$$A_n \cos\left(\omega_n t + \theta_n\right),\tag{4b}$$

the noise is representable as the Fourier integral

$$\frac{N}{\pi} \int \cos\left(\omega_n t + \theta_n\right)d\omega_n\tag{5b}$$

and the corresponding *noise power* for the frequency interval $\omega_1 < \omega_n < \omega_2$ is, by the Fourier integral energy theorem,

$$\frac{N^2}{\pi} \int_{\omega_1}^{\omega_2} d\omega_n = \frac{1}{\pi}(\omega_2 - \omega_1)N^2.\tag{6b}$$

The Fourier integral energy theorem states that, if in the epoch $0 \leq t \leq T$, the function $f(t)$ is representable as the Fourier integral

$$f(t) = \frac{1}{\pi} \int_0^\infty F(\omega) \cdot \cos\left(\omega t + \theta(\omega)\right)d\omega,\tag{7b}$$

then

$$\int_0^T f^2 dt = \frac{1}{\pi} \int_0^\infty F^2 d\omega.^7 \tag{8b}$$

Replacing (4b) by (5b) to take care of the distributed noise, the noise term of (3b) becomes

$$\cos\left(\lambda \int_0^t s dt\right) \cdot \frac{N}{\pi} \int (\omega_0 + \omega_n + \mu s) \cdot \cos(\omega_n t + \theta_n) d\omega_n$$

$$+ \sin\left(\lambda \int_0^t s dt\right) \cdot \frac{N}{\pi} \int (\omega_0 + \omega_n + \mu s) \cdot \sin(\omega_n t + \theta_n) d\omega_n. \tag{9b}$$

Now this noise in the low frequency circuit is passed through a low pass filter, which cuts off all frequencies above ω_a. ω_a is the maximum essential frequency in the signal $s(t)$.

It is therefore necessary to express (9b) as a frequency function before calculating the noise power. To this end we write the Fourier integrals

$$\cos\left(\lambda \int_0^t s dt\right) = \frac{1}{\pi} \int_0^\infty F_c \cos(\omega t + \theta_c) d\omega, \tag{10b}$$

$$\sin\left(\lambda \int_0^t s dt\right) = \frac{1}{\pi} \int_0^\infty F_s \sin(\omega t + \theta_s) d\omega. \tag{11b}$$

We note also that

$$\mu s \cdot \cos\left(\lambda \int_0^t s dt\right) = \frac{\mu}{\lambda} \frac{d}{dt} \sin\left(\lambda \int_0^t s dt\right)$$

$$= \frac{1}{\pi} \int_0^\infty \frac{\mu \omega}{\lambda} F_s \cos(\omega t + \theta_s) d\omega, \tag{12b}$$

$$\mu s \cdot \sin\left(\lambda \int_0^t s dt\right) = -\frac{\mu}{\lambda} \frac{d}{dt} \cos\left(\lambda \int_0^t s dt\right)$$

$$= \frac{1}{\pi} \int_0^\infty \frac{\mu \omega}{\lambda} F_c \sin(\omega t + \theta_c) d\omega. \tag{13b}$$

Substituting (10b), (11b), (12b) and (13b) in (9b) and carrying through straightforward operations, we find that the noise is given by

$$\frac{N}{2\pi^2} \int_0^\infty F_p d\omega \int_{\omega-\omega_a}^{\omega+\omega_a} \left(\omega_0 + \omega_n + \frac{\mu}{\lambda}\omega\right) \cos((\omega - \omega_n)t + \theta_p) d\omega_n$$

$$+ \frac{N}{2\pi^2} \int_0^\infty F_m d\omega \int_{-(\omega+\omega_a)}^{-(\omega-\omega_a)} \left(\omega_0 + \omega_n - \frac{\mu}{\lambda}\omega\right) \cos((\omega + \omega_n)t + \theta_m) d\omega_n, \tag{14b}$$

---

[7] See "Transient Oscillations in Electric Wave Filters," Carson and Zobel, *B. S. T. J.*, July, 1923.

where

$$F_p^2 = F_c^2 + F_s^2 + 2F_cF_s \cos(\theta_c - \theta_s), \qquad (15b)$$

$$F_m^2 = F_c^2 + F_s^2 - 2F_cF_s \cos(\theta_c - \theta_s). \qquad (16b)$$

The limits of integration of ω_n are determined by the fact that, $\omega - \omega_n$ in the first integral of (14b) and $\omega + \omega_n$ in the second, must lie between $\pm \omega_a$; all other frequencies are eliminated by the low pass filter.

From formula (14b) and the Fourier integral energy theorem, the *noise power* P_N is given by

$$P_N = \frac{N^2}{4\pi^3 T} \int_0^\infty F_p^2 d\omega \int_{\omega-\omega_a}^{\omega+\omega_a} \left(\omega_0 + \omega_n + \frac{\mu}{\lambda}\omega \right)^2 d\omega_n$$

$$+ \frac{N^2}{4\pi^3 T} \int_0^\infty F_m^2 d\omega \int_{-(\omega+\omega_a)}^{-(\omega-\omega_a)} \left(\omega_0 + \omega_n - \frac{\mu}{\lambda}\omega \right)^2 d\omega_n. \quad (17b)$$

Integrating with respect to ω_n, we have

$$P_N = \frac{N^2\omega_a}{2\pi^3 T} \int_0^\infty d\omega\{[(\omega_0 + (1 + \nu)\omega)^2 + \tfrac{1}{3}\omega_a^2]F_p^2$$

$$+ [(\omega_0 - (1 + \nu)\omega)^2 + \tfrac{1}{3}\omega_a^2]F_m^2\}, \quad (18b)$$

where $\nu = \mu/\lambda$.

Replacing F_p^2 and F_m^2 in (18b) by their values as given by (15b) and (16b), we get

$$P_N = \frac{\omega_a N^2}{\pi^3 T} \int_0^\infty (\omega_0^2 + (1 + \nu)^2\omega^2 + \tfrac{1}{3}\omega_a^2)(F_c^2 + F_s^2)d\omega$$

$$+ 4\frac{\omega_a N^2}{\pi^3 T} \int_0^\infty (1 + \nu)\omega_0\omega F_cF_s \cos(\theta_c - \theta_s)d\omega. \quad (19b)$$

To evaluate (19b) we make use of the formulas, derived below

$$\frac{1}{\pi T} \int_0^\infty (F_c^2 + F_s^2)d\omega = 1, \qquad (20b)$$

$$\frac{1}{\pi T} \int_0^\infty \omega^2(F_c^2 + F_s^2)d\omega = \lambda^2\overline{s^2} = P_S, \qquad (21b)$$

$$\frac{1}{\pi T} \int_0^\infty \omega F_cF_s \cos(\theta_c - \theta_s)d\omega \to 0 \text{ as } T \to \infty. \qquad (22b)$$

Substitution of (20b), (21b), (22b) in (19b) gives for large values of T

$$P_N = (\tfrac{1}{3}\omega_a^2 + \omega_0^2 + (1 + \nu)^2\lambda^2\overline{s^2})\omega_a N^2. \qquad (23b)$$

Here, for convenience, we have replaced N^2/π^2 of (19b) by N^2, so that N^2 of (23b) may be defined and regarded as the high frequency *noise power level.*

It remains to establish formulas (20*b*), (21*b*) and (22*b*). From the defining formulas (10*b*) and (11*b*) and the Fourier integral energy theorem, we have

$$\frac{1}{\pi T}\int_0^\infty F_c^2 d\omega = \frac{1}{T}\int_0^T \cos^2\left(\lambda\int_0^t s\,dt\right)dt,$$

$$\frac{1}{\pi T}\int_0^\infty F_s^2 d\omega = \frac{1}{T}\int_0^T \sin^2\left(\lambda\int_0^t s\,dt\right)dt. \qquad (24b)$$

Adding we get (20*b*).

Now differentiate (10*b*) and (11*b*) with respect to t and apply the Fourier integral energy theorem; we get

$$\frac{1}{\pi T}\int_0^\infty \omega^2 F_c^2 d\omega = \frac{1}{T}\int_0^T \lambda^2 s^2 \sin^2\left(\lambda\int_0^t s\,dt\right)dt,$$

$$\frac{1}{\pi T}\int_0^\infty \omega^2 F_s^2 d\omega = \frac{1}{T}\int_0^T \lambda^2 s^2 \cos^2\left(\lambda\int_0^t s\,dt\right)dt \qquad (25b)$$

and, by addition, we get (21*b*).

To prove (22*b*) we note that

$$(1 + \mu s)\cos\left(\lambda\int_0^t s\,dt\right)$$

$$= \cos\left(\lambda\int_0^t s\,dt\right) + \frac{\mu}{\lambda}\frac{d}{dt}\sin\left(\lambda\int_0^t s\,dt\right)$$

$$= \frac{1}{\pi}\int_0^\infty \left[F_c\cos(\omega t + \theta_c) + \frac{\mu}{\lambda}\omega F_s\cos(\omega t + \theta_s)\right]d\omega$$

$$= \frac{1}{\pi}\int_0^\infty \left[F_c^2 + \left(\frac{\mu}{\lambda}\right)^2\omega^2 F_s^2\right.$$

$$\left. + 2\frac{\mu}{\lambda}\omega F_c F_s\cos(\theta_c - \theta_s)\right]^{1/2}\cos(\omega t + \Phi)d\omega. \qquad (26b)$$

Consequently, by the Fourier integral energy theorem,

$$\frac{1}{T}\int_0^T (1 + \mu s)^2\cos^2\left(\lambda\int_0^t s\,dt\right)dt$$

$$= \frac{1}{\pi T}\int_0^\infty \left[F_c^2 + \left(\frac{\mu}{\lambda}\right)^2\omega^2 F_s^2 + 2\frac{\mu}{\lambda}\omega F_c F_s\cos(\theta_c - \theta_s)\right]d\omega \qquad (27b)$$

and

$$\frac{1}{T}\int_0^T \mu s\cdot\cos^2\left(\lambda\int_0^t s\,dt\right)dt$$

$$= \frac{1}{\pi T}\left(\frac{\mu}{\lambda}\right)\int_0^\infty \omega F_c F_s\cos(\theta_c - \theta_s)d\omega. \qquad (28b)$$

263

By simple transformations (28*b*) becomes

$$\frac{1}{\pi T}\int_0^\infty \omega F_c F_s \cos(\theta_c - \theta_s)d\omega$$
$$= \frac{1}{2T}\int_0^T \lambda s dt + \frac{1}{4T}\int_0^T \frac{d}{dt}\sin\left(2\lambda\int_0^t s dt\right)dt$$
$$= \frac{1}{2}\lambda\bar{s} + \frac{1}{4T}\sin\left(2\lambda\int_0^T s dt\right)$$
$$\to 0 \text{ as } T \to \infty, \tag{29b}$$

since by hypothesis $\bar{s} = 0$.

We note for reference that

$$-\frac{1}{\pi T}\int_0^\infty F_c F_s \sin(\theta_c - \theta_s)d\omega = \frac{1}{2T}\int_0^T \sin\left(2\lambda\int_0^t s dt\right)dt. \tag{30b}$$

BIBLIOGRAPHY

1. Carson, J. R., "Notes on the Theory of Modulation," *Proc. I. R. E.*, **10**, pp. 57–64, Feb., 1922.
2. Roder, H., "Über Frequenzmodulation," *Telefunken-Zeitung*, **10**, pp. 48–54, Dec., 1929.
3. Heilmann, A., "Einige Betrachtungen zum Problem der Frequenzmodulation," *Elek. Nach. Tech.*, **7**, pp. 217–225, June, 1930.
4. Van der Pol, B., "Frequency Modulation," *Proc. I. R. E.*, **18**, pp. 1194–1205, July, 1930.
5. Eckersley, T. L., "Frequency Modulation and Distortion," *Exp. Wireless and Wireless Engg.*, **7**, pp. 482–484, Sept., 1930.
6. Runge, W., "Untersuchungen an ampliduten- und frequenz-modulierten Sendern," *Elek. Nach. Tech.*, **7**, pp. 488–494, Dec., 1930.
7. Roder, H., "Amplitude, Phase and Frequency-Modulation," *Proc. I. R. E.*, **19**, pp. 2145–2175, Dec., 1931.
8. Andrew, V. J., "The Reception of Frequency Modulated Radio Signals," *Proc. I. R. E.*, **20**, pp. 835–840, May, 1932.
9. Barrow, W. L., "Frequency Modulation and the Effects of a Periodic Capacity Variation in a Non-dissipative Oscillatory Circuit," *Proc. I. R. E.*, **21**, pp. 1182–1202, Aug., 1933.
10. Barrow, W. L., "On the Oscillations of a Circuit Having a Periodically Varying Capacitance; Contribution to the Theory of Nonlinear Circuits with Large Applied Voltages," *Proc. I. R. E.*, **22**, pp. 201–212, Feb., 1934, also *M. I. T. Serial* 97, Oct., 1934.
11. Chaffee, J. G., "The Detection of Frequency-Modulated Waves," *Proc. I. R. E.*, **23**, pp. 517–540, May, 1935.
12. Armstrong, E. H., "A Method of Reducing Disturbances in Radio Signaling by a System of Frequency-Modulation," *Proc. I. R. E.*, **24**, pp. 689–740, May, 1936.
13. Crosby, M. G., "Frequency Modulation Propagation Characteristics," *Proc. I. R. E.*, **24**, pp. 898–913, June, 1936.
14. Crosby, M. G., "Frequency-Modulated Noise Characteristics," *Proc. I. R. E.*, **25**, pp. 472–514, April, 1937.
15. Roder, H., "Noise in Frequency Modulation," *Electronics*, **10**, pp. 22–25, 60, 62, 64, May, 1937.

Reprinted from *Journal IEE* (London), Vol. 93, Pt. 3, No. 23, pp. 153–158, May 1946

THE FUNDAMENTAL PRINCIPLES OF FREQUENCY MODULATION

By Prof. Balth. van der Pol, D.Sc.*

(Lecture *delivered before the* Radio Section, *25th April, 1945.*)

In his preliminary remarks Prof. van der Pol expressed the pleasure with which he accepted the invitation to lecture before the Radio Section, for this was his first visit to Great Britain since the liberation of the southern part of the Netherlands. He also expressed his great joy in meeting again, after such a long time of separation, his many English friends in the scientific and technical fields. Above all he expressed his sense of relief at being again in a country where one might freely say what one thought and utter what one felt, which was for so long denied to the Dutch.

---

(1) INTRODUCTION

It is not an uncommon experience that, while discussing problems of frequency modulation, even with radio engineers of several years of theoretical and practical training, one is struck by the fact that in the reasoning errors are made, which can often be traced to some technical rules that in themselves are perfectly right and serve a useful purpose as long as they are applied to amplitude modulation, but·which do *not* apply to frequency modulation. For example, the well-known methods of elementary alternating-current engineering in which $\cos \omega t$ is replaced by $e^{j\omega t}$, and the resultant complex circuit theory, can no longer by applied when the frequency itself is made a function of the time. Even the concept of "instantaneous frequency," which—it may at once be admitted—is of a somewhat arbitrary but nevertheless highly useful nature, is often misunderstood and even misrepresented. The same is true of the response of a linear passive network to a frequency-modulated e.m.f., where the question arises whether the parameters of the e.m.f. and the network are such that the network can "follow" the instantaneous frequency, or, in other words, whether the response is of a "quasi-stationary" or "adiabatic" nature.

The only way at present available to solve these and similar problems is to go back to the very first and fundamental principles. This implies a theoretical treatment beginning with the differential equations of the problem concerned. Unfortunately these equations can seldom be solved in terms of well-known functions, such as real or complex exponentials, to which the radio engineer is so much accustomed, and I think it is precisely to this fact that the main difficulties can be traced.

It therefore seems worth while in this Lecture to consider two simple but nevertheless very fundamental problems concerning frequency modulation. Notwithstanding the simplicity of these problems, their solutions—as will appear—require some care.

(2) TWO FUNDAMENTAL PROBLEMS

(*a*) The first problem is the following: *Given an ordinary circuit with L, C, r and condenser leak R, but of which all these four elements are each arbitrary functions of the time, so that $L = L(t)$, $C = C(t)$, $r = r(t)$, $R = R(t)$, it is required to find the possible current and voltage in the circuit.*

(*b*) The second problem is the following: *Let there be applied to a linear, passive, constant network an e.m.f. which is frequency-modulated with an arbitrary function of the time. It is required to find the expression for the current in this network.*

* Natuurkundig Laboratorium der N. V. Philips' Gloeilampenfabrieken, Eindhoven, Holland.

Before solving these problems we shall first have to consider the definitions of *amplitude*, *phase* and *frequency*, of which quite a collection is to be found in the literature.

(3) DEFINITIONS OF AMPLITUDE, PHASE AND FREQUENCY

The subject of harmonic oscillations has often been treated in mathematical, physical and technical texts. In order, therefore, to find the various definitions of amplitude, phase and frequency as given by different authors, I consulted some 50 books ranging from elementary technical expositions to such volumes as Thomson and Tait, "Natural Philosophy"; Rayleigh, "Sound"; Whittaker-Watson, "Modern Analysis"; several articles (by different authors) in the "Encyclopaedie der Mathematischen Wissenschaften," e.g. Study, "Algebra," I, 4; Stäckel, "Dynamik," IV, 6; Lamb, "Akoustik," IV, 26; Wangerin, "Optik," V, 21. This search brought to light the most varied definitions of some of these three fundamental concepts. Most authors agree to call A the *amplitude* in the expression

$$A \sin (\omega t + \psi) \quad \ldots \quad \ldots \quad (1)$$

but whereas some (Hort, Barkhausen, Orlich, Stäckel) call ψ in (1) the *phase*, others (Kalähne, Lamb, Weber-Gans, Helmholtz) call ψ the *phase constant* (*Phasenconstante*). Again others (Elias, Berliner-Scheel) call it the *phase angle*, whereas still other authors (Zenneck, Fleming) reserve the same nomenclature for ψ in the expression

$$A \sin (\omega t - \psi) \quad \ldots \quad \ldots \quad (2)$$

The term *phase* is reserved by Felix Klein ("Elementarmathematik vom höheren Standpunkt," **1**, p. 203) for quite a different quantity, namely the constant t_0 in the expression

$$A \cos [\omega(t - t_0)]$$

whereas one also encounters the term *phase* for the constant t_0 in the expressions

$$A \sin [\omega(t + t_0)] \text{ and } A \cos [\omega(t + t_0)]$$

Thomson and Tait in

$$A \cos (\omega t - \psi)$$

call the constant ψ the *epoch*, whereas Lamb calls it the *phase constant* in the expression

$$A \cos (\omega t + \psi)$$

I found another quite different definition in Max Planck ("Mechanik"), where the author, referring to $x = \sin \omega t$, says: "The angle which varies with time and which follows 'sin' is called the phase."

Finally, a similar, but still somewhat extended, definition to which I shall refer later *in extenso*, is to be found in Weber-Gans ("Repertorium der Physik") and Wangerin (*op. cit.*). It runs as follows: in the expression

$$A \sin (\omega t + \psi)$$

the angle $(\omega t + \psi)$ is called the *phase*. Similarly A B. Wood

("Sound") and Born ("Optik") call the same quantity the *phase* in the expression

$$A \cos (\omega t + \psi)$$

Thus a great variety of definitions of phase is to be found in the literature.

Most of the above definitions of phase are given in connection with physical or technical problems. But it is well known that a harmonic oscillation may be represented by a vector in an Argand diagram, and that exactly the same diagram is used in pure mathematics for the representation of a complex variable. Moreover, in the Argand diagram the addition of two oscillations of the same frequency is effected vectorially in exactly the same way as the addition of two complex quantities in the complex plane. Thus the same question of nomenclature arises in pure mathematics, where a complex variable z is represented by

$$z = x + iy = \rho e^{i\phi} \quad \rho > 0, \phi \, real \quad \cdot \quad \cdot \quad (3)$$

It is therefore of interest to note the fact that the following names for the quantity ϕ in (3) are to be found in different standard books on pure mathematics:

argument	angle (*Winkel*)
Abweichung	direction angle (*Richtungswinkel*)
anomaly	phase
azimuth	slope
arcus	amplitude

Referring especially to the last item in the list, I wish to stress the fact that what pure mathematicians occasionally call the *amplitude* is similar to the quantity which physicists and engineers call the *phase*.

In this matter of a somewhat confused nomenclature I would therefore strongly recommend the following definition of phase:
In the expression for a harmonic motion

$$y = A \cos (\omega t + \psi) \quad \cdot \quad \cdot \quad \cdot \quad \cdot \quad (4)$$

the whole argument of the cosine function, namely $(\omega t + \psi)$, is the phase. This definition has, among others, the advantage of enabling one to speak of a phase difference of two oscillations of different frequencies. This phase difference is then simply a linear function of the time, just as one phase by itself is already such a function of the time.

Now what is the *frequency*? Most commonly in harmonic motion, ω is called the *angular frequency*. But, considering amplitude modulation, we make A a function of the time, so that in (4) we have

$$A = a_0 [1 + mg(t)] \text{ (amplitude modulation)} \quad \cdot \quad \cdot \quad (5)$$

where $g(t)$ is the modulating audio signal, and still call this $A [= A(t)]$ the *instantaneous amplitude*. We have thus left the domain of a simple harmonic function, but continue to speak of the *amplitude*.

Similarly we may modulate the phase in (4) and thus get, for example,

$$\psi = \psi_0 [1 + mg(t)] \text{ (phase modulation)} \quad \cdot \quad \cdot \quad (6)$$

In order to obtain an expression for frequency modulation, it would, however, be erroneous simply to write in (4)

$$\omega = \omega_0 [1 + mg(t)] \quad \cdot \quad \cdot \quad \cdot \quad \cdot \quad (7)$$

for this would lead to a physical absurdity. Thus it is necessary first to rewrite (4) as

$$y = A \cos \left(\int_0^t \omega dt + \phi \right) \quad \cdot \quad \cdot \quad \cdot \quad (4a)$$

which is exactly the same expression as (4) but written in a different way, thus enabling one to extend this expression for a simple harmonic motion to the case of frequency modulation. For when in (4a) we substitute (7) we obtain

$$y = A \cos \left\{ \int_0^t \{ \omega_0 [1 + mg(t)] \} dt + \phi \right\} \quad \cdot \quad \cdot \quad (8a)$$

$$= A \cos \left[\omega_0 t + m\omega_0 \int_0^t g(t) dt + \phi \right] \quad \cdot \quad \cdot \quad (8b)$$

which is no doubt the proper expression for the frequency-modulated oscillation the engineer is aiming at.

These considerations bring us at once to the question of what we shall call the *instantaneous frequency*. When, as referred to already, we represent an ordinary harmonic motion by a vector in an Argand diagram, this vector is supposed to rotate with a constant *angular velocity* and then our simple harmonic oscillation is represented by its projection on, for example, the horizontal axis. This angular velocity then coincides with the *angular frequency* of the oscillation. The instantaneous angle the vector makes with the horizontal axis we have agreed to call the *phase*. Thus, for a simple harmonic motion the time differential of the *phase* is the *angular frequency*, and in the expression

$$A \cos (\omega t + \phi)$$

the phase is $(\omega t + \phi)$ and the angular frequency is d/dt (phase) $= d/dt(\omega t + \phi) = \omega$.

We can now extend our definition

$$\boxed{\text{angular frequency} = \frac{d}{dt} \text{ (phase)}} \quad \cdot \quad \cdot \quad \cdot \quad (9)$$

to a frequency-modulated oscillation such as is represented by (8a), where the vector now rotates with a variable angular velocity. Nothing prevents us also in this case from associating the concept of instantaneous frequency with that of *instantaneous velocity*, which is just the definition (9). For thus we find

$$\frac{d}{dt} \left\{ \int_0^t \omega_0 [1 + mg(t)] dt + \phi \right\} = \omega_0 [1 + mg(t)]$$

which is exactly the expression (7) for the instantaneous frequency.

It will further be clear that we can write (8a) also as

$$y = A \mathscr{R} e^{j\{ \int_0^t \omega_0 [1 + mg(t)] dt + \phi \}} \quad \cdot \quad \cdot \quad (10)$$

where $\mathscr{R}$ means "the real part of."

As to nomenclature, in all three cases (amplitude, phase and frequency modulation) we call the constant m the *modulation depth*. If, in particular, our audio modulation is a sinusoidal function of the time so that

$$g(t) = \cos pt$$

the frequency-modulated oscillation (10) becomes

$$y = A \mathscr{R} e^{j[\int_0^t \omega_0(1 + m \cos pt) dt + \phi]}$$

$$= A \mathscr{R} e^{j\left(\omega_0 t + \frac{m\omega_0}{p} \sin pt + \phi \right)}$$

In this case the maximum instantaneous angular frequency deviation $\Delta\omega_0$ is $m\omega_0$ so that in f.m. we can very well speak of a modulation depth m, meaning the ratio of the maximum frequency deviation, $m\omega_0 = \Delta\omega_0$, from the central frequency ω_0. This *modulation depth m* should not be confused with the *modulation index m'*, which is defined as

$$m' = m\frac{\omega_0}{p} = \frac{\Delta\omega_0}{p}$$

and therefore equals the ratio of the maximum frequency deviation $\Delta\omega_0$ to the (constant) frequency p.

In present practice we have

$$m = \frac{\Delta\omega_0}{\omega_0} \ll 1$$

and

$$m' = \frac{\Delta\omega_0}{p} \gg 1$$

so that the *modulation depth* is usually very small, whereas the *modulation index* is usually very large.

To simplify our language we will further call ω simply *frequency* instead of *angular frequency*, as in theoretical work the concept of $\omega (= 2\pi\nu)$ is of a much more fundamental nature than ν. This question in two dimensions is somewhat analogous to the Heaviside–Lorentz suggestion of dropping the 4π in three-dimensional electromagnetic theory (rationalized units).

(4) FIRST PROBLEM (as stated in Section 2)

With these preliminary definitions we are ready to tackle our first problem. For the current in an ordinary oscillatory circuit with constant L, C, r and R (condenser leak) we have the well-known differential equation for any potential difference v in the circuit

$$\frac{d^2v}{dt^2} + \left(\frac{r}{L} + \frac{1}{CR}\right)\frac{dv}{dt} + \frac{1}{LC}\left(1 + \frac{r}{R}\right)v = 0 \quad . \quad (11)$$

If, however, we want to study the current and p.d. in a circuit, where L, C, r and R are variable with the time, it would be quite erroneous simply to substitute in (11) $L = L(t)$, $C = C(t)$, $r = r(t)$, $R = R(t)$. Here, again, we have to go back to first principles. Thus (see Figure below), considering on the one

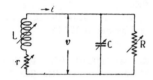

hand the potential difference v across the inductance-resistance branch, and on the other hand the total current in the leaking condenser, we have the two differential equations

$$\left. \begin{array}{l} v = \dfrac{d}{dt}(Li) + ri \\[2mm] -i = \dfrac{d}{dt}(Cv) + \dfrac{v}{R} \end{array} \right\} \quad . \quad . \quad . \quad . \quad (12)$$

We can, of course, eliminate i, for example, and thus arrive at the single differential equation for v

$$\frac{d}{dt}\left[L\frac{d}{dt}(Cv) + \frac{L}{R}v\right] + r\frac{d}{dt}(Cv) + \left(1 + \frac{r}{R}\right)v = 0 \quad . \quad (13)$$

which, for time-independent L, C, r and R, reduces to (11).

In the general case (variable elements) we have in (13) four functions of the time $L(t)$, $C(t)$, $r(t)$ and $R(t)$, and the question arises of how to reduce the number of these functions occurring in the final equations. To this end we return to (12) and notice that the common choice of variables, namely v and i in the usual circuit theory is simply governed by the fact that we have voltmeters and ammeters, whereas the physics of our problem induces us (with Maxwell and Heaviside) to consider as much more fundamental variables:

(*a*) the total electric flux in the condenser, which corresponds to its charge q, and

(*b*) the total magnetic flux in the coil, which we will denote by ϕ.

They are related to the common variables v and i by

$$\left. \begin{array}{l} \phi = Li \\[1mm] q = Cv \end{array} \right\} \quad . \quad . \quad . \quad . \quad (14)$$

Introducing the two time functions $\alpha_1(t)$ and $\alpha_2(t)$ defined by

$$\left. \begin{array}{l} 2\alpha_1(t) = \dfrac{r}{L} \\[2mm] 2\alpha_2(t) = \dfrac{1}{CR} \end{array} \right\} \quad . \quad . \quad . \quad . \quad (15)$$

which, in the case of constant elements, determine together the damping of the free oscillations, we can write (12) as

$$\left. \begin{array}{l} \dfrac{q}{C} = \phi' + 2\alpha_1\phi \\[2mm] -\dfrac{\phi}{L} = q' + 2\alpha_2 q \end{array} \right\} \quad . \quad . \quad . \quad (16)$$

Introducing further the derived variables ϕ_1 and q_1 defined by

$$\left. \begin{array}{l} \phi = e^{-2\int \alpha_1 dt}\,\phi_1 \\[1mm] q = e^{-2\int \alpha_2 dt}\,q_1 \end{array} \right\} \quad . \quad . \quad . \quad (17)$$

we obtain, instead of (16),

$$\left. \begin{array}{l} q_1 = C e^{-2\int(\alpha_1 - \alpha_2)dt}\,\phi_1' \\[1mm] -\phi_1 = L e^{+2\int(\alpha_1 - \alpha_2)dt}\,q_1' \end{array} \right\} \quad . \quad . \quad (18)$$

Finally, putting

$$2\int(\alpha_1 - \alpha_2)dt = A(t)$$

and

$$\left. \begin{array}{l} C e^{-A} = \gamma \\[1mm] L e^{+A} = \lambda \end{array} \right\} \quad . \quad . \quad . \quad . \quad (19)$$

we obtain our equations for the unknown q_1 and ϕ_1 in the canonical form

$$\boxed{\begin{array}{l} q_1 = \gamma\phi_1' \\[1mm] -\phi_1 = \lambda q_1' \end{array}} \quad . \quad . \quad . \quad . \quad (20)$$

where now only the *two* known time functions $\gamma(t)$ and $\lambda(t)$ are left in the coefficients.

Several practical consequences may now be derived from (20), for elimination of one of the variables in (20) leads to

$$\left. \begin{array}{l} \phi_1'' + \dfrac{\gamma'}{\gamma}\phi_1' + \dfrac{1}{\lambda\gamma}\phi_1 = 0 \\[2mm] q_1'' + \dfrac{\lambda'}{\lambda}q_1' + \dfrac{1}{\lambda\gamma}q_1 = 0 \end{array} \right\} \quad . \quad . \quad . \quad (21)$$

or, going back to the original parameters L, C, r and R,

$$\left. \begin{array}{l} \phi_1'' + \left[\dfrac{C'}{C} - 2\left(\dfrac{r}{L} - \dfrac{1}{CR}\right)\right]\phi_1' + \dfrac{1}{CL}\phi_1 = 0 \\[3mm] q_1'' + \left[\dfrac{L'}{L} + 2\left(\dfrac{r}{L} - \dfrac{1}{CR}\right)\right]q_1' + \dfrac{1}{CL}q_1 = 0 \end{array} \right\} \quad . \quad (22)$$

Incidentally, it is worth noticing that the coefficients of ϕ_1' and q_1' clearly show that the dimension of C' is the same as that of R^{-1} and that of L' the same as r, so that, for example, the change of an inductance in henrys per second is expressible in ohms. Further, if it happens that $\alpha_1 = \alpha_2$, i.e. that the time functions L, C, r and R are such that

$$\frac{r}{L} = \frac{1}{CR}$$

(which, for constant elements, would correspond to the distorsionless cable of Heaviside), then $A = 0$, $\gamma = C$, $\lambda = L$, and (22) assumes the simple form

$$\left.\begin{array}{c} \phi_1'' + \dfrac{C'}{C}\phi_1' + \dfrac{1}{CL}\phi_1 = 0 \\[2mm] q_1'' + \dfrac{L'}{L}q_1' + \dfrac{1}{CL}q_1 = 0 \end{array}\right\} \quad \cdots \quad (23)$$

where apparent damping terms occur with coefficients C'/C and L'/L. But it further follows from (23) that even when the product $1/CL$ also is independent of the time we are *not* led back to equations with constant coefficients, and therefore we cannot expect a pure harmonic oscillation to occur.

Again, it is a simple matter to assign time functions to C and L such that (23) becomes the differential equation of, say, a Bessel function, Legendre polynomial, Mathieu function, etc. In fact, the moment we enter the field of variable coefficients a great variety of mathematical possibilities and functions appear. It is, however, not our intention to exhaust this field completely.

Finally, it may be remarked that when C and L are constant but only α_1 and α_2 are taken to vary with the time (which amounts to r and R only being variable) and when moreover $\alpha_1 = \alpha_2$, then (22) reduces to

$$\left.\begin{array}{c} \phi_1'' + \dfrac{1}{CL}\phi_1 = 0 \\[2mm] q_1'' + \dfrac{1}{CL}q_1 = 0 \end{array}\right\} \quad \cdots \quad (24)$$

so that both ϕ_1 and q_1 become simple harmonic functions, and our original variables ϕ and q, with the aid of (20), can be written in the form

$$\left.\begin{array}{c} \phi = B\sqrt{L}e^{-2\int\alpha_1 dt}\sin(\omega_0 t + \psi) \\[2mm] q = B\sqrt{C}e^{-2\int\alpha_2 dt}\cos(\omega_0 t + \psi) \end{array}\right\} \quad \cdots \quad (25)$$

where B and ψ are constants of integration and $\omega_0^2 = 1/LC$.

Hence in this case we rigorously obtain pure amplitude modulation, without any trace of frequency modulation.

All the above seems sufficient indication that we should treat problems with variable parameters with much care, the common circuit theory not being of any value in this case.

I conclude this Section with the remark that it would perhaps be worth while to extend the common circuit theory where all the elements are taken to be independent of the time to a more general theory with variable elements.

Most likely, going back to the equations of Lagrange would here provide the necessary means. At any rate it would be advisable not to consider the usual variables v and i, but the more fundamental ones, namely the electric and magnetic fluxes q and ϕ, as was done above.

(5) FIRST PROBLEM (contd.)

Returning to the general problem as expressed by (20), let us now consider still another special case. When all losses can be ignored ($r = 0$, $R \to \infty$) we have $\alpha_1 = \alpha_2 = A = 0$, $\phi = \phi_1$, $q = q_1$, so that (23) becomes

$$\left.\begin{array}{c} \phi'' + \dfrac{C'}{C}\phi' + \dfrac{1}{CL}\phi = 0 \\[2mm] q'' + \dfrac{L'}{L}q' + \dfrac{1}{CL}q = 0 \end{array}\right\} \quad \cdots \quad (26)$$

If in the first of these equations C is considered constant and L variable, and if in the second equation L is considered constant and C variable, we obtain:

$$\left.\begin{array}{c} \phi'' + \dfrac{1}{CL(t)}\phi = 0 \\[2mm] q'' + \dfrac{1}{C(t)L}q = 0 \end{array}\right\} \quad \cdots \quad (27)$$

which are of the form

$$q'' + \Omega^2(t)q = 0 \quad \cdots \quad (28)$$

where

$$\Omega^2(t) = \dfrac{1}{LC} \quad \cdots \quad (29)$$

(28) represents the simplest special case of our problem, and the question arises how to solve (28). In this general form no solution is known, but *when $\Omega^2(t)$ does not vary too fast or too much with time* an approximate solution is

$$q = q_0 \cos\left[\int^t \Omega(t)dt + \psi\right] \quad \cdots \quad (30)$$

According to our definition of instantaneous frequency (Section 3) this is given by

$$\frac{d}{dt}\left[\int^t \Omega(t)dt + \psi\right] = \Omega(t)$$

so that we arrive at the result that, in the circumstances specified, the $\Omega(t)$ occurring in (28) actually represents the instantaneous frequency.

In the domain of quantum mechanics the approximate solution (30) of (28) is known as the Wenzel–Kramers–Brillouin solution, the so-called "W.K.B." form. I would point out, however, that this approximate solution had already been given by H. Jeffreys in 1923.[*]

The solution (30), with special reference to frequency modulation, was also derived by me in 1929.[†]

(6) SECOND PROBLEM: THE RESPONSE OF AN ARBITRARY, LINEAR, CONSTANT NETWORK TO A FREQUENCY-MODULATED E.M.F.

Consider an arbitrary, linear, constant network with admittance $Y(j\omega)$. When we apply an e.m.f. $v(t)$ of an arbitrary time function to this network the resulting current $i(t)$ is given symbolically by

$$i(t) = Y\left(\frac{d}{dt}\right)v(t) \quad \cdots \quad (31)$$

If, further, $v(t)$ is oscillatory but modulated in amplitude, phase or frequency, it may be represented by

$$v(t) = e^{j\omega t}f(t)$$

so that the current becomes

$$i(t) = Y\left(\frac{d}{dt}\right)[e^{j\omega t}f(t)] \quad \cdots \quad (32)$$

Now there is a well-known rule (used extensively by Heaviside) for moving in such an operational expression the factor $e^{j\omega t}$ in front of the operator $Y(d/dt)$. This rule enables us to write (32) as

$$i(t) = e^{j\omega t}Y\left(j\omega + \frac{d}{dt}\right)f(t) \quad \cdots \quad (33)$$

If next we develop the operator Y in a Maclaurin series as follows

$$Y\left(j\omega + \frac{d}{dt}\right) = \sum_0^\infty \frac{1}{n!}\left[\frac{d^n}{d(j\omega)^n}Y(j\omega)\right]\left(\frac{d}{dt}\right)^n$$

* JEFFREYS, H., *Proceedings of the London Mathematical Society*, 1923, **23**, p. 428.
† VAN DER POL, B., *Tijdschrift Nederlandsch Radio Genootschap*, 1929, **4**, p. 57, and *Proceedings of the Institute of Radio Engineers*, 1930, **18**, p. 1194.

we obtain at once from (33)

$$i(t) = e^{j\omega t}\sum_{0}^{\infty}\frac{1}{n!}Y^{(n)}(j\omega)f^{(n)}(t) \quad . \quad . \quad (34)$$

[where $Y^{(n)}(j\omega) = \dfrac{d^n}{d(j\omega)^n}Y(j\omega)$ and $f^{(n)}(t) = \dfrac{d^n}{dt^n}f(t)$], which expression was also derived, although in a more complicated way, by Carson and Fry;* on it they base their complete investigation of the problem at hand.

However, we shall here follow a different course, and to that end we first notice the complete symmetry of (34). For if in (34) we make the following changes

$$Y \rightarrow f$$
$$f \rightarrow Y$$
$$j\omega \rightarrow t$$
$$t \rightarrow j\omega$$

the right-hand side of (34) is left completely unaltered. We can therefore conclude that we are free to write instead of (33)

$$i(t) = e^{j\omega t}f\left\{t + \frac{d}{d(j\omega)}\right\}Y(j\omega) \quad . \quad . \quad (35)$$

and it is on *this* expression that we will base our present investigation.

Before proceeding, I wish to point out at the outset that in the above derivation, just as in the paper by Carson and Fry, the functions $Y(j\omega)$ and $f(t)$ are assumed to possess derivatives of any order. Further the series (34) in general does not converge, but as was pointed out by Dr. Stumpers (Philips Laboratory, Eindhoven) in a research not yet published, it is of an asymptotic nature. This latter fact need not prevent us from using it [or (35)] for numerical purposes; in fact the same is done in practically the whole domain of astronomy.

So far $f(t)$ has not yet been specified. It may be so chosen that either amplitude or frequency modulation emerges. For if we take

$$f(t) = f_1(t) = 1 + mg(t)$$

where again $g(t)$ is an audio signal, our e.m.f. becomes

$$v(t) = [1 + mg(t)]e^{j\omega t}$$

which represents amplitude modulation.

Again, taking

$$f(t) = f_2(t) = e^{j\Delta\omega\int^t g(t)dt}$$

our signal becomes frequency modulated as is shown by (8). It can also be made to represent phase modulation.

Hence, limiting ourselves to frequency modulation ($f = f_2$) and writing for short

$$\Delta\omega\int^t g(t)dt = s(t)$$

(35) becomes

$$i(t) = e^{j\omega t + js\left(t + \frac{d}{dj\omega}\right)}Y(j\omega) \quad . \quad . \quad . \quad (36)$$

and now, developing the exponent $s\left(t + \dfrac{d}{dj\omega}\right)$ in a Maclaurin series, we can write (36) as

$$i(t) = e^{j\left[\omega t + s + \frac{s'}{1!}\frac{d}{d(j\omega)} + \frac{s''}{2!}\left(\frac{d}{dj\omega}\right)^2 + \cdots\right]}Y(j\omega)$$
$$= e^{j(\omega t + s)}\left(e^{-\frac{js''}{2!}\frac{d^2}{d\omega^2} + \cdots}e^{s'\frac{d}{dj\omega}}\right)Y(j\omega)$$
$$= e^{j(\omega t + s)}e^{-\frac{js''}{2}\frac{d^2}{d\omega^2} + \cdots}Y[j(\omega + s')] \quad . \quad . \quad (37)$$

* *Bell System Technical Journal*, 1937, **16**, p. 513.

Now it will be clear that according to our definition in Section 2 the instantaneous frequency $\Omega(t)$ of the e.m.f. is given by

$$\Omega(t) = \frac{d}{dt}(\omega t + s) = \omega + s'$$

so that (37) may also be written

$$i(t) = e^{j\int^t\Omega dt}e^{-j\frac{s''}{2}\frac{d^2}{d\omega^2} + \cdots}Y(j\Omega) \quad . \quad . \quad (38)$$

If we may neglect the operator

$$e^{-j\frac{s''}{2}\frac{d^2}{d\omega^2}}$$

and replace it by unity, (38) is reduced to the extremely simple form

$$i(t) = e^{j\int^t\Omega dt}Y(j\Omega) \quad . \quad . \quad . \quad . \quad (39)$$

which is completely analogous to the well-known expression for a constant-frequency e.m.f.:

$$i(t) = e^{j\omega t}Y(j\omega)$$

Thus (39) represents what may be called a *quasi-stationary solution*, meaning that the circuit is completely capable of following through stationary states the variable frequency of the applied e.m.f.

This quasi-stationary solution obviously represents a limiting case, and in fact the circuit may be too sluggish or too "stiff" to follow the relatively rapid changes of frequency of the applied e.m.f. Exactly similar circumstances may arise with amplitude modulation, where, due to the "stiffness" of the circuit, sidebands may be cut off. This cutting off of sidebands in amplitude modulation can also be demonstrated in our present analysis if $f(t)$ is chosen to be $f_1(t)$ so as to represent amplitude modulation. It will have been noticed that, so far, our present attack of the problem completely eliminates any considerations of the spectral distribution of the applied e.m.f., but concentrates entirely on time functions rather than on frequency functions and sidebands. Further, the quasi-stationary solution is completely analogous to the case where with amplitude modulation the amplitude of the e.m.f. varies so slowly (or the damping of the circuit is so great) that the circuit can follow the variable amplitude through stationary states, which, in other words, can also be expressed by the fact that no sidebands are being cut off.

Returning to (39) and limiting ourselves provisionally therefore to the quasi-stationary solution, let us consider what became of our signal. To this end we write

$$Y(j\omega) = \rho(\omega)e^{j\phi(\omega)}$$

where $\rho(\omega)$ and $\phi(\omega)$ are the modulus and phase respectively of the admittance as functions of the frequency ω. (39) thus becomes

$$i(t) \simeq e^{j\int^t\Omega dt}\rho(\Omega)e^{j\phi(\Omega)}$$

and the instantaneous frequency of the current (representing what became of the signal) can, according to our definition in Section 2, be written as

$$\frac{d}{dt}\left[\int^t\Omega dt + \phi(\Omega)\right] = \Omega(t) + \frac{d}{dt}\phi(\Omega)$$
$$= \omega + \Delta\omega\left\{g(t) + g'(t)\frac{d}{d\omega}\phi[\omega + \Delta\omega g(t)]\right\} \quad . \quad (40)$$

so that *in the quasi-stationary approximation the total current signal (including its distortion) is completely determined by $\phi(\Omega)$, i.e. by the phase characteristic of the admittance only, and is therefore—at last explicitly—independent of the amplitude characteristic $\rho(\Omega)$ of the admittance.*

Thus the signal, which in the applied e.m.f. was $g(t)$ becomes in the current

$$g(t) + g'(t)\frac{d}{d\omega}\phi[\omega + \Delta\omega g(t)]$$
$$= g(t) + g'(t)[\phi'(\omega) + \Delta\omega g(t)\phi''(\omega) + \ldots]$$

The term $g'(t)\phi'(\omega)$ can, if desired, be interpreted as a retardation, so that the current signal becomes approximately

$$g[t + \phi'(\omega)] + \Delta\omega g(t)g'(t)\phi''(\omega)$$

where the first term represents the retardation and the second the distortion already present in the quasi-stationary approximation. This distortion is seen to be small when $\phi''(\omega) \to 0$, i.e. when one is working at an inflection point (or on a straight part) of the phase characteristic of the admittance.

As an illustration of (40) we consider a signal of the form $\omega + \Delta\omega \cos pt$, where p is an audio frequency, applied to an ordinary r, C, L circuit in resonance. Equation (40) then shows that the current signal becomes

$$\cos pt + \frac{\alpha p \sin pt}{\alpha^2 + (\Delta\omega)^2 \cos^2 pt}, \quad \left(\alpha = \frac{r}{2L}\right)$$

where the last term represents both the retardation and the distortion. Obviously in general the retardation of a distorted signal cannot be sharply defined.

(7) SECOND PROBLEM (contd.)

Finally, we investigate (a) in what circumstances the quasi-stationary solution is a good approximation, and (b) what is the influence upon the signal when a higher approximation than the quasi-stationary solution must be taken into account.

Considering (a), we return to (38) and remark that the first effect upon the current of the operator

$$e^{-j\frac{s''}{2}\frac{d^2}{d\omega^2}}$$

can be written as

$$i(t) \simeq e^{j\int^t \Omega dt}\left(1 - \frac{js''}{2}\frac{d^2}{d\omega^2}\right)Y(j\Omega)$$
$$= e^{j\int^t \Omega dt}\left(1 - \frac{j}{2}\frac{\partial^2}{\partial\omega\partial t}\right)Y(j\Omega) \quad . \quad . \quad . \quad (41)$$

$$\left(\text{where} \quad \Omega = \omega + s'(t) \quad \text{and therefore} \quad s''\frac{d}{d\omega} = \frac{d}{dt}\right)$$

so that the quasi-stationary solution is a sufficient approximation so long as

$$\left|\frac{1}{2}\frac{\partial^2}{\partial\omega\partial t}Y(j\Omega)\right| \ll |\dot{Y}(j\Omega)| \quad . \quad . \quad . \quad . \quad (42)$$

which may also be written approximately

$$\frac{\Delta\omega}{2}g'(t)|Y''(j\omega)| \ll |Y(j\omega)| \quad . \quad . \quad . \quad (43)$$

Applying this general result again to an ordinary r, C, L circuit in resonance and taking $g(t) = \cos pt$, we obtain as the condition for the validity of a quasi-stationary solution

$$p\Delta\omega \ll \omega_0\frac{r}{L}$$

which expresses a necessary limitation of the product of the audio frequency p and the maximum frequency deviation $\Delta\omega$ relative to the damping constant $r/2L$ of the circuit.

We obtain another illustration of (42) or (43) when an e.m.f., the frequency of which varies *linearly* with the time, is applied to a network of admittance $Y(j\omega)$. The e.m.f. may then be expressed by

$$e^{\frac{1}{2}j\omega_0^2 t^2}$$

so that the instantaneous frequency becomes

$$\Omega(t) = \frac{d}{dt}\left(\frac{1}{2}\omega_0^2 t^2\right) = \omega_0^2 t$$

and the speed with which this frequency is varied

$$\frac{d}{dt}\Omega(t) = \omega_0^2$$

With this e.m.f. the quasi-stationary solution is valid when

$$\frac{1}{2}\omega_0^2|Y''(j\Omega)| \ll |Y(j\Omega)|$$

which therefore puts a limit to ω_0^2, i.e. to the rapidity with which the frequency may be varied.

Considering next (b), the influence upon the signal when a quasi-stationary solution does not suffice, we have to calculate the instantaneous frequency associated with

$$e^{j\int^t \Omega dt}\left(1 - \frac{js''}{2}\frac{d^2}{d\omega^2}\right)Y(j\Omega)$$
$$= e^{j\int^t \Omega dt}\left(1 - \frac{js''}{2}\frac{d^2}{d\omega^2}\right)\rho(\Omega)e^{j\phi(\Omega)}$$
$$= e^{j\int^t \Omega dt + j\phi(\Omega)}\rho_1(\Omega)e^{j\phi_1(\Omega)} \quad \text{say.}$$

We are not concerned with ρ_1 but a little calculation shows that

$$\phi_1 \simeq \tan\phi_1 \simeq \frac{s''}{2}\left(\phi'^2 - \frac{\rho''}{\rho}\right) = \frac{d\Omega}{dt}\left(\phi'^2 - \frac{\rho''}{\rho}\right)$$

so that the instantaneous frequency in the current becomes

$$\Omega(t) + \frac{d}{dt}\phi(\Omega) + \frac{d}{dt}\left\{\frac{d\Omega}{dt}\left[\phi'^2(\Omega) - \frac{\rho''(\Omega)}{\rho(\Omega)}\right]\right\} \quad . \quad (44)$$

Here, in the first term $\Omega(t)$ represents the original signal.

The second term represents the distortion discussed above, which is already present in the quasi-stationary approximation, whereas the third term gives the distortion which must also be considered when the quasi-stationary solution does not suffice.

In most practical applications (44) may still further be reduced to

$$\Omega(t) + \frac{d}{dt}\phi(\Omega) + \left[\phi'^2(\omega) - \frac{\rho''(\omega)}{\rho(\omega)}\right]\frac{d^2\Omega}{dt^2} \quad . \quad (45)$$

Equation (45) shows that, whereas the quasi-stationary solution is determined by the phase characteristic of the admittance only, the last term also depends upon the amplitude characteristic of the admittance.

In conclusion, it may be remarked that, although the above investigation relates to the resulting current $i(t)$ for a given impressed e.m.f. leading therefore to a consideration of the admittance $Y(j\omega)$, a similar analysis can be used when, instead of the current, the potential difference across any circuit is required.

PAPER NO. 10

Reprinted from *Communications News* (Phillips), Vol. 9, No. 3, pp. 82–92, April 1948

DISTORTION OF FREQUENCY-MODULATED SIGNALS IN ELECTRICAL NETWORKS

F. L. H. M. STUMPERS.

621.396.619.13:621.392

Receiver filters introduce distortion in a frequencymodulated signal. This distortion can be calculated by F o u r i e r analysis of the signal or by the "quasi-stationary" method. C a r s o n and F r y derived the latter method from their series, the asymptotical character of which is shown here. The mathematical derivation is reviewed critically. The quasi-stationary solution is the first term of another also asymptotical series, and a method is given for judging the error. — The production of harmonics and intermodulation by single tuned and coupled circuits, as well as the effect of detuning are considered.

Introduction

1) The problem of calculating the current produced in an electrical network by a voltage of which the frequency is modulated attracted attention [1] at a very early stage and has since been many times discussed. My thesis [2] (chapters 4 and 5) contained a critical review of the mathematical problems involved and also discussed a number of applications. In the following a brief survey of these investigations is given.

For these calculations both the F o u r i e r and the asymptotic methods come into consideration. Which of the two methods deserves preference depends amongst other factors upon the magnitude of the modulation index. In the first mentioned method, the original signal is analysed into its F o u r i e r components which are then handled in accordance with the theory of alternating currents. An objection to the use of the F o u r i e r method is that it does not show clearly the relation between the distortion and the characteristic values of the network; this is much more clearly brought out by the asymptotic method developed by C a r s o n and F r y [3] which yields the result in the form of a power series of especially elegant type. When applied to frequency modulation, however, a different form of asymptotic expansion, which we are about to demonstrate, may be preferred.

W i l d e [4] has applied to frequency-modulation the methods of F e l d t k e l l e r [5] and G e n s e l [6] for approximating transient phenomena by means of a "cable harp", but in the case of sinusoidal modulation this method is somewhat cumbersome.

Fourier method

2) Modulation with a sinusoidal signal produces the following F o u r i e r spectrum

$$e^{i\omega_0 t + im \sin pt} = \sum_{-\infty}^{\infty} J_n(m) \, e^{i(\omega_0 t + npt)}; \quad m = \Delta\omega/p \quad (1)$$

Let the differential equation of the network be

$$\left(a_0 + a_1 \frac{d}{dt} + \dots + a_r \frac{d^r}{dt^r} \right) E(t) =$$
$$= \left(b_0 + b_1 \frac{d}{dt} + \dots + b_s \frac{d^s}{dt^s} \right) I(t) \quad (2)$$

in which the current is given by

$$I(t) = e^{i(\omega_0 t + m \sin pt)}$$

Abbreviating the equation (2), we write

$$f\left(\frac{d}{dt}\right) . E(t) = g\left(\frac{d}{dt}\right) . e^{i(\omega_0 t + m \sin pt)} \quad (3)$$

Ignoring initial phenomena, the steady state alternating current is obtained

$$E(t) = \sum_{-\infty}^{+\infty} J_n(m) . \frac{g(i\omega_0 + inp)}{f(i\omega_0 + inp)} e^{i\omega_0 t + inpt} \quad (4)$$

We shall confine the investigation to passive networks, so that all terms in the series remain finite. The latter converges in accordance with d'A l e m b e r t's test, since for n large $J_n(m)$ behaves in the same way as

$$\frac{1}{n!} \left(\frac{m}{2}\right)^n$$

whilst the behaviour of g/f for n large is given by the terms of the highest degree in both numerator and denominator. We now substitute Z for g/f and divide the former into real and imaginary parts, viz.

$$Z = Z(i\omega_0) = X_0 + iY_0 = X(i\omega_0) + iY(i\omega_0)$$
$$Z_n = X_n + iY_n = Z(i\omega_0 + inp)$$

Then the phase angle becomes

$$\varphi = \omega_0 t +$$
$$+ \text{arc tan} \frac{\Sigma J_n(m) . (X_n \sin npt + Y_n \cos npt)}{\Sigma J_n(m) . (X_n \cos npt - Y_n \sin npt)}$$
$$(5)$$

A detector tuned to the frequency ω_0 will then produce a voltage proportional to $d\varphi/dt - \omega_0$.

When

$A_{nm} = X_n X_m + Y_n Y_m$ and $B_{nm} = X_n Y_m - X_m Y_n$, the audio-frequency voltage is

$$\frac{d\varphi}{dt} - \omega_0 =$$

$$\frac{\Sigma_n \Sigma_m \, mp \, J_n J_m \, \{ A_{nm} \cos (n-m) \, pt +}{\Sigma \, \Sigma \, J_n J_m \quad \{ A_{nm} \cos (n-m) \, pt +}$$

$$\frac{+ \, B_{nm} \sin (n-m) \, pt \}}{+ \, B_{nm} \sin (n-m) \, pt \}} \qquad (6)$$

It is usual to determine the harmonics graphically. The method gives not only the instantaneous frequency, but also the instantaneous amplitude, as the square of the latter is identical with the denominator of expression (6).

The asymptotic method

3) Employing the differential equation (3) and assuming that $E(t) = E_0(t) \, e^{i\omega_0 t}$, the former expression can be converted to the following form

$$f \left(i\omega_0 + \frac{d}{dt} \right) . E_0(t) = g \left(i\omega_0 + \frac{d}{dt} \right) . e^{im \sin pt} \qquad (7)$$

The accuracy of this may be checked by means of formal development. The next step is to discover whether the following function is a particular solution of (7)

$$E_0(t) = \frac{g}{f} \left(i\omega_0 + \frac{d}{dt} \right) . e^{im \sin pt} \qquad (8)$$

Formally, this function meets the case and, as all the terms are periodic, it corresponds to the steady state. In place of g/f we write Z, the impedance of the network, and the formal solution then takes the form

$$E_0(t) = \left\{ Z \left(i\omega_0 \right) + Z' \, (i\omega_0) \frac{d}{dt} + \right.$$

$$\left. + \frac{Z'' \, (i\omega_0)}{2!} \frac{d^2}{dt^2} + \dots \right\} e^{im \sin pt} \qquad (8a)$$

It was roughly in this form (with evaluation of the individual differentiations) that the series was first derived by C a r s o n and F r y (l.c. p. 521), after whom the series will therefore be named in what follows. In the next section we shall review the derivation chosen by these authors.

Whereas the expansion of $g(i\omega_0 + d/dt)$ in expression (7) automatically terminates, this is, generally speaking not the case with the expansion of $Z(i\omega_0 + d/dt)$. The fact that it leads to the

correct result when breaking off, is immediately evident, and the question arises as to the convergence of the C a r s o n and F r y series when it produces an infinite number of terms; in order to see what happens we calculate the remainder after deducting the first $(k+1)$ terms of the series from the exact expression obtained by the F o u r i e r method. This remainder is

$$J_n(m) . Z \, (i\omega_0 + inp) \, e^{inpt} -$$

$$\left\{ Z + Z' \frac{d}{dt} + \frac{Z''}{2!} \frac{d^2}{dt^2} + \dots \frac{Z^{(k)}}{k!} \frac{d^k}{dt^k} \right\} \Sigma J_n(m) \, e^{inpt}$$

Since the result converges uniformly the F o u r i e r series may in this case be differentiated term by term; moreover, since it converges absolutely, it is permissible to change the sequence of the terms and thus place together the terms having the same factor e^{inpt}. Our remainder then becomes

$$J_n(m) . e^{inpt} \left\{ Z \, (i\omega_0 + inp) - Z \, (i\omega_0) - \right.$$

$$\left. - inp \, Z' \, (i\omega_0) \dots - \frac{(inp)^k}{k!} \, Z^{(k)} \, (i\omega_0) \right.$$

This takes the form of a F o u r i e r series, in which the amplitude of the component e^{inpt} is the product of $J_n(m)$ and the remainder of the expansion of $Z \, (i\omega_0 + inp)$ in the neighbourhood of $i\omega_0$, after $(k+1)$ terms. Now, if k be increased, leaving n constant, this amplitude will decrease to zero only when $i\omega_0 + inp$ lies within the circle of convergence of this expansion. In the other case the amplitude will in the long run increase. Since sinusoidal modulation leads to an infinite frequency spectrum, components will certainly occur for which $i\omega_0 + inp$ is outside the circle of convergence of Z around $i\omega_0$, at any rate when the radius of this circle is finite. If the radius is infinite (corresponding to resistance plus inductance) the series terminates. Apart from this instance the series of C a r s o n and F r y may certainly not be written without remainder; it is possible however to see from the remainder-for instance by splitting the impedance function into partial fractions that the series is asymptotic in the sence of P o i n c a r é, when $p \to 0$. It is the asymptotic series for phase modulation, as will be shown in section 6.

Calculation according to Carson and Fry

4) Let us now view in a more critical light the derivation put forward by C a r s o n and F r y and subsequently followed with very minor modifications by other authors [7].

It is first necessary to introduce the concept of the indicial impedance $A(t)$, this being the function of time represented by the voltage on the network when a unit-function current is passed through it at a time $t = 0$. The unit-function current $U(t)$ is 0 when $t < 0$ and 1 when $t < 0 . U(0) = \frac{1}{2}$. From the superposition principle it then follows that when an arbitrary current $F(t)$ is applied at a time $t = 0$, prior to which there was no current in the network, the voltage is represented by

$$E(t) = \int_0^t A(t - \tau) \, dF(\tau) \qquad (9)$$

Integration by parts then gives

$$E(t) = A(0)F(t) - A(t)F(0) - $$
$$- \int_0^t \frac{dA(t-\tau)}{d\tau} F(\tau) \, d\tau$$

In order to avoid singularities we shall confine the discussion to networks whose self-oscillations are damped and which at high frequency have a capacitive character. In this case $A(0)$ is 0, and the voltage is expressed by

$$E(t) = \int_0^t A'(\tau) F(t - \tau) \, d\tau \qquad (10)$$

When $F(t)$ is a periodic function of t, this integral can be split into two parts

$$E(t) = \int_0^\infty F(t - \tau) A'(\tau) \, d\tau - $$
$$- \int_t^\infty F(t - \tau) A'(\tau) \, d\tau \qquad (11)$$

The second integral represents a transient phenomenon which decays when t is large; the first expresses the steady state.

In their first derivation C a r s o n and F r y proceed as follows: Let the given signal be

$$F(t) = e^{i\omega t + i \int_0^t \mu d\tau}$$

Suppose that

$$F(t - \tau) = F(t) M(t, \tau) e^{-i\omega\tau}$$

(It follows that $M(t, 0) = 1$)

Incorporating this in equation (11) we obtain:

$$E(t) = F(t) \int_0^\infty M(t, \tau) A'(\tau) e^{-i\omega\tau} \, d\tau$$

C a r s o n and F r y then proceed to the expansion

$$M(t, \tau) = M(t, 0) + \tau \left\{ \frac{dM(t, \tau)}{d\tau} \right\}_{\tau=0} + $$
$$+ \frac{\tau^2}{2!} \frac{d^2 M(t, \tau)}{d\tau^2} \bigg\}_{\tau=0} + \cdots$$

which is used in the integral and integrated term by term

$$E(t) = F(t) \left\{ \int_0^\infty e^{-i\omega\tau} A'(\tau) \, d\tau + \right.$$
$$+ \left. \Sigma_1^\infty C_n \int_0^\infty \frac{\tau^n}{n!} e^{-i\omega\tau} A'(\tau) \, d\tau \right. \qquad (12)$$

$C_n(t)$ is determined by

$$C_n(t) = \left\{ \frac{d^n}{d\tau^n} M(t, \tau) \right\}_{\tau=0}$$

If we now introduce in equation (11) $F(t) = e^{i\omega t}$ for the current, it will be seen that

$$Z(i\omega) = \int_0^\infty e^{-i\omega\tau} A'(\tau) \, d\tau$$

Similarly

$$\frac{dZ(i\omega)}{di\omega} = - \int_0^\infty \tau e^{-i\omega\tau} A'(\tau) \, d\tau$$

it being permissible to differentiate under the integral symbol, since $A'(\tau)$ consists of a sum of damped oscillations. Substituting these results in expression (12) and taking into account that

$$C_n = (-1)_n \frac{d^n \{ F(t) e^{-i\omega_0 t} \}}{dt^n}$$

we obtain the desired expansion. The weak point is the integration term by term, which is not really permissible because the power series for the sine and cosine, although converging in the whole plane, do not do so uniformly at ∞.

The other derivation, also chosen by V e l l a t[7] proceeds from the F o u r i e r integral for the modulation

$$exp \left(i \int_0^t \mu d\tau \right) = \int_{-\infty}^\infty F(i\omega) \cdot e^{i\omega t} d\omega$$

(With sinusoidal modulation this is a F o u r i e r series).

The voltage is

$$E = e^{i\omega_0 t} \int_{-\infty}^\infty F(i\omega) Z(i\omega_0 + i\omega) e^{i\omega t} d\omega$$

We then develop

$$Z\,(i\omega_0 + i\omega) = Z\,(i\omega_0) +$$

$$+ \sum_1^\infty \frac{\omega^n}{n!} \left\{ \frac{d^n}{d\omega^n}\, Z\,(i\omega) \right\}_{\omega=\omega_0}$$

This is introduced into the integral and integrated term by term; with a slight modification we arrive again at the series of C a r s o n and F r y. Since the expansion of $Z\,(i\omega_0 + i\omega)$ with trivial exceptions does not hold good in the range from $-\infty$ to ∞, integration of the individual terms is again impermissible. The series may again not be written without a remainder.

Derivation of the asymptotic expansion

5) We shall now consider a signal, modulated both in amplitude and frequency. It will be assumed, further, that each modulation contains the same audiofrequency fundamental of p radians per second, the signal being thus represented by

$$a\,(pt)\,.\,exp\,\left(i\omega_0 t + i\int_0^t h\,(p\tau)\,d\tau\right)$$

The instantaneous amplitude of this input signal is then $a\,(pt)$, and employing the earlier abbreviations, the differential equation for the voltage is

$$f\left(\frac{d}{dt}\right) E = g\left(\frac{d}{dt}\right) a\,(pt)\,exp\left\{i\omega_0 t + i\int_0^t h\,(p\tau)\,d\tau\right\}$$

Now assuming that there are no continuous oscillations among the general integrals, introducing $E = E_0 e^{i\omega_0 t}$ and substituting u for pt, the differential equation for $E_0\,(u)$ will be

$$f\left(i\omega_0 + p\frac{d}{du}\right) E_0\,(u) =$$

$$= g\left(i\omega_0 + p\frac{d}{du}\right) a\,(u)\,exp\;\frac{i}{p}\int_0^u h\,(x)\,dx$$

The expression in the second term may be worked out by executing the s differentiations (See section 2). The result is

$$f\left(i\omega_0 + p\frac{d}{du}\right) E_0\,(u) = G\,(p,u)\,exp\left\{\frac{i}{p}\int_0^u h(x)\,dx\right\}$$

$$(12)$$

In this $G\,(p,u)$ is a polynomial in p; the second term is a periodic function of u. The solution of the differential equation now consists of the sum of a periodic function and r general integrals each multiplied by a suitable constant.

It is now possible to make use of a theorem of P e r r o n [8] concerning the dependence on a parameter of the integrals in a differential equation

of the n th order; from this it follows that there is an integral of the form

$$\left\{exp\frac{i}{p}\int_0^u h\,(x)\,dx\right\}\sum_0^\infty p^v\,w_v\,(u) \qquad (13)$$

asymptotic for $p \to 0$

The asymptotic expression

$$f\,(p) \approx g\,(p)\,\sum_0^\infty A_n p^n \text{ for } p \to 0$$

states that when

$$f\,(p) = g\,(p)\,\sum_0^m A_n p^n + R_m\,(p),\text{ if } m = 0,\,1,\,2,\text{ etc}$$

$$\lim_{p\to 0} p_m R_m\,(p) = 0$$

This then furnishes exactly the periodic solution of the original differential equation. By formal substitution of expression (13) in the differential equation and equating the terms of the same degree in p in the left and right hand side of the equation, the functions $w_v\,(u)$ can now be found. Thus

$$E_0(u) = \left\{ exp\frac{i}{p}\int_0^u h(x)dx \right\}\left[aZ + p\left(a'Z' + \frac{ih'Z''}{2}\right) + \right.$$

$$+ p^2\left\{\frac{a''Z''}{2} + \frac{ia'h'Z'''}{2} + \right.$$

$$\left.\left. + a\left(\frac{ih''Z'''}{6} - \frac{h'^2\,Z''''}{8}\right)\right\}+\dots\right] \qquad (14)$$

in which

$$a = a\,(u)\,;\,h = h\,(u)\,;\,Z = Z\,(i\omega_m)\,;\,Z' = \frac{dZ(i\omega_m)}{di\omega_m}$$

and so on.

$$\omega_m = \omega_0 + h\,(u)$$

Taking the particular case of a signal modulated in amplitude only, we have

$$E_0\,(u) = a\,(u)\,Z\,(i\omega_0) + pa'\,(u)\,Z'\,(i\omega_0) +$$

$$+ p^2\frac{a''\,(u)\,Z''\,(i\omega_0)}{2} + \dots \qquad (15)$$

and for a signal modulated only in frequency:

$$E_0\,(u) = \left\{ exp\frac{i}{p}\int_0^u h\,(x)\,dx \right\}\left[Z + \frac{iph'Z''}{2} + \right.$$

$$\left. + p^2\left(\frac{ih''Z'''}{6} - \frac{h'^2\,Z''''}{8}\right) + \dots\right] \qquad (16)$$

The instantaneous frequency and amplitude can be derived from these series without trouble; to this end the impedance is divided into modulus and phase $Z\,(i\omega) = M\,(\omega)\,e^{i\phi\,(\omega)}$. The series for the instantaneous frequency in (16) then commences with

274

$$\omega_m = \omega_0 + h(pt) + p\varphi'h' + p^2\left\{\frac{h''}{2}\left(-\frac{M''}{M}+\varphi'^2\right)+\right.$$

$$\left.+\frac{h'^2}{2}\left(\frac{M''M'-M'''M}{M^2}+2\varphi'\varphi''\right)\right\}+\ldots$$

where

$$h' = \frac{dh(pt)}{dpt}; \varphi' = \frac{d\varphi}{d\omega_m}; M' = \frac{dM(\omega_m)}{d\omega_m}; \text{ etc}$$

$$(16a)$$

For the instantaneous amplitude we have in the same case

$$A_m = M(\omega_m) + p\left(M'\varphi' + \frac{\varphi''M}{2}\right) \quad (16b)$$

In the series (16, 16a and 16b) we shall call the terms in p of zero degree the "static terms" and those of the first degree the "quasi-stationary" terms; together they constitute the "quasi-stationary approximation".

In contrast with expansion (16), that of C a r-s o n and F r y does not directly yield a series in powers of p for frequency-modulated signals. (It gives such a series in case of phase-modulation). V e l l a t took for his calculation five terms from the series of C a r s o n and F r y and discarded from the latter those terms in the derivatives of $e^{im\sin pt}$ in which p appears explicitly. In this way he obtains an approximation of the quasi-stationary term of series (16a). For frequency-modulation the series (16) may be preferred, the more so as the "quasi-stationary approximation" is usually sufficient.

Estimation of the error in the use of the asymptotic series

6) In this section the asymptotic series will be evolved in a different manner, so that the error may be easily estimated.

The following equation is taken as a basis

$$E(t) = \int_0^\infty F(t-\tau) A'(\tau) d\tau$$

in which $A(\tau)$ still retains the form

$$\Sigma^n (A_n + iB_n) e^{-(a_n+i\beta_n)\tau}$$

the terms either occur in complex pairs of are real.

In the case of frequency-modulation

$$F(t) = Re\, e^{i(\omega_0 t + m\sin pt)}$$

$$E(t) = Re\, e^{i(\omega_0 t + m\sin x)} \Sigma(A_n + iB_n)\int_0^\infty e^{-a_n\tau+ig(\tau)} d\tau$$

$$g(\tau) = -(\beta_n + \omega_0)\tau + m\sin(x-p\tau) -$$

$$- m\sin x; x = pt$$

When $e^{ig(\tau)}$ is expanded in terms of p and the integration carried out term by term, the result is the asymptotic series. In the integrand, the term $e^{-a_n\tau}$ decreases rapidly; the most significant contribution to the integral is therefore to be expected when $a_n\tau$ is small, e.g. less than 4.

Consider, for instance, the integral

$$\int_0^\infty e^{-a_n\tau}\cos(\beta_n+\omega_0)\,\tau\cos m\{\sin(x-p\tau)-\sin x\}\,d\tau$$

$$(17)$$

This is closely approximated by

$$\int_0^\infty e^{-a_n\tau}\cos(\beta_n+\omega_0)\,\tau\cos(\Delta\omega\tau\cos x)\,d\tau$$

as long as it holds good for $\tau_0 = 4/a_n$ that

$$\sin p\tau_0 \cong p\tau_0 \text{ and } \frac{\Delta\omega p\tau_0^2}{2} \ll 1$$

thus, for instance, when

$$p < 0.05\, a_n \text{ and } p\Delta\omega < 0.03\, a_n^2$$

If more terms in p are taken $e^{ig(\tau)}$ will certainly be more closely approximated for the same value of τ, but one obtains for the difference between the integral (17) and the approximation, integrals of the following form

$$\int_0^\infty R_n(\tau)\,\tau^n e^{-a_n\tau}\,d\tau$$

The centre of gravity of the integrand moves more and more to the right, so that the range in which the approximation is accurate becomes smaller and smaller.

Application of the theory to a tuned circuit

7) To a tuned circuit a current is applied of which the frequency is modulated, whilst the amplitude remains constant.

$$\left(LC\frac{d^2}{dt^2} + RC\frac{d}{dt} + 1\right)E =$$

$$= \left(R + L\frac{d}{dt}\right)e^{i(\omega_0 t + m\sin pt)}$$

See *fig. 1*.

Suppose that $LC = \omega_0^{-2}$ and $R/L = 2a$. We shall also introduce $E = E_0 e^{i\omega_0 t}$ and take it that a and $\Delta\omega\,(=mp)$ are both small with respect to ω_0.

In this case the differential equation may be given in the following simplified form

$$aE_0 + \frac{dE_0}{dt} = e^{im\sin pt} \quad (18)$$

Asymptotic development then gives the approximation for the square of the instantaneous amplitude

$$A^2 = \frac{1}{a^2 + \Delta\omega^2 \cos^2 pt} - \frac{2pa\Delta\omega^2 \sin 2pt}{(a^2 + \Delta\omega^2 \cos^2 pt)^3} \quad (19)$$

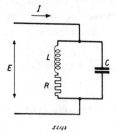

Fig. 1. Single tuned circuit.

The instantaneous frequency is

$$\omega_m = \omega_0 + \Delta\omega \cos pt +$$
$$+ \frac{pa\Delta\omega \sin pt}{a^2 + \Delta\omega^2 \cos^2 pt} + \frac{p^2\Delta\omega \cos pt}{(a^2 + \Delta\omega^2 \cos^2 pt)^3}$$
$$. \{ - a^4 - 6a^2\Delta\omega^2 \sin^2 pt + \Delta\omega^4 \cos^2 pt$$

$$(1 + \sin^2 pt)\} \quad (20)$$

If we require an expansion in harmonics

$$\omega_m = \omega_0 + \Delta\omega \cos pt + p \Sigma C_{2n+1} \sin (2n+1) pt +$$
$$+ p^2 \Sigma D_{2n+1} \cos (2n + 1) pt \quad (21)$$

F o u r i e r integration yields

$$C_{2n+1} = (-1)^n 2 c^{-(2n+1)} \{(1+c^2)^{\frac{1}{2}} - 1\}^{2n+1} ;$$
$$; c = \Delta\omega/a \quad (22)$$

and

$$D_{2n+1} = (-1)^n 2 (2n+1)^2 a^{-1} c^{-(2n+1)} (1+c^2)^{-\frac{1}{2}}$$
$$\{(1+c^2)^{\frac{1}{2}} - 1\}^{2n+1} \quad (23)$$

When the frequency deviation is small the third harmonic in the quasi-stationary term is proportional to $\Delta\omega^3$, the fifth to $\Delta\omega^5$ and so on.

$$C_3 = 6 \sum_0^\infty \frac{(-1)^{n+1} (2n+2)!}{(n+3)!n!} \left(\frac{\Delta\omega}{2a}\right)^{2n+3}$$

$$C_5 = 10 \sum_0^\infty \frac{(-1)^n (2n+4)!}{(n+5)!n!} \left(\frac{\Delta\omega}{2a}\right)^{2n+5} \text{ etc}$$

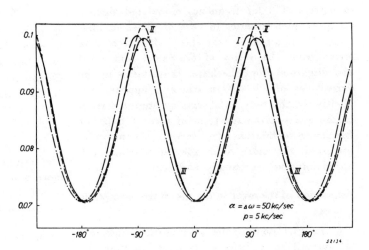

Fig. 2. Instantaneous amplitude: I Static approximation. II Quasi-stationary approximation. III Exact value, calculated by the Fourier method. XX Second order approximation from the asymptotic series.

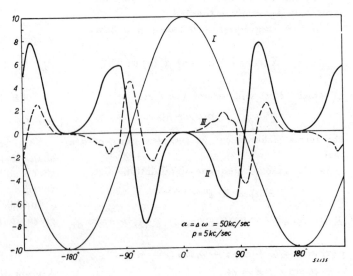

Fig. 3. Instantaneous frequency:
  I Exact value (Fourier method)
  II Difference between exact value and first approximation (enlarged 50 times) ($\sqrt{\overline{\Delta^2}} = 0.0793$)
  III [Difference between exact value and second approximation (enlarged 50 times) ($\sqrt{\overline{\Delta^2}} = 0.0341$)

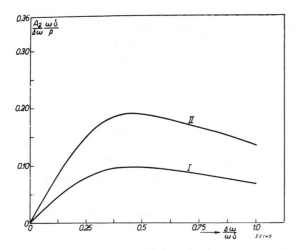

Fig. 4. Second harmonic produced by detuning a single circuit. Upper curve $\omega_1 - \omega_0 = 0.1\,\omega\delta$. Lower curve: $\omega_1 - \omega_0 = 0.05\,\omega\delta$. (Calculated from the quasi-stationary approximation).

Whereas the F o u r i e r method gives the result only point by point and involves a great deal of graphical work, the harmonics are now obtained immediately in compact form. By setting down an upper limit for the error in the manner employed in the preceding section a reasonable degree of accuracy may be anticipated for $p \leq 0.1\,a$ and $p\varDelta\omega \leq 0.1\,a^2$. In this instance the second order approximation would show a further improvement. We take as an example the case where $p = 0.1\,a$; $\varDelta\omega = a$. The results of the F o u r i e r method and the asymptotic approximation to the second order were calculated, for both the instantaneous amplitude and the instantaneous frequency. In *fig. 2* the exact value of the instantaneous amplitude and its approximations are given, and in *fig. 3* the exact value of the instantaneous frequency and also the difference between this function and its first and second order approximations enlarged 50 times.

Further a number of harmonics of the instantaneous frequency were calculated by the F o u r i e r method and compared with those derived from the quasi-stationary approximation:

So far we have assumed that the filter was tuned to the central frequency of the frequency-modulated signal. A small amount of detuning has no large effect on the odd harmonics but introduces the even harmonics. If the circuit is tuned to the frequency ω_1 and the central frequency of the signal is ω_0, we put

$$b = (\omega_1 - \omega_0)/a$$

The instantaneous frequency is derived in the same way as before and for the quasi-stationary term the result is

$$\frac{pa\varDelta\omega \sin pt}{a^2 + (ba - \varDelta\omega \cos pt)^2}$$

The harmonics are calculated by F o u r i e r integration. If A_n is the amplitude of the n th harmonic

$$A_n^2 = -p^2 (z_1{}^n - z_2{}^n)^2$$

where

$$z_1 = \frac{1}{c}(b - u + i - iv) \; ;$$

$$z_2 = \frac{1}{c}(b - u - i + iv) \; ; \quad c = \varDelta\omega/a \; ;$$

$$u = \left[\frac{b^2 - 1 - c^2}{2} + \frac{1}{2}\{(b^2 - 1 - c^2)^2 + 4b^2\}^{\frac{1}{2}}\right]^{\frac{1}{2}}$$

$$v = \left[\frac{1 + c^2 - b^2}{2} + \frac{1}{2}\{(1 + c^2 - b^2)^2 + 4b^2\}^{\frac{1}{2}}\right]^{\frac{1}{2}}$$

For $n = 2$ we get in this way $A_2 = \dfrac{4p}{c^2} u (1 - v)^2$

and for $n = 3$ $\quad A_3 = \dfrac{2p}{c^3} (1 - 3u^2)(1 - v)^3$

We can plot the percentage harmonic for a general single turned circuit by expressing the audio-frequency p and the frequency sweep $\varDelta\omega$ in the filter constant a (or as in our figures in $\omega\delta = 2a$). In *fig. 4* the lower curve is drawn for $b = 0.1$ and the upper one for $b = 0.2$. The percentage second harmonic is $A_2/\varDelta\omega$. The ordinate gives $A_2\omega\delta/p\varDelta\omega$ as a function of $\varDelta\omega/\omega\delta$. Thus when $\omega\delta$ and p are given one has only to multiply the ordinate of the curve in fig. 2 by $p/\omega\delta$ to get the percentage second harmonic for any given value of $\varDelta\omega/\omega\delta$. The odd harmonics are given

p/a	$\varDelta\omega/a$	$p\,\varDelta\omega/a^2$	Fourier		Quasi-stationary	
			% 3rd harm	% 5th harm	% 3rd harm	% 5th harm
0.1	1	0.1	1.42	0.24	1.38	0.24
0.33	1	0.33	4.00	0.64	4.74	0.81
0.28	1.08	0.30	4.7	0.87	5.53	1.06
0.28	1.79	0.50	8.7	3.9	8.08	2.78
0.28	2.51	0.70	9.8	6.3	8.91	3.53
0.28	3.59	1.01	10.1	6.55	8.74	4.59

in the same way in *figs.* 6 and 7. They apply when the filter is tuned to the central frequency but are not much dependent on exact tuning.

Application to the theory of coupled circuits

8) In *fig.* 5, M is the coefficient of mutual inductance. $M^2 = k^2 L_1 L_2$. It is assumed that both circuits are tuned to the frequency ω_0 and that

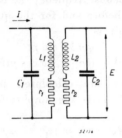

Fig. 5. Coupled circuits.

$r_1 = \omega_0 L_1 \delta_1$, $r_2 = \omega_0 L_2 \delta_2$. Now when the values of δ_1, δ_2 and k are small in relation to unity and $\Delta\omega$ and p small with respect to ω_0, the differential equation to be solved is

$$\omega_0^2 (k^2 + \delta_1\delta_2) E_0 + 2\omega_0 (\delta_1 + \delta_2) E_0' + 4E_0'' = i \, e^{im \sin pt} \qquad (24)$$

The instantaneous frequency is (quasi-stationary)

$$\omega_m = \omega_0 + \Delta\omega \cos pt + \frac{2pa\Delta\omega \sin pt \, (\beta^2 + \Delta\omega^2 \cos^2 pt)}{\beta^4 + (4a^2 - 2\beta^2) \Delta\omega^2 \cos^2 pt + \Delta\omega^4 \cos^4 pt} \qquad (25)$$

where

$$4\beta^2 = (k^2 + \delta_1\delta_2) \omega_0^2 \text{ and } 4a = \omega_0 (\delta_1 + \delta_2)$$

From this the harmonics can be extracted by F o u r i e r integration. When

$$\delta_1 = \delta_2 = \delta, \quad k/\delta = a, \text{ and } 2\Delta\omega/\omega\delta = x,$$

we find for the quasi-stationary approximation of the third harmonic

$$A_3 = \frac{4p}{x} \left[3 + \frac{4(1 - 3a^2)}{x^2} - \frac{1}{x^2} \{ (x^2 + 4 - 4a^2) A - 8aB \} \right]$$

and, similarly, for the fifth harmonic

$$A_5 = \frac{-4p}{x} \left[5 + \frac{20 - 60a^2}{x^2} + \frac{16 - 160a^2 + 80a^4}{x^4} - \left\{ 1 + \frac{12(1 - a^2)}{x^2} + \frac{16(1 - 6a^2 + a^4)}{x^4} \right\} A + \left\{ \frac{24a}{x^2} + \frac{64a(1 - a^2)}{x^4} \right\} B \right]$$

whilst in both cases

$$2A^2 = x^2 + 1 - a^2 + \{ (x^2 + 1 - a^2)^2 + 4a^2 \}^{\frac{1}{2}}$$
$$2B^2 = -x^2 - 1 + a^2 + \{ (x^2 + 1 - a^2)^2 + 4a^2 \}^{\frac{1}{2}}$$

By means of these expressions curves have been plotted for different values of a k/δ, from which the distortion as a function of the frequency swing can be read. This swing, and the audio-frequency must then be expressed in terms of the bandwidth $\omega\delta$, which is characteristic of the filter. For low values of x one should use the series expansion, which in the case of A_3 commences with

$$-\frac{px^3 (1 - 3a^2)}{2 (1 + a^2)^3} + \frac{3}{8} \frac{px^5 (1 - 10a^2 + 5a^4)}{(1 + a^2)^5}$$

and for A_5 with

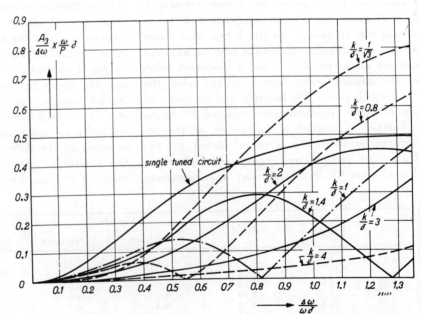

Fig. 6. Third harmonics produced by band-filter and single tuned circuit in general constants. (Calculated from the quasi-stationary approximation).

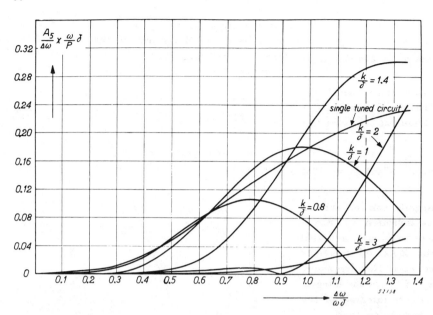

Fig. 7. Fifth harmonics produced by bandfilters and single tuned circuit, in general constants.
(Calculated from the quasi-stationary approximation).

$$\frac{px^5\,(1-10a^2+5a^4)}{8\,(1+a^2)^5} - \frac{5}{32}\;\frac{px^5\,(1-21a^2+35a^4-7a^4)}{(1-a^2)^7}$$

The following are a few comparative values of

of p and $\Delta\omega$ they give an indication of the result to be expected, but one should check the result by the exact F o u r i e r method.

k/δ	$\Delta\omega/\omega\delta$	$p/\omega\delta$	Fourier method		Quasi-stationary	
			% 3rd harm	% 5th harm	% 3rd harm	% 5th harm
1	0.5	0.1	1.31	0.53	1.37	0.39
1	0.4	0.2	2.20	0.45	2.30	0.32
2	1	0.1	1.65	2.10	1.73	1.79
2	1	0.1	3.86	1.13	3.99	0.24
1	1	0.2	2.53	3.06	3.46	3.59
1	0.5	0.5	4.91	0.39	6.88	1.93

exact results and quasi-stationary approximations:

The absolute values of A_3 and A_5 are graphed in *figs. 6* and *7*. When these pass through zero the phase is rotated 180°, which opens up the possibility of employing different filters in subsequent stages to obtain compensation. One could combine a filter of $k/\delta = 0.8$ with one of $k/\delta = 2$ aiming at a more attractive amplitude characteristic and reduced distortion. If $p = 0.1$ $\omega\delta$ and $\Delta\omega = 0.5\ \omega\delta$ the quasi-stationary approximation yields 1.249% third and 0.473% fifth harmonic. The exact theory gives respectively 1.106% and 0.451%. For $p = 0.1\ \omega\delta$ and $\Delta\omega = \omega\delta$ the quasi-stationary prediction is 0.060% third and 0.948% fifth harmonic, whereas the exact theory yields 0.658% and 1.264%. For lower values of $p/\omega\delta$ the correspondence is still better. The method is also applicable to filters of different δ.

The graphs of fig. 6 and 7 give no exact values, but the quasi-stationary approximation. Their use for calculations is justified up to $p/\omega\delta < 0.2$ and $p\Delta\omega/\omega^2\delta^2 < 0.1$. For larger values

Intermodulation

9) When a transmitter is modulated with different audiofrequency tones, sum- and difference-frequencies are produced by the receiver filters. Take the case of a single tuned circuit.

Let the instantaneous frequency of the transmitted signal be

$$\omega_0 + \Delta\omega_1 \cos pt + \Delta\omega_2 \cos qt$$

After the circuit the quasi-stationary term is then

$$\frac{d}{dt} \arctan \frac{\Delta\omega_1 \cos pt + \Delta\omega_2 \cos qt}{a}$$

Using the expression

$$\arctan a = \int_0^\infty \frac{e^{-u}}{u} \sin au\, du \qquad (24)$$

we insert

$a = 2x \cos pt + 2y \cos qt$, where $2xa = \Delta\omega_1$ and

$$2ya = \Delta\omega_2$$

Then

$$\arctan a = \int_0^\infty \frac{e^{-u}}{u} \sin (2ux \cos pt + 2yu \cos qt)\, du$$

By means of the well-known expansion in B e s-s e l functions, the coefficient of $\cos (2k + 1) pt + 2m\, qt$ in the expansion of the arc tan appears as

$$2\, (-1)^{k+m} \int_0^\infty \frac{e^{-u}}{u^2}\, J_{2k+1}(2ux)\, J_{2m}(2uy)\, du \qquad (25)$$

In many cases these integrals can be further simplified by the methods indicated by W a t-s o n [9]; for instance, whenever $k = m$, and also when the arguments of the B e s s e l functions are equal. In the latter case the frequency deviations are equal and $x = y$. Then the integral is reducable to complete elliptic integrals of the first and second kind, K and E, so that

$$C_{3p} = \frac{-(1 + 16x^2)^{\frac{1}{2}}}{36\pi x^3} \left\{ (11 - 2x^2)\, E - 5K \right\} -$$

$$-\frac{1}{2x} + \frac{1}{6x^3} \qquad (26)$$

The argument of K and E is $k = \dfrac{4x}{(1 + 16x^2)^{\frac{1}{2}}}$

(if K and E are tabulated as functions of a, where $\sin a = k$, $\tan a$ is equal to $4x$). C_{3p} is the coefficient of $\cos 3pt$ in the expansion of the arc tan. It must be remembered that the arc tan has to be differentiated in order to produce the frequency. In this way the amplitude corresponding to the frequency $3p$ after demodulation A_{3p} is given by

$$A_{3p} = 3p\, C_{3p}$$

For the frequency $2p \pm q$ we get analogously

$$C_{2p \pm q} = \frac{-(1 + 16x^2)^{\frac{1}{2}}}{12\pi x^3} \left\{ (-1 + 4x^2)\, E + K \right\} + \frac{1}{2x}$$

and

$$A_{2p \pm q} = (2p \pm q)\, C_{2p \pm q}$$

For small frequency deviations, the expansion of arc tan a can also be quite well employed; this gives us

$$C_{2p \pm q} =$$

$$= 2 \sum_0^\infty{}^m \sum_0^\infty{}^n (-1)^{m+n+1} \frac{(2m+2n+2)!\, x^{2n+2}\, y^{2m+1}}{(n+2)!\, n!\, (m+1)!\, m!}$$

and

$$C_{2p \pm 3q} = 2 \sum_0^\infty{}^m \sum_0^\infty{}^n (-1)^{m+n+2}$$

$$\frac{(2m + 2n + 4)!\, x^{2n+2}\, y^{2m+3}}{(n + 2)!\, n!\, (m + 3)!\, m!}$$

Fig. 8 gives the curves of A_{2p+q} and A_{3p} for equal frequency deviations of the fundamental tones. If the quasi-stationary approximation is to answer the purpose, it is necessary that p/a,

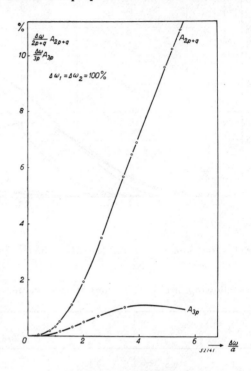

Fig. 8. [Intermodulation on applying a frequency-modulated signal. to a single tuned circuit. Two audio tones of equal swing in the original modulation. (Quasi-stationary approximation).

q/a, $p\Delta\omega/a^2$ and $q\Delta\omega/a^2$ shall not be too great for example p/a and $q/a < 0.4$ and

$$p\Delta\omega_1 + q\Delta\omega_2 < 0.8\, a^2$$

Intermodulation in the case of coupled circuits may be computed along similar lines.

Conclusion

10) Apart from the examples we have given here, many other applications are possible. The application to the frequency detector circuit was discussed in my thesis. Other examples and an alternative derivation of the asymptotic series were given by v a n d e r P o l [10].

For many practical purposes the condition that p the audio-frequency is small with respect to $\omega\delta$, is satisfied, and the use of the quasi-stationary approximation yields good results. In fact it may be the only way, as in this case the F o u r i e r spectrum has so many components, that this form of computation is excluded.

To Prof. B a l t h. v a n d e r P o l and Prof.

H. B r e m e k a m p the writer wishes to express his sincere thanks for many helpful suggestions.

References

[1.] R o d e r, H. "Effects of tuned circuits on a frequency modulated signal". Proc. Inst. Rad. Engrs **25**, 1617, 1937.

[2.] S t u m p e r s, F. L. H. M. „Eenige onderzoekingen over trillingen met frequentiemodulatie". Diss. Delft, May 1946.

[3.] C a r s o n, J. R. and F r y, T. C. "Variable frequency electric circuit theory with application to the theory of frequency modulation". Bell Syst. Tech. Journ. **16**, 513, 1937.

[4.] W i l d e, H. "Verzerrung frequenzmodulierter Schwingungen am Schwingungskreis". T.F.T. **32**, 150, 1942.

[5.] F e l d t k e l l e r, R. and W i l d e, H. "Gleitfrequenzen in Schwingungskreisen". T.F.T. **30**, 347, 1941.

[6.] G e n s e l, R. "Naherungsverfahren zur Berechnung von Einschwingvorgange in Siebschaltungen". T.F.T. **31**, 299, 1941.

[7.] V e l l a t, T. "Der Empfang frequenzmodulierter Wellen". Elek. Nach. Tech. **18**, 72, 1940.
J a f f e, D. L. "A theoretical and experimental investigation of tuned circuit distortion in frequency-modulation systems". Proc. Inst. Radio Engrs **33**, 318, 1945.

[8.] P e r r o n, O. "Uber die Abhangigkeit der Integrale eines Systems linearer Differential gleichungen von einem Parameter". Sitz. Ber. der Heidelberger Akad. der Wiss. 1918, Abh. 15.

[9.] W a t s o n, G. N. "A treatise on the theory of Bessel functions". 1st. ed. Cambridge 1922.

[10.] V a n d e r P o l B a l t h. "The fundamental principles of frequency modulation". Journ. Instit. Electr. Engrs **93**, III, 153, 1946.

PAPER NO. 11

Reprinted from *Proc. IEEE*, Vol. 53, No. 6, pp. 564–575, June 1965

The Quasi-Stationary Response of Linear Systems to Modulated Waveforms

D. D. WEINER, MEMBER, IEEE, AND B. J. LEON, MEMBER, IEEE

Abstract—A major consideration in communications is the transmission of modulated waveforms through linear systems. The engineer usually wants the modulation to be preserved, for it is the modulation that carries the transmitted information. This paper presents a straightforward method for relating distortion of the modulation to readily computed characteristics of the linear system and the input signal. Through a simple integration by parts, the response is broken up into 1) a quasi-stationary term that preserves the characteristics of the modulation, and 2) a correction term that represents the distortion. Examples of the application of the results to both AM and FM signals through linear time-invariant systems and FM waves through first order time-variant systems are presented. A fairly extensive bibliography of the basic problem is also presented.

I. INTRODUCTION

A MAJOR CONSIDERATION in the design of communication equipment is the transmission of modulated waveforms through linear systems. The engineer usually wants the modulation at the output to be the same as the modulation at the input, for it is the modulation that carries the transmitted information. Thus, he would like a straightforward way of establishing a relation between readily computed characteristics of the system he is to design and distortion of the waveforms to be transmitted through that system. Conventional analysis techniques can be applied to specific systems with specific signals but in these analyses the important design considerations often lose their identity. For example, in the analysis of FM signal processing, one may be able to compute the various frequency components of the input in terms of Bessel Function coefficients, operate on each of these with the appropriate transfer function, and then express the output as an infinite sum of frequency components. What is the modulation of the output i.e., what will a frequency discriminator detect? These are the type of questions dealt with in this paper.

Our purpose is to develop an approach in which the instantaneous amplitude and instantaneous frequency of the input signal retain their identity. The response is expressed as the sum of a quasi-stationary term plus a correction term. The quasi-stationary term is defined as the input signal multiplied by the appropriately modified version of the conventional sinusoidal steady-state

system function. This is illustrated in Table I.[1] For the time-invariant system with AM input, the carrier frequency is a constant equal to ω_c. Hence, the quasi-stationary response is defined as the input multiplied by $H(j\omega)$ evaluated at ω_c. In the time-invariant FM case, the instantaneous frequency of the signal $\omega(t)$ varies with time. The quasi-stationary response is obtained by multiplying the input by $H(j\omega)$ with ω replaced by $\omega(t)$. The resulting $H[j\omega(t)]$ is referred to as the quasi-stationary transfer function. Finally, in the case of a time-variant system excited by an FM signal, both the parameters of the system and the instantaneous frequency of the input vary with time. Here, the quasi-stationary response results through multiplying the input by $H(j\omega)$ in which the system parameters, e.g., R, L, C, μ, as well as the frequency variable ω, are replaced by $R(t)$, $L(t)$, $C(t)$, $\mu(t)$, and $\omega(t)$, respectively. Again, the resulting $H[j\omega(t); t]$ is referred to as the quasi-stationary transfer function.

Unfortunately, the quasi-stationary term does not comprise the total response. In addition, there is a correction term which in many applications represents the distortion. It arises because the system response is unable to build up and decay as fast as the quasi-stationary term would dictate. When the time-varying parameters vary "slowly enough," the correction term is negligible. Then it is possible to reason on a quasi-stationary basis provided it is known what is meant by "slowly enough." For faster modulations the correction term is appreciable and insight into the system behavior is obtained by interpreting the total response as an interference phenomenon between the quasi-stationary and correction terms.

In both the AM and FM cases, as documented in the following, there has been considerable engineering literature devoted to even the simplest of problems in this area (for example, the burst of carrier in the AM case and the step change in frequency in the FM case). The primary result of this paper is a demonstration that both these problems, plus more complicated AM and FM problems for all completely stable (no j axis poles) time-invariant lumped systems and for, at least, first-order time-variant systems, can readily be put into the desired form with the exact correction term as a closed-form integral. This result is obtained by the simple

Manuscript received October 30, 1964; revised February 2, 1965. The work reported in this paper was supported by the National Science Foundation Grant GP-581. Portions of this paper have been taken from a Ph.D. dissertation at Purdue University, Lafayette, Ind. The thesis is available as a technical report [40].

D. D. Weiner is with the Dept. of Electrical Engineering, Syracuse University, N. Y. He was formerly with Purdue University, Lafayette, Ind.

B. J. Leon is with Purdue University, Lafayette, Ind.

[1] In this paper it is to be understood that wherever signals are expressed as complex time functions, the real part is implied. For example, $A(t)e^{j\omega_c t}$ implies $R_e[A(t)e^{j\omega_c t}] = A(t) \cos \omega_c t$.

282

TABLE I
DEFINITION OF QUASI-STATIONARY RESPONSE

	Input Signal	Transfer Function	Quasi-Stationary Response
Time-invariant system with AM input	$A(t)e^{j\omega_c t}$	$H(j\omega_c)$	$A(t)H(j\omega_c)e^{j\omega_c t}$
Time-invariant system with FM input	$\exp\left[j\int_0^t \omega(\xi)d\xi\right]$	$H[j\omega(t)]$	$H[j\omega(t)]\left[e^{j\int_0^t \omega(\xi)d\xi}\right]$
Time-variant system with FM input	$\exp\left[j\int_0^t \omega(\xi)d\xi\right]$	$H[j\omega(t); t]$	$H[j\omega(t); t]\left[e^{j\int_0^t \omega(\xi)d\xi}\right]$

mathematical formalism of integration by parts—nothing more. Several examples are discussed to demonstrate application of the results.

II. QUASI-STATIONARY RESPONSE OF LINEAR TIME-INVARIANT SYSTEMS EXCITED BY AM SIGNALS

It is well known that if the frequency spectrum of a signal is restricted to a region within which the system gain is essentially constant and the system phase is essentially linear, then the output signal will be a delayed but undistorted replica of the input [1]. These conditions are frequently not satisfied in practice because of noise considerations, component bandwidth limitations, etc. The problem then becomes one of determining the distortion in the output signal. This can be accomplished using conventional frequency and/or time domain methods [2]–[21]. In spite of the wealth of literature referred to, the state of the art in analyzing linear time-invariant systems excited by amplitude modulated signals leaves much to be desired from the designer's point of view. The exact analyses are extremely tedious while the approximate ones are limited in their applications. None of the methods really give much insight into the dynamic behavior of the system. Furthermore, the instantaneous amplitude, the quantity of interest, usually becomes lost in the complexity of the mathematics. The analysis which follows is an attempt to correct some of these deficiencies.

A. Derivation of Response as Quasi-Stationary Term Plus Correction Term

Consider the stable system function

$$H(s) = \frac{a_m s^m + \cdots + a_1 s + a_0}{(s + s_1)(s + s_2) \cdots (s + s_n)},$$

$$m \leq n, \ s_i \neq s_j \quad (1)$$

in which the degree of the numerator polynomial is no more than the degree of the denominator polynomial and, in which, all poles are simple. Expanding in partial fractions gives

$$H(s) = \sum_{y=1}^{n} \frac{K_y}{s + s_y} + K_0. \quad (2)$$

The impulse response is simply

$$h(t) = \sum_{y=1}^{n} K_y e^{-s_y t} + K_0 \delta(t) \quad (3)$$

where $\delta(t)$ is the unit impulse occurring at $t=0$. Denote the system input by $u(t)$ and the system output by $v(t)$. Let $u(t)$ be given by

$$u(t) = \begin{cases} A(0_-)e^{j\omega_c t} & t < 0 \\ A(t)e^{j\omega_c t} & t \geq 0. \end{cases} \quad (4)$$

Hence, the input signal is a constant amplitude sinusoid for $t<0$ and an amplitude modulated wave with continuous envelope $A(t)$ for $t \geq 0$. (A discontinuity is allowed in the envelope at $t=0$.) The frequency of the carrier is a constant equal to ω_c. Using the convolution integral, the response of the system for $t \geq 0$ is

$$v(t) = \int_{-\infty}^{t} h(t - \tau)u(\tau)d\tau$$

$$= \int_{-\infty}^{0} \sum_{y=1}^{n} K_y e^{-s_y(t-\tau)} A(0_-)e^{j\omega_c \tau}d\tau$$

$$+ \int_{0}^{t} \sum_{y=1}^{n} K_y e^{-s_y(t-\tau)} A(\tau)e^{j\omega_c \tau}d\tau + K_0 A(t)e^{j\omega_c t}. \quad (5)$$

Evaluating the integrals from $-\infty$ to 0 and performing an integration by parts on the integrals from 0 to t yields

$$v(t) = A(t)H(j\omega_c)e^{j\omega_c t} - \sum_{y=1}^{n} \frac{K_y}{s_y + j\omega_c}[A(0_+) - A(0_-)]e^{-s_y t}$$

$$- \sum_{y=1}^{n} \int_{0_+}^{t} \frac{K_y}{s_y + j\omega_c} A'(\tau)e^{-s_y(t-\tau)}e^{j\omega_c \tau}d\tau \quad (6)$$

where 0_+ is used to denote the value of t immediately after $t=0$. Equation (6) is the desired result. The first term represents the quasi-stationary response. The second term, involving $A(0_-)$ and $A(0_+)$, gives that part of the correction term due to a step change in the envelope occurring at $t=0$. When $A(0_+)=A(0_-)$, this term is zero. The third term, involving the integrations from 0_+ to t reflects the inability of the system response to follow the amplitude variations of the input for $t>0$. Note that the integrands contain the first derivative of the envelope. When the envelope varies slowly enough, the last term is negligible. Hence, our notation will be

$$v(t) = v_q(t) + v_c(t) \qquad (7)$$

where the quasi-stationary term in this case is

$$v_q(t) = A(t)H(j\omega_c)e^{j\omega_c t} \qquad (8)$$

and the correction term is

$$v_c(t) = -\sum_{y=1}^{n} \frac{K_y}{s_y + j\omega_c}[A(0_+) - A(0_-)]e^{-s_y t}$$
$$- \sum_{y=1}^{n} \int_{0_+}^{t} \frac{K_y}{s_y + j\omega_c} A'(\tau)e^{-s_y(t-\tau)}e^{j\omega_c \tau}d\tau. \qquad (9)$$

For many engineering purposes an estimate, in the form of an upper bound on the departure of the system performance from quasi-stationary behavior, is adequate. Since the absolute value of a sum is less than or equal to the sum of the absolute values,

$$|v_c(t)| \le \sum_{y=1}^{n}\left|\frac{K_y}{s_y + j\omega_c}[A(0_+) - A(0_-)]e^{-s_y t}\right|$$
$$+ \sum_{y=1}^{n}\left|\int_{0_+}^{t} \frac{K_y}{s_y + j\omega_c} A'(\tau)e^{-s_y(t-\tau)}e^{j\omega_c \tau}d\tau\right|. \qquad (10)$$

If $A'(t)$ is sufficiently simple, convenient bounds can be obtained by straight forward substitution into (10) and evaluation of the resulting integrals. For those applications in which this is too laborious or unfeasible, (10) can be simplified to

$$|v_c(t)| \le B(t)$$
$$= \sum_{y=1}^{n}\left|\frac{K_y}{s_y + j\omega_c}\right|\left|A(0_+) - A(0_-)\right|e^{-\alpha_y t}$$
$$+ \sum_{y=1}^{n}\left|\frac{K_y}{s_y + j\omega_c}\right|\left|A'(\tau)\right|_{\max} \frac{1}{\alpha_y}(1 - e^{-\alpha_y t}) \qquad (11)$$

where α_y equals the real part of s_y. As time increases, $B(t)$, the upper bound in (11), asymptotes to B_∞ where

$$B_\infty = \left|A'(\tau)\right|_{\max} \sum_{y=1}^{n} \frac{1}{\alpha_y}\left|\frac{K_y}{s_y + j\omega_c}\right|. \qquad (12)$$

The bound given by (12) is particularly convenient for the case in which $A(t)$ is a sample function from a stochastic process, the process being characterized by $|A'(t)|_{\max}$. In general, the choice between (10), (11), or (12) will depend upon the particular application. These equations should not be interpreted as giving the tightest possible bounds. They are presented mainly because of their simplicity. By using ingenuity and the detailed properties of $A(t)$, considerably tighter bounds can be achieved. A tradeoff exists between simplicity of the bounds and improved accuracy.

The analysis is readily extended to systems having multiple-order poles. If there is a pole of order p at frequency s_i, then the partial fraction (2) has a set of terms of the form $K_{ij}/(s+s_i)^j$ with j taking on integer values from 1 to p. The time response (3) then has terms $(K_{ij}/(j-1)!)t^{j-1}e^{-s_i t}$. Each of these terms requires j integrations by parts to get the desired form. The quasi-stationary term is defined from the system function as before. The correction term becomes

$$v_c(t) = -\sum_{y=1}^{n} \frac{K_{y1}}{s_y + j\omega_c}[A(0_+) - A(0_-)]e^{-s_y t}$$
$$- \sum_{q=2}^{p}\sum_{y=1}^{q} \frac{K_{iq}}{(q-y)!}\frac{t^{q-y}}{(s_i + j\omega_c)^y}[A(0_+) - A(0_-)]e^{-s_i t}$$
$$- \sum_{y=1}^{n} \int_{0_+}^{t} \frac{K_{y1}}{s_y + j\omega_c} A'(\tau)e^{-s_y(t-\tau)}e^{j\omega_c \tau}d\tau$$
$$- \sum_{q=2}^{p}\sum_{y=1}^{q} \int_{0_+}^{t} \frac{K_{iq}}{(q-y)!}\frac{(t-\tau)^{q-y}}{(s_i + j\omega_c)^y} A'(\tau)$$
$$\cdot e^{-s_i(t-\tau)}e^{j\omega_c \tau}d\tau. \qquad (13)$$

Bounds on this correction term can be obtained by the same procedures as in the simple pole case.

B. Application of Results

The instantaneous envelope of the system response equals $|v(t)|$ and is constrained to lie between the bounds $|v_q(t)| \pm |v_c(t)|$. By comparing $|v_c(t)|$ with $|v_q(t)|$, it is possible to determine to what extent the system is behaving in a quasi-stationary manner. When the correction term is not negligible, the expressions for $|v_q(t)|$ and $B(t)$ [the upper bound on $|v_c(t)|$], are often significantly simpler than the corresponding expression for $|v(t)|$. In such a situation much useful information can easily be obtained by plotting $|v_q(t)| \pm B(t)$ as illustrated in the following examples.

The first case considered is the response of a high-Q, single-tuned circuit (with half-bandwidth α as measured between the 0.707 points and resonant frequency ω_r) to a burst of carrier frequency ω_c with trapezoidal leading edge. The excitation is shown in Fig. 1. Application of (6) yields

$$|v(\tau)| = \begin{cases} \frac{1}{\tau_0}\frac{1}{1+x^2}[(1-\tau)^2 - 2(1-\tau)e^{-\tau}\cos x\tau \\ \quad - 2x\tau e^{-\tau}\sin x\tau + e^{-2\tau} + x^2\tau^2]^{1/2} \quad 0 \le \tau \le \tau_0 \\[2mm] \frac{1}{\tau_0}\frac{1}{1+x^2}[(1+x^2)\tau_0^2 \\ \quad - 2\tau_0 e^{(\tau_0-\tau)}\cos x(\tau_0 - \tau) \\ \quad + 2\tau_0 e^{-\tau}\cos x\tau - 2x\tau_0 e^{(\tau_0-\tau)}\sin x(\tau_0 - \tau) \\ \quad - 2x\tau_0 e^{-\tau}\sin x\tau + e^{-2\tau} + e^{2(\tau_0-\tau)} \\ \quad - 2e^{(\tau_0-2\tau)}\cos x\tau_0]^{1/2} \quad \tau \ge \tau_0 \end{cases} \qquad (14)$$

where $\tau = \alpha t$, $\tau_0 = \alpha t_0$, and $x = (\omega_c - \omega_r)/\alpha$. Using (10) an upper bound on the magnitude of the correction term is found to be

$$|v_c(\tau)| \le B(\tau) = \begin{cases} \dfrac{1}{[1+x^2]^{1/2}}\dfrac{1}{\tau_0}(1 - e^{-\tau}) \\ \qquad\qquad 0 \le \tau \le \tau_0 \\[3mm] \dfrac{1}{[1+x^2]^{1/2}}\dfrac{1}{\tau_0}[e^{-(\tau-\tau_0)} - e^{-\tau}] \\ \qquad\qquad \tau \ge \tau_0. \end{cases} \qquad (15)$$

The bounds on the envelope of the exact response are given by $|v_q(\tau)| \pm B(\tau)$. These are plotted in Fig. 2, along with the envelope responses for two different values of ω_c. When $\omega_c = \omega_r$, the lower bound exactly equals the envelope response, while for $\omega_c = \omega_r + 2\alpha$, the bounds hug the first overshoot and first undershoot very nicely. Note the relative simplicity of (15) as compared with (14).

A second example involves the response of high-Q single and double-tuned circuits to a sudden burst of carrier for which $A(0_-) = 0$ and $A(t) = 1$. Equations (6) and (11) were used [40] in obtaining the curves of Fig. 3 (a), (b), and (c), while a corresponding upper bound for

(13) was used in Fig. 3 (d). In each instance, the bounds yield reasonable estimates to the exact response. The tightness of these bounds is appreciated all the more if the complexity of the exact expression for $v(t)$ is compared to the relative simplicity of the expressions for the bounds.

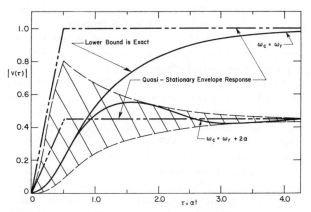

Fig. 2.   Envelope response and bounds for single-tuned circuit excited by a burst of carrier with trapezoidal leading edge. The poles are located at $s = -\alpha \pm j\omega_r$.

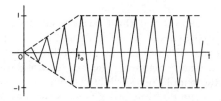

Fig. 1.   Burst of carrier with trapezoidal leading edge.

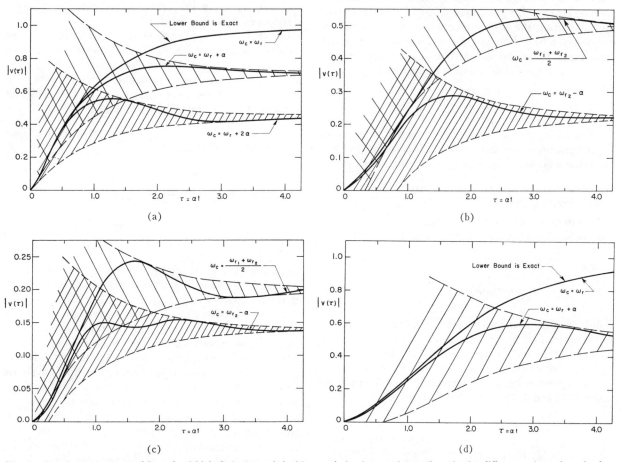

(a)   (b)

(c)   (d)

Fig. 3.   Envelope response and bounds of high-Q single- and double-tuned circuits to a burst of carrier for different values of carrier frequency. (a) Single-tuned circuit with poles located at $s = -\alpha \pm j\omega_r$. (b) Double-tuned Butterworth filter with poles located at $s = -\alpha \pm j\omega_{r_1}$ and $-\alpha \pm j\omega_{r_2}$, $(\omega_{r_1} - \omega_{r_2}) = 2\alpha$. (c) Overcoupled double-tuned filter with poles located at $s = -\alpha \pm j\omega_{r_1}$ and $-\alpha \pm j\omega_{r_2}$, $(\omega_{r_1} - \omega_{r_2}) = 4\alpha$. (d) Synchronously tuned double-tuned circuit with second-order poles located at $s = -\alpha \pm j\omega_r$.

III. Quasi-stationary Response of Linear Time-Invariant Systems Excited by FM Signals

The first published results on the quasi-stationary response of linear time-invariant systems with FM inputs are due to Carson and Fry [22]. They presented the correction term as an infinite series expansion without investigating the convergence properties of their expansion. Van der Pol [23] and Stumpers [24], in a more straightforward manner, also expanded the response into a quasi-stationary term plus an infinite series for the correction term. Stumpers pointed out the asymptotic behavior of his series as well as that of Carson and Fry. No clear-cut bounds on the errors involved in truncating the series and using only a few of the leading terms were presented. Baghdady [25] unified the approaches of Carson and Fry, Van der Pol, and Stumpers and attempted to provide bounds on the error due to truncation of the series. Rowe [26] showed Baghdady's analysis to be defective and supported his claim with several counterexamples. Hess [27], feeling that it was impossible to obtain an exact expression in the desired form, derived a result which supposedly approximates the output closely and keeps a tight bound on the error term between the approximate and actual response. Hupert [28] obtained a closed form result by approximating the instantaneous frequency of the excitation as a limit function of a series of small steps and then using the principle of superposition. His results are incorrect due to the omission of the factor

$$\exp\left[j\int_0^t \omega(\xi)d\xi\right]$$

from the integrand in his (14). Since Hupert's derivation is not presented in sufficient detail, it is speculated that the error was caused by improper normalization.

A. Derivation of Response as Quasi-Stationary Term Plus Correction Term

The derivation proceeds in a manner very similar to that in Section II-A. Let the FM input be given by

$$u(t) = \begin{cases} e^{j\omega_v t} & t < 0 \\ e^{j\theta(t)} = e^{j\int_0^t \omega(\xi)d\xi} & t \geq 0. \end{cases} \quad (16)$$

Thus, the instantaneous frequency of the input signal is constant for $t < 0$, jumps from ω_v to ω_0 at $t = 0$, and varies in some arbitrary but specified manner for $t \geq 0$. For a system with n distinct poles and the ith pole of order p, the integration by parts technique (see Weiner and Leon [29] and [40] for more detail) gives

$$v(t) = H[j\omega(t)]e^{j\theta(t)}$$

$$+ \sum_{y=1}^{n} K_{y1}\left[\frac{1}{(s_y + j\omega_v)} - \frac{1}{(s_y + j\omega_0)}\right]e^{-s_y t}$$

$$+ \sum_{q=2}^{p}\sum_{y=1}^{q}\frac{K_{iq}}{(q-y)!}$$

$$\cdot\left[\frac{1}{(s_i - j\omega_v)^y} - \frac{1}{(s_i + j\omega_0)^y}\right]t^{q-y}e^{-s_i t} \quad (17)$$

$$+ \sum_{y=1}^{n}\int_{0_+}^{t}K_{y1}\frac{j\omega'(\tau)}{[s_y + j\omega(\tau)]^2}e^{-s_y(t-\tau)}e^{j\theta(\tau)}d\tau$$

$$+ \sum_{q=2}^{p}\sum_{y=1}^{q}\int_{0_+}^{t}\frac{yK_{iq}}{(q-y)!}$$

$$\cdot\frac{j\omega'(\tau)}{[s_i + j\omega(\tau)]^{(y+1)}}(t-\tau)^{q-y}e^{-s_i(t-\tau)}e^{j\theta(\tau)}d\tau.$$

The first term in (17) is the desired quasi-stationary term. The second and third terms result from the frequency step at $t = 0$. The fourth and fifth terms arise from the inability of the system to follow the input in a quasi-stationary manner. The single summations are due to the simple poles while the double summations are due to the multiple-order pole. Note that only the first derivative of $\omega(\tau)$ appears. This is significant since previous analyses involving infinite series expansions of the correction term [22]–[25] had indicated that higher derivatives were also involved.

In most FM applications the modulation can be classified as abrupt (as with FSK) or continuous (as with commercial FM broadcast.) For the continuous case, $\omega_v = \omega_0$, and the second and third terms drop out of (17). Now the correction term (designated by the subscript cc for continuous correction) is bounded by

$$|v_{cc}(t)| \leq B = \sum_{y=1}^{n}\frac{|K_{y1}|}{\alpha_y}\frac{|\omega'(\tau)|_{max}}{|[s_y + j\omega(\tau)]^2|_{min}}$$

$$+ \sum_{q=2}^{p}\sum_{y=1}^{q}y\frac{|K_{iq}|}{(\alpha_i)^{q-y+1}}$$

$$\cdot\frac{|\omega'(\tau)|_{max}}{|[s_i + j\omega(\tau)]^{y+1}|_{min}}. \quad (18)$$

Equation (18) is especially convenient for the case in which $\omega(t)$ is a sample function from a stochastic process characterized by $|\omega'(t)|_{max}$.

If the quasi-stationary and correction terms are thought of as rotating phasors, the response can be interpreted as the interference between these two phasors. This interference phenomenon provides an effective pictorial explanation for the behavior of linear systems. For the AM case [40], it offers a straightforward interpretation of such measures as the percentage envelope overshoot and undershoot, their time of occurrence, conditions under which the instantaneous frequency of the response equals that of the input, and the severity of frequency modulations in the response. For the FM case, the phasor interpretation even more dramatically explains many unusual features of the instantaneous frequency of the response, as is shown in Section III-B. Finally, the interference mechanism explains the response of linear first-order time-variant circuits as discussed in Section IV.

B. Response of High-Q Single-Tuned Circuit to a Frequency Step

Many investigators have studied the response of a single-tuned circuit to a sinusoidal input that is frequency modulated by a step. Salinger [30], after idealizing the filter characteristics, used contour integration in the complex plane to evaluate the response. Hatton [31] solved the differential equation directly while Clavier [32], McCoy [33], and Gumowski [34] used transform techniques. These approaches all involved unwieldy mathematical expressions that shed little light on what is actually going on. Weiner and Baghdady [35] conducted both an experimental and theoretical study of the FM transient problem which resulted in an extensive library of FM transient oscillograms. Their theoretical work yielded the first simple explanation of the FM transient response as an interference phenomenon [36], [37].

Consider the input to the circuit of Fig. 4 to be

$$i_s(t) = \begin{cases} e^{j\omega_\nu t} & t < 0 \\ e^{j\omega_0 t} & t \geq 0. \end{cases} \quad (19)$$

The instantaneous frequency is ω_ν for $t < 0$, and jumps to ω_0 for $t > 0$. The problem is to determine the instantaneous frequency of the output voltage $e_0(t)$.

Oscillograms of the instantaneous frequency of the response of a single-tuned circuit to a series of frequency steps are shown in Fig. 5 [35]–[37]. The normalized variables x_ν and x_0 give the initial and final frequency deviations of the input from the resonant frequency of the filter in terms of the number of half-bandwidths of the filter. For example, $x_\nu = 0$ and $x_0 = 1$ indicate a frequency step from the resonant frequency of the filter to a frequency one half-bandwidth above resonance. Figure 5(f) demonstrates that during transitions in which an overshoot becomes an undershoot, the instantaneous amplitude of the response drops to zero. In Figs. 5(1) and (m), the leading edge of the square wave corresponds to a frequency deviation away from resonance ($x_\nu = 0$); the trailing edge corresponds to a deviation towards resonance ($x_0 = 0$). When $x_0 = 0$, the response never overshoots, regardless of the values of x_ν. Figure 5(n) shows a response in which the overshoots have no inflection point in the leading and trailing edges. Figure 5(o) indicates that when $x_\nu = -x_0$, the leading and trailing edges of the response are symmetrical.

As is obvious from Fig. 5, exact expressions describing the waveforms would be extremely complicated. Nevertheless, the behavior observed in the oscillograms can be explained in a very simple and straightforward manner. Dividing (17) by the quasi-stationary term, the normalized response is

$$e_{0N}(t) = \frac{e_0(t)}{Z(j\omega_0)e^{j\omega_0 t}} = 1 + A_\nu e^{-\alpha t} e^{-j(\omega_0 - \omega_r)t} \quad (20)$$

where

$$A_\nu = |A_\nu| e^{j\Phi_\nu}$$

and

$$|A_\nu| = \frac{|x_0 - x_\nu|}{[1 + x_\nu^2]^{1/2}} \qquad \Phi_\nu = \text{ctn}^{-1} x_\nu. \quad (21)$$

Given a phasor interpretation to the normalized response, the resultant phasor $e_{0N}(t)$ is the sum of a constant unit phasor plus a phasor of magnitude $|A_\nu| e^{-\alpha t}$ rotating clockwise at a frequency of $(\omega_0 - \omega_r)$ radians/s. The magnitude of the resultant phasor is the envelope of $e_{0N}(t)$, while the time derivative of the phase angle gives the instantaneous frequency of $e_{0N}(t)$. The envelope and instantaneous frequency of $e_0(t)$ are obtained by multiplying the envelope of $e_{0N}(t)$ by $|Z(j\omega_0)|$ and adding ω_0 radians/s to the instantaneous frequency of $e_{0N}(t)$.

The phasor model easily explains all the phenomena observed in the oscillograms of Fig. 5. Consider, for example, the situation shown in Fig. 6. As the rotating phasor decays exponentially, the resultant phasor $e_{0N}(t)$ wobbles back and forth with a period that is approximately equal to $2\pi/\omega_0 - \omega_r$. At points A, C, E, and G, $e_{0N}(t)$ is at rest, and the instantaneous frequency of $e_0(t)$ equals ω_0. Local maxima occur at B and F and local minima at D and H. Undershoots in the response are explained by Fig. 7. Here, the rotating phasor is sufficiently large at $t = 0$, and the frequency $(\omega_0 - \omega_r)$ is sufficiently high to enable the tip of the $e_{0N}(t)$ phasor to encircle the origin twice. Since the instantaneous frequency overshoots in the direction of the frequency of the stronger signal [38], the instantaneous frequency of $e_{0N}(t)$ undershoots at points to the left of 0, and overshoots at points to the right of 0. Note that undershoots must always precede overshoots; also, successive undershoots have increasing magnitudes whereas successive overshoots have decreasing magnitudes. When the tip of the $e_{0N}(t)$ phasor passes through the origin, the resultant is zero, and the instantaneous frequency of the response is transitional between an overshoot and an undershoot.

When $\omega_0 = \omega_r$, $(\omega_0 - \omega_r) = 0$ and the rotating phasor is stationary. Therefore, the response to a frequency step

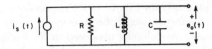

Fig. 4.   Single-tuned circuit.

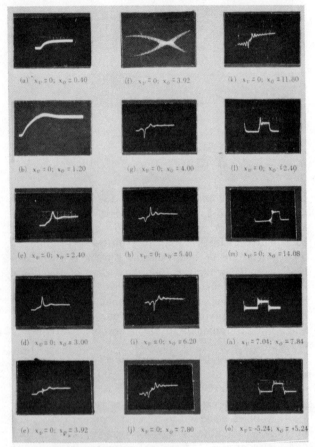

Fig. 5.   Oscillograms of the instantaneous frequency of the response of a single-tuned circuit (with half-bandwidth α as measured between the 0.707 points and resonant frequency ω_r) to a series of frequency steps.

$$x = \frac{(\omega - \omega_r)}{\alpha} \quad \text{and} \quad x_0 = \frac{(\omega_0 - \omega_r)}{\alpha}.$$

that terminates at ω_r consists of a smooth exponential rise caused by the decay of $|A_\nu|e^{-\alpha t}$. For certain values of $(\omega_0 - \omega_r)$ an inflection point occurs because the response is alternately influenced by the decay and rotation of the rotating phasor. If $(\omega_0 - \omega_r)$ is small, only the decay predominates, and if $(\omega_0 - \omega_r)$ is very large, only the rotation predominates. In these two cases no inflection point occurs.

Quantitative information concerning the FM transients can also be obtained from the phasor model. For example, when $x_\nu = 0$ and $x_0 > 0$, $|A_\nu| = x_0$ and the initial phase angle of the rotating phasor is $\pi/2$ radians. Since the rotating phasor rotates at $(\omega_0 - \omega_r)$ radians/sec., the

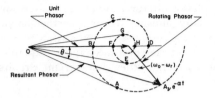

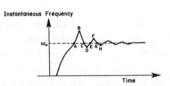

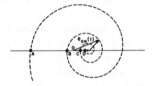

Fig. 6.   Behavior of the resultant phasor $e_{0N}(t)$ that results in the occurrence of instantaneous-frequency and instantaneous-amplitude transients in the single-tuned circuit response.

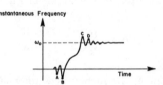

Fig. 7.   Behavior of $e_{0N}(t)$ that results in the occurrence of both undershoots and overshoots in the instantaneous frequency of $e_0(t)$.

normalized times of peak undershoots and overshoots are given by

$$t_n = \frac{\left(2n\pi - \dfrac{\pi}{2}\right)}{(\omega_0 - \omega_r)}. \qquad n = 1, 2, 3 \cdots . \qquad (22)$$

The value of the instantaneous frequency at these instances is obtained as follows: at t_n the resultant and rotating phasors are colinear but pointing in opposite directions. The tangential velocity of the rotating phasor tip is $x_0 e^{-\alpha t_n}(\omega_0 - \omega_r)$ and equals that of the resultant phasor which is $(1 - x_0 e^{-\alpha t_n})\omega_N(t_n)$. $\omega_N(t_n)$ is the instantaneous frequency of the resultant phasor. Solving for $\omega_N(t_n)$ yields

$$\omega_N(t_n) = \frac{x_0(\omega_0 - \omega_r)}{e^{\alpha t_n} - x_0}. \qquad (23)$$

Since the instantaneous frequency of $e_0(t)$ equals that of $e_{0N}(t)$ plus ω_0, the fractional overshoot a_n, or the frac-

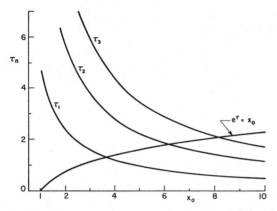

Fig. 8.   Plot of

$$\tau_n = \frac{\left(2n\pi - \dfrac{\pi}{2}\right)}{x_0} \qquad \text{for } x_v = 0 \quad \text{and} \quad x_0 > 0.$$

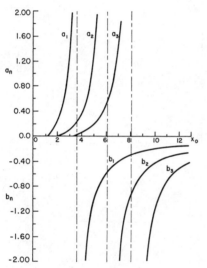

Fig. 9.   Plot of theoretical fractional overshoots and undershoots for $x_v = 0$ and $x_0 > 0$.

tional undershoot b_n, at the nth peak is given by

$$a_n = \frac{[\omega_N(t_n) + \omega_0] - \omega_0}{\omega_0 - \omega_v} = \frac{x_0}{e^{\tau_n} - x_0} \qquad (24)$$

$$b_n = \frac{[\omega_N(t_n) + \omega_0] - \omega_v}{\omega_0 - \omega_v} = \frac{x_0}{e^{\tau_n} - x_0} + 1 \qquad (25)$$

where $\tau_n = \alpha t_n$. Equations (22), (24) and (25), are plotted in Figs. 8 and 9, respectively. The transition between an overshoot and undershoot occurs when $1 = x_0 e^{-\tau_n}$. Hence, the curve $e^\tau = x_0$ is also plotted in Fig. 8. The intersections of this curve with the τ_1, τ_2, and τ_3 curves indicate that the first three overshoots should become undershoots when x_0 reaches the values 3.63, 6.12, and 8.12, respectively.

C. Conditions for Quasi-Stationary Response of Several Filters of Interest

There are many applications in which it is desirable to have a system respond in a quasi-stationary manner to an FM input. Assume that the instantaneous phase angle of the input is

$$\theta(t) = \omega_c t + \Delta\omega \int_0^t f(\xi) d\xi. \qquad (26)$$

The instantaneous frequency of the input is

$$\omega(t) = \omega_c + \Delta\omega f(t). \qquad (27)$$

Assume $|f(t)| < 1$ so that the maximum frequency deviation is $\Delta\omega$. Assuming the filters to be high Q, (18) yields

$$|v_c(t)| \le B = \Delta\omega |f'(\tau)|_{\max} \sum_{y=1}^{n} \frac{|K_{y1}|}{(\alpha_y)^3}$$

$$+ \Delta\omega |f'(\tau)|_{\max} \sum_{q=2}^{p} \sum_{y=1}^{q} y \frac{|K_{iq}|}{(\alpha_i)^{q+2}}. \qquad (28)$$

The results for several filters whose gain characteristics have been normalized are summarized in Table II. In each case, the bound is proportional to the frequency deviation and the maximum rate of change of the instantaneous frequency of the input, while being inversely proportional to the square of the filter bandwidth. Clearly, by making the bandwidth large enough for a given input or by choosing a suitable input for a given filter, the correction term can be made as small as desired.

When the conditions for quasi-stationary behavior are violated, it is of interest to investigate the nature of the response. Figure 10 shows oscillograms of the response of a single-tuned circuit to a sinusoidally modulated FM wave. These oscillograms have many similarities with those of Fig. 5. Figure 10(a) shows the filter behaving in a quasi-stationary manner. With an increase in the modulating frequency and the size of the deviation, distortion becomes evident [see Fig. 10(b)]. Violating quasi-stationary conditions even further, the response starts to ring and the first overshoot grows in magnitude. Eventually, the first overshoot becomes an undershoot and at the instant of transition the instantaneous amplitude of the response dips to zero [see Figs. 10(c)–(f)]. Other distorted waveforms are shown in Figs. 10(g)–(j).

The violent distortion in the frequency response may at first appear to be somewhat surprising in view of the smooth nature of the modulation. However, if the response is interpreted as the result of an interference phenomenon, the severe distortion is easily explained.

TABLE II
BOUND ON CORRECTION TERM FOR SEVERAL FILTER TYPES

Filter	Pole-Pattern	Bandwidth (BW)	Bound on Correction Term*
First-Order Butterworth	$j\omega_r$, $-\alpha$	2α	$\dfrac{4D}{(BW)^2}$
Second-Order Butterworth	$j(\omega_r + \alpha)$, $j\omega_r$, $j(\omega_r - \alpha)$, $-\alpha$	$2\sqrt{2}\,\alpha$	$\dfrac{16D}{(BW)^2}$
Third-Order Butterworth	$j\left(\omega_r + \dfrac{\sqrt{3}}{2}\alpha\right)$, $j\omega_r$, $j\left(\omega_r - \dfrac{\sqrt{3}}{2}\alpha\right)$, $-\alpha$, $\dfrac{-\alpha}{2}$	2α	$\dfrac{41D}{(BW)^2}$
n-Identical Cascaded Single-Tuned Circuits	$j\omega_r$, nth-order pole, $-\alpha$	$2\alpha\sqrt{2^{1/n}-1}$	$\dfrac{2n(1+n)(2^{1/n}-1)D}{(BW)^2}$

* Note: $D = \Delta\omega \, |f'(t)|_{\max}$.

IV. QUASI-STATIONARY RESPONSE OF LINEAR FIRST-ORDER TIME-VARIANT SYSTEMS EXCITED BY FM SIGNALS

For the most general first-order linear differential equation

$$\frac{dv(t)}{dt} + g(t)v(t) = u(t). \tag{29}$$

The standard solution usual given (see, for example, Ince [39]) is

$$v = C \exp\left[-\int g\,dt\right] + \exp\left[-\int g\,dt\right]$$

$$\int u \exp\left[\int g\,dt\right] dt \tag{30}$$

where C is a constant. The difficulty with the aforementioned solution is that it fails to describe the system behavior in terms of familiar sinusoidal steady-state circuit concepts. As a result, even for first-order, time-variant systems, it is not known under what conditions such useful quantities as impedance, bandwidth, etc., are meaningful. The results presented in this section clarify the situation.

Response of a Time-Variant Parallel RC Circuit to an FM Wave

The circuit to be analyzed is shown in Fig. 11. The quasi-stationary impedance $Z[j\omega(t);t]$ is defined as the conventional steady-state impedance in which the parameters R, C, and ω are replaced by $R(t)$, $C(t)$, and $\omega(t)$, respectively.[2] Hence,

$$Z[j\omega(t);t] = \frac{\dfrac{1}{C(t)}}{\alpha(t) + j\omega(t)} \tag{31}$$

where

$$\alpha(t) = 1/[R(t)C(t)].$$

[2] $Z[j\omega(t);t]$ should not be confused with Zadeh's system function [41]. The latter is defined as $H[j\omega;t] = y(t)/e^{j\omega t}$ where $y(t)$ is the system response to the input $e^{j\omega t}$.

For $t<0$, $R(t)$ and $C(t)$ are constants R_0 and C_0, respectively. At $t=0$, the resistance and capacitance begin to vary. The circuit elements are always assumed to be positive. The FM input $i(t)$ is the same as $u(t)$ in (16). The problem is to determine the output voltages $e_0(t)$ as a quasi-stationary term plus a correction term. Using the principle of superposition, the output voltage equals

$$e_0(t) = \int_{-\infty}^{t} h(t, \tau) i(\tau) d\tau \qquad (32)$$

where $h(t, \tau)$ is the system response at time t due to a unit impulse applied at time τ. For $t>0$, it is readily shown that

$$h(t, \tau) = \begin{cases} \dfrac{1}{C(t)} \exp[\alpha_0 \tau] \exp\left[-\int_0^t \alpha(\xi) d\xi\right] & \tau < 0 \\[2ex] \dfrac{1}{C(t)} \exp\left[-\int_\tau^t \alpha(\xi) d\xi\right] & \tau \geq 0 \end{cases} \qquad (33)$$

where $\alpha_0 = 1/R_0 C_0$. Subsitution of (16) and (33) into (32) and integrating by parts results in

$$e_0(t) = Z[j\omega(t); t] \exp\left[j \int_0^t \omega(\xi) d\xi\right]$$

$$+ \frac{C_0}{C(t)} [Z(j\omega_\nu; 0) - Z(j\omega_0; 0)]$$

$$\cdot \exp\left[-\int_0^t \alpha(\xi) d\xi\right]$$

$$- \frac{1}{C(t)} \int_0^t \frac{d}{d\tau} \{C(\tau) Z[j\omega(\tau); \tau]\}$$

$$\cdot \exp\left[-\int_\tau^t \alpha(\xi) d\xi + j \int_0^t \omega(\xi) d\xi\right] d\tau. \qquad (34)$$

Just as with the time-invariant AM and FM cases, the response consists of a quasi-stationary term plus a correction term. The first part of the correction term involving ω_ν and ω_0 results from a frequency step in the input at $t=0$. The latter part of the correction term reflects the inability of the system to respond to the variations of $R(t)$, $C(t)$, and $\omega(t)$.

As was the situation in sections II and III, a trade off exists between the tightness and the simplicity of the bound on $|v_c(t)|$. If the minimum value of $\alpha(t)$ is denoted by $\alpha_{\min}$ and the maximum value of $d/d\tau\{C(\tau) Z[j\omega(\tau); \tau]\}$ by $D_{\max}$, a simple bound on $|v_c(t)|$ is given by

$$|v_c(t)| \leq B(t) = \frac{C_0}{C(t)} |Z(j\omega_v; 0) - Z(j\omega_0; 0)| e^{-\alpha_{\min} t}$$

$$+ \frac{1}{C(t)} \frac{D_{\max}}{\alpha_{\min}} [1 - e^{-\alpha_{\min} t}]. \qquad (35)$$

For large values of time the bound asymptotes to

$$|v_c(t)| \leq B_\infty = \frac{1}{C(t)} \frac{D_{\max}}{\alpha_{\min}}. \qquad (36)$$

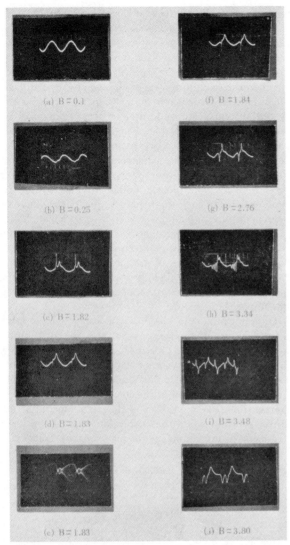

(a) B=0.1   (f) B=1.84

(b) B=0.25   (g) B=2.76

(c) B=1.82   (h) B=3.34

(d) B=1.83   (i) B=3.48

(e) B=1.83   (j) B=3.80

Fig. 10. Oscillograms of the instantaneous frequency of the response of a single-tuned circuit to a series of frequency steps.

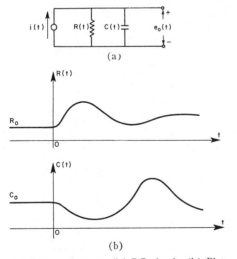

Fig. 11. (a) Time-variant parallel RC circuit. (b) Plots of $R(t)$ and $C(t)$ vs. time.

Using the previous analysis, it is possible to determine under what conditions the conventional sinusoidal steady-state concepts have meaning in the response of a time-variant parallel RC circuit.

When the instantaneous frequency of the input is constant for all time, (34) becomes

$$e_0(t) = Z(j\omega; t)e^{j\omega t}$$
$$- \frac{1}{C(t)} \int_0^t \frac{d}{d\tau} [C(\tau)Z(j\omega; \tau)]$$
$$\cdot e^{j\omega\tau} \exp\left[- \int_\tau^t \alpha(\xi)d\xi\right] d\tau. \qquad (37)$$

The correction term in (37) shows that it is possible to vary $R(t)$ and $C(t)$ simultaneously so that $v_c(t)$ is identically zero. Since $[C(\tau)Z(j\omega; \tau)]$ equals $1/\alpha(\tau)+j\omega$, $v_c(t)=0$ when α is a constant, (i.e., when the resistance varies inversely proportional to the capacitance). Under these conditions the denominator of $Z(j\omega; t)$ is constant. However, the numerator varies inversely proportional to $C(t)$. Therefore, it is possible to vary the magnitude of the impedance without changing the pole-zero pattern by appropriately changing R and C as continous functions of time. This is the only known case in time-variant circuits in which the correction term is zero and the quasi-stationary term gives the total response.

In a similar manner the response of other first-order, time-variant circuits can be expressed as the sum of a quasi-stationary term plus a correction term. In many situations a phasor interpretation can be extremely effective in explaining the response. The resultant phasor is the sum of a quasi-stationary phasor plus a correction phasor. In general, the correction phasor will rotate with respect to the quasi-stationary phasor. Thus, the wobbling of the resultant phasor readily explains the frequency fluctuations in the output of a time-variant circuit to a constant frequency sinusoid. Other phenomena such as nulls and the instantaneous envelope of the output are also readily explained by the phasor model.

V. Conclusion

This paper has been concerned with the response of linear systems to modulated waveforms. It has been shown that an exact closed form solution for time-invariant and first-order, time-variant systems can be obtained as the sum of a quasi-stationary term plus a correction term. Two situations arise depending upon whether or not the magnitude of the correction term is negligible with respect to that of the quasi-stationary term.

When the maximum rate of change of the modulation is slow compared to the speed with which the system can respond, the quasi-stationary term predominates and the system can be analyzed on a "dynamic" basis by appropriately modifying the conventional sinusoidal steady-state system function. This approach does not require a spectral analysis of the modulated input. The extent to which a system is behaving in a quasi-stationary manner can be estimated by evaluating relatively simple expressions for the bounds on the correction term.

When the maximum rate of change of the modulation is comparable to or faster than the speed with which the system response can build up and decay, the correction term becomes appreciable. Insight into the system behavior is obtained by representing the quasi-stationary and correction terms as phasors and interpreting the total response as an interference phenomenon between these phasors. The phasor model provides an intuitive explanation for such quantities as percentage envelope overshoot and undershoot; the time at which such overshoots and undershoots occur; the conditions under which the instantaneous frequency of the response equals the instantaneous frequency of the input; and the severity and detailed properties of any frequency modulations which may occur in the instantaneous frequency of the response.

Of major importance is the realization that the fundamental nature of the response is the same for both AM and FM excitations to either time-invariant or first-order, time-variant systems. This is due to the interaction of the quasi-stationary and correction terms. The quasi-stationary term is the logical extension from the conventional system theory of the particular (or steady-state) solution while the correction term is the logical extension of the homogeneous (or transient) solution.

References

[1] Mason, S. J., and H. J. Zimmerman, *Electronic Circuits, Signals and Systems.* New York: Wiley, 1960, pp 366–369.
[2] Schwartz, M., *Information Transmission, Modulation, and Noise.* New York: McGraw-Hill, 1959, ch 2.
[3] Gent, A. W., The transient response of RF and IF filters to a wave pocket, *Proc. IEE*, pt 4, vol 99, Nov 1952, pp 326–335.
[4] Talkin, A. I., Transient response of cascaded-tuned circuits, *IRE Trans. on Circuit Theory*, vol CT-1, Sep 1954, pp 65–68.
[5] Wait, J. R., An approximate method of obtaining the transient response from the frequency response, *Canad. J. Tech.*, vol 31, Jun 1953, pp 127–131.
[6] Rosenbrock, H. H., Approximate relations between transient and frequency response, *J. British IRE*, vol 18, Jan 1958, pp 57–64.
[7] Hupert, J. J., Envelope and angle response of asymmetrical narrow-band networks, *IRE Trans. on Circuit Theory*, vol CT-6, Sep 1959, pp 292–295.
[8] Valley, G. E., and H. Wallman, *Vacuum Tube Amplifiers.* New York: McGraw-Hill, 1948, ch 7.
[9] Tucker, D. G., Transient response of filters, *Wireless Engineer*, vol 23, Feb–Mar 1946, pp 36, 84.
[10] Eaglefield, C. C., Transient response filters, *ibid.*, pp 67–74.
[11] Cherry, E. C., The transmission characteristics of asymmetric sideband communication networks, pt I, *J. IEE*, pt 3, vol 89, Mar 1942, pp 19–42; and pt II, *J. IEE*, pt 3, vol 90, Jun 1943, pp 75–88.
[12] Aigrain, P. R., B. R. Teare, and E. M. Williams, Generalized theory of the band-pass low-pass analogy, *Proc. IRE*, vol 37, Oct 1949, pp 1152–1155.
[13] Mayne, D. Q., Transient response of band-pass filters to modulated signals, *Proc. IEE*, vol 47, pt C, vol 106, Sep 1959, pp 144–152.
[14] Kallman, H. E., R. E. Spencer, and C. P. Singer, Transient response, *Proc. IRE*, vol 33, Mar 1945, pp 169–195.
[15] Henderson, K. W., and W. H. Kautz, Transient responses of conventional filters, *IRE Trans. on Circuit Theory*, vol CT-5, Dec 1958, pp 333–347.
[16] Mulligan, J. H., Jr., The effect of pole and zero locations on the

transient response of linear dynamic systems, *Proc. IRE*, vol 37, May 1949, pp 516–529.

[17] Zemanian, A. H., Bounds existing on the time and frequency responses of various types of networks, *Proc. IRE*, vol 42, May 1954, pp 835–839.

[18] ——, Further bounds existing on the transient response of various types of networks, *Proc. IRE*, vol 43, Mar 1955, pp 322–326.

[19] ——, Restrictions on the shape factors of the step response of positive real system functions, *Proc. IRE*, vol 44, Sep 1956, pp 1160–1165.

[20] Zemanian, A. H., and N. Chang, Behavior of attenuation for systems having monotonic step and impulse responses, *IEEE Trans. on Circuit Theory*, vol CT-10, Jun 1963, pp 252–255.

[21] Zemanian, A. H., and P. E. Fleischer, On the transient responses of ladder networks, *IRE Trans. on Circuit Theory*, vol CT-5, Sep 1958, pp 197–201.

[22] Carson, J. R., and T. C. Fry, Variable-frequency electric circuit theory, *Bell Sys. Tech. J.*, vol 16, Oct 1937, pp 513–540.

[23] Van der Pol, B., The fundamental principles of frequency modulation, *J. IEE (London)*, pt 3, vol 93, May 1946, pp 153–158.

[24] Stumpers, F. L. H. M., Distortion of frequency-modulated signals in electrical networks, *Commun. News*, vol 9, Apr 1948, pp 82–92.

[25] Baghdady, E. J., Theory of low-distortion reproduction of FM signals in linear systems, *IRE Trans. on Circuit Theory*, vol CT-5, Sep 1958, pp 202–214.

[26] Rowe, H. E., Distortion of angle-modulated waves by linear networks, *IRE Trans. on Circuit Theory (Correspondence)*, vol CT-9, Sep 1962, pp 286–290.

[27] Hess, D. T., Transmission of FM signals through linear filters, *Proc. NEC*, vol 18, Oct 1962, pp 469–476.

[28] Hupert, J. J., Transient response of narrow-band networks to angle-modulated signals, *Proc. NEC*, vol 18, Oct 1962, pp 458–468.

[29] Weiner, D. D., and B. J. Leon, On the quasi-stationary response of linear time-invariant filters to arbitrary FM signals, *IEEE Trans. on Circuit Theory (Correspondence)*, vol CT-11, Jun 1964, pp 308–309.

[30] Salinger, H., Transients in frequency modulation, *Proc. IRE*, vol 30, Aug 1942, pp 378–383.

[31] Hatton, W. L., Simplified FM transient response, Tech Rept 196, Research Lab. of Electronics, M.I.T., Cambridge, Mass., Apr 1951.

[32] Clavier, A. G., Application of Fourier transforms to variable-frequency circuit analysis, *Proc. IRE*, vol 37, Nov 1949, pp 1287–1290.

[33] McCoy, R. E., FM transient response of band-pass circuits, *Proc IRE*, vol 42, Mar 1954, pp 574–579.

[34] Gumowski, I., Transient response in FM, *Proc. IRE*, vol 42, May 1954, pp 819–822.

[35] Weiner, D. D., Experimental study of FM transients and quasi-static response, M. S. Thesis, M.I.T., Cambridge, Mass., Jan 1958.

[36] Weiner, D. D., and E. J. Baghdady, FM transients, *M.I.T. Research Lab. of Electronics Quarterly Progress Rept.*, Apr 1958, pp 36–42.

[37] Baghdady, E. J., Analog modulation systems, in *Lectures on Communication Theory*, E. J. Baghdady, Ed. New York: McGraw-Hill, 1961, ch 19, pp 460–466.

[38] ——, Theory of stronger-signal capture in FM reception, *Proc. IRE*, vol 46, Apr 1958, pp 729–738.

[39] Ince, E. L., *Ordinary Differential Equations*. New York: Dover, 1956.

[40] Weiner, D. D., and B. J. Leon, The quasi-stationary response of linear systems to modulated waveforms, Tech Rept TR-EE64-10 under NSF Grant GP-581, School of Electrical Engineering, Purdue University, Lafayette, Ind., May 1964.

[41] Zadeh, L. A., Frequency analysis of variable networks, *Proc. IRE*, vol 38, Mar 1950, pp 291–299.

PAPER NO. 12

Reprinted from *Proc. IEEE*, Vol. 56, No. 1, pp. 2–13, Jan. 1968

Distortion and Crosstalk of Linearly Filtered, Angle-Modulated Signals

EDWARD BEDROSIAN, SENIOR MEMBER, IEEE, AND STEPHEN O. RICE, FELLOW, IEEE

Abstract—An important problem in the theory and practice of receiving angle-modulated signals is the design of the filtering elements which must be employed. It has long been known that filtering introduces distortion and crosstalk into the signal. However, the computation of these effects is difficult. The methods customarily used employ approximations of one kind or another, and the equations used do not apply to all cases of practical interest. Here formulas are presented which enlarge somewhat the domain of cases amenable to calculation.

In this analysis, an angle-modulated signal having an arbitrary phase function is applied to a general linear filter, and the phase of the output is expanded in a series having the linearly filtered input as the leading term. The expansion is then specialized to the case of a narrowband signal applied to a narrow, symmetrical, bandpass filter.

A spectral analysis is performed by assuming a Gaussian input phase and examining terms through fifth order in the output phase expansion. This leads to the main results of the paper, namely expressions for the leading terms in the output spectrum. It is argued that these terms represent the principal contribution in the case where the distortion is small.

To demonstrate their application to a practical problem, the formulas are used to calculate the distortion and crosstalk produced when an FM signal, having a flat baseband spectrum, is passed through a single-pole filter. This example is of some current interest because such a filter has been employed in the forward path of a feedback FM receiver used for satellite communication. A number of cases are considered, and the results of the computations are plotted.

I. INTRODUCTION

ONE OF THE most intriguing problems relating to the theory of angle (i.e., frequency or phase) modulation is that of obtaining a useful relationship between the properties of the input and output modulation when an angle-modulated signal is passed through a linear network. From the earliest work by Roder[1] and by Carson and Fry[2] in 1937 to the present, interest has persisted as others have extended results and added new approaches. Many of these are cited by Baghdady, Panter, Downing, and Rowe.[3]-[6]

Aside from the exact, but laborious, computational methods, the output phase angle (or its derivative) is generally obtained as a series in which the leading term is the principal one. For example, the leading term may be the input phase angle, with suitable delay. Again, it may be an approximation to the output phase angle obtained by quasi-stationary analysis.

The basic series derived here for the output differs from earlier ones in that the leading term is simply the linearly

Manuscript received April 27, 1967; revised November 6, 1967. The part of this research performed by E. Bedrosian is an extension of work reported earlier in RAND Memo. RM-4888-NASA, sponsored by the National Aeronautics and Space Administration under Contract NASr-21(02).

E. Bedrosian is with The RAND Corporation, Santa Monica, Calif.
S. O. Rice is with Bell Telephone Laboratories, Inc., Murray Hill, N. J.

filtered input phase angle. The subsequent terms constitute the higher-order contributions to the output. The series converges when the input phase angle or its time derivative is not too large.

In Section II a general expansion is derived for the output when a carrier with arbitrary angle modulation is applied to a general linear filter. The expansion is then specialized in Section III to the case of a narrowband signal applied to a narrow, symmetrical, bandpass filter. Section IV presents the leading terms in the spectrum of the output phase when the angle modulation is Gaussian and the filter is symmetrical about the carrier frequency. Elements of this spectrum are identified as the linear-signal, cross-power, and intermodulation components of the output phase. In Section V, their values are computed numerically for the specific example of a signal frequency-modulated by a flat spectrum of Gaussian noise and passed through a single-pole filter. This example is of interest in the design of FM systems with frequency feedback.[7],[8]

II. INPUT-OUTPUT PHASE RELATIONSHIP

Consider a modulated signal $s(t)$ of unit amplitude formed by phase-modulating a harmonic carrier of frequency f_o by a real signal $\phi(t)$. In complex form, $s(t)$ becomes

$$s(t) = e^{i2\pi f_o t + i\phi(t)}. \tag{1}$$

Let $s(t)$ be applied to a linear filter having a steady-state transfer function $G(f)$ and an impulse response $g(t)$. These are then a Fourier pair

$$g(t) = \int_{-\infty}^{\infty} G(f)e^{i2\pi f t}df, \quad G(f) = \int_{-\infty}^{\infty} g(t)e^{-i2\pi f t}dt \tag{2}$$

where $g(t)$ is real and, for a physically realizable filter, vanishes when $t < 0$. $G(f)$ is complex in general with an even real part and an odd imaginary part; therefore,

$$G(f) = G^*(-f) \tag{3}$$

where * denotes the complex conjugate.

The filter output is given by

$$s_o(t) = \int_0^{\infty} g(\tau)s(t - \tau)d\tau$$

$$= \left[\int_0^{\infty} g(\tau)e^{-i2\pi f_o \tau}e^{i\phi(t-\tau)}d\tau\right]e^{i2\pi f_o t} \tag{4}$$

where the bracketed factor indicates a simultaneous amplitude and phase modulation brought about by the filtering. This factor may be interpreted as the output obtained when an input $\exp\left[i\phi(t)\right]$ is applied to a filter with impulse response

$$g(t)e^{-i2\pi f_o t} \qquad (5)$$

and transfer function

$$\int_0^\infty g(t)e^{-i2\pi(f_o+f)t}dt = G(f_o+f). \qquad (6)$$

It is convenient to introduce normalized transfer and impulse response functions defined by

$$\Gamma(f) = \frac{G(f_o+f)}{G(f_o)} \qquad (7)$$

and

$$\gamma(t) = \frac{g(t)e^{-i2\pi f_o t}}{G(f_o)}, \qquad (8)$$

respectively. Then

$$\int_0^\infty \gamma(\tau)d\tau = 1. \qquad (9)$$

The output phase is given by the argument of the modulation factor in (4). Upon using the definition of $\gamma(t)$ given by (8), it is found that

$$\text{output phase} = \text{Im} \log \int_0^\infty G(f_o)\gamma(\tau)e^{i\phi(t-\tau)}d\tau$$

$$= \text{Im} \log \int_0^\infty \gamma(\tau)e^{i\phi(t-\tau)}d\tau + \text{Im} \log\ G(f_o)$$

$$= \theta(t) - \beta_o \qquad (10)$$

where $\theta(t)$ is the time-dependent component of the output phase and β_o is the carrier phase shift.[1] It is shown in Appendix I that the output phase can be expanded as

$$\theta(t) = \text{Re}\ \Phi(t) + \sum_{n=2}^\infty \frac{1}{n!}\ \text{Im}\ (i^n f_n). \qquad (11)$$

Here

$$\Phi(t) = \int_0^\infty \gamma(\tau)\phi(t-\tau)d\tau \qquad (12)$$

and the coefficients f_n to $n=7$ are given by

$$\begin{array}{ll} f_2 = F_2 & f_4 = F_4 - 3F_2^2 \\ f_3 = F_3 & f_5 = F_5 - 10F_3F_2 \\ f_6 = F_6 - 15F_4F_2 - 10F_3^2 + 30F_2^3 \\ f_7 = F_7 - 21F_5F_2 - 35F_4F_3 + 210F_3F_2^2 \end{array} \qquad (13)$$

where

$$F_n = \int_0^\infty \gamma(\tau)[\phi(t-\tau) - \Phi(t)]^n d\tau. \qquad (14)$$

A sufficient condition for the series in (11) to converge (absolutely) is that

$$F(\zeta) = \int_0^\infty \gamma(\tau)e^{\zeta[\phi(t-\tau)-\Phi(t)]}d\tau \qquad (15)$$

be regular, and have no zeros in or on the circle $|\zeta|=1$. The output phase rate $\theta(t)$ is given by the derivative of (11), provided the resulting series converges uniformly.

The foregoing expansion is quite general in that it places no restriction, except for convergence of the series, on either the modulated signal or the filter to which it is applied. Its basic form is particularly desirable in that the leading term, $\text{Re}\ \Phi(t)$, is simply the linearly filtered input. The remaining terms then constitute the various orders of distortion.

III. APPLICATION TO BANDPASS FILTER

It has been pointed out by a number of writers that the study of FM distortion produced by a narrowband filter is simplified by consideration of an equivalent low-pass filter. In the approach used here, this simplification is associated with the neglect of the rapidly varying components in the impulse response $\gamma(t)$ defined by (8). The details are given in the following discussion. The essential idea is that when the input is written in complex form, its frequency components are negligibly small over the image passband around $f=-f_o$.

The transfer function $G(f)$ for a filter passing a narrow band of frequencies around f_o has passbands around both $\pm f_o$. Since

$$\Gamma(f) = G(f_o + f)/G(f_o),$$

$\Gamma(f)$ has passbands at $f=0$ and $-2f_o$. Choose some convenient function, $\Gamma_a(f)$, which approximates $\Gamma(f)$ in the passband around $f=0$ and is negligibly small outside this band. The function $\Gamma_a(f)$ will be called the normalized, low-pass equivalent transfer function.[2]

Let $\gamma_a(t)$ be the Fourier transform of $\Gamma_a(f)$. The desired simplification consists of replacing $\gamma(t)$ by $\gamma_a(t)$ in the integrals of Section II. To examine the error introduced, note that the integral in the fundamental relation given by (10) is a convolution integral and may be written as

$$\int_0^\infty \gamma(\tau)e^{i\phi(t-\tau)}d\tau = \int_{-\infty}^\infty e^{i2\pi ft}\Gamma(f)\Omega(f)df \qquad (16)$$

where $\gamma(t)$, $\Gamma(f)$, and $\exp\left[i\phi(t)\right]$, $\Omega(f)$ are Fourier pairs. If the Fourier integral giving $\Omega(f)$ does not converge, a suitable convergence factor or truncation is tacitly assumed.

Upon using $\Gamma(f)=\Gamma_a(f)+[\Gamma(f)-\Gamma_a(f)]$, (16) becomes

$$\int_0^\infty \gamma(\tau)e^{i\phi(t-\tau)} = \int_0^\infty \gamma_a(\tau)e^{i\phi(t-\tau)}d\tau + v(t)$$

where $v(t)$ is the contribution of $\Gamma(f)-\Gamma_a(f)$ to the integral

---

[1] Let α_o and β_o denote, respectively, the attenuation and phase shift of the filter at the carrier frequency f_o. Then $G(f_o)=\exp\ (-\alpha_o-i\beta_o)$, and $\beta_o = -\text{Im} \log G(f_o)$.

[2] In any particular case, the method of choosing $\Gamma_a(f)$ usually depends upon how $G(f)$ is specified. $G(f)$ may be given as a rational function of f, as a partial-fraction expansion, as curves of attenuation and phase shift, or in still other ways.

on the right-hand side of (16). The absolute value of $v(t)$ satisfies the inequality

$$|v(t)| \le \int_{-\infty}^{\infty} |\Gamma(f) - \Gamma_a(f)| \; |\Omega(f)| df.$$

In cases of practical interest the spectrum $\Omega(f)$ of $\exp\left[i\phi(t)\right]$ is significant only in the region where $\Gamma(f) - \Gamma_a(f)$ is small, i.e., where $\Gamma_a(f)$ is a good approximation to $\Gamma(f)$. Thus, the product $|\Gamma(f) - \Gamma_a(f)| \; |\Omega(f)|$ is usually small (the delta functions appearing in $\Omega(f)$ when $\phi(t)$ is periodic may be avoided by using Fourier series instead of integrals) for all values of f and, if $\Omega(f)$ decreases with sufficient rapidity as $|f| \to \infty$, $v(t)$ itself will be small. Therefore, the relative error introduced when $\gamma(t)$ is replaced by $\gamma_a(t)$ in the second integral in (10) is small except at the infrequent instants when the integral itself is small. Indeed, when the convergence condition associated with (15) is satisfied, the integral is bounded away from zero. This follows from the fact that the expression

$$F(i) = e^{-i\Phi(t)} \int_{0}^{\infty} \gamma(\tau) e^{i\phi(t-\tau)} d\tau$$

cannot then be zero, as explained in connection with (47).

For most physical narrowband filters symmetrical about f_o, $G(f_o - f)$ is approximately equal to $G^*(f_o + f)$, and the phase shift β_o appearing in the last equation of (10) is nearly zero. It will be assumed that the approximation $\Gamma_a(f)$ to $\Gamma(f)$ is chosen so that $\Gamma_a(-f) = \Gamma_a^*(f)$. Then $\gamma_a(t)$ is real, and so are the quantities $\Phi(t)$, F_n, and f_n computed by using $\gamma_a(t)$ in the integrals of Section II. Only odd-order terms are nonzero in (11), and the expansion becomes

$$\theta(t) = \Phi(t) + \sum_{n=1}^{\infty} \frac{(-1)^n}{(2n+1)!} f_{2n+1}, \qquad (17)$$

the first terms of which are

$$\theta(t) = \Phi(t) - \frac{1}{3!} F_3 + \frac{1}{5!} (F_5 - 10 F_3 F_2) + O(\phi^7) \quad (18)$$

where $O(\phi^7)$ denotes terms of order seven and higher in $\phi(t)$. A necessary and sufficient condition for (17) to be valid is that $F(\zeta)/F(-\zeta)$ have no zeros or singularities in or on the unit circle, where $F(\zeta)$ is given by (15). The derivation of this condition and some of its consequences are given in Appendix I.

In all of the following work, except Appendix I, which deals with general filters, the functions $\Gamma(f)$, $\gamma(t)$ will denote $\Gamma_a(f)$, $\gamma_a(t)$, respectively, the subscript a being omitted for simplicity. To repeat, the only filters which will be considered are those which are narrowband and symmetrical about the carrier frequency. Henceforth (in the simplified notation), $\Gamma(-f) = \Gamma^*(f)$, $\gamma(t)$ will be real, and (17) and (18) will apply.

IV. SPECTRAL ANALYSIS

In communication system analysis, the effect of operations encountered in signal processing is frequently better described in terms of the statistical properties of an ensemble of representative signals than by the explicit behavior of a specific signal. Typical of the situations in which an angle-modulated signal can suffer significant distortion due to bandpass filtering is one in which the phase or instantaneous frequency of the carrier is made to vary in accordance with the amplitude of a multichannel, frequency-division-multiplexed speech or data baseband signal. For many purposes, such a composite signal is adequately approximated by a random time function having a Gaussian distribution of values.

Under these conditions, an expression for the spectral density of the output signal in terms of the filter transfer function and the spectral density of the imput signal, i.e., a second-moment analysis, can provide useful engineering information. This technique will be applied to the input-output relationship for a symmetric filter given by (17). The convergence of the series (17) is studied in Appendix I. One set of sufficient conditions includes the requirement that $|\phi(t)|$ be less than $\pi/2$ for all values of t. Although this requirement is violated by any Gaussian $\phi(t)$, the looseness of the condition and the plausibility of the computed results suggest that the series is valid for phase functions well in excess of this bound.

Let $\phi(t)$ denote a sample function of a stationary, differentiable, zero-mean, Gaussian process having an autocorrelation function $R_\phi(t)$ and a spectral density $W_\phi(f)$, where R and W are a Fourier pair:

$$R(t) = \int_{-\infty}^{\infty} df \, W(f) e^{i2\pi ft}, \quad W(f) = \int_{-\infty}^{\infty} dt \, R(t) e^{-i2\pi ft}. \quad (19)$$

The autocorrelation function is defined by

$$R_\phi(\tau) = E\left[\phi(t)\phi(t+\tau)\right] \qquad (20)$$

where the expectation operator E denotes an ensemble average. The output spectral density is determined formally from (17) by using (20), with θ for ϕ, to form the autocorrelation function and then applying (19). That is,

$$W_\theta(f) = \mathscr{F} E\left[\theta(t)\theta(t+\tau)\right] \qquad (21)$$

where $\mathscr{F}$ denotes the Fourier transform operator and where the operations are applied term by term to the pairings which result from using the expansion for θ. If it is desired to express the results in terms of the input and output phase rates, the relationships

$$W_{\dot{\theta}}(f) = (2\pi f)^2 W_\theta(f), \qquad W_\phi(f) = W_{\dot{\phi}}(f)/(2\pi f)^2 \quad (22)$$

may be used. Note that if the angles are measured in radians and frequency in hertz then the units of $W_\theta(f)$ are rad^2/Hz and those of $W_{\dot{\theta}}(f)$ are $(\text{rad/s})^2/\text{Hz}$.

To identify the various components of the output spectrum, it is necessary to examine the diverse pairings which result when the expansion for $\theta(t)$ given by (17) is substituted in (21). To facilitate this process, let the terms in (17) be denoted by Φ, f_3, f_5, etc., and use the notation

$$\Phi \times f \equiv \mathscr{F} E\left[\Phi(t) f(t+\tau)\right].$$

When these operations are performed, as shown in Appendix II, it is seen that terms of three distinct types emerge. These are presented in schematic form as

$$\Phi \times \Phi \Rightarrow W_\phi(f)|\Gamma(f)|^2$$
$$= W_\theta^L(f)$$

$$\left.\begin{array}{c}\Phi \times f_i\\f_i \times f_i\end{array}\right\} \Rightarrow W_\phi(f)\left[\int d\rho\, W_\phi(\rho) + \int d\rho\, W_\phi(\rho)W_\phi(f-\rho) + \cdots\right]$$
$$= W_\theta^C(f)$$

$$f_i \times f_i \Rightarrow \int d\rho \int d\sigma\, W_\phi(\rho)W_\phi(\sigma)W_\phi(f-\rho-\sigma) + \cdots$$
$$= W_\theta^I(f).$$

The first type is unique and results from the $\Phi \times \Phi$ term. It is a linear function of the input spectrum $W_\phi(f)$, since Φ is simply the linearly filtered input. For this reason, it will be denoted by $W_\theta^L(f)$ and referred to as the "linear-signal" component of the output spectrum. The term "signal" is deliberately avoided because an unambiguous specification of the signal content in the output spectrum appears difficult when the interference consists of a distortion of the input signal rather than the familiar independent random noise.

The $\Phi \times f_i$ terms and parts of the $f_i \times f_i$ terms yield contributions of the second type which have the form of the input spectrum multiplied by convolutions of the input spectrum with itself. The presence of the $W_\phi(f)$ multiplier indicates that this component of the output spectrum comes primarily from a cross correlation between the linear and nonlinear terms in the output phase expansion. It will therefore be denoted by $W_\theta^C(f)$ and referred to as the "cross-power" component of the output spectrum. It has significance mainly as a measure of the correlation between the distorted and undistorted elements of the output.

The third type of contribution to the output spectrum arises from the balance of the $f_i \times f_i$ terms. The result is a spectrum of intermodulation products constituting true nonlinear distortion; this type of term will be called the "intermodulation" component of the ouput spectrum and will be denoted by $W_\theta^I(f)$.

The leading terms of these types are, from Appendix II,

$$W_\theta^L(f) = W_\phi(f)|\Gamma(f)|^2$$

$$W_\theta^C(f) = 2W_\phi(f)\int_{-\infty}^{\infty} d\rho\, W_\phi(\rho)\{\mathrm{Re}\,\Gamma(\rho)\Gamma(f)\Gamma(-\rho-f)$$
$$-|\Gamma(\rho)|^2\,|\Gamma(f)|^2\}$$

$$W_\theta^I(f) = \frac{1}{6}\int_{-\infty}^{\infty} d\rho \int_{-\infty}^{\infty} d\sigma\, W_\phi(\rho)W_\phi(\sigma)W_\phi(f-\rho-\sigma)$$

$$\times |2\Gamma(\rho)\Gamma(\sigma)\Gamma(f-\rho-\sigma) - \Gamma(f-\rho-\sigma)\Gamma(\rho+\sigma)$$
$$-\Gamma(\rho)\Gamma(f-\rho) - \Gamma(\sigma)\Gamma(f-\sigma) + \Gamma(f)|^2. \quad (23)$$

The cross-power component given is the result of only the $\Phi \times f_3$ operation; the higher-order contributions from the $\Phi \times f_5$ and $f_3 \times f_3$ terms are assumed to be negligible in comparison. The principal intermodulation component comes from the remainder of the $f_3 \times f_3$ term.

Limiting forms of the cross-power and intermodulation spectral densities for narrowband signals can be obtained by assuming that the highest effective frequency in $\phi(t)$ tends to zero and that $\Gamma(f)$ can be expanded as

$$\Gamma(f) = 1 + a(if) + b(if)^2 + c(if)^3 + d(if)^4 + \cdots$$

where, from $\Gamma(-f) = \Gamma^*(f)$, the coefficients $a, b, c, d, \cdots$ are real. Then $W_\phi(f)$ becomes negligibly small except near $f=0$ and it may be shown from (23) that $W_\theta^C(f)$ and $W_\theta^I(f)$ approach the values

$$W_\theta^C(f) = 2(6d - 6ac - b^2 + 4a^2b - a^4)W_\phi(f)f^2$$
$$\int_{-\infty}^{\infty} d\rho\, W_\phi(\rho)\rho^2$$

$$W_\theta^I(f) = \frac{2}{3}(3c - 3ab + a^3)^2 \int_{-\infty}^{\infty} d\rho \int_{-\infty}^{\infty} d\sigma\, W_\phi(\rho)W_\phi(\sigma)$$
$$W_\phi(f-\rho-\sigma)\rho^2\sigma^2(f-\rho-\sigma)^2.$$

In general, the linear-signal and cross-power components cover the same spectral range as the input spectrum, but the intermodulation spectrum contains newly created frequency components and therefore covers a greater frequency range. Naturally, only that portion of the intermodulation spectrum which coincides with the signal bandwidth need be considered when computing the output distortion.

An additional consideration develops when the input signal consists of two or more channels in frequency-division multiplex. The total intermodulation appearing in a given channel is given by $W_\theta^I(f)$ integrated over the appropriate frequency interval, as before. However, it now becomes meaningful to inquire as to the source of the intermodulation products appearing in the channel. Those arising from the interactions of frequency components within the channel, both between themselves and with frequency components in other channels, can be termed "self-distortion" since they are related to the presence of the signal in the channel of interest. Those arising from interactions between frequency components in the other channels can be termed "crosstalk" or "interchannel interference" as is occasionally experienced on idle telephone channels.

The crosstalk can be determined by computing the intermodulation using a slotted input spectrum, i.e., one in which $W_\phi(f)$ is made zero in the channel of interest while in the remaining channels it retains the original values corresponding to active talkers. The resulting intermodulation is then pure crosstalk since the self-distortion is made to vanish by slotting the input. The self-distortion is then given by the difference between the intermodulation spectra for the slotted and unslotted cases. If the channel under consideration is narrow in comparison with the signal bandwidth, then the effect on the intermodulation spectrum of slotting the input is small, so that, for all practical purposes, the crosstalk is well approximated by the unslotted intermodulation spectrum. The self-distortion is then negligible in comparison with the crosstalk.

V. Numerical Example—FM with Uniform Baseband Through a Single-Pole Filter

As mentioned in Section I, an important application of the foregoing analysis is to the case of the single-pole filter.[3] The normalized equivalent low-pass transfer function of this filter is given by

$$\Gamma(f) = \frac{1}{1 + if}, \qquad f = \frac{f}{f_c} \qquad (24)$$

where f_c is the 3-dB cutoff frequency (which corresponds to the 3-dB half-bandwidth of the original bandpass filter). Substituting (24) into (23) yields, after some tedious but straightforward algebra,

$$W_\theta^L(f) = \frac{W_\phi(f)}{1 + f^2}$$

$$W_\theta^C(f) = \frac{4f^2 W_\phi(f)}{1 + f^2} \times \int_{-\infty}^{\infty} \frac{d\rho \, \rho^2 W_\phi(\rho)}{(1 + \rho^2)[1 + (\rho + f)^2][1 + (\rho - f)^2]}$$

$$W_\theta^I(f) = \frac{1}{6(1 + f^2)} \int_{-\infty}^{\infty} d\rho \int_{-\infty}^{\infty} d\sigma W_\phi(f - \rho - \sigma) W_\phi(\rho) W_\phi(\sigma)(f - \rho - \sigma)^2 \rho^2 \sigma^2 \qquad (25)$$

$$\times \frac{4 + 4f^2 + 4f(\rho + \sigma)(f - \rho)(f - \sigma) + (\rho + \sigma)^2(f - \rho)^2(f - \sigma)^2}{[1 + (f - \rho - \sigma)^2](1 + \sigma^2)(1 + \rho^2)[1 + (\rho + \sigma)^2][1 + (f - \rho)^2][1 + (f - \sigma)^2]}$$

where the boldface quantities are normalized to f_c.

An input signal of particular interest is one which is frequency-modulated by a Gaussian signal having a uniform spectral density in (A, B) Hz, where $A \ll B$, since such a baseband closely approximates a frequency-division-multiplexed, multichannel communication signal. The spectral density of the input phase rate can then be written

$$W_{\dot\phi}(f) = \begin{cases} \dfrac{(2\pi D)^2}{2(B - A)}, & A \le |f| \le B \\ 0, & f \text{ elsewhere} \end{cases} \qquad (26)$$

where D is the rms deviation of instantaneous frequency in hertz and where, from (22), the spectral density of the input phase becomes

$$W_\phi(f) = \begin{cases} \dfrac{D^2}{2(B - A)f^2}, & A \le |f| \le B \\ 0, & f \text{ elsewhere.} \end{cases} \qquad (27)$$

It is convenient, for simplicity, to set the lowest baseband frequency A equal to zero despite the fact that the rms phase (i.e., the square root of the integral of $W_\phi(f)$ over all f) is then infinite. The expansion for the output phase is meaningless in that case, but the corresponding expansion for the output phase rate apparently remains valid. However, the only justification given here is that if the spectral density of the output phase rate is computed from (22) and (25) by using (27) for $W_\phi(f)$, and if the limiting values are taken as A approaches zero, then the results have meaning and are identical to those obtained by simply setting A equal to zero. Consequently, the spectra of the input phase rate and phase will be taken as

$$\left. \begin{aligned} W_{\dot\phi}(f) &= \frac{2\pi^2 D^2}{B} \\ W_\phi(f) &= \frac{D^2}{2Bf^2} \end{aligned} \right\} |f| \le B \qquad (28)$$

and zero elsewhere.

Substituting $W_\phi(f)$ from (28) into (25) and using (22) to obtain the spectra for the output phase rate then yields

$$\frac{W_\theta^L(f)}{D^2/f_c} = \frac{2\pi^2}{B(1 + f^2)}, \qquad |f| \le B$$

$$\frac{W_\theta^C(f)}{D^4/f_c^3} = \frac{8\pi^2 f^2}{B^2(1 + f^2)} \int_0^B \frac{d\rho}{(1 + \rho^2)[1 + (\rho + f)^2][1 + (\rho - f)^2]}, \qquad |f| \le B$$

$$\frac{W_\theta^I(f)}{D^6/f_c^5} = \frac{\pi^2 f^2}{12B^3(1 + f^2)} \int d\rho \int d\sigma$$

$$\times \frac{4 + 4f^2 + 4f(\rho + \sigma)(f - \rho)(f - \sigma) + (\rho + \sigma)^2(f - \rho)^2(f - \sigma)^2}{[1 + (f - \rho - \sigma)^2](1 + \sigma^2)(1 + \rho^2)[1 + (\rho + \sigma)^2][1 + (f - \rho)^2][1 + (f - \sigma)^2]} \qquad (29)$$

where it is understood that the terms vanish in those frequency intervals in which they are not defined and where boldface indicates normalization to f_c. The region of integration for the intermodulation term is shown in Fig. 1.

---

[3] It should be noted that the apparent validity of using a Gaussian process for the phase $\phi(t)$ in the spectral analysis of Section IV is considerably enhanced for this particular filter. As noted after (51) in Appendix I, a bound $|\dot\phi(t)| < \omega_c/3$ on the phase rate rather than on the phase itself will suffice to insure convergence of the series in (17).

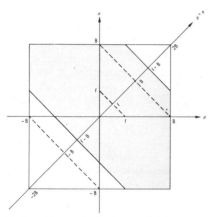

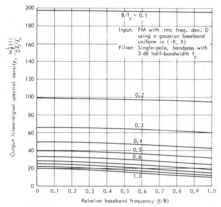

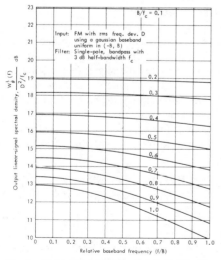

Fig. 1. Region of integration for intermodulation computation.

Fig. 2. Output linear-signal spectral density.

Fig. 3. Output linear-signal spectral density, dB.

Since these spectral densities are even functions of frequency, they will be plotted subsequently only for positive frequencies.

The spectral density of the linear-signal component of the output phase rate given by the first equation of (29) is plotted in Figs. 2 and 3 as a function of the relative baseband frequency f/B, for a number of baseband-to-filter half-bandwidth ratios B/f_c. As might be expected, the output spectral density is fairly uniform when the filter is wide, i.e., when B/f_c is small. As the filter is narrowed, the output takes the shape of the filter power response and, indeed, the spectral density is down 3 dB at the highest baseband frequency for a filter for which $B/f_c = 1$.

The spectral density of the cross-power component of the output phase rate given by the second equation (29) can be integrated by expanding the integrand in partial fractions. The result is

$$\frac{W_{\dot\theta}^C(f)}{D^4/f_c^3} = \frac{8\pi^2}{B^2(f^2+1)(f^2+4)}$$

$$\left\{\tan^{-1} B + \frac{3f}{8(f^2+1)}\log\left[\frac{(B+f)^2+1}{(B-f)^2+1}\right]\right.$$

$$\left. + \frac{f^2-2}{4(f^2+1)}\left[\tan^{-1}(B+f) + \tan^{-1}(B-f)\right]\right\},$$

$$|f| \leq B, \quad (30)$$

which is plotted in Figs. 4 and 5. The cross-power increases with baseband frequency in general, changing as the square of the frequency for wide filters. A drop-off at the higher frequencies, resembling the skirt of the filter power response, occurs as the filter is narrowed, i.e., as f_c is decreased, thereby increasing B/f_c.

In principle, at least one of the integrations for the intermodulation can be performed by the same procedure of expanding by partial fractions. However, the algebra required appears too formidable to warrant the effort. Even if it were done, it is highly likely that the second integration would not be possible analytically. Therefore, numerical results were obtained by machine computation on the RAND IBM 7044, using the technique of iterated integration to evaluate the double integral.[4] The results are plotted in Fig. 6, and in decibels in Fig. 7.

Like the cross power, the intermodulation tends to zero at the lower end of the baseband and, in general, increases with frequency within the baseband. Also, the effect of the filter power response is again evident for narrow filters. In the manner typical of a third-order spectral density, the intermodulation spectrum extends to three times the highest baseband frequency and falls smoothly to zero. The linear-signal and cross-power spectra both vanish abruptly beyond the highest baseband frequencies since they both contain the input signal spectrum as a multiplicative factor.

---

[4] Romberg integration was used. The Romberg approximation sequence was halted when a relative difference of 10^{-3} between successive approximations was attained.

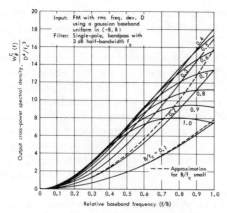

Fig. 4. Output cross-power spectral density.

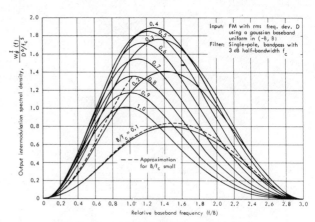

Fig. 6. Output intermodulation spectral density.

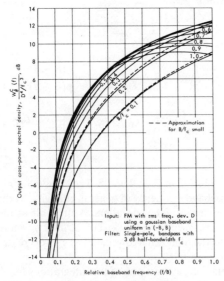

Fig. 5. Output cross-power spectral density, dB.

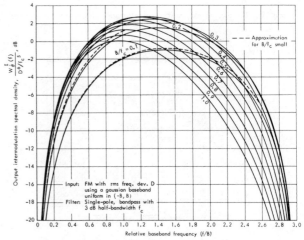

Fig. 7. Output intermodulation spectral density, dB.

Useful approximations to these spectra can be obtained by assuming the highest baseband frequency B to be sufficiently small in comparison with the filter 3-dB half-bandwidth f_c that variations of the denominator terms in (25) over the regions of integration can be neglected. Simple integration then yields

$$\frac{W_\theta^L(f)}{D^2/f_c} \approx \frac{2\pi^2}{B}, \qquad |f| \le B$$

$$\frac{W_\theta^C(f)}{D^4/f_c^3} \approx \frac{8\pi^2 f^2}{B}, \qquad |f| \le B$$

$$\frac{W_\theta^I(f)}{D^6/f_c^5} \approx \begin{cases} \dfrac{\pi^2 f^2}{3B^3}(3B^2 - f^2), & |f| \le B \\[2mm] \dfrac{\pi^2 f^2}{6B^3}(3B - |f|)^2, & B \le |f| \le 3B. \end{cases} \quad (31)$$

These spectra exceed the correct values slightly but do not differ significantly for values of B/f_c up to about 0.2. They are shown in Figs. 4–7 as dashed lines for $B/f_c = 0.1$ and 0.2. The approximation for $W_\theta^C(f)$ and $W_\theta^I(f)$ can also be ob-

tained by starting with the limiting forms of $W_\theta^C(f)$ and $W_\theta^I(f)$ given in Section IV.

The ratio of the linear-signal spectral density to the intermodulation spectral density is computed from (29) and plotted in Fig. 8. For a multichannel modulating signal, this quantity approximates the signal-to-crosstalk ratio, SCR, in a narrow channel as a function of its location within the baseband (see discussion at the end of Section IV). As is the case with thermal noise, the interference is greatest in the highest channel. The approximations of (31) can be used to obtain a simple formulation of this ratio valid for small B/f_c. Thus,

$$\text{SCR} \cong \frac{W_\theta^L(f)}{W_\theta^I(f)} = \frac{6B^2}{f^2 D^4 (3B^2 - f^2)}, \qquad |f| \le B, \quad (32)$$

which is shown by dashed lines in Fig. 8 for the two lower values of B/f_c.

Finally, the ratio of the total linear signal to the total intermodulation within the baseband, i.e., the ratio of the integrals, from $-B$ to B, of the first and third equations of (29), is plotted in Fig. 9. When the modulating signal consists of a single channel, this quantity gives the output signal-to-

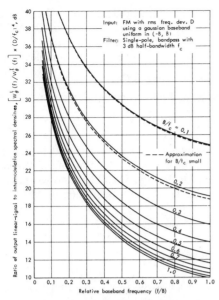

Fig. 8. Ratio of output linear signal to intermodulation spectral densities, dB. Approximates the signal-to-crosstalk ratio in a narrow channel as a function of its location within the baseband.

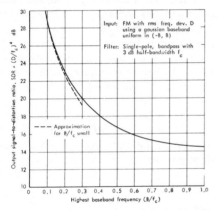

Fig. 9. Output signal-to-distortion ratio in total baseband, dB.

distortion ratio, SDR. It also serves as a measure of the validity of using only the leading terms in the spectral analysis given by (29). If the SDR ratio is large, then it can be assumed that the higher-order contributions to the distortion can be safely neglected. As before, an approximation valid for small B/f_c can be made by integrating (31) and forming the ratio

$$\text{SDR} \cong \frac{15}{2B^2 D^4} \tag{33}$$

which is shown in Fig. 9 as a dashed line.

To illustrate the use of these curves, consider the case for which the filter half-bandwidth f_c equals 2 MHz, the rms frequency deviation D equals 1.4 MHz, and the highest baseband frequency B equals 1 MHz, i.e., $D/f_c = 0.7 \Rightarrow -1.5$ dB and $B/f_c = 0.5$. Then, from Fig. 9, the SDR is $16.7 + 6 = 22.7$ dB, which is large enough to ensure the accuracy of

the approximation, while, from Fig. 8, the signal-to-cross-talk ratio at the upper end of the baseband (the worst channel) is $12.6 + 6 = 18.6$ dB. This is, of course, a somewhat extreme example of a badly distorted channel but serves nevertheless to indicate the range of applicability of the results.

As a more practical example, consider an FM signal modulated by a uniform baseband extending to 2.4 MHz and having an rms deviation of 20 MHz in the RF channel. Assume a frequency-feedback receiver having a feedback factor of 20 dB, uniform across the baseband, and a single-pole IF filter with a 3-dB half-bandwidth of 4 MHz. The rms frequency deviation in the IF channel is then 2 MHz, and the system parameters become $B = 2.4$ MHz, $D = 2$ MHz, and $f_c = 4$ MHz, yielding the ratios $B/f_c = 0.6$ and $D/f_c = 0.5 \Rightarrow -3$ dB. Since the distortion in a linear feedback network is reduced by the feedback factor, it follows from Fig. 9 that the signal-to-distortion ratio $\text{SDR} = 15.8 + 12 + 20 = 47.8$ dB. The SCR at the upper end of the baseband is $11.6 + 12 + 20 = 43.6$ dB from Fig. 8.

It is seen, in this case, that neither the distortion, if the modulation consists of a single wideband signal such as a TV video output, nor the crosstalk in the worst channel, if the modulation consists of a multichannel telephone base, is negligible in terms of the criteria for high-fidelity transmission. Further improvements can be obtained by increasing the feedback, which poses severe practical problems in such wideband circuitry, or by widening the IF filter, which vitiates the threshold-reducing advantage of the feedback process.

APPENDIX I

EXPANSION OF OUTPUT PHASE

The time-dependent component of the output phase is given by (10) as

$$\theta(t) = \text{Im} \log \int_0^\infty \gamma(\tau) e^{i\phi(t-\tau)} d\tau. \tag{34}$$

The expression on the right may be expanded in a number of different ways. Here the linear portion of $\theta(t)$ will be extracted first and then the remaining nonlinear portion expanded as a functional power series. This procedure yields the higher-order terms in a form suited to the spectral analysis of Appendix II.

To obtain the linear portion of $\theta(t)$, take as a guide the case of small $\phi(t)$ and let

$$z(\zeta, t) = \log \int_0^\infty \gamma(\tau) e^{\zeta\phi(t-\tau)} d\tau. \tag{35}$$

Since ζ and $\phi(t)$ occur only in the product $\zeta\phi(t-\tau)$, the part of $z(\zeta, t)$ which is linear in $\phi(t)$ is also the part linear in ζ. The linear portion of $z(\zeta, t)$ is the linear term in its Maclaurin series in powers of ζ:

$$\frac{\zeta}{1!} \left[\frac{\partial}{\partial \zeta} z(\zeta, t) \right]_{\zeta=0} = \zeta \int_0^\infty \gamma(\tau)\phi(t-\tau) d\tau \tag{36}$$

$$= \zeta\Phi(t)$$

where (9) has been used to show that the denominator in the derivative of the logarithm becomes unity at $\zeta=0$. Since $\theta(t) = \text{Im } z(i, t)$, the linear portion of $\theta(t)$ equals the linear portion of Im $z(i, t)$. From (36) this is Im $[i\Phi(t)] = \text{Re } [\Phi(t)]$. Thus, the linear portion of $\theta(t)$ is Re $\Phi(t)$ where

$$\Phi(t) = \int_0^\infty \gamma(\tau)\phi(t - \tau)d\tau. \tag{37}$$

Adding and subtracting $i\Phi(t)$ in the exponent carries (34) into an expression for $\theta(t)$ as the sum of its linear and non-linear parts:

$$\theta(t) = \text{Re } \Phi(t) + \text{Im log} \int_0^\infty \gamma(\tau)e^{i[\phi(t-\tau)-\Phi(t)]}d\tau. \tag{38}$$

To expand the nonlinear component of $\theta(t)$, i.e., the second term on the right in (38), consider

$$f(\zeta) = \log F(\zeta) \tag{39}$$

where

$$F(\zeta) = \int_0^\infty \gamma(\tau)e^{\zeta[\phi(t-\tau)-\Phi(t)]}d\tau \tag{40}$$

and expand $f(\zeta)$ in a Maclaurin series in ζ. Then

$$f(\zeta) = \sum_{n=0}^\infty \frac{\zeta^n}{n!} f_n \tag{41}$$

where f_n is the nth derivative of $f(\zeta)$ evaluated at $\zeta=0$. The coefficients f_n are obtained by first differentiating (39) to obtain

$$F(\zeta) \frac{d}{d\zeta} f(\zeta) = \frac{d}{d\zeta} F(\zeta), \tag{42}$$

then differentiating $(n-1)$ more times. Setting $\zeta=0$ in the result yields

$$\sum_{j=0}^{n-1} \binom{n-1}{j} F_j f_{n-j} = F_n, \qquad n \geq 1 \tag{43}$$

where

$$F_n = \left[\left(\frac{d}{d\zeta}\right)^n F(\zeta)\right]_{\zeta=0} = \int_0^\infty \gamma(\tau)[\phi(t-\tau) - \Phi(t)]^n d\tau. \tag{44}$$

In particular, from (9) and (37), $F_0=1$ and $F_1=0$. Then, from $F_0=1$ and (39), f_0 is zero[5] and from $F_0=1$, $F_1=0$, and (42), $f_1=0$. Thus, (43) goes into the recurrence relation

$$f_n = F_n - \sum_{j=2}^{n-2} \binom{n-1}{j} F_j f_{n-j}, \qquad n \geq 2 \tag{45}$$

from which the higher-order coefficients may be computed. The coefficients are found to be

$$f_2 = F_2, \, f_3 = F_3, \, f_4 = F_4 - 3F_2^2, \cdots \tag{46}$$

as stated in (13). The desired expansion of (38) is then given by evaluating (41) at $\zeta=i$, thereby yielding (11).

[5] This f_0 is not to be confused with carrier frequency f_o.

It is interesting to note that f_n is related to $F_n, F_{n-1}, \cdots$, 0, 1 in the same way that the nth semi-invariant[9],[10] of a distribution is related to its moments $\mu_n, \mu_{n-1}, \cdots, 0, 1$ about the mean. However, f_n and F_n may be complex here.

A. Convergence: General Case

A necessary and sufficient condition for $f(i)$ to be given by

$$f(i) = \sum_{n=2}^\infty \frac{i^n}{n!} f_n, \tag{47}$$

where the series converges absolutely, is that $f(\zeta)$ be regular in and on the circle $|\zeta|=1$.[11] In terms of $F(\zeta)$ this condition requires that $F(\zeta)$ be regular and have no zeros in or on the unit circle. This condition is then sufficient for

$$\text{Im } f(i) = \sum_{n=2}^\infty \frac{1}{n!} \text{Im } (i^n f_n) \tag{48}$$

to hold, and hence for

$$\theta(t) = \text{Re } \Phi(t) + \sum_{n=2}^\infty \frac{1}{n!} \text{Im } (i^n f_n) \tag{49}$$

to be a valid expansion for (38).

Note that the condition is only sufficient for the convergence of (49). It is not a necessary condition because the divergence of a complex series does not imply divergence of its imaginary part. However, if the imaginary part of the series converges (and the real part diverges), it may not converge to Im $f(i)$. Consequently the sufficient condition cannot be strengthened to include zeros in or on the unit circle just because the imaginary part of the series converges.

When $\phi(\tau) \equiv 0$ for $0 \leq \tau \leq \infty$, (40) for $F(\zeta)$ and (37) for $\Phi(t)$ show that $F(\zeta)=1$ for all ζ. This suggests that $F(\zeta)$ will not vanish in or on the unit circle if $|\phi(\tau)|$ remains small enough. More precisely, it can be shown that $\theta(t)$ is given by (49) if $\phi(\tau)$ and $\gamma(\tau)$ are continuous and if there exists a number A such that

$$|\phi(t - \tau) - \Phi(t)| < A, \qquad 0 \leq \tau \leq \infty$$
$$\int_0^\infty |\gamma(\tau)|d\tau < 2A^{-2}e^{-A}. \tag{50}$$

B. Convergence: Bandpass Case

Slightly more relaxed conditions result from the expansion of the output phase for the bandpass case of Section III. A necessary and sufficient condition for (17) to be valid can be obtained by first writing it as

$$\theta(t) = \Phi(t) + \frac{1}{2i}[f(i) - f(-i)] \tag{51}$$

and observing that the power series for $f(\zeta) - f(-\zeta)$ $= \log [F(\zeta)/F(-\zeta)]$ will converge absolutely at $\zeta=i$ if $f(\zeta) - f(-\zeta)$ is regular in and on the unit circle. Therefore, for the expansion for $\theta(t)$ to hold, it is necessary and sufficient for $F(\zeta)/F(-\zeta)$ to have no zeros or singularities in or on the unit circle.

Simple, but rather loose, sufficient conditions for the validity of the expansion in (17) for all values of t can be given.

They are: 1) that the integral of $\gamma(t)$ from $\tau = 0$ to ∞ converge; 2) that, for all real values of τ, $\phi(\tau)$ and $\gamma(\tau)$ be continuous; 3) that $\gamma(\tau) \geq 0$; and 4) that $|\phi(\tau)| < \pi/2$. Under these conditions, the integral defining $F(\zeta)$ is regular for all finite values of ζ, its nth derivative is given by (44) for F_n, and its real part remains positive inside and on the unit circle. Consequently, $F(\zeta)/F(-\zeta)$ has no zeros or singularities inside or on $|\zeta| = 1$ and the validity of (17) follows.

However, the series may converge even when $\phi(t)$ is not small. If the derivative $\phi'(t)$ is bounded, the inequality $|\phi(t - \tau) - \phi(t)| \leq |\tau| \max |\phi'(t)|$ may be used to show that, for the filter $\gamma(t) = \omega_c \exp(-\omega_c t)$, a sufficient condition for the series to converge is $\max |\phi'(t)| < \omega_c/3$.

The series given by (17) was used to compute values of $\theta(t)$ for the single-pole filter of the preceding paragraph. Computations for the case $\phi(t) = A \sin \omega_c t$ show that the conditions $\max |\phi(t)| < \pi/2$ and $\max |\phi'(t)| < \omega_c/3$ are unduly restrictive. Even though these conditions require $A < \pi/2$ and $A < 1/3$, respectively, the computations show that the series for $\theta(0)$ converges when $A = 3$. However, when $A = 5$, the function $F(\zeta)$ has zeros at $\zeta = 0.572 \pm i\,0.587$ inside the unit circle, and the series diverges.

APPENDIX II

SPECTRAL ANALYSIS FOR GAUSSIAN $\phi(t)$

As suggested in Section IV, let the output phase of (18) be written

$$\theta(t) = \Phi + f_3 + f_5 + \cdots \qquad (52)$$

where Φ, f_3, and f_5 denote the linear-signal, and the third- and fifth-order nonlinear components of the output, respectively. From (21), the output spectral density $W_\theta(f)$ is given by the Fourier transform of $E[\theta(t + \tau)\theta(t)]$,

$$W_\theta(f) = \int_{-\infty}^{\infty} d\tau\, e^{-i\omega\tau} E[\theta(t + \tau)\theta(t)], \qquad \omega = 2\pi f. \qquad (53)$$

Throughout this Appendix, ω will denote $2\pi f$.

From (52) it follows that $W_\theta(f)$ may be written as the sum of contributions of terms of the form $\Phi \times \Phi$, $\Phi \times f_3$, $f_3 \times f_3$, etc.

The linear-signal component $W_\theta^L(f)$ of $W_\theta(f)$ is obtained immediately from the $\Phi \times \Phi$ term by noting that $\Phi(t)$ is the response when $\phi(t)$ is applied to a filter having the impulse response $\gamma(t)$ and the steady-state transfer function $\Gamma(f)$:

$$W_\theta^L(f) = W_\phi(f)|\Gamma(f)|^2. \qquad (54)$$

This is the first equation of (23).

The $\Phi \times f_3$, $\Phi \times f_5$, and $f_3 \times f_3$ terms, which are the only higher-order ones considered here, are more difficult to compute. Expressions for the expected values of powers and products of terms similar to $[\Phi(t) - \phi(t - u)]$ are needed. One method of obtaining these expected values is to start with the relation, for Gaussian variables,

$$E(x_1 x_2 \cdots x_{2N-1} x_{2N})$$

$$= \sum_{\text{all pairs}} E(x_i x_j) E(x_k x_l) \cdots E(x_m x_n) \qquad (55)$$

where (i, j), (k, l), $\cdots$, (m, n) are N pairs of integers selected from $1, 2, \cdots, 2N$ and the summation extends over all possible pairings. The expectations on the right may be expressed in terms of the autocorrelation function $R_\phi(\tau)$. A second method, the one used here, is based on the following result. Let L be a linear operator (operating on functions of t) such that

$$L(a e^{i\omega t}) = a A(\omega) e^{i\omega t} \qquad (56)$$

and let $\phi(t)$ be a Gaussian process with spectral density $W_\phi(f)$. The process $L[\phi(t)]$ is then also Gaussian. Representing $\phi(t)$ by the series

$$\sum_{1}^{N} (a_n \cos \omega_n t + b_n \sin \omega_n t) \qquad (57)$$

where $\omega_n = 2\pi f_n$, $f_n = n\Delta f$, $\Delta f \to 0$, $N \to \infty$, and a_n and b_n are independent Gaussian random variables with mean zero and variance

$$E(a_n^2) = E(b_n^2) = 2W_\phi(f)\Delta f \qquad (58)$$

and considering the characteristic function of $L[\phi(t)]$ lead to

$$E \exp\{iL[\phi(t)]\}$$

$$= \exp\left[-\frac{1}{2}\int_{-\infty}^{\infty} W_\phi(f)A(\omega)A(-\omega)df\right]. \qquad (59)$$

This result will be used to obtain expressions for the required expected values.

Equation (53) for $W_\theta(f)$ contains two $\Phi \times f_3$ terms. From (18), the first one is the Fourier transform of

$$F(\tau) = E\left\{\Phi(t + \tau)\frac{1}{3!}\int_0^\infty du\,\gamma(u)[\Phi(t) - \phi(t - u)]^3\right\} \qquad (60)$$

and the second one can be written as the Fourier transform of $F(-\tau)$. Since $F(\tau)$ is real, the contribution of the two $\Phi \times f_3$ terms to $W_\theta^C(f)$ is

$$W_\theta^C(f) = \int_{-\infty}^{\infty} d\tau\, e^{-i\omega\tau}[F(\tau) + F(-\tau)]$$

$$= \text{Re}\,2\int_{-\infty}^{\infty} d\tau\, e^{-i\omega\tau}F(\tau). \qquad (61)$$

The expected value of

$$\frac{1}{3!}\Phi(t + \tau)[\Phi(t) - \phi(t - u)]^3 \qquad (62)$$

is the coefficient of $(ix_1)(ix_2)^3$ in the expansion of $E \exp(iM)$ where

$$M = x_1\Phi(t + \tau) + x_2[\Phi(t) - \phi(t - u)]$$

$$= x_1\int_0^\infty ds\,\gamma(s)\phi(t + \tau - s)$$

$$+ x_2\left[\int_0^\infty ds\,\gamma(s)\phi(t - s) - \phi(t - u)\right]. \qquad (63)$$

303

The quantity M is the result of a linear operation on $\phi(t)$ and may be taken as the $L[\phi(t)]$ appearing in (59). The corresponding $A(\omega)$ is obtained by replacing the functions $\phi(v)$ in (63) by $\exp(i\omega v)$, where v denotes $t+\tau-s$, $t-s$, and $t-u$, in turn, and by noting that the Fourier transform of $\gamma(s)$ is $\Gamma(f)$. It then follows from (56) and (63) that

$$A(\omega)e^{i\omega t} = L(e^{i\omega t}) = x_1 e^{i\omega t}e^{i\omega \tau}\Gamma(f)$$
$$+ x_2 e^{i\omega t}[\Gamma(f) - e^{-i\omega u}].$$

Therefore

$$A(\omega) = x_1\Gamma(f)e^{i\omega\tau} + x_2 H_u(f) \qquad (64)$$

where

$$H_u(f) = \Gamma(f) - e^{-i\omega u}. \qquad (65)$$

Consequently

$$Ee^{iM} = \exp\left[ax_1^2 + bx_2^2 + cx_1 x_2\right] \qquad (66)$$

in which a, b, and c are integrals obtainable from (59).

The coefficient of $x_1 x_2^3$ in the power series expansion of the right-hand side of (66) is bc where

$$b = -\frac{1}{2}\int_{-\infty}^{\infty} df\, W_\phi(f)H_u(f)H_u(-f)$$

$$c = -\frac{1}{2}\int_{-\infty}^{\infty} df\, W_\phi(f)[\Gamma(f)H_u(-f)e^{i\omega\tau}+\Gamma(-f)H_u(f)e^{-i\omega\tau}]$$

$$= -\int_{-\infty}^{\infty} df\, W_\phi(f)\Gamma(f)H_u(-f)e^{i\omega\tau}.$$

The step from the second equation to the third makes use of the fact that $W_\phi(f)$ is an even function of f.

Equation (60) for $F(\tau)$ may now be written as

$$F(\tau) = \int_0^\infty du\gamma(u)bc$$

$$= \frac{1}{2}\int_0^\infty du\gamma(u)\int_{-\infty}^\infty d\rho\, W_\phi(\rho)H_u(\rho)H_u(-\rho)$$

$$\int_{-\infty}^\infty d\sigma\, W_\phi(\sigma)\Gamma(\sigma)H_u(-\sigma)e^{i2\pi\sigma\tau}. \qquad (67)$$

Equation (65) for $H_u(f)$ shows that

$$H_u(\rho)H_u(-\rho) = |\Gamma(\rho)|^2 + 1 - \Gamma(\rho)e^{i2\pi\rho u} - \Gamma(-\rho)e^{-i2\pi\rho u}.$$

When this is multiplied by $\gamma(u)H_u(-\sigma)$ and the integration with respect to u is performed [with the help of the Fourier relation between $\Gamma(f)$ and $\gamma(t)$], the contribution of $|\Gamma(\rho)|^2 + 1$ vanishes, leaving

$$\int_0^\infty du\gamma(u)H_u(\rho)H_u(-\rho)H_u(-\sigma)$$

$$= \Gamma(\rho)\Gamma(-\rho-\sigma) + \Gamma(-\rho)\Gamma(\rho-\sigma) - 2\Gamma(\rho)\Gamma(-\rho)\Gamma(-\sigma).$$

The triple integral (67) for $F(\tau)$ reduces to a double integral in ρ and σ. Changing the sign of the variable of integration

ρ in the portion containing $\Gamma(-\rho)\Gamma(\rho-\sigma)$ shows that its contribution to $F(\tau)$ is equal to that of the portion containing $\Gamma(\rho)\Gamma(-\rho-\sigma)$.

Therefore

$$F(\tau) = \int_{-\infty}^\infty d\rho \int_{-\infty}^\infty d\sigma\, W_\phi(\rho)W_\phi(\sigma)e^{i2\pi\sigma\tau}$$

$$\cdot[\Gamma(\rho)\Gamma(\sigma)\Gamma(-\rho-\sigma) - 2|\Gamma(\rho)\Gamma(\sigma)|^2].$$

Substitution of the expression for $F(\tau)$ in the integral (61) for $W_\theta^C(f)$ and use of the Fourier integral theorem removes the integrations with respect to τ and σ, replaces σ by f, and gives the final expression

$$W_\theta^C(f) = 2W_\phi(f)\int_{-\infty}^\infty d\rho\, W_\phi(\rho)$$

$$\cdot\{\operatorname{Re}\Gamma(\rho)\Gamma(f)\Gamma(-\rho-f) - |\Gamma(\rho)|^2|\Gamma(f)|^2\}. \qquad (68)$$

This is the second equation of (23).

The terms of next higher order arise from $\Phi \times f_5$ and $f_3 \times f_3$. The $\Phi \times f_5$ terms are of the same nature as the $\Phi \times f_3$ terms, and their contribution to $W_\theta(f)$ may be shown to contain $W_\phi(f)$ as a multiplicative factor. They will be neglected since they furnish higher-order contributions to the cross power $W_\theta^C(f)$ given by (68).

From (18) the $f_3 \times f_3$ contribution to $W_\theta(f)$ is the Fourier transform of

$$F(\tau) = \int_0^\infty du\gamma(u)\int_0^\infty dv\gamma(v)E\{[\Phi(t+\tau) - \phi(t+\tau-r)]^3$$

$$\cdot[\Phi(t) - \phi(t-v)]^3/36\}. \qquad (69)$$

The expectation appearing in the integrand is equal to the coefficient of $(ix_1)^3(ix_2)^3$ in the expansion of $E\exp(iN)$, where

$$N = x_1[\Phi(t+\tau) - \phi(t+\tau-u)] + x_2[\Phi(t) - \phi(t-v)]$$

$$= x_1\left[\int_0^\infty ds\gamma(s)\phi(t+\tau-s) - \phi(t+\tau-u)\right]$$

$$+ x_2\left[\int_0^\infty ds\gamma(s)\phi(t-s) - \phi(t-v)\right]. \qquad (70)$$

This expression for N is similar to (63) for M and, as before, replacing the functions $\phi(t+\tau-s)$, $\cdots$ by the corresponding exponentials $\exp[i\omega(t+\tau-s)]$, $\cdots$ shows that the $A(\omega)$ corresponding to N is

$$A(\omega) = x_1 H_u(f)e^{i\omega\tau} + x_2 H_v(f) \qquad (71)$$

where the H's are given by (65). Then

$$Ee^{iN} = \exp\left[ax_1^2 + bx_2^2 + cx_1 x_2\right] \qquad (72)$$

where a, b, and c are integrals obtainable from (59) and (71).

The coefficient of $x_1^3 x_2^3$ in the expansion of the exponential in (72) is $abc+c^3/6$, and its negative is the value of the expectation in (69) for $F(\tau)$. Taking the Fourier transform of $F(\tau)$ shows that the $f_3 \times f_3$ contribution to $W_\theta(f)$ is

$$- \int_{-\infty}^{\infty} d\tau e^{-i\omega\tau} \int_0^{\infty} du\gamma(u) \int_0^{\infty} dv\gamma(v)(abc + c^3/6). \quad (73)$$

The integrals a and b are independent of τ, and

$$c = -\frac{1}{2} \int_{-\infty}^{\infty} df W_\phi(f)[H_u(f)H_v(-f)e^{i\omega\tau}$$
$$+ H_u(-f)H_v(f)e^{-i\omega\tau}]$$
$$= - \int_{-\infty}^{\infty} df W_\phi(f)H_u(f)H_v(-f)e^{i\omega\tau} \quad (74)$$

contains τ only through the factor $\exp(i\omega\tau)$ in the integrand. From the Fourier integral theorem it follows that the contribution of the a, b, c term to $W_\theta(f)$ contains the factor $W_\phi(f)$. It is thus a higher-order contribution to the cross power than is $W_\theta^C(f)$, given by (68), and will be neglected.

The c^3 term in (73) gives the leading term in the power spectral density of the intermodulation

$$W_\theta^I(f) = -\frac{1}{6} \int_{-\infty}^{\infty} d\tau e^{-i\omega\tau} \int_0^{\infty} du\gamma(u) \int_0^{\infty} dv\gamma(v)c^3. \quad (75)$$

Replacing the variable of integration f in (74) for c by ρ, σ, and v, in turn, leads to a triple integral for c^3, and to a sixfold integral for $W_\theta^I(f)$. In the latter, the integral with respect to u contains the product $H_u(\rho)H_u(\sigma)H_u(v)$. Introducing the definition (65) of $H_u(f)$ and multiplying out gives eight terms. Two of these cancel after integration with respect to u, leaving

$$\int_0^{\infty} du\gamma(u)H_u(\rho)H_u(\sigma)H_u(v) = -2\Gamma(\rho)\Gamma(\sigma)\Gamma(v) + \Gamma(v)\Gamma(\rho + \sigma)$$
$$+ \Gamma(\rho)\Gamma(\sigma + v) + \Gamma(\sigma)\Gamma(v + \rho) - \Gamma(\rho + \sigma + v) = S(\rho, \sigma, v). \quad (76)$$

The value of the corresponding integral with respect to v is the complex conjugate $S(-\rho, -\sigma, -v)$ of $S(\rho, \sigma, v)$.

Integration with respect to u and v reduces the integral for $W_\theta^I(f)$ to the fourfold integral

$$W_\theta^I(f) = \int_{-\infty}^{\infty} d\tau e^{-i\omega\tau} \frac{1}{6} \int_{-\infty}^{\infty} d\rho \int_{-\infty}^{\infty} d\sigma \int_{-\infty}^{\infty} dv e^{i2\pi(\rho + \sigma + v)\tau}$$
$$\times W_\phi(\rho)W_\phi(\sigma)W_\phi(v)|S(\rho, \sigma, v)|^2.$$

Integration with respect to τ yields the delta function $\delta(f - \rho - \sigma - v)$, thereby permitting a simple integration on

v. Finally, therefore,

$$W_\theta^I(f) = \frac{1}{6} \int_{-\infty}^{\infty} d\rho \int_{-\infty}^{\infty} d\sigma W_\phi(\rho)W_\phi(\sigma)W_\phi(f - \rho - \sigma)$$
$$\times |2\Gamma(\rho)\Gamma(\sigma)\Gamma(f - \rho - \sigma) - \Gamma(f - \rho - \sigma)\Gamma(\rho + \sigma)$$
$$- \Gamma(\rho)\Gamma(f - \rho) - \Gamma(\sigma)\Gamma(f - \sigma) + \Gamma(f)|^2 \quad (77)$$

which is the third equation of (23).

ACKNOWLEDGMENT

The suggestions and criticisms of W. Sollfrey and M. B. Marcus of The RAND Corporation and I. S. Reed of the University of Southern California, Los Angeles, Calif., contributed greatly to the successful completion of this analysis and are acknowledged with gratitude. The programming for the numerical computations of Section V was performed by J. E. Rieber of The RAND Corporation. Helpful comments by L. H. Enloe, H. E. Rowe, and C. L. Ruthroff of the Bell Telephone Laboratories, Inc., are also acknowledged with thanks.

Before learning of one another's work, the authors had independently derived substantially the same results for the third-order spectral components. They became aware of this through C. L. Ruthroff, who had published material in which FM distortion was one item of concern. It was decided, after some correspondence, that the best way to present the results was through a joint paper.

REFERENCES

[1] H. Roder, "Effects of tuned circuits upon a frequency modulated signal," Proc. IRE, vol. 25, pp. 1617–1647, December 1937.
[2] J. R. Carson and T. C. Fry, "Variable frequency electric circuit theory with application to the theory of frequency modulation," Bell Sys. Tech. J., vol. 16, pp. 513–540, October 1937.
[3] E. J. Baghdady, Lectures on Communication System Theory. New York: McGraw-Hill, 1961.
[4] P. F. Panter, Modulation, Noise and Spectral Analysis. New York: McGraw-Hill, 1965.
[5] J. J. Downing, Modulation Systems and Noise. Englewood Cliffs, N. J.: Prentice-Hall, 1964.
[6] H. E. Rowe, Signals and Noise in Communication Systems. Princeton, N. J.: Van Nostrand, 1965.
[7] J. C. Chaffee, "The application of negative feedback to frequency modulation systems," Bell Sys. Tech. J., vol. 18, pp. 404–437, July 1939.
[8] L. H. Enloe, "Decreasing the threshold in FM by frequency feedback," Proc. IRE, vol. 50, pp. 18–30, January 1962.
[9] H. Cramér, Mathematical Methods of Statistics. Princeton, N. J.: Princeton University Press, 1946, p. 186.
[10] J. Riordan, An Introduction to Combinatorial Analysis. New York: Wiley, 1958, p. 37.
[11] E. G. Phillips, Functions of a Complex Variable, 6th ed. Edinburgh: Oliver and Boyd, 1949, p. 19.

III

FM Threshold Reduction

PAPER NO. 13

Reprinted from *Proc. IRE*, Vol. 50, No. 1, pp. 18–30, Jan. 1962

Decreasing the Threshold in FM by Frequency Feedback*

L. H. ENLOE†, MEMBER, IRE

Summary—The "frequency feedback demodulator" or "frequency compression demodulator" can be used to extend the threshold of signal-to-noise improvement in large index frequency-modulation systems. Previous papers have advanced the argument that the threshold occurs in the usual manner when noise peaks exceed the carrier envelope at the input to the frequency detector of a feedback demodulator. However, correlation between calculated values and carefully measured experimental values has been poor. The calculated values have been incorrect by orders of magnitude in many typical cases. In this paper it is shown that the threshold can also occur because of the feedback action of the demodulator. When this is taken into account, the threshold can be calculated accurately. It is shown that the threshold cannot be improved by the often suggested scheme of inserting a carrier of the proper phase at the input to the frequency detector. The phase-locked loop, a related device, is shown to have a threshold which is equal to or poorer than the threshold of the feedback demodulator for large modulation indices. It is shown that the stability of the feedback loop (and consequently the threshold) of the feedback demodulator is a function of modulation, but that by following the procedure given, the effect can be almost entirely eliminated. The results of this paper allow one to design feedback demodulators for the first time which will extend the threshold in a predictable manner.

INTRODUCTION

THERE is widespread interest at the present time in modulation techniques which trade bandwidth of the transmitted signal for improved baseband signal-to-noise and signal-to-interference ratios. These techniques are of special interest in the field of space communications where distances are so large that even the use of masers and large low-noise antennas results in a signal-to-noise ratio, assuming single-sideband modulation, too small for commercial telephone service by a factor of about a hundred. In addition, the most efficient operation requires the use of a single-frequency channel by many satellites and ground stations simultaneously. This mode of operation demands a modulation technique which can increase the baseband signal-to-mutual-interference ratio considerably over that of single-sideband modulation. Large index frequency modulation satisfies both of these requirements and could be used but for the fact that it has a relatively poor threshold. *The threshold is defined as the minimum carrier-to-noise ratio yielding an FM improvement which is not significantly deteriorated from the value predicted by the usual small noise signal-to-noise formulas.* This paper discusses special demodulators using frequency feedback to extend the threshold and bring frequency modulation into an attractive position.

The frequency feedback demodulator originated

with Chaffee [1] in the early thirties. His paper and a companion paper by Carson [2] contain interesting results concerning the distortion and small noise signal-to-noise characteristics of the demodulator. In the past few years various papers have been published [3], [4] to show that a form of this demodulator (a limiter was added) could be used to decrease the threshold of signal-to-noise improvement in large index FM. More recently, demodulators of this type were used in the Project Echo experiment for the same purpose [5]. In these demodulators the frequency deviation of the large index wave is compressed by using feedback, so that it may be passed through a narrow-band-pass IF filter before demodulation. Previous papers [3], [4] have advanced the theory that the system threshold occurs in the usual manner at the frequency detector of the feedback demodulator when noise peaks exceed the carrier, and is equal to that of a conventional FM demodulator having the same narrow-band-pass IF filter. This implies that, if the system were above threshold with no feedback applied, an unlimited signal-to-noise ratio could theoretically be obtained by increasing the frequency deviation of the carrier and the amount of feedback simultaneously. In practice it has been found that this behavior cannot be obtained, and it has also been found that the threshold can be different from the predicted value by orders of magnitude. The tendency has been to attribute these inconsistencies to the difficulties involved in obtaining a stable feedback system for large amounts of feedback. It will be shown in this paper that the more fundamental reason is that the threshold deteriorates as a function of the feedback action of the system. A mathematical analysis of the feedback demodulator when it is operating *below* threshold is a difficult and unsolved problem. The determination of where the threshold occurs, however, is a considerably less difficult problem, and it is this problem which is considered. In the past it has been suggested [4] that the threshold of the feedback demodulator could be eliminated by properly inserting a carrier in front of the frequency detector. In this paper it is shown that in a properly designed system the threshold would not be improved by carrier insertion. It is also shown that the feedback stability and the threshold of the feedback demodulator are both functions of modulation, but that by following the suggested procedure the effect can be almost entirely eliminated. The results of this paper allow one, for the first time, to design feedback demodulators which will extend the threshold in a predictable manner.

* Received by the IRE, August 11, 1961; revised manuscript received, October 31, 1961.
† Bell Telephone Laboratories, Inc., Holmdel, N. J.

309

I. The Concept of Frequency Feedback

The block diagram of the feedback demodulator is shown in Fig. 1. In order to understand its operation, imagine for the moment that the voltage controlled oscillator (VCO) is removed from the circuit and the feedback loop left open. Assume that a wide index FM wave is applied to the input of the mixer, and a second FM wave, from the same source but whose index is a fraction smaller, is applied to the VCO terminal of the mixer. The output of the mixer would consist of the difference frequency, since the sum frequency components are removed by the bandpass filter. The frequency deviation of the mixer output would be small, although the frequency deviation of both input waves is large, since the difference between their instantaneous deviations is small. Hence, the indexes of modulation would subtract and the resulting wave would have a smaller index of modulation. The reduced index wave may be passed through a filter, whose bandwidth need be only a fraction of that required for either large index wave, and frequency detected. It is now apparent that the second FM wave could be obtained by feeding the output of the frequency detector back to the VCO.

ing the band-pass IF filter is small compared to unity.[1] Under this condition the IF filter may be represented by its low-pass equivalent.[2] The VCO and frequency detector may be replaced by an ideal integrator and an ideal differentiator respectively. The mixer is a frequency subtractor. The open-loop transfer function $K_v K_f A_L(j\omega) H(j\omega)$ must, of course, satisfy the usual Nyquist and other stability criteria.[2] This eliminates any possibility of using a so-called "rectangular" IF filter. Let us define the closed-loop transfer function as the function which relates the phase of the wave generated by the VCO to the phase of the wave at the signal input to the mixer, *i.e.,* $\phi(j\omega)\phi_1(j\omega)$ in Fig. 2. *From linear feedback theory it follows that the closed-loop bandwidth is unavoidably larger than the open-loop bandwidth.* In later sections we shall see that this fact plays an important role in the determination of the threshold.

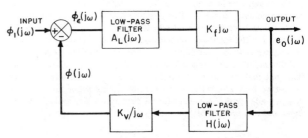

Fig. 2—Baseband analog of the feedback demodulator when the wave entering the IF band-pass filter has a modulation index much smaller than unity, *i.e.,* when $|\phi_\epsilon(t)| \ll 1$. $A_L(j\omega)$ is the low-pass equivalent of the band-pass IF filter $A(j\omega)$. $\phi_1(j\omega)$, $\phi_\epsilon(j\omega)$, $\phi(j\omega)$ and $e_0(j\omega)$ are the Fourier transforms of $\phi_1(t)$, $\phi_\epsilon(t)$, $\phi(t)$ and $e_1(t)$, respectively (shown in Fig. 1).

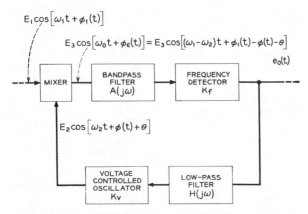

Fig. 1—Block diagram for a frequency feedback demodulator. K_f and K_v are the gain constants of the frequency detector and the VCO respectively. They relate radian frequency to voltage. $A(j\omega)$ and $H(j\omega)$ are the transfer functions of the band-pass IF filter and low-pass baseband filter in the feedback path respectively. Although the output is indicated as being immediately behind the frequency detector, it might equally well be considered to be any other point in the loop.

II. Small Index Response and Signal-to-Noise Ratio Above the Threshold

The "moving finger" or quasi-stationary behavior of the feedback demodulator is more or less obvious, and so our attention will be concentrated on the less obvious small index behavior. It is the small index response which may be used to determine the system stability, the effective or closed-loop noise bandwidth, the baseband signal-to-noise ratio and the threshold. The baseband analog of the demodulator, shown in Fig. 2, is valid as long as the modulation index of the wave enter-

It will now be shown that the signal-to-noise ratio of a feedback demodulator is the same as that of a conventional FM demodulator receiving the same signal and noise density if the carrier-to-noise power ratio is sufficiently large. Assume for the moment that there is no feedback, *i.e.,* that $K_v = 0$. Let the unmodulated carrier and noise at the input to the frequency detector be given by [6]

$$e(t) = \cos \omega_0 t + I_c(t) \cos \omega_0 t - I_s(t) \sin \omega_0 t$$

$$= R_e[1 + I_c(t) + j I_s(t)]\epsilon^{j\omega_0 t}$$

$$= R_e \sqrt{[1 + I_c(t)]^2 + I_s{}^2(t)}$$

$$\epsilon \left[j\omega_0 t + j \tan^{-1} \frac{I_s(t)}{1 + I_c(t)} \right],$$

---

[1] It should be noted that the modulation indexes of the waves at the VCO output and mixer signal input may be large. These waves have many sidebands for each baseband frequency component, but they are correlated in such a manner that when they are multiplied together in the mixer a cancellation process takes place which results in a carrier and a single pair of quadrature sidebands per baseband component.

[2] This will be discussed further in Section V and in Appendix C.

where $R_e(\ )$ stands for the real part of $(\ )$. $I_c(t)$ and $I_s(t)$ are random Gaussian variables in phase quadrature with each other, normalized by the carrier envelope. The last equation places in evidence the envelope and phase of the composite wave.[3] The constant frequency term ω_0 can be subtracted out by a balanced frequency detector or eliminated by filtering,. and so will be dropped. If the envelope of the noise is much smaller than the carrier envelope except for a negligible portion of time, *i.e.*, if

$$| I_c(t) + jI_s(t) | \ll 1,$$

then the phase and envelope of the composite wave are given by

$$\Psi_1(t) \overset{\circ}{=} I_s(t)$$

and

$$E_1(t) \overset{\circ}{=} 1 + I_c(t).$$

These expressions show that the composite wave at the input to the frequency detector is small index phase-modulated by the term $I_s(t)$, which is derived from the component of noise which is in phase quadrature with the carrier. The wave is also small index envelope-modulated by the term $I_c(t)$, which is derived from the component of noise which is in phase with the carrier. When feedback is applied, the VCO generates a wave which reduces the angle-modulation index of the wave in the IF, *i.e.*, the quadrature component of noise $I_s(t)$. Thus it is seen that, as long as the carrier-to-noise ratio is sufficiently large, the demodulator does not respond to the in-phase noise (envelope of the composite wave), but that it demodulates the quadrature noise in exactly the same fashion as it would demodulate signal modulation. Signal and quadrature noise are reduced in the same proportion by feedback, with the result that the baseband signal-to-noise ratio is independent of feedback. For large carrier-to-noise ratios the baseband signal-to-noise ratio of a feedback demodulator is then the same as that of a conventional FM demodulator [8].

III. Qualitative Discussion of the Threshold and Experimental Results

The conventional FM receiver accepts a band of noise at the input to its frequency detector equal to the bandwidth required by the large index transmitted wave. When noise peaks exceed the carrier for any significant portion of time, a threshold occurs. This threshold limits, for a given carrier power and noise density, the baseband signal-to-noise improvement obtainable by increasing the deviation of the transmitted wave. It was shown in Section II that the frequency feedback demodulator reduced the quadrature noise

[3] The phase and envelope of any general wave are uniquely defined by its pre-envelope [7].

$I_s(t)$ reaching the frequency detector by the use of feedback. While the in-phase noise $I_c(t)$ was not reduced by feedback, it was also not detected by the ideal frequency detector and consequently was not fed around the loop. Feedback then decreases the envelope of the noise at the input to the frequency detector. Based upon this observation, it has been *wrongly* suggested [3], [4] that the threshold is determined by the envelope *of* the noise at the frequency detector and hence should not be degraded by feedback. If this were true, the baseband signal-to-noise ratio could (in theory) be increased indefinitely by increasing the deviation of the transmitted wave and the amount of feedback simultaneously, as long as the system were above threshold with no feedback applied. Unfortunately, we shall see later that the threshold is degraded by the feedback action of the demodulator because of effects which are not at all obvious from the above oversimplified argument.

Let us study briefly the reasons why the feedback demodulator is able to extend the threshold. The conventional large index FM receiver fails to use a very important piece of *a priori* information, that even though the carrier frequency will have large frequency deviations, its rate of change will be at the baseband rate. Both the feedback FM demodulator and the phase-locked loop use this *a priori* information to extend the threshold. They are essentially "tracking filters" which can track only the slowly varying frequency of large index waves, and they consequently respond to only a narrow band of noise centered about the instantaneous carrier frequency. We would expect intuitively that the threshold of an ideal tracking filter would be equal to that of any other frequency detector receiving the same carrier and narrow band of noise. The bandwidth of noise to which the feedback demodulator and phase-locked loop respond is precisely the band of noise which the VCO tracks. The noise bandwidth of both systems is that of the closed-loop response function defined in Section II. While the thresholds of the phase-locked loop and feedback demodulator occur because of the same basic mechanism, the details by which they occur are, of course, different. Let us concentrate now on the feedback demodulator. Assume for the moment that the feedback loop is open and that the demodulator input consists of an unmodulated carrier wave and a band of Gaussian noise. The noise can be separated into components which are in phase and in phase quadrature with the carrier, as discussed in Section II. If the carrier-to-noise ratio at the input to the frequency detector is large, then the composite wave at that point is small index phase-modulated by the quadrature noise and small index envelope-modulated by the in-phase noise. For the purpose of this paper the threshold will be said to occur when the phase of the composite wave at the input to the frequency detector contains a significant amount of noise in addition to that derived from the quadrature noise, *i.e.*, in addition to the noise predicted by the FM improvement formula.

Now close the feedback loop. The baseband noise angle modulates the VCO. If the root-mean-square phase deviation of the VCO wave is small, the spectrum consists primarily of a carrier "spike" and the first-order spectral sideband zone. These first-order sidebands are in phase quadrature with the VCO carrier spike. The mixer forms the product of the VCO wave with the demodulator input carrier and noise. The product of the VCO carrier spike and incoming carrier yields a new carrier in the IF. The product of the VCO carrier spike and incoming quadrature and in-phase noise terms yields new quadrature and in-phase noise terms in the IF. The quadrature noise becomes angle noise and the in-phase noise becomes envelope noise. The product of the quadrature term (first-order sidebands) of the VCO with the incoming carrier yields a second quadrature term in the IF which tends to cancel the first quadrature term. It is because of this cancellation that noise in the baseband is reduced by feedback. It is also this cancellation that allows the VCO wave to have a frequency deviation which may be wider than the IF filter bandwidth. If the products mentioned so far are the only ones of any significance, the system behaves in a linear fashion. Two products remain, however. The product of the quadrature term of the VCO wave with the incoming quadrature and in-phase noise terms yields additional angle and envelope noise. *The threshold occurs when this additional angle noise becomes significant.* This happens when the root-mean-square phase deviation of the VCO wave caused by noise is no longer small compared to unity. The exact relations are discussed quantitatively in Section IV. Experimentally it is found that, soon after the additional noise becomes noticeable, noise impulses appear in the baseband. Their onset serves as a convenient measure of the threshold in practical systems. They appear suddenly and increase rapidly in number as the carrier-to-noise ratio decreases. They first appear when the root-mean-square phase deviation of the VCO wave (caused by noise) is $\phi_{\mathrm{rms}} = 1/3.11$ radian, *i.e.*, approximately one-third of a radian. They are apparently the result of higher-order nonlinearities which occur in this region.

The threshold should be expressed in terms of the input carrier-to-noise ratio in the closed-loop noise bandwidth in order for its full implication to be appreciated. This is easily done by using the small index baseband analog shown in Fig. 2. The mean-square phase of the wave generated by the VCO, ϕ_{rms}^2, can be obtained by integrating the product of the input phase spectral power density and the square of the absolute value of the closed-loop response function. Above threshold the input phase spectral power density caused by noise is equal to the spectral power density of the normalized quadrature noise. Thus

$$\phi_{\mathrm{rms}}^2 = \sigma_s^2 \int_{-\infty}^{+\infty} \left| \frac{K_v K_f A_L(j\omega) H(j\omega)}{1 + K_v K_f A_L(j\omega) H(j\omega)} \right|^2 df,$$

where σ_s^2 is the spectral power *density* of the normalized quadrature noise $I_s(t)$, which is also equal to the total input normalized white noise density. The closed-loop noise bandwidth[4] is

$$B_c = \left[\frac{1 + K_v K_f}{K_v K_f} \right]^2 \int_{-\infty}^{+\infty} \left| \frac{K_v K_f A_L(j\omega) H(j\omega)}{1 + K_v K_f A_L(j\omega) H(j\omega)} \right|^2 df,$$

and the carrier-to-noise ratio in front of the mixer in a bandwidth equal to the closed-loop noise bandwidth is $\rho = 1/2 B_c \sigma_s^2$. Consequently, the input carrier-to-noise ratio in a bandwidth equal to the closed-loop noise bandwidth is given, in terms of the mean-square phase of the VCO wave, by

$$\rho = \frac{1}{2\phi_{\mathrm{rms}}^2} \left[\frac{K_v K_f}{1 + K_v K_f} \right]^2.$$

As determined by experiment, the threshold occurs at

$$\rho_T = \frac{1}{2(1/3.11)^2} \left[\frac{K_v K_f}{1 + K_v K_f} \right]^2 = 4.8 \left[\frac{F-1}{F} \right]^2, \quad (1)$$

where $F = 1 + K_v K_f =$ amount of feedback or frequency compression. For large feedback $\rho_T = 4.8$ or 6.8 db.

Let us dwell upon the full implications of (1), for they are probably the most important results in this paper. This equation tells us where the *feedback threshold* occurs in the demodulator as a function of the feedback factor and the input carrier-to-noise ratio in a bandwidth equal to the closed-loop noise bandwidth of the system. Recall that the development of this equation assumed that the system was above threshold on an open-loop basis. Therefore (1) must be used only after assurance that this condition is satisfied. For instance, the result $\rho_T = 0$ when $F = 1$ means only that a system without feedback does not have a threshold produced by feedback. *In order for a feedback demodulator to be above threshold, it must be above both the open-loop and the feedback thresholds independently.* Observe that the closed-loop noise bandwidth is not uniquely related to the IF filter bandwidth, baseband bandwidth or the bandwidth required by the large index wave in front of the mixer. However, once the open-loop transfer function $K_v K_f A_L(j\omega) H(j\omega)$ is specified, the closed-loop noise bandwidth is determined uniquely. It will always be larger than the open-loop bandwidth, and for large feedback it will be much larger. Consequently, if the system were designed by placing all the selectivity in the IF filter, then the carrier-to-noise ratio in the IF band in front of the frequency detector would be much larger than that in a bandwidth equal to the closed-loop noise bandwidth (by the ratio of the bandwidths). As a result, the system threshold would be caused entirely by the feedback threshold.

---

[4] Notice that power is assumed to reside in both positive and negative frequencies, and that the closed-loop bandwidth is two sided.

Since the closed-loop noise bandwidth does not depend on whether the filtering is done at IF or baseband or a combination of both, it is immediately clear how to improve the performance of the demodulator. First, note that full feedback should be maintained over all baseband frequencies to be transmitted so that the frequency deviation in the IF filter is fully compressed. With this restriction, the closed-loop noise bandwidth should be as small as possible for a given amount of feedback, in order to minimize the feedback threshold. The filtering should be proportioned so that the IF filter is as wide as possible without allowing the open-loop threshold to predominate. This allows the largest possible frequency deviation in the IF and therefore yields the largest possible baseband signal-to-noise ratio for a fixed system threshold carrier-to-noise ratio. The remaining filtering required to realize the desired closed-loop response function is done at baseband in the feedback path.

The concepts discussed in this Section have been thoroughly tested experimentally. Demodulators were deliberately designed to have the feedback threshold predominate. A typical set of data is shown in Table I. The threshold was defined as the point where the noise impulses first appear (actually, in order to obtain a consistent measuring point, the value where the impulses were occurring at an average rate of one per second was used). The closed-loop noise bandwidths were varied by using different IF filter shapes and/or by placing different filters in the baseband feedback path. The IF filter 3-db bandwidths were 6 to 7 kc in all cases. The mean value is $\phi_{rms}=1/3.11$. All values fall within a spread of pulse or minus 7 per cent of the mean value, which is within the estimated accuracy of the measurements. Although this data is for a 3-kc baseband system, C. L. Ruthroff (to be published) has obtained similar results for a 1-Mc baseband system.

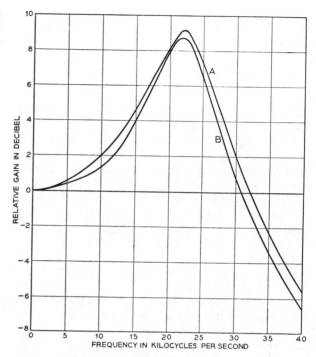

Fig. 3—Measured closed-loop system response. Curve A: 6-kc single-pole IF filter, 15-kc single-pole baseband filter. Curve B: 30-kc single-pole IF filter, 3-kc single-pole baseband filter.

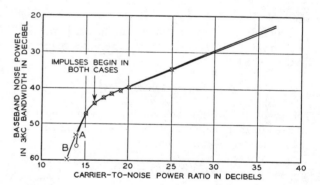

Fig. 4—Plot of the relative baseband noise power in a 3-kc bandwidth vs the input carrier-to-noise power ratio in an arbitrary but fixed bandwidth for the two cases in Fig. 3. The input noise was white for all practical purposes.

TABLE I

THE ROOT-MEAN-SQUARE PHASE OF THE VCO WAVE AT THE THRESHOLD OF TWO FEEDBACK DEMODULATORS FOR DIFFERENT AMOUNTS OF FEEDBACK AND DIFFERENT CLOSED-LOOP NOISE BANDWIDTHS

	$F=1+K_vK_f$ (Feedback) DB	B_c (Closed-Loop Noise Bandwidth) KC	ϕ_{rms} Radians
Receiver No. 1	20	172	1/2.89
	20	147	1/2.92
	20	39	1/3.20
	14	68	1/3.20
Receiver No. 2	20	45	1/3.27
	30	135	1/3.20

A typical set of experimental curves in Figs. 3 and 4 illustrate that the IF bandwidth of a feedback demodulator designed using the old theory can indeed be greatly increased without increasing the over-all system threshold. The closed-loop frequency response functions for the two different filter arrangements are shown in Fig. 3. In case A the IF filter was a single pole with a 6-kc, 3-db bandwidth, and the baseband filter in the feedback path was a single pole with a 15-kc 3-db bandwidth. In case B the 3-db IF bandwidth was increased to approximately 30 kc, and the 3-db baseband bandwidth was decreased to approximately 3 kc. Curves of the output baseband noise in a 3-kc bandwidth vs the input carrier-to-noise ratio in an arbitrary but fixed bandwidth are shown in Fig. 4. It can be seen that the system threshold was the same in both cases. The important point is that while the thresholds were equal, the wideband IF system could receive an FM signal with an index approximately 5 times larger than that of the narrow IF system, yielding an increase in signal-to-noise ratio of 25.[5]

IV. Analysis of the Threshold (Carrier Unmodulated)

A mathematical analysis of the feedback demodulator when it is operating *below* threshold is a difficult and unsolved problem. The determination of where the threshold occurs, however, is a considerably less difficult problem, and it is this problem which will now be considered. We have seen that when the feedback demodulator is above threshold, the baseband noise is derived entirely from the quadrature component of the input noise. This condition must be satisfied in order for the FM improvement formulas to be valid. The threshold will be assumed to occur when a significant amount of additional noise appears in the baseband. A series approach will be used to determine where this additional noise becomes significant. In order to assure the validity of the result, we will then have to check the envelope in front of the frequency detector to make sure that the probability of zero crossings is negligibly small.

Let the carrier and noise at the input to the mixer be given by

$$e_1 = \cos \omega_1 t + N_c(t) \cos \omega_1 t - N_s(t) \sin \omega_1 t,$$

where $N_c(t)$ and $N_s(t)$ are Gaussian random variables in time quadrature with each other, normalized with respect to the carrier magnitude. Let the wave generated by the VCO be given by

$$e_2 = 2 \cos [\omega_2 t + \phi(t) + \theta].$$

The problem is to find the conditions which must be satisfied so that the phase of the wave entering the frequency detector will be essentially

$$\Psi_1(t) \doteq \overline{N}_s(t) - \bar{\phi}(t),$$

where $\overline{N}_s(t)$ and $\bar{\phi}(t)$ represent the output of the baseband equivalent of the narrow-band IF filter to $N_s(t)$ and $\phi(t)$ respectively. If only the difference frequency terms are retained, the mixer output is

$$e_3(t) = e_1 e_2 = [\cos \phi(t) + N_c(t) \cos \phi(t)$$
$$+ N_s(t) \sin \phi(t)] \cos (\omega_0 t - \theta)$$
$$- [-\sin \phi(t) - N_c(t) \sin \phi(t)$$
$$+ N_s(t) \cos \phi(t)] \sin (\omega_0 t - \theta),$$

where $\omega_0 = \omega_1 - \omega_2$. When this wave is passed through the IF filter and $\sin \phi(t)$ and $\cos \phi(t)$ are expressed in series form, then

[5] The system parameters should not be taken as optimum in any sense. This design served only to verify theory and nothing more.

$$e_3(t) = \left\{ \sum_{m=0}^{\infty} (-1)^m \frac{\overline{\phi^{2m}}}{(2m)!} + \sum_{m=0}^{\infty} (-1)^m \frac{\overline{N_c \phi^{2m}}}{(2m)!} \right.$$
$$\left. + \sum_{m=0}^{\infty} (-1)^m \frac{\overline{N_s \phi^{2m+1}}}{(2m+1)!} \right\} \cos (\omega_0 t - \theta)$$
$$- \left\{ - \sum_{m=0}^{\infty} (-1)^m \frac{\overline{\phi^{2m+1}}}{(2m+1)!} - \sum_{m=0}^{\infty} (-1)^m \frac{\overline{N_c \phi^{2m+1}}}{(2m+1)!} \right.$$
$$\left. + \sum_{m=0}^{\infty} (-1)^m \frac{\overline{N_s \phi^{2m}}}{(2m)!} \right\} \sin (\omega_0 t - \theta).$$

The envelope and phase of a wave in the form $A(t) \cos \omega_0 t - B(t) \sin \omega_0 t$ are given respectively by

$$E_1(t) = \sqrt{A^2(t) + B^2(t)}$$

and

$$\Psi_1(t) = \tan^{-1} \frac{B(t)}{A(t)}.$$

Neglecting all third-order and higher terms, we have

$$\Psi_1(t) \doteq \overline{N}_s(t) - \bar{\phi}(t) - \overline{N_c(t)\phi(t)}$$
$$- \overline{N_c(t)\overline{N}_s(t)} + \overline{N_c(t)\bar{\phi}(t)}$$

and

$$E_1(t) \doteq 1 + \overline{N}_c(t) + \overline{N_s(t)\phi(t)} - \tfrac{1}{2}\overline{\phi^2(t)} + \tfrac{1}{2}\overline{N}_c^2(t)$$
$$+ \tfrac{1}{2}\overline{N}_s^2(t) + \tfrac{1}{2}\bar{\phi}^2(t) - \overline{N}_s(t)\bar{\phi}(t). \qquad (2)$$

Pay particular attention to the difference between terms which were multiplied and then filtered, such as $\overline{N_s(t)\phi(t)}$, and terms which were filtered and then multiplied, such as $\overline{N}_s(t)\bar{\phi}(t)$. The relative importance of these second-order terms will now be discussed. Assume that there is no feedback and that the resulting open-loop system is above threshold. $\phi(t)$ is then zero, and only the first and fourth terms remain in the expression for the phase $\Psi_1(t)$. The first term represents the "above threshold noise" or quadrature noise. The fourth term is negligible because the system is assumed above threshold. Now assume that feedback is applied, and that the closed-loop system is also above threshold. If there is a significant amount of feedback, $\bar{\phi}(t)$ approaches $\overline{N}_s(t)$. The first two terms represent the "above-threshold" noise of the feedback demodulator. The fifth term, which is introduced by feedback, is almost equal to and tends to cancel the fourth term, which is present with or without feedback. They may both be neglected. The third term, which is also introduced by feedback, is considerably larger than the other second-order terms. It represents phase noise which adds to the above-threshold noise, *i.e.*, the first two terms. The threshold is assumed to occur when this additional noise becomes significant. The third term is a distortion term since it is a function of the "signal" $\phi(t)$. In the analysis of feedback systems having distortion, it is often possible to represent the complete system as a linear one in which the distortion components are introduced by generators [9]. This approach is valid as long as the fed-back distortion components are small; *i.e.*, all significant distortion must be caused by the signal. An important aspect of such an analysis is that one can tell to a fair degree of approximation where the results of the linearized model cease to be accurate. This general approach proves to be a useful one in the present problem. Let $\phi_s(t)$ be the VCO phase contributed by the quadrature noise $N_s(t)$, and let $\phi_d(t)$ be the phase contributed by the distortion component. Then if $\phi_s(t) \gg \phi_d(t)$, the significant distortion would be caused by $\phi_s(t)$, and the linearized baseband model shown in Fig. 5 is valid. The quadrature noise and distortion are introduced into the system in parallel and are treated identically by feedback. Consequently $\phi_s(t) \gg \phi_d(t)$ if the spectral density of the distortion input $\phi_s(t)N_c(t)$ is much smaller than the spectral density of the quadrature term $N_s(t)$. In Appendix A it is shown that for large feedback the ratio of the spectral density of $\phi_s(t)N_c(t)$ to the density of $N_s(t)$ is equal to the mean-square value of $\phi_s(t)$, *i.e.*, ϕ_{rms}^2. Thus, the linear model in Fig. 5 is valid as long as $\phi_{\mathrm{rms}}^2 \ll 1$. In this region of operation the phase term in (2) becomes approximately

$$\Psi_1(t) \doteq \{\overline{N}_s(t) - \bar{\phi}_s(t)\} - \{\overline{N_c(t)\phi_s(t)} - \bar{\phi}_d(t)\},$$

where the second bracketed term is much smaller than the first. This second bracketed term represents the excess noise which is not derived from the quadrature input noise. One would expect intuitively that $\phi_{\mathrm{rms}}^2 \doteq 1/10$ would be close to the upper limit of validity for the model of Fig. 5, for much beyond that the non-linear action on the feedback distortion would produce sufficient additional distortion to result in a cumulative situation. This result agrees quite well with the measured threshold value of $\phi_{\mathrm{rms}} = 1/3.11$.

In order for the results in the above paragraph to be correct, it is necessary that the number of zero crossings of the envelope be negligibly small for $\phi_{\mathrm{rms}}^2 \leq 1/10$. Calculation of the actual probability of zero crossings is prohibitively difficult. In lieu of this calculation it will be shown that the root-mean-square fluctuation of the envelope is negligibly small compared to its mean value. The first term in the envelope expression of (2) is the normalized carrier envelope. The second, fifth, and sixth terms represent the envelope noise when the loop is open. Since the system is above threshold on an open-loop basis these terms are sufficiently small. The last three terms tend to cancel each other and can be neglected if $\bar{\phi}(t) \doteq \overline{N}_s(t)$. Demonstration that the third and fourth terms are small compared to the normalized carrier envelope, *i.e.*, unity, requires the specification of a particular closed-loop transfer function. We shall select the two-pole low-pass function, having a damping ratio less than unity, because of its practical importance. It is felt that the results are typical of most other transfer functions of interest. From (2) it can be seen that the third and fourth terms are formed by multiplying two variables together and then passing the result through the low-pass equivalent of the IF filter. In Appendix B it is shown that the total power lying within a rectangular band of frequencies equal to *twice* the noise bandwidth of the closed-loop response is $P = 11/4\phi_{\mathrm{rms}}^4$. Since the power passed by the IF filter is smaller still,[6] the envelope noise of the composite wave at the input to the frequency detector is small compared to unity for $\phi_{\mathrm{rms}}^2 \leq 1/10$.

In summary, the mean-square phase deviation of the VCO wave ϕ_{rms}^2 caused by noise must be small compared to unity in order for the noise appearing in the baseband to be primarily derived from the quadrature input noise, *i.e.*, predictable from the FM improvement formula. This, then, is also the condition which must be satisfied for a system to be above the feedback threshold.

V. Modulation

In Section IV the carrier was assumed to be unmodulated. Here it is shown that the system *stability*, and hence the closed-loop bandwidth and feedback threshold, are all functions of *modulation*. It is then shown that the dependence may be virtually eliminated by proper design.

Assume that the carrier frequency is varying slowly so that it may be represented quasi-statically. Noise can then be separated into components in phase and in quadrature with the carrier. The quadrature component is the angle noise, and it is the response of the system to this noise that is of interest. The nature of this re-

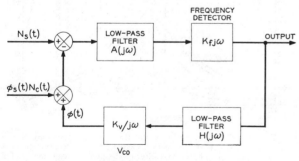

Fig. 5—Block diagram for representing the effect of excess phase noise or "noise distortion."

---

[6] Notice that, for a fixed closed-loop noise bandwidth, increasing the resonant peak of the closed-loop response function allows more envelope noise to pass through the IF filter.

sponse may be determined from the small index response of the system. It is desired to find a transfer function which will relate the signal component of the instantaneous phase at the output of a mistuned high-Q band-pass filter to the phase of the small index modulated carrier applied to its input. The mistuning, of course, represents the deviation of the quasi-statically varying carrier frequency caused by signal modulation. Distortion will not be considered since it does not affect stability. In Appendix C[7] it is shown that the baseband analog of the mistuned high-Q band-pass filter is

$$Y(j\omega) = \frac{1}{2R(\omega_d)} \left[Y_x(j\omega) + Y_x^*(-j\omega) \right],$$

where

$$Y_x(j\omega) = Y_L[j(\omega + \omega_d)]\epsilon^{-j\theta(\omega_d)},$$

$$Y_L[j(\omega + \omega_d)] = R(\omega + \omega_d)\epsilon^{j\theta(\omega+\omega_d)},$$

> ω_d = difference between the carrier frequency and the filter center frequency,
>
> $Y_L(j\omega)$ = conventional low-pass equivalent of the band-pass circuit.

In Figs. 6 and 7 are shown magnitude and phase plots for a single-pole and Bode-type filter [10]. The conventional low-pass equivalent is given by the $\omega_d = 0$ curve. Modulation manifests itself as a variation in ω_d, and so its effect on the phase margin of the system is very apparent. A slow roll-off filter such as the single-pole filter is obviously required in the IF. The single-pole filter introduces only a slight excess phase shift for carrier frequency deviations, *i.e.*, variations in ω_d, within the 3-db bandwidth. Consequently, only slight variations occur in the closed-loop bandwidth and therefore in the feedback threshold.

This instability can have real nuisance value because of the likelihood that it will be overlooked in routine tests on experimental systems. The reason for this is that the system is unstable only on the the peaks of the modulation. Further, the oscillations are generally at a frequency which is high compared to the baseband bandwidth and consequently do not appear *after* the baseband filter which usually follows FM demodulators to remove noise lying outside of the baseband bandwidth. The effect which is noticed is a degradation of the threshold as a function of modulation. The carrier frequency in the IF breaks into oscillation at the modulation peaks. The IF filter behaves like a slope circuit and produces envelope modulation on the carrier, making it much easier for noise peaks to exceed the carrier envelope in front of the frequency detector and produce the threshold.

[7] The equation developed can also be obtained from a much more general analysis of distortion presented in an unpublished memorandum in 1954 by S. Doba.

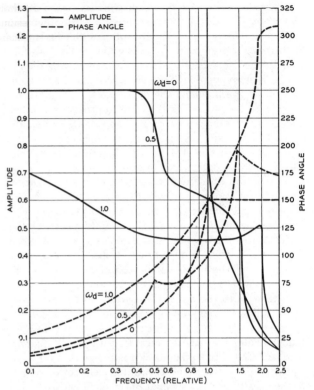

Fig. 6—Amplitude and phase of the transfer function relating the signal component of the instantaneous phase at the output of a Bode-type high-Q band-pass filter to the phase of the small index modulated carrier applied to its input. The carrier is mistuned by ω_d.

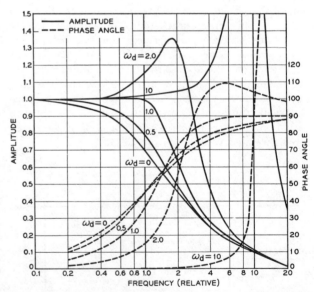

Fig. 7—Amplitude and phase of the transfer function relating the signal component of the instantaneous phase at the output of a single-pole high-Q band-pass filter to the phase of the small index modulated carrier applied to its input. The carrier is mistuned by ω_d.

The optimum IF filter has not yet been determined; however, it is felt that a single-pole filter will be close. Additional filtering is done at baseband in order to accomplish the following objectives:

1) Full feedback should be maintained over all baseband frequencies to be transmitted, so that the frequency deviation in the IF is fully compressed.
2) The closed-loop noise bandwidth should be as small as possible for a given amount of feedback in order to keep the feedback threshold at a minimum.
3) The IF filter should be as wide as possible without allowing the open-loop threshold to predominate. This allows maximum frequency deviation in the IF and thus maximum baseband signal-to-noise ratio.

VI. Sample Design

The material presented above will now be illustrated by means of an example. This example is not offered as THE way to design a system; it is meant only to consolidate thinking. In particular, we shall use the experimental feedback threshold value given in Table I. In practice the "operating threshold" would depend upon the intended use, *i.e.*, television, data, multiplex telephony, etc.

Assume the following data:

$f_b =$ 3 Mc = baseband bandwidth,
$\Delta f = \mp$ 60 Mc = peak frequency deviation.

The design will use a Bode-type [10] open-loop characteristic with a phase margin of 50°. It can be seen from Fig. 8 that this phase margin yields a closed-loop noise bandwidth very close to the minimum value for values of feedback less than 30 db. The characteristic will be realized in two parts: A single-pole filter in the IF and the rest at baseband. Now we must determine the required amount of feedback. Fig. 8 illustrates graphically that we must use as little as possible in order to keep the closed-loop bandwidth minimized. However, we must use enough to compress the deviation of the IF wave sufficiently for it to pass through the single-pole IF filter. The IF filter must be narrow enough to prevent the open-loop threshold from being dominant. Thus, we must know the feedback threshold, which can not be calculated until the amount of feedback is determined. We see that we have gone in a full circle. The result is that we consider only the baseband bandwidth as being known, estimate a value of feedback, and then solve for the frequency deviation. If we do not obtain the required deviation, we make another estimate and try again.

Let us choose 20 db of feedback for our first estimate, *i.e.*, $F = 10$. The closed-loop bandwidth is, from Fig. 8, $B_c \approx 11.6\ f_b = 34.8$ Mc. From (1), the carrier-to-noise

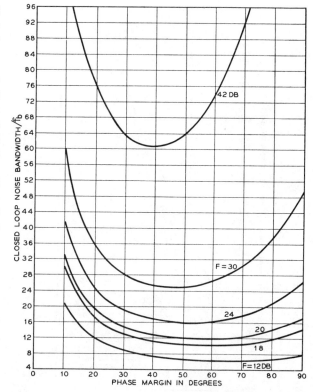

Fig. 8—The closed-loop noise bandwidth of systems having a Bode-type open-loop characteristic for different amounts of feedback and phase margin.

ratio at the input to the mixer in this bandwidth at the feedback threshold is

$$\rho_T = 4.8 \left[\frac{9}{10} \right]^2 = 3.92 \quad \text{or} \quad 5.94 \text{ db.}$$

The single-pole filter in the IF must have a noise bandwidth small enough so that the open-loop threshold does not predominate. The threshold of a frequency detector depends upon the ratio of the input noise bandwidth to baseband bandwidth. See Fig. 9.[8] Therefore we are forced to guess and check our guess.

Guess at a value of 6 for the ratio of the IF filter noise bandwidth to baseband bandwidth. From Fig. 9 it is seen that for $B/f_b = 6$ the signal-to-noise ratio is proportional to the carrier-to-noise ratio for all carrier-to-noise ratios greater than 8.5 db. 8.5 db will be used as the open-loop threshold. Now for the check. The noise density which yields a carrier-to-noise ratio of 5.94 db in a 34.8-Mc bandwidth (closed-loop noise

---

[8] These curves were derived for an ideal frequency detector with a carrier and a narrow band of Gaussian noise at its input. Since in the feedback demodulator the quadrature noise has been reduced by feedback, the curves do not apply exactly. However, experience has shown that they may be used to estimate the frequency detector threshold for systems typified by those in Table I.

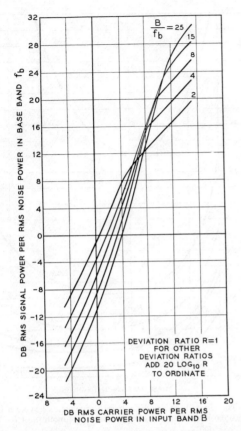

Fig. 9—Threshold curves. (Presented by F. J. Skinner in an unpublished memorandum in 1954; derived from the results of Rice [6]. The threshold phenomenon generates a more or less flat power spectrum which is superimposed on the parabolic power spectrum (triangular voltage spectrum) at baseband. Thus, the threshold effects are noticed sooner in the lower frequencies, *i.e.*, for large B/f_b ratios. For similar curves see Stumpers [11].

bandwidth) also yields a carrier-to-noise ratio of 8.5 db in a 19.35-Mc bandwidth (IF noise bandwidth). Using 19.35 Mc for the IF bandwidth means that B/f_b is slightly greater than 6 but not enough to matter. The 3-db bandwidth of a single-pole filter with a noise bandwidth of 19.35 Mc is $B = 19.35/1.57 = 12.3$ Mc. The compressed wave could have an index[9] of 2 in this filter and not have sufficient distortion to be detectable on an oscilloscope (this distortion is reduced by feedback) [1]. The peak frequency deviation in front of the mixer would be

$$\Delta f = 10 \times 2 \times 3 = 60 \text{ Mc.}$$

From Fig. 9 a conventional demodulator receiving the same transmitted wave would have a threshold greater than 12 db in the required 120-Mc rf bandwidth. Thus,

[9] In practice one could not realize a closed-loop noise bandwidth as small as this example suggests. We have selected a rather idealized open-loop characteristic and have completely neglected excess linear phase shift caused by parasitics. It is for closed-loop noise bandwidths much greater than the baseband bandwidth where the ability to widen the IF bandwidth pays dividends.

the feedback demodulator has a threshold which is better by at least

$$12 - 5.94 + 10 \log_{10} \frac{120}{34.8} \doteq 11.5 \text{ db.}$$

VII. CARRIER INSERTION AND THE PHASE-LOCKED LOOP

In amplitude modulation systems a locally generated sine wave, whose phase and frequency are the same as the carrier, may be inserted to extend the threshold indefinitely, at least theoretically [12]. It has been proposed [4] that such a scheme be incorporated with the feedback demodulator. If the large index transmitted wave were phase modulated instead of frequency modulated, feedback could be used to reduce the deviation in the IF until only the first-order sidebands were important. A locally generated sine wave of the proper phase and frequency could then be inserted to eliminate the threshold of the phase detector. It will be shown in the next few paragraphs that the thresholds of the phase-locked-loop and the feedback demodulator with carrier insertion are equal to or poorer than that of a feedback demodulator without carrier insertion. The reason for this stems from the fact that in the first two cases the instantaneous phase deviation of the wave behind the mixer (multiplier) must be small compared to unity in order to prevent severe distortion. For large transmitter modulation indexes this requires more feedback than would be required by a straightforward feedback demodulator, and consequently results in a larger closed-loop noise bandwidth. The larger closed-loop noise bandwidth yields a poorer feedback threshold. First let us consider the carrier insertion scheme. Let the carrier be inserted directly after the mixer. The combined output is then

$$\begin{aligned} e_3(t) = &\{ C_2 + \cos\phi(t) + N_c(t)\cos\phi(t) \\ &+ N_s(t)\sin\phi(t) \} \cos(\omega_0 t - \theta) \\ &- \{ -\sin\phi(t) - N_c(t)\sin\phi(t) \\ &+ N_s(t)\cos\phi(t) \} \sin(\omega_0 t - \theta), \quad (3) \end{aligned}$$

where $C_2 \cos(\omega_0 t - \theta)$ is the inserted carrier, normalized by the incoming carrier magnitude.

The envelope and phase may be found exactly as in the development of (2). They are

$$\begin{aligned} E_1(t) = 1 + &\frac{1}{1 + C_2} \{ \overline{N_c}(t) + \overline{N_s(t)\phi(t)} - \tfrac{1}{2}\overline{\phi^2(t)} \} \\ + &\frac{1}{(1 + C_2)^2} \{ \tfrac{1}{2}\overline{N_s}^2(t) + \tfrac{1}{2}\overline{N_c}^2(t) + \tfrac{1}{2}\overline{\phi}^2(t) \\ &- \overline{N_s(t)\phi(t)} \}, \end{aligned}$$

$$\begin{aligned} \Psi_1(t) = &\frac{1}{1 + C_2} \{ \overline{N_s}(t) - \overline{\phi}(t) - \overline{N_c(t)\phi(t)} \} \\ &- \frac{1}{(1 + C_2)^2} \{ \overline{N_c}(t)\overline{N_s}(t) - \overline{N_c(t)\phi(t)} \}. \end{aligned}$$

The terms proportional to $1/(1+C_2)^2$ tend to cancel just as did the corresponding terms in (2).

There are two points to notice:

1) The inserted carrier reduces the percentage envelope modulation by the factor $1+C_2$, decreasing system susceptibility to envelope noise.

2) The above-threshold phase noise $\overline{N}_s(t) - \bar{\phi}(t)$ and the "threshold-producing" phase noise $\overline{N_c(t)\phi(t)}$ are reduced by the same factor $1+C_2$, as modulation would be if it were present. This reduction produces the same effect on the baseband as a decrease in the gain constant of the phase detector. Hence, the effect is simply to reduce the amount of feedback. The feedback threshold of this demodulator is not affected by carrier insertion.

Now consider the phase-locked loop. In the phase-locked loop the IF frequency ω_0 is zero and $\theta = \pi/2$. The output of the mixer (multiplier) is a baseband signal proportional to the quadrature component of (3). Therefore, there is no need for a separate phase detector or carrier insertion. Notice that the quadrature term is $Q(t) \doteq N_s(t) - \phi(t) - N_c(t)\phi(t)$. Again the threshold producing noise is present and undiminished, yielding the feedback threshold.

In summary, both the phase-locked loop and the feedback demodulator with carrier insertion require the instantaneous phase deviation of the wave behind the mixer (multiplier) to be small compared to unity. For large transmitted modulation indexes this requires more feedback on the highest baseband frequency than would be required by a straightforward feedback demodulator, and consequently a larger closed-loop noise bandwidth is obtained. The larger closed-loop noise bandwidth yields a poorer feedback threshold.

VIII. Conclusions

The over-all threshold of the frequency feedback demodulator was determined. The results show that two thresholds can occur. The first occurs in the usual manner at the frequency detector; the second occurs when the root-mean-square phase of the VCO wave (caused by noise) is no longer small compared to unity. Experimentally it was found that the second threshold occurs when $\phi_{rms} = 1/3.11 \pm 7$ per cent radians. This was related quantitatively to the carrier-to-noise ratio in the closed-loop noise bandwidth. It was pointed out that for maximum baseband signal-to-noise ratio the feedback threshold and open-loop threshold should occur in the same neighborhood. It was also shown that sharp cutoff filters in the IF can cause a serious loss in threshold under modulated conditions, and for this reason slow roll-off filters such as the single-pole filter should be used. Realizing these facts can improve the baseband signal-to-noise ratio by one or two orders of magnitude, for a given carrier power and noise density, over that which would be obtained in systems designed be previous theories [3], [4]. It was shown that the phase-locked loop and the feedback demodulator with carrier insertion have a poorer threshold than the standard feedback demodulator for large index modulation.

The author would like to point out several problems that need further study. The "operational" threshold, which determines the limit of usefulness of a demodulator, varies according to the service for which it is used, *i.e.*, multiplex telephony, television, etc. This threshold needs to be determined subjectively for each of these services. It has been pointed out in this paper that feedback demodulators should be designed to have the feedback and open-loop thresholds occur in the same neighborhood. Experimental results for the various services are also needed here in order to be more specific. Intermodulation distortion produced by the IF filter should be studied to determine the maximum allowable frequency deviation. Last but not least, it would be useful if a general analysis of the behavior of the feedback demodulator below threshold were available. However, the fact that the threshold of an ideal frequency detector, in the absence of feedback, is not understood to a satisfactory degree gives some insight into the difficulties.

Appendix A

It is desired to show that the ratio of the spectral power density of the term $\phi_s(t)N_c(t)$ to the density of $N_s(t)$ is equal to the mean-square value of $\phi_s(t)$ for large feedback. $N_c(t)$ and $N_s(t)$ are the in-phase and quadrature components of the input noise, normalized by the carrier magnitude. $\phi_s(t)$ is the component of the phase of the VCO wave derived from the quadrature component of the input noise.

The spectral density of the term $\phi_s(t)N_c(t)$ is found by determining the covariance and then taking the Fourier transform. The covariance is

$$R_1(\tau) = E[\phi_s(t_1)N_c(t_1)\phi_s(t_2)N_c(t_2)],$$

where $E( )$ is the statistical average or expectation of $( )$. Now $\phi_s(t)$ is derived from $N_s(t)$ by linear filtering; it follows that $\phi_s(t)$ and $N_c(t)$ are independent processes as long as the RF and IF filters are symmetrical [13]. Thus

$$R_1(\tau) = R_{\phi_s}(\tau)R_{N_c}(\tau),$$

where R_{ϕ_s} and R_{N_c} are the covariances of ϕ_s and N_c, respectively. The spectral density can now be found by convolving the individual densities.

$$S_1(f) = \int_{-\infty}^{+\infty} S_{N_c}(f_1 - f)S_{\phi_s}(f_1)df_1.$$

We are interested in the power density only within the bandwidth of the closed-loop response function. If the feedback is equal to or greater than 15 db or so, the bandwidth of the RF band of noise is equal to or greater

319

than twice the 3-db bandwidth of the closed-loop response function. Under these conditions

$$S_{N_c}(f_1 - f) \doteq S_{N_c}(f)$$

for f and f_1 within the desired range. Thus

$$\frac{S_1(f)}{S_{N_s}(f)} \doteq \int_{-\infty}^{+\infty} S_{\phi_s}(f_1)df_1 \doteq \phi_{\text{rms}}^2,$$

since

$$S_{N_c}(f) = S_{N_s}(f).$$

Appendix B

It is desired to determine an upper bound to the total power passed by an ideal rectangular filter, whose bandwidth is equal to twice the closed-loop noise bandwidth, when the input is the first two second-order terms in the envelope expression of (2), *i.e.*,

$$E_2(t) = N_s(t)\phi(t) - \tfrac{1}{2}\phi^2(t).$$

A two-pole low-pass transfer function will be chosen as representative of the closed-loop transfer function. Let the impulse and transfer functions be, respectively,

$$h(t) = \frac{\omega_n}{\sqrt{1 - \zeta^2}} \epsilon^{-\omega_n \zeta t} \sin \left\{ \omega_n \sqrt{1 - \zeta^2} t \right\}$$

and

$$H(j\omega) = \frac{\omega_n^2}{\omega_n^2 - \omega^2 + 2j\omega\omega_n\zeta},$$

where $\zeta < 1$.

The covariance can be expanded in the following form [14]:

$$R_{E_2}(\tau) = R_{N\phi}^2(0) + R_{NN}(\tau)R_{\phi\phi}(\tau) + R_{N\phi}(\tau)R_{\phi N}(\tau)$$
$$- R_{N\phi}(0) R_{\phi\phi}(0) - R_{N\phi}(\tau)R_{\phi\phi}(\tau) - R_{\phi N}(\tau)$$
$$R_{\phi\phi}(\tau) + \tfrac{1}{4}R_{\phi\phi}^2(0) + \tfrac{1}{2}R_{\phi\phi}^2(\tau),$$

where $R_{N\phi}(\tau) \equiv E[N_s(t_1)\phi(t_2)]$, etc. Now the individual terms must be calculated. The term $R_{\phi\phi}(\tau)$ may be found in Middleton [15]:

$$R_{\phi\phi}(\tau) = \frac{\sigma_s^2\omega_n}{4\zeta} \epsilon^{-\zeta\omega_n|\tau|} \cos\left\{ \omega_n\sqrt{1 - \zeta^2}\tau \right\} +$$
$$+ \frac{\zeta}{\sqrt{1 - \zeta^2}} \sin\left\{ \omega_n\sqrt{1 - \zeta^2} | \tau | \right\},$$

where σ_s^2 is the spectral density of $N_s(t)$. The term for $R_{NN}(\tau)$ is

$$R_{NN}(\tau) = \sigma_s^2\delta(\tau),$$

where $\delta(\tau)$ is the "delta" function. The term $R_{\phi N}(\tau)$ is simply the product of the impulse function $h(\tau)$ and σ_s^2.

$$R_{\phi N}(\tau) \left\{ \begin{array}{l} = \dfrac{\sigma_s^2\omega_n}{\sqrt{1 - \zeta^2}} \epsilon^{-\omega_n\zeta\tau} \sin \omega_n \sqrt{1 - \zeta^2}\tau, \quad \tau \geq 0, \\[2ex] 0, \tau \leq 0. \end{array} \right.$$

The spectral density corresponding to the inverse Fourier transform of $R_{E_1}(\tau)$ contains a dc term, a term of uniform density and a number of terms whose energy is concentrated within a bandwidth twice that of the function $H(j\omega)$. The total power which would be passed by a rectangular filter whose bandwidth is equal to twice the closed-loop noise bandwidth $(B_c = \omega_n/4\zeta)$ is smaller than

$$P = \frac{11}{4}\left(\frac{\sigma^4 s^{\omega^2} n}{16\zeta^2}\right) = \frac{11}{4} \phi_{\text{rms}}^4.$$

P is calculated from $R_{E_1}(\tau)$ by letting $\tau = 0$ in all terms except the one containing $R_{NN}(\tau)$ and adding them. The term $R_{NN}(\tau)$ yields a uniform spectral density. Only the power within a bandwidth equal to twice the closed-loop noise bandwidth was included.

Appendix C

Mistuned Angle-Modulated Carrier

It is desired to find a transfer function which will relate the signal component of the instantaneous phase at the output of a mistuned high-Q band-pass filter to the phase of the small index modulated carrier applied to its input. Distortion will not be considered since it is only the signal transfer function which is important to stability. Let the small index signal be

$$e_1(t) = \cos\left[\omega_0 t + \phi(t)\right] = R_e\left(\epsilon^{j\omega_0 t}\sum_0^\infty \frac{[j\phi(t)]^n}{n!}\right)$$
$$\doteq R_e([1 + j\phi(t)]\epsilon^{j\omega_0 t}).$$

The result is a "baseband" signal which has been translated in frequency by ω_0. If $Y_L(j\omega)$ is the transfer function of the conventional low-pass equivalent of the filter, then $Y_L[j(\omega + \omega_d)]$ is the transfer function seen by the baseband signal because of the carrier being mistuned by ω_d. The output signal will, in general, be both envelope and phase modulated. Only the signal component of the phase modulation is desired. Write

$$Y_L[j(\omega + \omega_d)] \equiv R(\omega + \omega_d)\epsilon^{j\theta(\omega + \omega_d)}$$
$$= \epsilon^{j\theta(\omega_d)} Y_x(j\omega),$$

where

$$Y_x(j\omega) \equiv R(\omega + \omega_d)\epsilon^{j\theta(\omega + \omega_d) - j\theta(\omega_d)}.$$

$Y_x(j\omega)$ is not the transfer function of a physically realizable network, and consequently its impulse function is complex, *i.e.*,

$$h(\tau) = h_1(\tau) + jh_2(\tau),$$

where $h_1(\tau)$ and $h_2(\tau)$ are real. The filter output may be found by convolution:

$$e_0(t) = \epsilon^{j\theta(\omega_d)} \int_{-\infty}^{+\infty} [h_1(\tau) + jh_2(\tau)][1 + j\phi(t - \tau)d\tau$$
$$= \epsilon^{j\theta(\omega_d)}R(\omega_d) + \int_{-\infty}^{+\infty} [h_1(\tau) + jh_2(\tau)][j\phi(t - \tau)]d\tau.$$

The signal component of phase is

$$\Psi(t) = \text{Im} \ln e_0(t)$$

$$= \theta(\omega_d) + \frac{1}{R(\omega_d)} \int_{-\infty}^{+\infty} h_1(\tau)\phi(t - \tau)d\tau$$

assuming

$$\left| \int_{-\infty}^{+\infty} \phi(t - \tau)[h_1(\tau) + jh_2(\tau)]d\tau \right| \ll R(\omega_d),$$

i.e., if the distortion is not too severe. $\theta(\omega_d)$ is a constant and can be ignored. The desired transfer function has an impulse function of $h_1(\tau)/R(\omega_d)$. $h_1(\tau)$ is the real part of $h(\tau)$, and consequently its transfer function consists of the even real part and odd imaginary part of the minimum phase function $Y_x(j\omega)$. The desired transfer function is then

$$Y(j\omega) = \frac{1}{2R(\omega_d)} \left[Y_x(j\omega) + Y_x^*(-j\omega) \right].$$

ACKNOWLEDGMENT

The author wishes to thank C. L. Ruthroff for innumerable discussions, and for provision of an experimental demodulator which greatly facilitated the work reported in this paper. He also wishes to thank Mrs. C. L. Beattie for programming the problems represented by Figs. 6, 7 and 8, and F. J. Skinner for allowing the use of Fig. 9.

REFERENCES

[1] J. G. Chaffee, "The application of negative feedback to frequency-modulation systems," *Bell Sys. Tech. J.*, vol. 18, pp. 403–437; July, 1939.

[2] J. R. Carson, "Frequency modulation—theory of the feedback receiving circuit," *Bell Sys. Tech. J.*, vol. 18, pp. 395–403; July, 1939.

[3] M. O. Felix and A. J. Buxton, "The performance of FM scatter systems using frequency compression," *Proc. Natl. Electronics Conf.*, vol. 14, pp. 1029–1043; 1958.

[4] M. Morita and S. Ito, "High sensitivity receiving system for frequency modulated waves," 1960 IRE INTERNATIONAL CONVENTION RECORD, pt. 5, pp. 228–237.

[5] C. L. Ruthroff, "Project Echo: FM demodulators with negative feedback," *Bell Sys. Tech. J.*, vol. XL, pp. 1149–1157; July, 1961.

[6] S. O. Rice, "Properties of a sine wave plus random noise," *Bell Sys. Tech. J.*, vol. 27, pp. 109–157; January, 1948.

[7] J. Dugunji, "Envelopes and pre-envelopes of real waveforms," IRE TRANS. ON INFORMATION THEORY, vol. IT-4, pp. 53–57; March, 1958.

[8] H. S. Black, "Modulation Theory," D. Van Nostrand Co., Inc., New York, N. Y., pp. 218–234; 1953.

[9] J. C. West, J. L. Douce and B. G. Leary, "Frequency Spectrum Distortion of Random Signals in Non-linear Feedback Systems," IEE Monograph No. 419M; November, 1960.

[10] H. W. Bode, "Network Analysis and Feedback Amplifier Design," D. Van Nostrand Co., Inc., New York, N. Y., pp. 451–529; 1945.

[11] F. L. H. M. Stumpers, "Theory of frequency modulation noise," PROC. IRE, vol. 36, pp. 1081–1092; September, 1948.

[12] D. B. Harris, "Selective modulation," PROC. IRE, vol. 33, pp. 565–572; June, 1945.

[13] D. Middleton, "Introduction to Statistical Communication Theory," McGraw-Hill Book Co., Inc., New York, N. Y., p. 399; 1960.

[14] *Ibid.*, p. 343.

[15] *Ibid.*, p. 168.

PAPER NO. 14

Reprinted from *Proc. IRE*, Vol. 51, No. 2, pp. 349–356, Feb. 1963

A Threshold Criterion for Phase-Lock Demodulation*

JEAN A. DEVELET, JR.†, MEMBER, IRE

Summary—An analytical threshold criterioₗ has been developed for the general phase-lock receiver utilizing Booton's quasi-linearization technique. This criterion is established for arbitrary information and noise spectral densities. The information is assumed phase- or frequency-encoded on the received signal. Explicit results are centered around the case of additive white Gaussian noise.

The principal nonlinearity is assumed to be the phase detector which is represented as a product device.

Threshold curves are derived for three types of signals:

1) Bandlimited phase-encoded white Gaussian signals, optimal receiver;

2) Bandlimited phase-encoded white Gaussian signals, second-order receiver;

3) Frequency-encoded white signals, optimal receiver.

The phase-encoded white Gaussian signal threshold is then compared with Shannon's results. It was found that the optimal receiver threshold occurs $10 \log_{10} (e)$ or 4.34 db above Shannon's limit.

The second-order receiver was found to threshold 2 to 3 db above the optimal receiver in the region of normally encountered output signal-to-noise power ratios.

Frequency-encoded white signals represent the character of residual noise in a communication link oscillator system. Residual frequency noise is induced by the ever present thermal noise in oscillator circuits. This particular thermal-induced noise cannot be removed entirely. For this fundamental noise process maximum receiver sensitivities are derived.

An interesting result of quasi-linearization is that, for the signals considered, previous solutions of the Wiener-Hopf equation may be applied with only slight modifications.

I. INTRODUCTION

PREVIOUS ANALYSES of phase-lock receiver performance have been based on linearized models of the actual transfer function.[1-4] Since the phe-

* Received June 25, 1962; revised manuscript received, September 26, 1962.

† Aerospace Corporation, El Segundo, Calif. Formerly with Space Technology Laboratories, Inc., a Subsidiary of Thompson Ramo Wooldridge, Inc., Redondo Beach, Calif.

[1] W. J. Gruen, "Theory of AFC synchronization," PROC. IRE, vol. 53, pp. 1043–1048; August, 1953.

[2] B. D. Martin, "A Coherent Minimum-Power Lunar Probe Telemetry System," Jet Propulsion Lab., California Inst. Tech., Pasadena, Calif., External Publication No. 610, pp. 1–72; August, 1959.

[3] C. E. Gilchriest, "Application of the phase-locked loop to telemetry as a discriminator or tracking filter," IRE TRANS. ON TELEMETRY AND REMOTE CONTROL, vol. TRC-4, pp. 20–35; June, 1958.

[4] R. Jaffe and E. Rechtin, "Design and performance of phase-lock circuits capable of near-optimum performance over a wide range of input signal and noise levels," IRE TRANS. ON INFORMATION THEORY, vol. IT-1, pp. 66–76; March, 1955.

nomenon of loop threshold is caused by the effects of nonlinearities, it is understandable that the linear models referenced cannot yield a threshold criterion.

This paper describes the application of the *quasi-linearization* technique of Booton[5] to the determination of an analytic threshold criterion. The analysis will be restricted to the case of Gaussian signals corrupted by additive white Gaussian noise. Extension to other situations is possible but not treated in this paper.

II. Analysis

Consider the phase-lock receiver of Fig. 1.

The low-frequency output of the product demodulator can be shown to be

$$v_e(t) = - E \sin \epsilon(t) - X(t) \sin [m(t) + \epsilon(t)]$$
$$+ Y(t) \cos [m(t) + \epsilon(t)] \qquad (1)$$

where

$E = \sqrt{2 S_{if}}$, received signal amplitude, volts;

$\epsilon(t) =$ instantaneous loop error, radians;

$m(t) =$ instantaneous signal modulation, a Gaussian random variable, radians;

$X(t)$, $Y(t) =$ uncorrelated white Gaussian noise variables of average power $\Phi_{if} B W_{if}$, watts.

For the normal phase-lock receiver situation the predetection bandwidth, BW_{if}, may be assumed much larger than the phase-lock loop bandwidth. In this situation $Y(t)$ and $X(t)$ have a correlation time[6] much shorter than $m(t)$ or $\epsilon(t)$. Under these conditions, it can be shown that the noise portion of (1) can be represented by a Gaussian variable $N(t)$ which has the same correlation function as $Y(t)$ or $X(t)$. Thus (1) becomes

$$v_e(t) = - E \sin \epsilon(t) + N(t). \qquad (2)$$

Eq. (2) may now be utilized to obtain an analytical representation of the general receiver of Fig. 1. Fig. 2 portrays this representation.

The representation of Fig. 2 can be seen to be a simple servomechanism with the exception of the nonlinear element $E \sin [\ ]$. Booton[5] has provided an approximation technique which can be used to replace the nonlinear element by an equivalent gain, K_A. This technique essentially determines the *average* gain of the nonlinear device under the expected operating conditions. K_A may be found utilizing an averaging procedure which is a slight variation of Booton's equation (58).[7] This variation was obtained by an integration by parts and is in a

[5] R. C. Booton, Jr., "Nonlinear Control Systems with Statistical Inputs," Mass. Inst. Tech., Cambridge, Mass., Rept. No. 61, pp. 1–35; March 1, 1952.
[6] The time in seconds it takes the correlation function to drop to a small value compared to the value at zero seconds.
[7] *Ibid.*, p. 21.

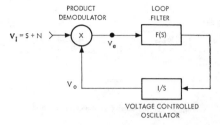

$$V_i = E \sin [wt + m(t)] + x(t) \sin wt + y(t) \cos wt$$
$$V_0 = 2 \cos [w(t) + m(t) + \epsilon(t)]$$

Fig. 1—General phase-lock receiver.

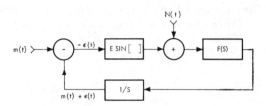

Fig. 2—Analytical receiver representation.

more convenient form:

$$K_A = \int_{-\infty}^{\infty} g'(x) p_1(x) dx \qquad (3)$$

where

$K_A =$ equivalent element gain,

$g'(x) = E \cos x$,

$p_1(x) =$ probability density of $\epsilon(t)$ which must be Gaussian to conform to Booton's criteria.

Letting $\overline{\epsilon(t)^2} = \sigma^2$, substituting into (3), and integrating yields

$$K_A = E \exp (-\sigma^2/2). \qquad (4)$$

The *quasi-linear* receiver representation obtained by linearizing the $E \sin [\ ]$ transfer function is now shown in Fig. 3.

Denoting the signal and noise *one-sided* power-spectral densities of $m(s)$ and $N(s)$ as $\Phi_m(\omega)$ rad²/cps and $\Phi_n(\omega)$ watts/cps respectively, it is a simple matter to show

Modulation Error

$$\sigma^2 = \int_0^{\infty} \Phi_m(\omega) \left| 1 - \frac{\phi_0}{\phi_i} (\omega) \right|^2 df$$

Noise Error

$$+ \int_0^{\infty} \frac{\exp (\sigma^2) \Phi_n(\omega)}{2 S_{if}} \left| \frac{\phi_0}{\phi_i} (\omega) \right|^2 df \qquad (5)$$

where

$$\frac{\phi_0}{\phi_i} (\omega) = \frac{E \exp \left(-\dfrac{\sigma^2}{2} \right) F(s)/s}{1 + E \exp (-\sigma^2/2) F(s)/s}.$$

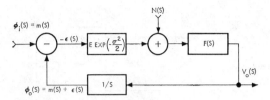

$$\phi_i(S) = m(S)$$
$$-\epsilon(S)$$
$$N(S)$$
$$V_o(S)$$
$$\phi_o(S) = m(S) + \epsilon(S)$$

Fig. 3—Quasi-linear phase-lock receiver.

Eq. (5) is a key relationship. As a consequence of the exp (σ^2), it implicitly contains a threshold criterion for the receiver model of Fig. 3. It will be shown more explicitly in what follows how this criterion is obtained.

Eq. (5) can be solved for the received signal power S_{if} as follows:

$$S_{if} = \frac{\dfrac{\exp \sigma^2}{2} \int_0^\infty \Phi_n(\omega) \left|\dfrac{\phi_0}{\phi_i}(\omega)\right|^2 df}{\sigma^2 - \int_0^\infty \Phi_m(\omega) \left|1 - \dfrac{\phi_0}{\phi_i}(\omega)\right|^2 df} . \quad (6)$$

It is desired now to choose a function ϕ_0/ϕ_i such that for a specified σ, $\Phi_m(\omega)$, and $\Phi_n(\omega)$, S_{if} is minimized. This function will yield maximum receiver sensitivity. In order to find this optimum function let

$$\frac{\phi_0}{\phi_i}(\omega) = A(\omega)e^{i\theta(\omega)}. \quad (7)$$

Substitution of (7) in (6) yields

$$S_{if} = \frac{\dfrac{\exp(\sigma^2)}{2} \int_0^\infty \Phi_n(\omega) A^2(\omega) df}{\sigma^2 - \int_0^\infty \Phi_m(\omega)[1 - 2A(\omega)\cos\theta(\omega) + A^2(\omega)]df} . \quad (8)$$

Clearly for minimum S_{if}, $\theta(\omega)$ should be chosen equal to zero. Thus,

$$S_{if} = \frac{\dfrac{\exp(\sigma^2)}{2} \int_0^\infty \Phi_n(\omega) A^2(\omega) df}{\sigma^2 - \int_0^\infty \Phi_m(\omega)[1 - A(\omega)]^2 df} . \quad (9)$$

$A(\omega)$ may be found by standard variational methods. That is, add to $A(\omega)|_{\text{opt}}$ the function $\epsilon\eta(\omega)$. Eq. (9) is then differentiated with respect to ϵ and the derivative and ϵ set equal to zero. Since $\eta(\omega)$ is arbitrary, an equation for $A(\omega)|_{\text{opt}}$ for all ω is obtained. The result of this mathematical manipulation is

$$A(\omega)\Big|_{\text{opt}} = \frac{\Phi_m(\omega)}{\Phi_m(\omega) + \dfrac{\Phi_n(\omega)\exp(\sigma^2)}{2S_{if_{\min}}}} . \quad (10)$$

Eq. (10) is an interesting result. It is the identical result one obtains if the minimization of σ^2 in (5) is

carried out. Since minimization of S_{if} and σ^2 is achieved by the same transfer function, the general results of Wiener are applicable with the exception of the fact that the noise power-spectral density is replaced by

$$\frac{\Phi_n(\omega)\exp(\sigma^2)}{2S_{if_{\min}}} .$$

Further consideration of Wiener-Hopf solutions will use results of previous workers in the field. In particular only the case of white Gaussian noise will be treated. That is $\Phi_n(\omega) = 2\Phi_{if}$ where Φ_{if} is the *one-sided* predetection power-spectral density of the receiver.

General Solution for White Gaussian Noise

Yovits and Jackson in their important paper 1955[8] found a particularly useful form of the optimum *realizable* transfer function for the situation of white noise.

Their results for *linear* filters are repeated below utilizing the terminology previously developed,[9] as follows:

$$\left|1 - \frac{\phi_0}{\phi_i}(\omega)\right|_{\text{opt}}^2 = \frac{\Phi_n}{\Phi_n + \Phi_m(\omega)} . \quad (11)$$

The corresponding minimum following error is

$$\sigma^2 = \Phi_n \int_0^\infty \log_\epsilon \left(1 + \frac{\Phi_m(\omega)}{\Phi_n}\right) df. \quad (12)$$

In order to treat the performance of the quasi-linear receiver model one need just replace Φ_n in (11) and (12) by $\Phi_{if} \exp(\sigma^2)/S_{if}$. The following fundamental relations are thereby obtained:

$$\left|1 - \frac{\phi_0}{\phi_i}(\omega)\right|_{\text{opt}}^2 = \frac{\Phi_{if}\exp(\sigma^2)}{\Phi_{if}\exp(\sigma^2) + S_{if}\Phi_m(\omega)}, \quad (13)$$

$$\sigma^2 = \frac{\Phi_{if}\exp(\sigma^2)}{S_{if}} \int_0^\infty \log_\epsilon \left(1 + \frac{S_{if}\Phi_m(\omega)}{\Phi_{if}\exp(\sigma^2)}\right) df. \quad (14)$$

Eq. (14) is one of the principal results of this paper. Given a receiver output quality constraint and a receiver noise density Φ_{if}, (14) will not yield a bounded solution for σ^2 if the received signal power S_{if} is too small. The value of S_{if} below which the solution for σ^2 ceases to exist is the threshold for phase-lock demodulation obtained from the quasi-linear model.

III. Applications

The previous results will now be applied to three situations which are important in communication engineering. In all cases the background noise will be assumed white and Gaussian. The signal modulations and transfer functions to be considered are the following:

[8] M. C. Yovits and J. L. Jackson, "Linear filter optimization with game theory considerations," 1955 IRE NATIONAL CONVENTION RECORD, pt. 4, pp. 193–199.
[9] *Ibid.*, pp. 195–196.

1) Reception of a bandlimited phase-encoded white Gaussian signal spectrum with the optimal transfer function,
2) Reception of a bandlimited phase-encoded white Gaussian signal spectrum with a second-order transfer function,
3) Reception of a frequency-encoded white signal spectrum with the optimal transfer function.[10]

The first case is important to illustrate optimal communication of information by use of a phase-lock receiver.

The second situation is of practical interest since a second-order loop is easily realized and is amenable to measurements verifying the theory which has been developed.

The last situation is important in that oscillators in a communication link are not perfectly stable and must be "tracked" in phase in order to continue receiving the signal. The results here point to fundamental sensitivity limitations of phase-lock receivers given the oscillator system is perturbed by a random walk phase process.

Bandlimited Phase-Encoded White Gaussian Signal, Optimal Receiver

Let the signal power-spectral density in (14) be given by

$$\Phi_m(\omega) = \Phi_m \qquad 0 \le f \le f_m,$$
$$\Phi_m(\omega) = 0 \qquad f_m < f. \tag{15}$$

Integration of (14) yields

$$\sigma^2 = \frac{\Phi_{if} \exp(\sigma^2)}{S_{if}} f_m \log_e \left[1 + \frac{\Phi_m S_{if}}{\Phi_{if} \exp(\sigma^2)} \right]. \tag{16}$$

Since $\Phi_{if} \exp(\sigma^2)/S_{if}$ is the *one-sided* phase noise power-spectral density in the receiver output,[11] and $\Phi_m f_m$ is the mean-square signal power in the receiver output, one may rewrite (16) in a more meaningful form exhibiting signal-to-noise power ratios input and output. Thus,

$$\left(\frac{S}{N}\right)_i = \frac{\exp(\sigma^2)}{2\sigma^2} \log_e \left[1 + \left(\frac{S}{N}\right)_0 \right] \tag{17}$$

where

$$\left(\frac{S}{N}\right)_i = \frac{S_{if}}{2\Phi_{if}f_m};$$

the input signal-to-noise power ratio referred to twice the information bandwidth.

$$\left(\frac{S}{N}\right)_0 = \frac{\Phi_m f_m}{\dfrac{\Phi_{if}}{S_{if}} \exp(\sigma^2) f_m} = 2\left(\frac{S}{N}\right)_i \frac{\sigma_m^2}{\exp(\sigma^2)};$$

[10] This situation corresponds to a random walk in signal phase.
[11] Inspection of the transfer function $V_0(s)/N(s)$ in Fig. 3 reveals the fact exp (σ^2) is a factor in the output noise also.

the output signal-to-noise power ratio referred to the information bandwidth. σ_m = modulation index, radians.

Eq. (17) has a minimum value for $(S/N)_i$ at $\sigma = 1.0$ for a fixed system output quality $(S/N)_0$. Substitution of $\sigma = 1.0$ radian in (17) yields the threshold result depicted in Fig. 4.

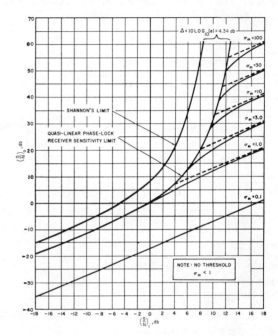

Fig. 4—Quasi-linear receiver performance for the situation of bandlimited white Gaussian phase-encoded signals with the optimal transfer function.

The curves above threshold represent system output quality vs signal-to-noise power ratio input asymptotically converging on the conventional high $(S/N)_i$ relation. These are constructed with the modulation index $\sigma_m = \sqrt{\Phi_m f_m}$ as a parameter. Since

$$\left(\frac{S}{N}\right)_0 = \frac{2\sigma_m^2 \left(\dfrac{S}{N}\right)_i}{\exp(\sigma^2)},$$

and σ is a function of $(S/N)_i$ governed by (17), a simultaneous solution of this relation and (17) yields σ and hence $(S/N)_0$ vs $(S/N)_i$. It is interesting to note the curvature as one approaches threshold. Note also for $\sigma_m < 1.0$ the quasi-linear model yields no threshold.

The threshold criterion depicted by Fig. 4 is significantly below a standard FM or PM discriminator as would be expected since more optimal demodulation is employed. It is, however, higher than that predicted by use of Shannon's results. This is as it should be, for above threshold information may be conveyed by the phase-lock communication system and the rate of communication should be bounded by a communication theoretical result such as Shannon's. More discussion of this appears in Section IV.

Bandlimited Phase-Encoded White Gaussian Signal, Second-Order Receiver

The previous section treated the case of the optimal receiver which requires an infinite number of elements to realize. The second-order receiver though not optimum is easy to realize in hardware.

As with the optimal receiver, let the signal power-spectral density be given by (15), as follows

$$\Phi_m(\omega) = \Phi_m \qquad 0 \le f \le f_m,$$
$$\Phi_m(\omega) = 0 \qquad f_m < f.$$

Return to the fundamental equation (5) is necessary since we are not dealing with an optimal receiver for the signal given by (15).

The transfer function $\phi_0/\phi_i(\omega)$ has been derived by Gruen for the second-order receiver.[1] Gruen's results are restated below for convenience.

$$\frac{\phi_0}{\phi_i}(s) = \frac{\omega_n^2\left[1 + \left(\dfrac{2\zeta}{\omega_n} - \dfrac{1}{K}\right)s\right]}{\omega_n^2 + 2\zeta\omega_n s + s^2} \qquad (18)$$

where

ω_n = loop natural frequency, rad/sec,
ζ = loop damping ratio,
K = loop gain, sec.$^{-1}$.

Eq. (18) may be obtained from (5) by substituting for $F(s)$ the transfer function of a lag-lead filter.

$$F(s) = K_0 \frac{1 + \tau_1 s}{1 + \tau_2 s} \qquad (19)$$

where

K_0 = a dc gain,
$\tau_1; \tau_2$ = time constants of the lag-lead filter, seconds.

Various methods of realizing $F(s)$ have been discussed thoroughly by Martin.[2] In addition to the above substitution for $F(s)$, the following parameters are defined after Gruen:[12]

$$\omega_n^2 = \frac{K}{\tau_2},$$

$$\tau_1 = \frac{2\zeta}{\omega_n} - \frac{1}{K},$$

$$K = E \exp\left(-\frac{\sigma^2}{2}\right)K_0. \qquad (20)$$

where

ω_n = loop natural frequency, rad/sec,
ζ = loop damping factor.

Since (20) shows that the loop natural frequency, and damping vary with σ the loop error, all analysis to fol-

[12] Gruen, *op. cit.*, p. 1045.

low will be carried out in terms of the zero σ (high $(S/N)_i$) values of these quantities, ω_{n0} and ζ_0 respectively. The receiver design will center around optimization of ω_{n0} to achieve maximum sensitivity.

Eqs. (15) and (18) may now be substituted in (5) and the two integrals evaluated. The following simplifying assumptions will be made to yield meaningful results without resort to computer integration of the modulation error in (5):

$$\frac{2\zeta}{\omega_n} \gg \frac{1}{K},$$

$$\frac{2\pi f_m}{\omega_n} \ll 1.$$

The first assumption is usually the case in practical receivers. The second assumption restricts the validity of our calculations to the more interesting region of high signal-to-noise ratios in the receiver output. With the above substitutions and assumptions, it can be shown that (5) becomes, for the second-order receiver

$$\sigma^2 = \frac{(2\pi)^4 \Phi_m f_m^5 \exp(\sigma^2)}{5\omega_{n0}^4}$$

$$+ \frac{\Phi_{if} \exp(\sigma^2)\left[1 + 4\zeta_0^2 \exp\left(-\dfrac{\sigma^2}{2}\right)\right]}{8S_{if}\zeta_0}\omega_{n0}. \qquad (21)$$

Assuming a fixed damping ζ_0, Φ_{if}, Φ_m and f_m, (21) may be solved for S_{if} and minimized with respect to ω_{n0}. Thus, maximum receiver sensitivity is achieved. Performing this manipulation one obtains as the minimum receiver input signal to noise-power ratio defined as in the optimal receiver,

$$\left(\frac{S}{N}\right)_i = \frac{S_{if}}{2\Phi_{if}f_m} = \frac{\exp\left(\dfrac{6}{5}\sigma^2\right)}{2\sigma^2}$$

$$\left\{\frac{5\pi\left[1 + 4\zeta_0^2 \exp\left(-\dfrac{\sigma^2}{2}\right)\right]}{16\zeta_0}\right\}^{4/5}\left(\frac{S}{N}\right)_0^{1/5}, \qquad (22)$$

where the relation for signal-to-noise output power ratio is given by the same relation as in the optimal receiver, *i.e.*,

$$\left(\frac{S}{N}\right)_0 = \frac{2\sigma_m^2\left(\dfrac{S}{N}\right)_i}{\exp(\sigma^2)}. \qquad (23)$$

As in (17), (22) has a minimum value at a particular value of σ which depends on system output quality $(S/N)_0$ and receiver damping ratio ζ_0. Only one particular value of $\zeta_0 = 1/\sqrt{2}$ will be chosen in what follows.[13]

[13] Since the previous section has shown the second-order receiver is not optimal for the type of signal treated here, the author does not feel justified in using other than a commonly encountered value for ζ_0.

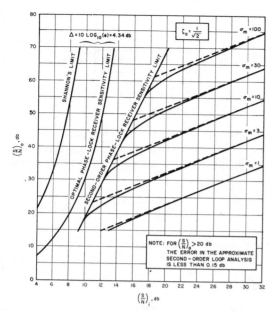

Fig. 5—Quasi-linear receiver performance for the situation of band-limited white Gaussian phase-encoded signals with a second-order transfer function.

For $\zeta_0 = 1/\sqrt{2}$ the value of σ which yields minimum $(S/N)_i$ and hence minimum receiver threshold is 1.01 radians. Substitution of $\sigma = 1.01$ radians into (22) yields the following threshold relation:

$$\left(\frac{S}{N}\right)_i = 4.08 \left(\frac{S}{N}\right)_0^{1/5}. \qquad (24)$$

Fig. 5 graphs (24). Note that only values for $(S/N)_0$ >20 db are considered. This is consistent with the approximate evaluation of the modulation error portion of (5).

As with the optimal receiver, the curves above threshold represent system output quality vs signal-to-noise power ratio input asymptotically converging on the conventional high $(S/N)_i$ relation. These are constructed with the modulation index σ_m as a parameter. Loop error, σ, may be eliminated for purposes of the graphs in Fig. 5 by simultaneous solution of (22) and (23).

Shannon's lower limit and the optimal phase-lock receiver sensitivity are shown for comparison in Fig. 5. It is interesting to note that for output signal-to-noise ratios of practical interest ($\approx$ 20 to 40 db) that the second-order receiver is only 2 to 3 db poorer than the optimal receiver or 6 to 7 db poorer than Shannon's limit.

Frequency-Encoded White Signals, Optimal Receiver

In a coherent communication system, oscillator (clock) stability is of great importance especially when accurate Doppler measurements or low information rates are to be conveyed through the link.

In general it is not easy to describe the effect of oscillator noise on coherent systems because the shape of the oscillator system phase power-spectral density is not known. In one specific instance, however, the effect is amenable to calculation, that is, when the resulting random process imposed on the received signal phase is caused by thermal noise. In this special instance Edson[14] has shown that the *frequency* modulation [$\dot{m}(t)$ Fig. 2] has a white power-spectral density. Develet[15] then obtained by simple integration and Fourier transformation the *one-sided* phase power-spectral density, Φ_m, given by

$$\Phi_m = \frac{2}{\tau_c \omega^2}, \qquad \frac{\text{rad}^2}{\text{cps}} \qquad (25)$$

where

τ_c = coherence time of the oscillator system; the time in seconds it takes the phase drift to build up to one (1) radian rms,

ω = radian frequency, rad/sec.

This power-spectral density can be identified with $m(t)$ a random walk process.

One seeks now the fundamental receiver sensitivity in the presence of additive white Gaussian noise given that the received signal has a phase power-spectral density governed by (25). Substitution of (25) in (14) yields

$$\sigma^2 = \frac{\Phi_{if} \exp(\sigma^2)}{S_{if}} \int_0^\infty \log_e \left[1 + \frac{2S_{if}}{\omega^2 \tau_c \Phi_{if} \exp(\sigma^2)} \right] df. \quad (26)$$

Integration gives the simple result

$$S_{if} = \frac{\Phi_{if}}{2\tau_c} \left[\frac{\exp(\sigma^2)}{\sigma^4} \right]. \qquad (27)$$

Eq. (27) may be differentiated with respect to σ to find the minimum value of S_{if}. Thus,

$$S_{if} \big|_{\min} = \frac{\Phi_{if}}{\tau_c} \left(\frac{e^2}{8} \right) \qquad (28)$$

where

$$\sigma_{\min} = \sqrt{2} \text{ radian.}$$

Considering that the receiver noise density is given by Boltzmann's constant times the equivalent receiver temperature, (28) may be restated as

$$S_{if} \big|_{\min} = \frac{KT_{eq}}{\tau_c} \left(\frac{e^2}{8} \right) \qquad (29)$$

[14] W. A. Edson, "Noise in oscillators," *Proc. IRE*, vol. 48, pp. 1454–1466; August, 1960.

[15] J. A. Develet, Jr., "Fundamental Sensitivity Limitations for Second Order Phase-Lock Receivers," presented at URSI Spring Meeting, Washington, D.C., May 4, 1961, STL Tech. Note 8616-0002-NU-000; June 1, 1961.

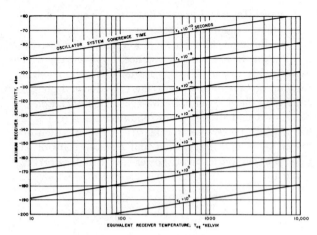

Fig. 6—Receiver sensitivity vs equivalent receiver temperature.

where

$$K = 1.38 \times 10^{-23}, \quad \text{Joules}/^\circ\text{K},$$

T_{eq} = equivalent receiver temperature, °K.

Fig. 6 plots phase-lock receiver sensitivity as a function of T_{eq} with τ_c as a parameter. It represents a fundamental sensitivity limitation for phase-lock reception given by the *quasi-linear* receiver model, for reception of signals disturbed by a random walk phase process.

IV. Comparison with Shannon's Results

Let us now derive Shannon's limit as depicted in Fig. 4. Consider the communication channel of Fig. 7. It is clear that the error-free information rate in bits/sec of the video portion of the link cannot be greater than the maximum error-free information rate which can be passed through the RF channel. This fact allows one to use Shannon's theorem to bound the output signal to noise-power ratio $(S/N)_0$ for a given input signal to noise-power ratio $(S/N)_i$. The phase-lock demodulator cannot enhance the information transfer process.

The RF signal modulation considered in the derivation of Fig. 4 was bandlimited, phase-encoded, white, and Gaussian. Since it is difficult to specify RF bandwidth occupancy for such a phase modulation at intermediate modulation indices ($\sigma_m \approx 1.0$), only an extreme case will be considered which yields an absolute lower bound on RF signal-to-noise power ratio for a given output signal-to-noise power ratio. It will be assumed that the RF channel has infinite bandwidth. Shannon has shown that this condition corresponds to the case of maximum information flow per watt of RF channel signal power in the presence of white noise.

For this case, the use of Shannon's result[16] gives the upper bound on information flow in the RF channel as

$$C_{\text{max}} = \frac{S_{if}}{\Phi_{if} \log_e 2}, \qquad \text{bits/sec.} \qquad (30)$$

<hr>

[16] C. E. Shannon, "Communication in the presence of noise," Proc. IRE, vol. 37, pp. 10–21; January, 1949.

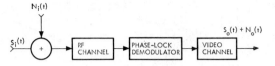

Fig. 7—Communication channel.

Since the signals are bandlimited and the noise is white and additive in the quasi-linear receiver model output, the information rate passing out of the demodulator cannot exceed

$$C_0 = f_m \frac{\log_e \{1 + (S/N)_0\}}{\log_e 2}, \qquad \text{bits/sec.} \qquad (31)$$

In addition, $C_{\text{max}} \geq C_0$ because the phase-lock demodulator cannot enhance the RF channel's ability to transfer information, but can only equal or degrade this ability. Therefore

$$\left(\frac{S}{N}\right)_i = \frac{S_{if}}{2\Phi_{if}f_m} \geq \frac{1}{2} \log_e \left\{1 + \left(\frac{S}{N}\right)_0\right\}. \qquad (32)$$

Eq. (32) represents a lower bound on required signal-to-noise power input and is graphed in Fig. 4. A significant point is the consistency of Booton's quasi-linear approximate model and Shannon's result. Note the constant difference factor, $10 \log_{10}(e)$, for all values of $(S/N)_0$. A difference factor is to be expected, for in real-time phase-lock demodulation no *a priori* knowledge of the signals to be transmitted is available as in Shannon demodulation.

V. Conclusions

For the first time a simple approximate theory for phase-lock threshold has been developed.

The quasi-linearization technique of Booton was the principal analytic tool in developing this threshold criterion.

It was shown that all previous Wiener-Hopf solutions in the linear filter domain apply in this situation with one minor adjustment. The noise power-spectral density is replaced by $\Phi_{if} \exp (\sigma^2)/S_i$ wherever it appears in previous results. It must be kept in mind that only one particular servomechanism (the phase-lock loop) and nonlinearity (the product demodulator) was treated in this analysis.

Certain interesting examples were shown to illustrate the usefulness of the procedure. In particular for the special case of communication with white Gaussian phase-encoded signals in the presence of additive white Gaussian noise the optimal phase-lock receiver threshold performance was found to be 4.34 db poorer than Shannon's limit. This degradation in performance is explained by the fact that no *a priori* knowledge of the signals to be transmitted is available in real-time phase-lock demodulation. This contrasts to the Shannon demodulator in which knowledge of the set of transmitted signals is available to the receiver. Apparently *a priori* knowledge is worth 4.34 db in signal power.

Limited measurements which have been conducted on a wideband second-order loop at Space Technology Laboratories, Inc., Redondo Beach, California, indicate very close agreement with predictions based on the quasi-linear theory for loop rms phase error, σ, out to the predicted threshold.[17]

It is important to note that the threshold analysis in this paper was based on a mean-square signal, noise, and loop error criterion. Threshold is defined as the input signal-to-noise power ratio at which, for a given quality constraint, the loop error becomes unbounded. If, however, short-term statistics of the receiver output are important to the observer, an entirely new criterion may require development. In any event, it is considered this paper presents the *lower bound* on sensitivity, because with any new criterion a bounded mean-square loop error will certainly be a prerequisite.

Acknowledgment

The author wishes to acknowledge the stimulating discussions with Dr. R. C. Booton, Jr., of Space Technology Laboratories, Inc., Redondo Beach, Calif., and Dr. C. R. Cahn of the Bissett-Berman Corp., Santa Monica, Calif., which aided in my solution of this problem.

---

[17] R. C. Booton, Jr., "Demodulation of wideband frequency modulation utilizing phase-lock techniques," *Proc. Nat'l Telemetering Conf.*, Washington, D. C., May 23–25, 1962, vol. 2, Fig. 11, p. 14.

PAPER NO. 15

Reprinted from *Proc. IEEE*, Vol. 51, No. 12, pp. 1737–1753, Dec. 1963

Phase-Locked Loop Dynamics in the Presence of Noise by Fokker-Planck Techniques*

ANDREW J. VITERBI†, SENIOR MEMBER, IEEE

Summary—Statistical parameters of the phase-error behavior of a phase-locked loop tracking a constant frequency signal in the presence of additive, stationary, Gaussian noise are obtained by treating the problem as a continuous random walk with a sinusoidal restoring force. The Fokker-Planck or diffusion equation is obtained for a general loop and for the case of frequency-modulated received signals. An exact expression for the steady-state phase-error distribution is available only for the first-order loop, but approximate and asymptotic expressions are derived for the second-order loop. Results are obtained also for the expected time to loss of lock and for the frequency of skipping cycles. Some of the results are extended to tracking loops with nonsinusoidal error functions. Validity thresholds of widely accepted approximate models of the phase-locked loop are obtained by comparison with the exact results available for the first-order loop.

PHASE-LOCKED LOOP DYNAMICS

THE PHASE-LOCKED LOOP is a communication receiver which operates as a coherent detector by continuously correcting its local oscillator frequency according to a measurement of the phase error. A block diagram of the device is shown in Fig. 1 with the pertinent input and output signals indicated.

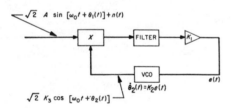

Fig. 1—Phase-locked loop.

dicated. The output of the voltage-controlled oscillator (VCO) is a sinusoid whose frequency is controlled by the input voltage, $e(t)$; that is,

$$\dot{\theta}_2(t) = \frac{d\theta_2(t)}{dt} = K_2 e(t),\qquad(1)$$

so that when $e(t) = 0$, the oscillator frequency is ω_0. The received signal is a sinusoid of power A^2 watts and of arbitrary frequency and phase which may be either fixed or time-varying because of frequency or phase

* Received April 17, 1963; revised manuscript received August 2, 1963. This paper presents the results of one phase of research carried out at the Jet Propulsion Laboratory, California Institute of Technology, under Contract No. NAS 7-100, sponsored by the National Aeronautics and Space Administration.

† University of California, Los Angeles, Calif., and consultant, Jet Propulsion Laboratory, Pasadena, Calif.

modulation at the transmitter. Thus, it may be represented by the expression

$$\sqrt{2}\, A \sin\left[\omega_0 t + \theta_1(t)\right].\qquad(2)$$

If the signal is a pure sinusoid with constant frequency ω and an initial phase θ, then

$$\theta_1(t) = (\omega - \omega_0)t + \theta.\qquad(3)$$

Although we shall limit ourselves to this case in several of the following sections, we shall continue at present with the general formulation in which $\theta_1(t)$ is an arbitrary and possibly random process.

The received noise is assumed to be a stationary white Gaussian process of one-sided spectral density N_0 w/cps. We shall assume that the phase-locked loop is preceded by a band-pass filter with center frequency ω_0, a bandwidth β, which is very wide compared to the frequency region of interest, and a transfer function which is flat over this region. The only restriction on β is that it be less than or equal to $2\omega_0$. If we let $\beta = 2\omega_0$, the band-pass filter becomes a low-pass filter with this bandwidth. The noise process $n(t)$ over an arbitrary period of duration T can be expanded in a Fourier series whose coefficients are Gaussian variables which become independent in the limit as T approaches infinity [1]. By collecting the sine and cosine terms of the series, we can represent the noise process of infinite duration by the expression

$$n(t) = \sqrt{2}\, n_1(t) \sin \omega_0 t + \sqrt{2}\, n_2(t) \cos \omega_0 t,\qquad(4)$$

where $n_1(t)$ and $n_2(t)$ are independent stationary wide-band Gaussian processes with flat spectra over the frequency range from zero to $\beta/2$. If we choose β to be $2\omega_0$ and restrict our interest to frequencies below ω_0 rad/sec, $n_1(t)$ and $n_2(t)$ may be regarded essentially as independent white Gaussian processes of one-sided spectral density N_0 w/cps.

Thus the product of input and reference signals is

$$2\big\{ A \sin\left[\omega_0 t + \theta_1(t)\right] + n_1(t) \sin \omega_0 t + n_2(t) \cos \omega_0 t\big\}$$
$$\cdot \big\{ K_3 \cos\left[\omega_0 t + \theta_2(t)\right]\big\}$$
$$= K_3\big\{ A \sin\left[\theta_1(t) - \theta_2(t)\right] - n_1(t) \sin \theta_2(t) + n_2(t) \cos \theta_2(t)$$
$$+ A \sin\left[2\omega_0 t + \theta_1(t) + \theta_2(t)\right] + n_1(t) \sin\left[2\omega_0 t + \theta_2(t)\right]$$
$$+ n_2(t) \cos\left[2\omega_0 t + \theta_2(t)\right]\big\}.$$

The double-frequency terms may be neglected since neither the filter nor the VCO will respond to these for reasonable large ω_0. Then from Fig. 1 we see that

$$e(t) = K_1 K_3 F(s)\{A \sin [\theta_1(t) - \theta_2(t)] - n_1(t) \sin \theta_2(t)$$
$$+ n_2(t) \cos \theta_2(t)\} \quad (5)$$

where $F(s)$ is a rational function which represents in operational notation the effect of the linear filter in the loop. If we let

$$\phi(t) = \theta_1(t) - \theta_2(t) \quad (6)$$

and

$$K = K_1 K_2 K_3 \quad (7)$$

we obtain from (1) and (6)

$$\phi(t) = \theta_1(t) - K_2 e(t),$$

and from (5) and (7) we have

$$\phi(t) = \theta_1(t) - KF(s)[A \sin \phi(t) - n_1(t) \sin \theta_2(t)$$
$$+ n_2(t) \cos \theta_2(t)]. \quad (8)$$

$\phi(t)$ defined by (6) is the instantaneous phase error or the phase difference between the received signal and the reference signal at the output of the phase-locked loop. If we let

$$n'(t) = - n_1(t) \sin \theta_2(t) + n_2(t) \cos \theta_2(t), \quad (9)$$

(8) reduces to

$$\phi(t) = \theta_1(t) - KF(s)[A \sin \phi(t) + n'(t)]. \quad (10)$$

This differential equation in operational form represents the dynamic operation of the phase-locked loop. It may also be written as the operational equation

$$\phi(t) = \theta_1(t) - \frac{KF(s)}{s} [A \sin \phi(t) + n'(t)], \quad (11)$$

which is represented by the block diagram or model of Fig. 2. It should be noted that (11) and the model of

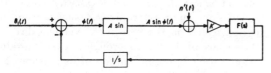

Fig. 2—Model of the phase-locked loop.

Fig. 2 are exact in all respects except for the fact that the terms centered about the double frequency, $2\omega_0$, have been assumed to be eliminated completely by the combination of the filter and the VCO.

Before we can proceed we must determine the statistics of the noise process $n'(t)$ defined by (9), which is the random driving function in the model. We can show that this is a white Gaussian process whenever the original noise process is Gaussian and white. We have

taken $n(t)$ to be essentially white at least for frequencies less than twice the VCO quiescent frequency, ω_0, and have shown that consequently $n_1(t)$ and $n_2(t)$ are essentially white for frequencies up to ω_0. We have from above

$$n' = - n_1 \sin \theta_2 + n_2 \cos \theta_2, \quad (9)$$

and let us define similarly

$$n'' = n_1 \cos \theta_2 + n_2 \sin \theta_2. \quad (9a)$$

The process $\theta_2(t)$ depends on $\theta_1(t)$ and the noise, as is clear upon inspection of Fig. 2. Thus $\theta_2(t)$ may be nonstationary as in the case, for example, when $\theta_1(t)$ is given by (3). Although $\theta_2(t)$ is a function of the input noise, it can depend only on its past [i.e., $n(t-\delta)$ where $\delta > 0$], and since the noise is an essentially white process,[1] $n(t)$ is independent of $n(t-\delta)$. Therefore, $\theta_2 = \theta_2(t)$ is independent of $n(t)$ and consequently also of $n_1(t)$ and $n_2(t)$, and we have as the joint probability density function of the three independent random variables: n_1, n_2, and θ_2,

$$p(n_1, n_2, \theta_2) = \frac{1}{2\pi\sigma^2} \exp (-n_1^2/2\sigma^2) \exp (-n_2^2/2\sigma^2) p(\theta_2),$$

where $\sigma^2 = N_0 \omega_0 / 2\pi$ and $p(\theta_2)$ is an arbitrary nonstationary distribution. From this we can obtain the joint-probability density function of n', n'', and θ_2 by performing the linear transformation of (9) and (9a). The result is

$$p(n', n'', \theta_2) = p(n_1, n_2, \theta_2) \left| J \left(\frac{n_1, n_2, \theta_2}{n', n'', \theta_2} \right) \right|$$
$$= \frac{1}{2\pi\sigma^2} \exp (-n'^2/2\sigma^2) \exp (-n''^2/2\sigma^2) p(\theta_2),$$

since the absolute value of the Jacobian is unity. Hence we conclude that $n'(t)$ is a stationary process with exactly the same statistics as $n_1(t)$ and $n_2(t)$. It is Gaussian and essentially white at least over the frequency region up to ω_0 rad/sec, and its one-sided spectral density is N_0 w/cps.

The model of Fig. 2 first appeared without proof in a paper by Develet [2]. In the absence of noise, the model has been known for some time and analyzed by several authors beginning with Gruen [3]. Solutions of (11) in the absence of noise have been obtained for a number of filter-transfer functions and also for the case of received sinusoids with linearly time-varying fre-

---

[1] Actually, since $n(t)$ is "white" only for radian frequencies below $2w_0$, $n(t-\delta)$ is essentially independent of $n(t)$ only for $\delta > k\pi/w_0$, where k is a sufficiently large constant. On the other hand, since the combination of filter and VCO has a low-pass transfer function which is extremely narrow compared to $2w_0$, $\theta_2(t)$ is essentially independent of the input, $n(t)$, for $\delta < k\pi/w_0$, so that the net result is the same.

quency [4]. The general case in which additive noise is present has been treated by a variety of approximations. The first approach, by Jaffe and Rechtin [5], essentially replaced the sinusoidal nonlinearity of the model of Fig. 2 by a linear amplifier of gain A, which is the gain for arbitrarily small ϕ. Margolis [6] first analyzed the nonlinear operation in the presence of noise by perturbation methods obtaining a series solution of the differential equation of operation. By using the first few terms of the series he determined approximate moments of the phase error. Develet [2], who first proposed the operational model, applied Booton's quasi-linearization technique [7] to replace the sinusoidal nonlinearity by a linear amplifier whose gain is the expected gain of the device. Most recently, Van Trees [8] obtained a Volterra series representation of the closed-loop response by a perturbation method similar to the method employed by Margolis [6], but with the advantage of the simplified model he obtained more extensive results.

Unlike these analyses, Fokker-Planck or continuous random-walk techniques yield exact expressions for the statistics of the random process, $\phi(t)$. Unfortunately, expressions in closed form are available only for the first-order loop (*i.e.*, when the filter is omitted). For the general case a partial differential equation in ϕ and linear combinations of its time derivatives is derived, but a solution cannot be obtained in general. These techniques were first applied to this problem by Tikhonov [9], [10], who determined the steady-state probability distribution of ϕ for the first-order loop and an approximate expression for the distribution when the loop contains a one-stage RC filter.

All the analyses of this device have been concerned with the steady-state behavior. In this paper we shall obtain for the first-order loop not only its steady-state probability distribution and variance, but also the mean time to loss of lock, which is a transient phenomenon equivalent to a random-walk problem with absorbing boundaries. Also, we shall derive the Fokker-Planck equation for the general loop filter which produces zero mean error. We shall specialize this equation to the second-order loop and determine the form of the solution for the steady-state probability distribution of its phase error. In later sections we shall treat the effect of random modulation and tracking loops with other than sinusoidal error function. Finally we shall compare some of the results of the exact analysis with previous approximate results to determine the degree of validity of the approximate models.

First of all, in the next section a simple mechanical analog of the phase-locked loop is presented which provides a qualitative description of the operation of the device and an understanding of the nature of the statistical parameters required for its quantitative description.

The First-Order Loop and its Mechnical Analog

If the filter is omitted in Fig. 1, then $F(s) = 1$ in (10). Furthermore, if we take the received signal to be a sinusoid of constant frequency and phase so that $\theta_1(t)$ is given by (3), we obtain the first-order differential equation

$$\dot{\phi}(t) = (\omega - \omega_0) - K[A \sin \phi(t) + n'(t)]. \quad (12)$$

Hence the term "first-order loop." Since $n'(t)$ is a white Gaussian process, the instantaneous change in ϕ represented by its derivative depends only on the present value of ϕ and the present value of the noise. Hence, $\phi(t)$ is a continuous Markov process, and we may use random-walk techniques to determine its probability distribution.

A mechanical analog is conducive to understanding the mechanism of this "random walk." Consider the pendulum of Fig. 3 which consists of a weightless ball

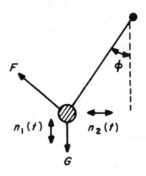

Fig. 3—Mechanical analog of the first-order loop.

attached by an infinitesimally thin weightless rod to a fixed point, and let the apparatus lie horizontally on a table top which is being randomly agitated. The pendulum is free to turn a full revolution about the point. Let the rod be initially at an angle ϕ with respect to the vertical axis and let an external force (such as a constant wind) be exerted on the ball in the vertical direction. Let the surface of the table be rough so that it produces a frictional force opposing motion of magnitude $f\dot{\phi}$. In addition, let the ball be equipped with an internal engine which exerts a constant force F along the axis of motion. The random agitation of the table produces a force on the ball which may be represented by two components which are stationary white Gaussian processes of zero means: $n_1(t)$ in the vertical direction, and $n_2(t)$ in the horizontal direction. Then by equating forces along the instantaneous axis of motion we obtain

$$f\dot{\phi} + G \sin \phi = F - n_1(t) \sin \phi - n_2(t) \cos \phi. \quad (13)$$

If we divide both sides of this equation by f and let

$$n_1 \sin \phi + n_2 \cos \phi = n',$$

$$F/f = \omega - \omega_0,$$

$$G/f = AK,$$

$$1/f = K,$$

we find that (13) is the same as (12). Also, the process $n'(t)$ defined here can be shown to be white and Gaussian by the same argument that was used in connection with the $n'(t)$ defined by (9) in the previous section. Thus the massless pendulum is the mechanical analog of a first-order loop with constant received frequency.

In the absence of the random forces it is clear that the pendulum approaches the equilibrium position

$$\phi_0 \sin^{-1}(F/G) = \sin^{-1}(\omega - \omega_0)/(AK), \qquad (14)$$

at which point the velocity is zero. Because this is a first-order system, there can be no overshoot. If $F > G$ or $(\omega - \omega_0) > AK$, there can be no equilibrium position, and the pendulum continues to revolve indefinitely, which corresponds to a loop that cannot achieve lock. When the random or noise forces are applied as well as the constant ones, the motion becomes a random walk, but when the noise variance is small, there is a strong tendency for the angle ϕ to approach and remain about this equilibrium position.

The complete statistical description of the random walk of the angle ϕ is given by its probability density as a function of time, $p(\phi, t)$. To understand qualitatively the behavior of this function, let us assume for the moment that the constant force $F = 0$ and that initially (at $t = 0$) the pendulum is at rest in the vertical position. Thus,[2] $p(\phi, 0) = \delta(\phi)$. With the passage of time, the effect of the random forces will be felt in the movement of the pendulum from the equilibrium position. The qualitative behavior of the probability density function is sketched in Fig. 4. Of course, the condition

$$\int_{-\infty}^{\infty} p(\phi, t) d\phi = 1$$

must always be met. After a sufficient amount of time, the random forces will push the pendulum around by more than half a revolution so that it will tend to return to the equilibrium position after a full cycle of rotation in either direction. This corresponds to the reference signal of the phase-locked loop advancing or retreating one cycle relative to the received signal. The average time for this occurrence depends on the signal-to-noise ratio. Thus after a sufficiently long period, the probability density will appear as a multimodal function, each mode being centered about equilibrium positions spaced 2π radians apart, the central mode being the largest with each successive maximum progressively smaller. After an even longer period equal to several times the aver-

[2] In certain standard treatments of the continuous random walk problem, the probability density function is denoted $p(\phi, t|\phi_0, 0)$ which signifies that we are dealing with the function at time t given that $\phi = \phi_0$ at $t = 0$. We avoid this cumbersome notation by specifying the initial condition on $p(\phi, t)$.

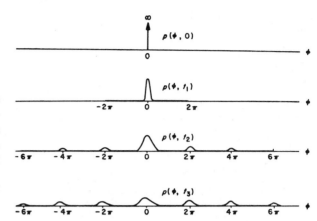

Fig. 4—Qualitative behavior of the probability density function for the first-order loop ($\omega = \omega_0$).

age time between revolutions, the central mode of the probability density will have diminished, the modes to either side will have become almost as large, and more modes of significant magnitude will have appeared. The central mode will remain the largest since the pendulum may have revolved in either direction with equal probability. Finally, in the steady state an arbitrary number of revolutions will have occurred.

In the case for which F is not zero, or equivalently $\omega \neq \omega_0$, then clearly the pendulum will have a greater tendency to swing around in the sense corresponding to the direction of the force. Hence, the density function $p(\phi, t)$ will not be symmetrical. In either case, we are led to realize that the significant parameter, at least in the steady state, is the angle (or phase error) ϕ modulo 2π, since the number of revolutions of the pendulum which have occurred does not affect the present state of the system. In fact, although $p(\phi, t)$ yields a complete description of the statistical behavior, it would appear that a combination of the steady-state distribution of ϕ modulo 2π, and the frequency or average time between revolutions would yield a simpler and nearly complete representation. In the next two sections these parameters will be obtained quantitatively.

THE STEADY-STATE PHASE-ERROR PROBABILITY DENSITY FOR THE FIRST-ORDER LOOP

A continuous random walk which is a Markov process is described by the statistical parameters of the incremental change of position as a function of the present position. Thus from (12) we note that in the infinitesimal increment of time Δt, the phase will change by an amount[3]

$$\Delta \phi = \int_{t}^{t+\Delta t} \dot{\phi}(t) dt = (\omega - \omega_0)\Delta t - (AK \sin \phi)\Delta t$$

$$- K \int_{t}^{t+\Delta t} n'(u) du.$$

[3] This assumes that $\phi(t)$ is a continuous process, which is justified by physical considerations.

Thus, since $n'(t)$ is a white Gaussian process with zero mean and

$$\overline{n'(u)n'(v)} = (N_0/2)\delta(u - v),$$

it follows that for a given position ϕ, $\Delta\phi$ is a Gaussian variable with mean

$$\overline{\Delta\phi} = [(\omega - \omega_0) - AK \sin \phi]\Delta t \qquad (15)$$

and variance

$$\sigma_{\Delta\phi}^2 = \overline{(\Delta\phi)^2} - (\overline{\Delta\phi})^2$$

$$= K^2 \int_t^{t+\Delta t} \int_t^{t+\Delta t} \overline{n'(u)n'(v)} \, dudv$$

$$= K^2(N_0/2)\Delta t. \qquad (16)$$

With the knowledge of the statistical parameters of the increment $\Delta\phi$, we may proceed to obtain $p(\phi, t)$. It was shown by Uhlenbeck and Ornstein [11] and Wang and Uhlenbeck [12] that for a continuous Markov process described by a first-order differential equation with a white Gaussian input, the instantaneous probability density $p(\phi, t)$ must satisfy the partial differential equation

$$\frac{\partial p(\phi, t)}{\partial t} = -\frac{\partial}{\partial \phi}[A(\phi)p(\phi, t)] + \frac{1}{2}\frac{\partial^2}{\partial \phi^2}[B(\phi)p(\phi, t)], \quad (17)$$

with the appropriate initial condition, where

$$A(\phi) = \lim_{\Delta t \to 0} \frac{1}{\Delta t} \overline{\Delta\phi}$$

$$B(\phi) = \lim_{\Delta t \to 0} \frac{1}{\Delta t} \overline{(\Delta\phi)^2},$$

provided

$$\lim_{\Delta t \to 0} \left(\frac{1}{\Delta t}\right)\overline{(\Delta\phi)^n} = 0 \qquad \text{for } n > 2.$$

Eq. (17) is known as the Fokker-Planck equation or the diffusion equation because it is a generalization of the equation for heat diffusion. From (15) and (16) we obtain for the first-order loop

$$A(\phi) = (\omega - \omega_0) - AK \sin \phi,$$
$$B(\phi) = K^2 N_0/2,$$

and it may be readily verified (using the property that the product of a set of Gaussian variables is the sum of products of pairs taken over all pairs of the variables) that

$$\lim_{\Delta t \to 0} \left(\frac{1}{\Delta t}\right)\overline{(\Delta\phi)^n} = 0 \qquad \text{for } n > 2.$$

Thus the Fokker-Planck equation holds for this case, and inserting the coefficients into (17) we obtain

$$\frac{\partial p}{\partial t} = \frac{\partial}{\partial \phi}[(AK \sin \phi + \omega_0 - \omega)p] + \frac{K^2 N_0}{4}\frac{\partial^2 p}{\partial \phi^2}. \quad (18)$$

If we take the initial value of ϕ to be ϕ_0, we have

$$p(\phi, 0) = \delta(\phi - \phi_0). \qquad (19)$$

As was pointed out in the previous section, we are really interested in $p(\phi, t)$, where ϕ is taken modulo 2π. Therefore, we are tempted to solve (18) in the region $-\pi \leq \phi \leq \pi$ with appropriate boundary conditions. To do this properly we must first realize that since the coefficients of the Fokker-Planck equation in this case are periodic in ϕ, if $p(\phi, t)$ is a solution, then so is $p(\phi + 2\pi n, t)$ where n is any integer. Let us define the function

$$P(\phi, t) = \sum_{n=-\infty}^{\infty} p(\phi + 2\pi n, t).$$

Since each term of the series is a solution of (18), then $P(\phi, t)$ must also satisfy the Fokker-Planck equation

$$\frac{\partial P(\phi, t)}{\partial t} = \frac{\partial}{\partial \phi}[(AK \sin \phi + \omega_0 - \omega)P(\phi, t)]$$

$$+ \frac{K^2 N_0}{4}\frac{\partial^2 P(\phi, t)}{\partial \phi^2}, \qquad (20)$$

with the initial condition

$$P(\phi, 0) = \sum_{n=-\infty}^{\infty} \delta(\phi - \phi_0 - 2\pi n).$$

Also, $P(\phi, t)$ must be periodic in ϕ since for any integer m

$$P(\phi + 2\pi m, t) = \sum_{n=-\infty}^{\infty} p[\phi + 2(m + n), t]$$

$$= \sum_{k=-\infty}^{\infty} p(\phi + 2k\pi, t) = P(\phi, t).$$

Therefore, we may solve (20) over the interval of just one period $(-\pi \leq \phi \leq \pi)$ with the initial condition

$$P(\phi, 0) = \delta(\phi - \phi_0), \quad -\pi \leq \phi \leq \pi, \qquad (21)$$

the boundary condition

$$P(\pi, t) = P(-\pi, t) \quad \text{for all } t, \qquad (22)$$

and the normalizing condition

$$\int_{-\pi}^{\pi} P(\phi, t) \, d\phi = 1 \quad \text{for all } t. \qquad (23)$$

Although in principle the linear partial differential (20) with the conditions (21), (22), and (23) can be solved for $P(\phi, t)$, the procedure is somewhat complicated by the nonlinear behavior of the variable coefficients, and a closed-form solution cannot be obtained. On the other hand, the result of greatest interest is the steady-state distribution

$$P(\phi) = \lim_{t \to \infty} P(\phi, t). \qquad (24)$$

By definition, the steady-state distribution is stationary.

Therefore,

$$\frac{\partial P(\phi)}{\partial t} = \lim_{t \to \infty} \frac{\partial P(\phi, t)}{\partial t} = 0. \tag{25}$$

Thus, in the steady state, the partial differential (20) reduces to an ordinary differential equation in $P(\phi)$. Letting

$$\alpha = (4A)/(KN_0) \tag{26}$$

and

$$\beta = [4(\omega - \omega_0)]/(K^2 N_0), \tag{27}$$

we obtain

$$0 = \frac{d}{d\phi}\left[(\alpha \sin \phi - \beta)P(\phi) + \frac{dP(\phi)}{d\phi}\right]. \tag{28}$$

If we integrate once with respect to ϕ, we obtain a first-order linear differential equation which is readily solved as[4]

$$P(\phi) = C \exp(\alpha \cos \phi + \beta\phi)$$
$$\left[1 + D \int_{-\pi}^{\phi} \exp - (\alpha \cos x + \beta x) dx\right]$$
$$-\pi \le \phi \le \pi. \tag{29}$$

To evaluate the constants, we must utilize the limiting form of the conditions (22) and (23); *i.e.*,

$$P(\pi) = P(-\pi) \tag{30}$$

and

$$\int_{-\pi}^{\pi} P(\phi)d\phi = 1. \tag{31}$$

Then using (30), we obtain

$$D = \frac{\exp(-2\beta\pi) - 1}{\int_{-\pi}^{\pi} \exp - (\alpha \cos x + \beta x) dx}, \tag{32}$$

and by means of (31), the constant C can be evaluated. In the special case $\beta = 0$ (which requires $\omega = \omega_0$; *i.e.*, when the frequency of the received signal is determined beforehand and the VCO quiescent frequency is tuned to this frequency so that the problem consists only of acquiring and tracking phase), we note from (32) and (27) that $D = 0$ so that

$$P(\phi) = \frac{\exp(\alpha \cos \phi)}{2\pi I_0(\alpha)} \quad -\pi \le \phi \le \pi, \tag{33}$$

[4] The results of (29), (32), and (33) were first obtained by V. I. Tikhonov [9]. Actually, these are a special case of an expression for a random-walk problem with arbitrary nonlinear restoring forces derived in A. A. Andronov, L. S. Pontryagin, and A. A. Witt [13].

since

$$C = \frac{1}{\int_{-\pi}^{\pi} \exp(\alpha \cos \phi)d\phi} = \frac{1}{2\pi I_0(\alpha)}.$$

The parameter α plays a very important role. From (26) we have

$$\alpha = \frac{(4A)}{(KN_0)} = \frac{(A^2)}{[N_0(AK/4)]}. \tag{34}$$

But A^2 is the received signal power, while $AK/4$ is an important parameter defined for the linearized model of the loop. If we replace the sinusoidal nonlinearity in the model of Fig. 2 by its gain A about $\phi = 0$, we obtain the linearized model. Then the variance of ϕ is obtained by using Parseval's theorem as:

$$\sigma_\phi^2 = \frac{1}{2\pi}\int_{-\infty}^{\infty} \frac{N_0}{2} \frac{K^2/\omega^2}{[1 + (A^2 K^2/\omega^2)]} d\omega$$
$$= \frac{N_0(AK/4)}{A^2} = \frac{1}{\alpha}. \tag{35}$$

The variance of ϕ is the same as the noise power at the output of an ideal low-pass filter of bandwidth $AK/4$ when the input is white noise of one-sided spectral density N_0. Hence, for the first-order filter, the loop bandwidth is defined as

$$B_L = AK/4, \tag{36}$$

so that (34) becomes

$$\alpha = (A^2)/(N_0 B_L), \tag{37}$$

which is the SNR in the bandwidth of the loop.

Eq. (33) is plotted in Fig. 5 (p. 1744) for several values of α. It resembles a Gaussian distribution when the SNR, α, is large and becomes flat as α approaches zero. The asymptotic behavior of (33) for large α is of interest. Since for large α

$$I_0(\alpha) \sim (\exp \alpha)/(2\pi\alpha)^{1/2},$$
$$P(\phi) = \frac{[\exp(\alpha \cos \phi)]}{[2\pi I_0(\alpha)]} \sim \frac{\{\exp[\alpha(\cos \phi - 1)]\}}{(2\pi/\alpha)^{1/2}}.$$

Expanding $\cos \phi$ in a Taylor series, we obtain

$$P(\phi) \sim \frac{\exp\left[\frac{-\alpha\phi^2}{2}\left(1 - \frac{2\phi^4}{4!} + \frac{2\phi^6}{6!} \cdots\right)\right]}{(2\pi/\alpha)^{1/2}}$$
$$-\pi \le \phi \le \pi. \tag{38}$$

When α is large, $P(\phi)$ decays rapidly, so that the function is very small for all but very small values of ϕ. Thus the higher-order terms of the series representation of $\cos \phi$ have very little effect for moderate values of $P(\phi)$. Hence, the graph of $P(\phi)$ will appear to be nearly Gaussian for large α and, in this case, the results of the linear model are quite accurate.

The cumulative steady-state probability distribution

$$\text{Prob}\,(|\,\phi\,| \,<\phi_1) = \int_{-\phi_1}^{\phi_1} p(\phi)d\phi \qquad 0 < \phi_1 < \pi$$

is also of interest since it indicates the percentage of time during which the absolute value of the loop phase error ϕ is less than a given magnitude ϕ_1. This may be calculated when $\omega = \omega_0$ in the following manner. Expanding $P(\phi)$ of (33) in a Fourier series, we have

$$P(\phi) = \frac{\exp\,(\alpha\cos\phi)}{2\pi I_0(\alpha)}$$

$$= \frac{1}{2\pi I_0(\alpha)}\Bigg[I_0(\alpha) + 2\sum_{n=1}^{\infty} I_n(\alpha)\cos n\phi\Bigg]$$

Then

$$\text{Prob}\,(|\,\phi\,| \,<\phi_1) = 2\int_0^{\phi_1} P(\phi)d(\phi)$$

$$= \frac{\phi_1}{\pi} + \frac{2}{\pi}\sum_{n=1}^{\infty}\frac{I_n(\alpha)\sin n\phi_1}{nI_0(\alpha)}$$

for

$$0 < \phi_1 < \pi \quad \text{and} \quad \omega = \omega_0. \tag{39}$$

This series converges so rapidly that (39) could be calculated for several values of α without the use of a large-scale digital computer. The results are shown in Fig. 6. The variance of ϕ can be similarly obtained:

$$\sigma_\phi^2 = \int_{-\pi}^{\pi} \phi^2\exp\,(\alpha\cos\phi)d\phi/2\pi I_0(\alpha)$$

$$= \frac{1}{2\pi I_0(\alpha)}\int_{\pi}^{\pi}\phi^2\Bigg[I_0(\alpha) + 2\sum_{n=1}^{\infty} I_n(\alpha)\cos n\phi\Bigg]d\phi$$

$$= \frac{\pi^2}{3} + 4\sum_{n=1}^{\infty}\frac{(-1)^n I_n(\alpha)}{n^2 I_0(\alpha)}. \tag{40}$$

This series converges even more rapidly than that of (39). It is plotted in Fig. 7 as a function of $1/\alpha$. Note that as the SNR, α, approaches zero, the variance approaches $\pi^2/3$, which is the variance of a random variable that is uniformly distributed from $-\pi$ to $+\pi$. Also shown in Fig. 7 is the variance of the phase error as determined from the linear model (35). It is evident that the linear model for the first-order loop with $\omega = \omega_0$ and no signal modulation produces results of reasonable accuracy (within 20 per cent) for $1/\alpha < 1/4$ or when the loop SNR, $\alpha > 4$.

For the general case $(\omega \neq \omega_0)$, (29), (31), and (32) yield the entire distribution. However, analog or digital computation is required to evaluate the pertinent integrals. The case for which $(\beta/\alpha) = (\omega - \omega_0)/(AK) = \sin(\pi/4)$ is shown in Fig. 8. The constants as well as the distribution were obtained by means of the analog computer.

MEAN TIME TO LOSS OF LOCK AND FREQUENCY OF SKIPPING CYCLES

Since we have obtained only solutions for steady-state probabilities, a valuable statistic is the expected time required for the absolute value of the phase error to exceed some value ϕ_l when it is initially zero. When this occurs the loop will be said to have lost lock. Of particular interest is the case for which $\phi_l = 2\pi$, which represents a loss or gain of a complete cycle, or for the mechanical analog, a complete revolution of the pendulum.

In the framework of the mechanical analog of the first-order loop, we may represent the out-of-lock boundaries by two knife edges at angles $+\phi_l$ and $-\phi_l$ relative to the vertical (Fig. 9). The pendulum is initially at an angle ϕ_0 relative to the vertical when the random external forces are applied. The first time that the angle reaches $\pm\phi_l$ the knife edges cut the rod and the process terminates.

This so-called first-passage time problem for Markov processes has been treated extensively in the literature [13]–[15]. It is possible to determine not only the first moment but all moments and even the distribution of the first-passage time for a Markov process described by a first-order differential equation with a white Gaussian driving function. However, computational difficulties render the form of the solution rather complex in all but the simplest cases. We shall employ a somewhat different method than previously used to obtain a simple expression for the expected time to the first occurrence of loss of lock. Closely related to the mean time to loss of lock is the frequency of skipping cycles. For the mechanical analog, this is the inverse of the expected time for the pendulum to swing a complete revolution in either direction. For the phase-locked loop, this represents the frequency of occurrence of the event that the loop VCO gains or drops a cycle relative to the received signal. In either case this corresponds to letting $\phi_l = 2\pi$ in the determination of expected time. This is a very important parameter for tracking applications in which the phase-locked loop is used to measure the received Doppler frequency which is then integrated to determine relative range. An error of a full cycle will cause a significant error in the result.

We treat only the case of the first-order loop for which the VCO quiescent frequency is tuned to the received frequency $(\omega_0 = \omega)$ so that the equilibrium position is at $\phi = 0$. This is also a good approximation to the steady-state behavior of the second-order loop with any value of $\omega - \omega_0$, but with very small integrator gain a, as will be discussed in a later section. For the first-order loop, when $\omega \neq \omega_0$, the same approach can be used, measuring phase error from the equilibrium position rather than from zero, but the results are in the form of integrals which require numerical calculation.

Let us assume that the loop is initially in lock so that

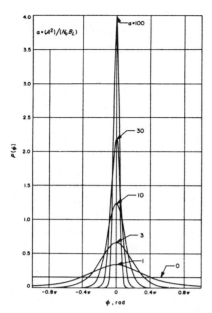

Fig. 5—First-order loop steady-state probability densities for $\omega = \omega_0$.

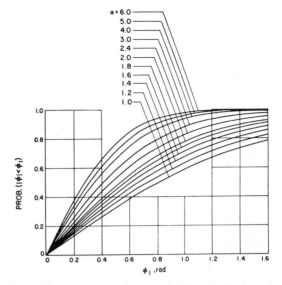

Fig. 6—Steady-state cumulative probability distributions of first-order loop for $\omega = \omega_0$.

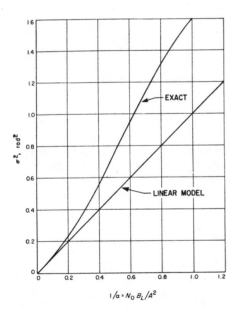

Fig. 7—Variance of phase-error for first-order loop where $\omega = \omega_0$.

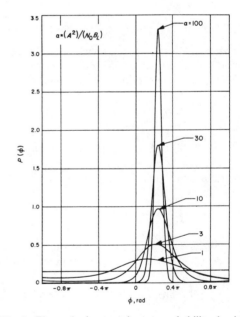

Fig. 8—First-order loop steady-state probability densities for $(\omega - \omega_0)/AK = \sin(\pi/4)$.

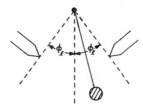

Fig. 9—Mechanical analog of first-passage time problem for first-order loop.

$\phi_0 = 0$. As long as the loop phase error (or the pendulum angle) remains within the limits $|\phi| < \phi_l$, the probability density of ϕ, which we denote here by $q(\phi, t)$, is described in the same manner as before by the Fokker-Planck equation

$$\frac{\partial q}{\partial t} = \frac{\partial}{\partial \phi}(AK \sin \phi q) + \frac{N_0 K^2}{4} \frac{\partial^2 q}{\partial \phi^2}, \qquad (41)$$

with the initial condition

$$q(\phi, 0) = \delta(\phi) \qquad \text{for } |\phi| < \phi_l,$$

when

$$\omega = \omega_0.$$

We have denoted the probability density function of the phase error by $q(\phi, t)$ to distinguish it from the corresponding function $p(\phi, t)$ for the previous case without boundaries. As soon as $|\phi|$ reaches ϕ_l for the first time, the pendulum is removed from action so that

$$q(\phi, t) = 0 \qquad \text{for all } |\phi| \geq \phi_l. \qquad (42)$$

Thus, in addition to the initial condition we have the boundary condition

$$q(\phi_l, t) = q(-\phi_l, t) = 0 \qquad \text{for all } t. \qquad (43)$$

Solution of (41) over the interval $-\phi_l < \phi < \phi_l$ with the boundary conditions of (43) would yield the probability density $q(\phi, t)$. Its integral over the interval

$$\psi(t) = \int_{-\phi_l}^{\phi_l} q(\phi, t)d\phi \qquad (44)$$

gives the probability that ϕ has not yet reached ϕ_l at time t. Note that as a consequence of (42) the limits on the integral of (44) could just as well be infinite. In fact,

$$\int_{-\infty}^{\infty} q(\phi, t)d\phi = \psi(t) \leq 1,$$

and this points out the fundamental difference between $q(\phi, t)$ and $p(\phi, t)$ of the previous section for which

$$\int_{-\infty}^{\infty} p(\phi, t)d\phi = 1 \qquad \text{for all } t.$$

In other words, $q(\phi, t)$ is not strictly a probability density function. In order to render it such we would have to normalize it by the probability that $|\phi|$ has never exceeded ϕ_l by the time $t[i.e., \psi(t)]$.

$\psi(t)$ must be a monotonically nonincreasing function of t, and from its definition it follows that the probability density function of the time required for ϕ to reach ϕ_l for the first time is $-[\partial\psi(t)/\partial t]$. Thus the expected time to reach the out-of-lock position ϕ_l is

$$T = \int_0^{\infty} -t \frac{\partial\psi(t)}{\partial t} dt = -[t\psi(t)]_0^{\infty} + \int_0^{\infty} \psi(t)dt. \qquad (45)$$

If the nonincreasing function $\psi(t)$ approaches zero faster than $0(1/t)$, the first term on the right side of (45) is zero. This must be the case, for otherwise the integral of the second term would not exist. Then using (44) we obtain the expression for the expected time

$$T = \int_0^{\infty} \int_{-\phi_l}^{\phi_l} q(\phi, t)d\phi dt. \qquad (46)$$

Now if we integrate both sides of (41) with respect to t over the infinite interval, we obtain

$$q(\phi, \infty) - q(\phi, 0)$$
$$= \frac{\partial}{\partial \phi}[AK \sin \phi Q(\phi)] + \frac{N_0 K^2}{4} \frac{\partial^2 Q(\phi)}{\partial \phi^2}, \qquad (47)$$

where

$$Q(\phi) = \int_0^{\infty} q(\phi, t)dt.$$

Clearly $q(\phi, \infty) = 0$, and since ϕ is assumed initially at zero, $q(\phi, 0) = \delta(\phi)$. Therefore, we have

$$-\delta(\phi) = \frac{\partial}{\partial \phi}[AK \sin \phi Q(\phi)] + \frac{N_0 K^2}{4} \frac{\partial^2 Q(\phi)}{\partial \phi^2}, \qquad (48)$$

with the boundary conditions

$$Q(\phi_l) = \int_0^{\infty} q(\phi_l, t)dt = 0$$

$$Q(-\phi_l) = \int_0^{\infty} q(-\phi_l, t)dt = 0, \qquad (49)$$

which follow from (43). The solution of (48) may then be integrated with respect to ϕ over the interval $[-\phi_l, \phi_l]$ to obtain T, the expected time of (46). Taking the indefinite integral of both sides of (48), we obtain

$$C - u(\phi) = AK \sin \phi Q(\phi) + \frac{N_0 K^2}{4} \frac{\partial Q(\phi)}{\partial \phi}, \qquad (50)$$

where $u(\phi)$ is the unit step function and C is a constant to be evaluated from the boundary conditions. The solution to the first-order differential equation is

$$Q(\phi) = D \exp(\alpha \cos \phi)$$
$$+ \exp(\alpha \cos \phi) \int_{-\phi_l}^{\phi} \frac{\exp(-\alpha \cos x)}{\gamma}[C - u(x)]dx, \qquad (51)$$

where

$$\alpha = \frac{A^2}{N_0(AK/4)}$$

and

$$\gamma = \frac{N_0 K^2}{4} = \frac{AK}{\alpha} = \frac{4B_L}{\alpha}.$$

Application of the boundary conditions of (49) yields the values of the constants as $D=0$ and $C=1/2$. Thus,

$$Q(\phi) = \frac{\exp(\alpha\cos\phi)}{\gamma} \int_{-\phi_l}^{\phi} \exp(-\alpha\cos x)[\tfrac{1}{2}-u(x)]dx. \quad (52)$$

Integrating with respect to ϕ over the interval $[-\phi_l, \phi_l]$, we obtain the expression for the mean time to lose lock:

$$T = \int_{\phi_l}^{\phi_l} Q(\phi)d\phi = \frac{1}{\gamma}\int_{-\phi_l}^{\phi_l} d\phi$$

$$\times \int_{-\phi_l}^{\phi} \exp\alpha(\cos\phi - \cos x)[\tfrac{1}{2} - u(x)]dx$$

$$= \frac{1}{\gamma}\int_0^{\phi_l}\int_{\phi}^{\phi_l} \exp\alpha(\cos\phi - \cos x)dxd\phi. \quad (53)$$

The domain of integration is the right isosceles triangle shown in Fig. 10. We can obtain a series representation of this double integral by expanding the integrands in Fourier series:

$$\exp(\alpha\cos\phi) = I_0(\alpha) + 2\sum_{m=1}^{\infty} I_m(\alpha)\cos m\phi$$

$$\exp(-\alpha\cos x) = I_0(\alpha) + 2\sum_{n=1}^{\infty}(-1)^n I_n(\alpha)\cos nx. \quad (54)$$

Then

$$T = \frac{1}{\gamma}\int_0^{\phi_l}\int_{\phi}^{\phi_l}$$

$$\times \begin{bmatrix} I_0^2(\alpha) + 4I_0(\alpha)\sum_{n=2,4,6\cdots}(-1)^n I_n(\alpha)\cos n\phi \\ + 4\sum_{m=1}^{\infty}\sum_{n=1}^{\infty} -(1)^n I_m(\alpha)I_n(\alpha)\cos m\phi\cos nx \end{bmatrix} d\phi dx$$

$$= \frac{1}{\gamma}\begin{bmatrix} \frac{I_0^2(\alpha)\phi_l^2}{2} + 4I_0(\alpha)\sum_{n=1}^{\infty}\frac{I_{2n}(\alpha)}{2n}\sin 2n\phi_l \\ + 4\sum_{m=1}^{\infty}\sum_{n=1}^{\infty}(-1)^n I_m(\alpha)I_n(\alpha) \end{bmatrix}$$

$$\times \int_0^{\phi_l}\int_{\phi}^{\phi_l}\cos m\phi\cos nxdxd\phi \Bigg], \quad (55)$$

where

$$\int_0^{\phi_l}\int_{\phi}^{\phi_l}\cos m\phi\cos nxdxd\phi$$

$$= \begin{cases} \cos(n-m)\phi_l\left[\dfrac{1}{nm} + \dfrac{1}{n(n-m)}\right] \\ \qquad - \dfrac{4\cos n\phi_l}{nm} - \dfrac{1}{(n-m)n} \end{cases} \quad \text{when } n = m$$

$$= \left(\frac{1}{m^2} - \frac{\cos m\phi_l}{m^2}\right) \quad \text{when} \quad n = m. \quad (56)$$

This expression may be computed without the aid of a large-scale digital computer because the sequence

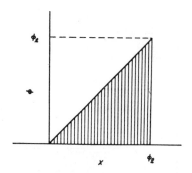

Fig. 10—Domain of integral T.

$I_n(\alpha)$, and consequently the above series, converges quite rapidly.

However, the most important result which we seek can be obtained in closed form. This is the frequency of skipping cycles, or, in other words, the inverse of the expected time between skipping cycles, which is $T(\phi_l = 2\pi)$. It is clear from (55) and (56) that when $\phi_l = 2\pi$,

$$T(2\pi) = \frac{2\pi^2}{\gamma}I_0^2(\alpha) = \frac{\pi^2\alpha I_0^2(\alpha)}{2B_L}, \quad (57)$$

where we have used

$$\gamma = \frac{N_0 K^2}{4} = \frac{AK}{\alpha} = \frac{4B_L}{\alpha}$$

so that

$$\text{frequency of skipping cycles} = (2B_L)/\pi^2\alpha I_0^2(\alpha). \quad (58)$$

This parameter normalized by B_L is shown as a function of α in Fig. 11.

For large SNR, α,

$$I_0(\alpha) \sim (e^\alpha)/(2\pi\alpha)^{1/2},$$

so that for large α,

$$\text{frequency of slipping cycles} \simeq [(4B_L)/\pi]e^{-2\alpha}. \quad (59)$$

Another parameter which is equally significant is the frequency of dropping or advancing half cycles ($\phi_l = \pi$). In the mechanical analog this corresponds to the pendulum arriving at the unstable equilibrium position and returning to the stable equilibrium position, either by the same route or by going around the full revolution. It is nearly intuitive that for a Markov process the frequency of this event is exactly double the frequency of skipping cycles. However, to show this rigorously, we note that the expected time for the pendulum to go from the equilibrium position $\phi = 0$ to $\phi = \pi$ and to return is $T(\pi) + T'(\pi)$, where $T(\pi)$ is the expected time to go from 0 to $\pm\pi$, and $T'(\pi)$ is the expected time to go from π to either 0 or 2π. $T(\pi)$ is given by (55) with $\phi_l = \pi$, while we can show that

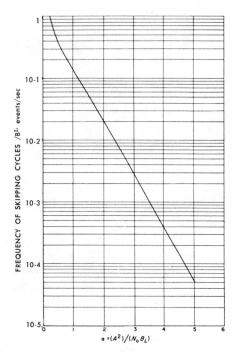

Fig. 11—Frequency of skipping cycles normalized by loop bandwidth for first-order loop where $\omega = \omega_0$.

$$T'(\pi) = \frac{1}{\gamma} \int_0^{\phi_l} \int_0^{\phi} \exp \alpha(\cos \phi - \cos x) dx d\phi.$$

The integrand is the same as that for $T(\pi)$, but the domain of integration is its complement with respect to the square of side π (Fig. 12). Therefore,

$$T(\pi) + T'(\pi) = \frac{1}{\gamma} \int_0^{\pi} \int_0^{\pi} \exp \alpha(\cos \phi - \cos x) dx d\phi$$

$$= (\pi^2/\gamma) I_0^2(\alpha) = [T(2\pi)/2], \qquad (60)$$

and

frequency of slipping half cycles $= (4B_L)/[\pi^2 \alpha I_0^2(\alpha)]$. (61)

THE FOKKER-PLANCK EQUATION FOR HIGHER-ORDER LOOPS

Consider the phase-locked loop whose filter has the rational transfer function

$$F(s) = G(s)/H(s),$$

where $G(s)$ and $H(s)$ are polynomials such that $G(0) = 1$, $H(0) = 0$ and

$$\deg G(s) \leq \deg H(s) = n - 1.$$

Then

$$G(s) = \sum_{k=0}^{n-1} a_k s^k; \qquad a_0 \neq 0$$

$$H(s) = \sum_{k=1}^{n-1} b_k s^k; \qquad b_{n-1} \neq 0. \qquad (62)$$

This will be referred to as an nth-order loop. In this

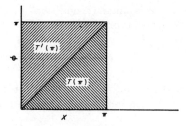

Fig. 12—Domains of integration for $T(\pi)$ and $T'(\pi)$.

case, (11) which describes the operation of the loop becomes

$$sH(s)\phi = -KG(s)(A \sin \phi + n'), \qquad (63)$$

since

$$s^k(\omega - \omega_0) = \frac{d^k}{dt}(\omega - \omega_0) = 0 \qquad \text{for } k \geq 1.$$

The reason for the pole at the origin of $F(s)$ is now clear. It eliminates the constant $(\omega - \omega_0)$ which causes the steady-state phase error in the first-order loop. Now let us define the random variable ϵ by the relation[5]

$$\phi = G(s)\epsilon. \qquad (64)$$

Inserting this in (63), we obtain

$$sH(s)\epsilon = -K(A \sin \phi + n'), \qquad (65)$$

which is an nth-order differential equation. Now let us define the n-random variables $x_0, x_1, \cdots, x_{n-1}$ as

$$x_k = \frac{d^k \epsilon}{dt^k} \quad (k = 0, 1, \cdots, n - 1.) \qquad (66)$$

Inserting these for the derivatives of ϵ in (65) and using (62), we obtain

$$b_{n-1}\dot{x}_{n-1} + \sum_{k=1}^{n-2} b_k x_{k+1} = -K(A \sin \phi + n').$$

Also, we have

$$x_k = \frac{d}{dt}\frac{d^{k-1}\epsilon}{dt^{k-1}} = \dot{x}_{k-1},$$

so that we may express the derivatives $\dot{x}_k$ in terms of the variables x_k by the n differential equations

$$\dot{x}_{n-1} = -\sum_{k=1}^{n-2} \frac{b_k}{b_{n+1}} x_{k+1} - \frac{K(A \sin \phi + n')}{b_{n-1}}$$

$$\dot{x}_{n-2} = x_{n-1}$$

$$\vdots \qquad \vdots$$

$$\dot{x}_0 = x_1 \qquad (67)$$

[5] This substitution which leads to the representation of ϕ as sum of the components of a Markov vector (67) was suggested by J. N. Franklin.

It follows also from (62), (64), and (66) that

$$\phi = \sum_{k=0}^{n-1} a_k x_k. \tag{68}$$

The random vector $(x_0, \cdots, x_{n-1})$ is a Markov vector since an incremental change depends only on the present state of the vector.

Wang and Uhlenbeck [12] have shown that for a vector Markov process, $x = (x_0, x_1, \cdots, x_{n-1})$, the Fokker-Planck equation is

$$\frac{\partial p(x)}{\partial t} = - \sum_{k=0}^{n-1} \frac{\partial}{\partial x_k} [A_k(x) p(x)]$$
$$+ \frac{1}{2} \sum_{k=0}^{n-1} \sum_{l=0}^{n-1} \frac{\partial^2}{\partial x_k \partial x_l} [B_{kl}(x) p(x)],$$

where

$$A_k(x) = \lim_{\Delta t \to 0} \frac{1}{\Delta t} \overline{(\Delta x_k)},$$

and

$$B_{kl}(x) = \lim_{\Delta t \to 0} \frac{1}{\Delta} \overline{(x_k)(x_l)} \tag{69}$$

with the initial condition

$$p(x, 0) = \prod_{k=0}^{n-1} \delta(x_k - x_{k,0}).$$

In our case,

$$A_k(x) = x_{k+1} \quad \text{for } k = 0, 1, \cdots, n-2$$

$$A_{n-1}(x) = - \sum_{k=1}^{n-2} \frac{b_k}{b_{n-1}} x_{k+1} - \frac{KA}{b_{n-1}} \sin \phi$$

$$B_{n-1,n-1}(x) = \lim_{\Delta t \to 0} \frac{1}{\Delta t} \frac{K^2}{b_{n-1}^2} \int_t^{t+\Delta t} \int_t^{t+\Delta t} \overline{n'(u) n'(v)} \, du \, dv$$
$$= \frac{K^2 N_{0/2}}{b_{n-1}^2}$$

$$B_{k,l}(x) = 0 \quad \text{for all } k \neq n-1 \text{ and } l \neq n-1.$$

Thus, the Fokker-Planck equation for the nth-order loop is

$$\frac{\partial p(x, t)}{\partial t} = - \sum_{k=0}^{n-2} x_{k+1} \frac{\partial p(x, t)}{\partial x_k} + \frac{1}{b_{n-1}} \frac{\partial}{\partial x_{n-1}}$$
$$\times \left[\left(\sum_{k=0}^{n-2} b_k x_{k+1} + KA \sin \phi \right) p(x, t) \right]$$
$$+ \frac{K^2 N_0}{4 b_{n-1}^2} \frac{\partial^2 p(x, t)}{\partial x_{n-1}^2}, \tag{70}$$

where

$$\phi = \sum_{k=0}^{n-1} a_k x_k.$$

Solution of this general case does not appear possible. However, in the next section some results are obtained for the second-order loop.

STEADY-STATE PROBABILITY DISTRIBUTION FOR THE SECOND-ORDER LOOP

The loop filter of greatest interest[6] is

$$F(s) = 1 + (a/s) = (s + a)/(s),$$

which requires a single integrator with gain a. In terms of the parameters of (62), $n = 2$, $a_0 = a$, $a_1 = 1$, $b_1 = 1$. Substituting these parameters in (67) and (70), we obtain the differential equations for the random variables

$$\dot{x}_1 = - K[(A \sin \phi + n'], \tag{71a}$$
$$\dot{x}_0 = x_1, \tag{71b}$$

and the Fokker-Planck equation

$$\frac{\partial p}{\partial t} = - x_1 \frac{\partial p}{\partial x_0} + \frac{\partial}{\partial x_1} [(AK \sin \phi) p] + \frac{K^2 N_0}{4} \frac{\partial^2 p}{\partial x_1^2}, \tag{72}$$

where

$$\phi = a x_0 + x_1. \tag{73}$$

If we restrict our attention to the steady-state probability distribution

$$p(x_0, x_1) = \lim_{t \to \infty} p(x_0, x_1, t).$$

Since

$$\lim_{t \to \infty} \frac{dp}{dt} (x_0, x_1, t) = 0,$$

we obtain

$$x_1 \frac{\partial p}{\partial x_0} = AK \frac{\partial}{\partial x_1} [(\sin \phi) p] + \frac{K^2 N_0}{4} \frac{\partial^2 p}{\partial x_1^2}.$$

With the substitution

$$z = a x_0, \tag{74}$$

we obtain an equation in $p(\phi, z)$ (note that the Jacobian of the transformation is a),

$$a(\phi - z) \left(\frac{\partial p}{\partial \phi} + \frac{\partial p}{\partial z} \right) = AK \frac{\partial}{\partial \phi} (\sin \phi p) + \frac{K^2 N_0}{4} \frac{\partial^2 p}{\partial \phi^2}. \tag{75}$$

Even this partial differential equation cannot be solved directly. However, since we are interested only in the density function of ϕ,

$$p(\phi) = \int_{-\infty}^{\infty} p(\phi, z) dz,$$

we may integrate both sides of (75) with respect to z

---

[6] Tikhonov [9], considered the RC low-pass filter whose transfer function is $1/(s+b)$. Its value is limited, however, since it does not reduce the mean phase-error to zero, as the perfect integrator does.

over the infinite line and obtain an ordinary differential equation in $p(\phi)$

$$a\left\{\frac{d(\phi p)}{d\phi} - \frac{d}{d\phi}\left[\int_{-\infty}^{\infty} zp(\phi, z)dz\right]\right\}$$
$$= AK\frac{d}{d\phi}(\sin\phi p) + \frac{K^2N_0}{4}\frac{d^2p}{d\phi^2}. \quad (76)$$

But

$$\int_{-\infty}^{\infty} zp(\phi, z)dz = p(\phi)\int_{-\infty}^{\infty} zp(z \mid \phi)dz = p(\phi)E(z \mid \phi),$$

so that (76) becomes

$$0 = \frac{d}{d\phi}\left\{[AK\sin\phi - a\phi + aE(z \mid \phi)]p + \frac{K^2N_0}{4}\frac{dp}{d\phi}\right\}. \quad (77)$$

Unfortunately, it is not possible to determine exactly $E(z \mid \phi)$, which is a function of ϕ, without knowing $p(z, \phi)$, which would require solution of (75). However, its general form can be obtained as follows: from (73) and (74) we have $z = \phi - x_1$, so that

$$E[z(t) \mid \phi(t)] = E[\phi(t) - x_1(t) \mid \phi(t)]$$
$$= \phi(t) - E[x_1(t) \mid \phi(t)]. \quad (78)$$

Integrating (71a) using (74), we have

$$x_1(\infty) - x_1(t) = -AK\int_{t}^{\infty}\sin\phi(\xi)d\xi$$
$$- K\int_{t}^{\infty} n'(\xi)d\xi. \quad (79)$$

The expectation of the second term on the right side of (79) is zero, since $n'(t) = 0$ for all t. Also,

$$E[x_1(\infty) \mid \phi(t)] = E[x_1(\infty)] = 0,$$

since it is clear that the mean of the process is zero. Therefore,

$$E[x_1(t) \mid \phi(t)] = AK\int_{j}^{\infty} E[\sin\phi(\xi) \mid \phi(t)]d\xi. \quad (80)$$

Combining (77), (78), and (80), and letting $\xi = t + \tau$, we obtain

$$0 = \frac{d}{d\phi}\left\{\frac{4A}{KN_0}\left(\sin\phi - a\int_{0}^{\infty} E[\sin\phi(t+\tau) \mid \phi(t)]d\tau\right)p(\phi)\right.$$
$$\left. + \frac{dp(\phi)}{d\phi}\right\}. \quad (81)$$

At this point we are dealing with the random phase process ϕ. We may once again deal with ϕ modulo 2π, or equivalently define the function

$$P(\phi) = \sum_{n=-\infty}^{\infty} p(\phi + 2\pi n)$$

as we did for the first-order loop. Since the coefficients of (81) are periodic in ϕ, if $p(\phi)$ is a solution then so is

$p(\phi + 2\pi n)$ for all integers n, and hence so is $P(\phi)$. Thus we may replace $p(\phi)$ in (81) by $P(\phi)$. The magnitude of the expectation is always less than one, and becomes negligible for values of τ several times the inverse bandwidth of the spectrum of $\phi(t)$. This bandwidth is proportional to AK, as we found for the first-order loop. Therefore, the order of magnitude of the integral is inversely proportional to AK, and if $a \ll AK$, the second term in the coefficient of $P(\phi)$ is much smaller than the first. Neglecting this second term reduces (81) to the steady-state Fokker-Planck equation for the first-order loop (28) with $\omega = \omega_0$, whose solution is (33). Thus when the second integrator gain $a \ll AK$,

$$P(\phi) \simeq \frac{\exp(\alpha\cos\phi)}{2\pi I_0(\alpha)} \quad -\pi \le \phi \le \pi. \quad (82)$$

On the other hand, for any value of a, when the SNR is large enough, $\phi(t)$ will be small for all time, so that $\sin\phi(t) \simeq \phi(t)$ and both $\phi(t)$ and $\sin\phi(t)$ will be nearly Gaussian processes. In this case, the expectation can be approximated by

$$\int_{0}^{\infty} E[\sin\phi(t+\tau) \mid \phi(t)]d\tau \simeq \int_{0}^{\infty} E[\phi(t+\tau) \mid \phi(t)]d\tau$$
$$= \left[\int_{0}^{\infty}\rho_\phi(\tau)d\tau\right]\sin\phi, \quad (83)$$

where $\rho_\phi(\tau)$ is the normalized autocorrelation function of the stationary process $\phi(t)$. The integral can be obtained by using Parseval's theorem:

$$\int_{0}^{\infty}\rho_\phi(\tau)dt = \frac{1}{2\sigma^2}\int_{-\infty}^{\infty} R_\phi(\tau)d\tau = \frac{S_\phi(0)}{2\sigma^2},$$

where $R_\phi(\tau)$ is the unnormalized autocorrelation function, σ^2 the variance of ϕ, and $S_\phi(\omega)$ the spectral density. Since we have approximated $\sin\phi$ by ϕ, we may use the linearized version of Fig. 2 with the loop filter $F(s) = 1 + (a/s)$ inserted. Then

$$S_\phi(\omega) = \frac{N_0K^2}{2}\left|\frac{s+a}{s^2 + AKs + aAK}\right|^2,$$

so that $S_\phi(0) = (N_0)/(2A^2)$.

$$\sigma^2 = \frac{1}{2\pi}\int_{-\infty}^{\infty} S_\phi(\omega)d\omega = \frac{N_0}{4A^2}(AK + a)$$

and

$$\int_{0}^{\infty}\rho_\phi(\tau)d\tau = 1/(AK + a).$$

Inserting this integral in (83) and substituting in (81) with $p(\phi)$ replaced by $P(\phi)$, we obtain

$$0 = \frac{d}{d\phi}\left\{\frac{4A}{KN_0}\left[\sin\phi\left(\frac{AK}{AK+a}\right)\right]P(\phi) + \frac{dP(\phi)}{d\phi}\right\},$$

whose solution with the boundary conditions of (30) and (31) is

$$P(\phi) \simeq \frac{\exp{(\alpha' \cos \phi)}}{2\pi I_0(\alpha')} \qquad \text{for large } \alpha' \qquad (84)$$

where the effective SNR, α', is given by

$$\alpha' = (A^2)/[N_0(AK + a)/(4)].$$

If we let $B_L = (AK + a)/4$, this is the same expression as that for the first-order loop with $\omega = \omega_0$. As would be expected, this expression for loop bandwidth for the second-order loop is that obtained from the linear model of the loop.

Random Modulation

Thus far we have considered only sinusoidal signals of constant frequency and phase. However, the exact method may be applied to signals with random frequency or phase modulation, provided the modulating process is stationary and Gaussian. We shall now derive the Fokker-Planck equation for a first-order loop and a specific random modulation, which will demonstrate the procedure in general.

The differential equation of operation for a first-order loop with arbitrary modulation is given by (10) as

$$\dot\phi(t) = \dot\theta_1(t) - K[A \sin \phi(t) + n'(t)]. \qquad (10)$$

We shall assume that the VCO quiescent frequency is equal to the carrier frequency ($\omega = \omega_0$), and that the carrier signal is frequency modulated by the stationary Gaussian process $m(t)$ with a modulation index

$$K_F \frac{\text{rad/sec}}{\text{volt}}$$

Then

$$\dot\theta_1(t) = K_F m(t), \qquad (85)$$

and (10) becomes

$$\dot\phi(t) = -AK \sin \phi(t) - Kn'(t) + K_F m(t). \qquad (86)$$

Note that the modulation is an additional term in the driving function of the differential equation. Specifically we shall consider a Gaussian modulating signal whose spectrum[7] is

$$S_m(\omega) = \frac{M_0/2}{\omega^2 + \beta^2}. \qquad (87)$$

With this driving function $\phi(t)$ as given by (86) is no longer a Markov process, since the present value of $m(t)$ is correlated with the past and consequently $\dot\phi(t)$ is no longer independent of the past states of the sys-

---

[7] A Gaussian process with this spectrum is necessarily a Markov process which may be generated by driving a first-order linear system with white Gaussian noise; *cf.*, Wang and Uhlenbeck [12].

tem. However, we may proceed to express $m(t)$ in terms of an auxiliary white process, which will allow us to treat the problem in terms of a two-dimensional Markov process whose components are $\phi(t)$ and $m(t)$.

A stationary Gaussian process whose spectrum is given by (87) has the same statistics as the output of a first-order linear system (such as an RC low-pass filter) excited by white Gaussian noise. That is, we may represent $m(t)$ in terms of the white Gaussian process $\eta(t)$ of zero mean and of one-sided spectral density M_0, by the operational equation

$$m(t) = \frac{1}{s + \beta} \eta(t),$$

or equivalently by the differential equation

$$\dot m(t) = -\beta m(t) + \eta(t). \qquad (88)$$

We may now treat the two first-order differential equations (86) and (88) as the defining equations for the two-dimensional Markov process $[\phi(t), m(t)]$ with white Gaussian driving functions $n'(t)$ and $\eta(t)$. Thus we may determine the Fokker-Planck equation for the two-dimensional probability density function in ϕ and m, $p(\phi, m, t)$, by evaluating the normalized moment coefficients of (69) and inserting these in the Fokker-Planck equation in the same way as was done for higher-order loops. The result is

$$\frac{\partial p}{\partial t} = \frac{\partial}{\partial \phi}\left[(AK \sin \phi - K_F m)p + \frac{K^2 N_0}{4}\frac{\partial p}{\partial \phi}\right]$$
$$+ \frac{\partial}{\partial m}\left[\beta m p + \frac{K^2 M_0}{4}\frac{\partial p}{\partial m}\right] \qquad (89)$$

The method used to derive (89) is easily generalized to the case of an nth-order loop with a received signal frequency modulated by a random Gaussian process with a rational spectrum, whose denominator polynominal is of degree $2k$, and whose numerator polynominal is of lower degree. The result in the general case is a Fokker-Planck equation for an $n + k$-dimension vector Markov process. The difficulty lies only in the solution of the partial differential equation. Even (89) which represents the simplest case ($n = k = 1$) appears formidable. For the steady-state or stationary case in which the time derivative is set equal to zero, we have a problem of the same magnitude as for the second-order loop without modulation, for which we could obtain only approximate results.

Other Error Functions

While the majority of closed-loop tracking systems employ sinusoidal carriers and reference oscillators, occasionally for low frequencies and specific applications, square wave carriers are employed. In such cases the VCO is replaced by a multivibrator which, when the loop is locked, generates a square wave which is displaced by exactly a quarter period relative to the received signal. If the received signal power is A^2, the amplitude of the square wave must be A. Let the reference square wave be of amplitude K_3 and let all the other components and parameters of the loop be the same as for the sinusoidal case (Fig. 1). Then by taking the Fourier series expansion of the square waves and reproducing the analysis for sinusoidal loops, we find that the equation of operation becomes

$$\dot\phi(t) = \dot\theta_1(t) - KF(s)[Ah(\phi) + n'(t)], \qquad (90)$$

where $n'(t)$ has the same statistics as for sinusoidal loops and $h(\phi)$ is the triangular wave, one period of which is shown in Fig. 13. Thus the model of Fig. 2 describes this case when the sinusoidal nonlinearity is replaced by the function $h(\phi)$.

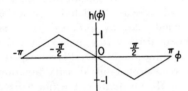

Fig. 13—Triangular error function.

Another system which has received much attention in a variety of applications including radar ranging is a tracking device often referred to as the delay-lock loop [16]–[18]. In one version of this device [18] the received signal alternates between $+A$ volts and $-A$ volts according to a binary code with a switching interval corresponding to the half period of the square wave in the above-mentioned case. The code is generated by a maximum-length linear shift register at the transmitter. The VCO in the receiver tracking loop consists of an identical shift register whose switching period is controlled by the loop filter output. The reference signal is derived by delaying the VCO-coder output by exactly two switching intervals and subtracting it from the undelayed coder output. The result of multiplying this by the received signal is an error function identical to Fig. 13 over the interval $-\pi < \phi < \pi$ and zero elsewhere [18]. However, the noise density is doubled by this procedure [so that the one-sided spectral density of $n'(t)$ is $2N_0$ in this case] and some self-noise is introduced into the system by the randomness of the code. However, the self-noise is negligible if the ratio N_0/A^2 is very large compared to the switching period [18]. We shall take this to be the case in the analysis which follows.

The various results obtained for sinusoidal loops can be generalized to arbitrary error functions $h(\phi)$ which are odd functions of ϕ, of which the cases just mentioned are particular examples. We shall consider only the re-

sults for the first-order loop. The steady-state probability density for a nonlinearity which is a periodic odd function of ϕ can be obtained by the same method as was used for the sinusoidal case [*cf.* (33)], and is

$$P(\phi) = C \exp[-\alpha g(\phi)] \qquad -\pi \le \phi \le \pi,$$

where

$$g(\phi) = \int^\phi h(x)dx, \qquad (91)$$

and

$$C = \frac{1}{\displaystyle\int_{-\pi}^{\pi} \exp[-\alpha g(\phi)]d\phi},$$

provided the timing periods of the received signal and the reference oscillator are initially synchronized.

The mean time to loss of lock has particular significance for the case of the coded signals of the delay-lock loop just discussed. For, assuming that the loop is initially in lock, if the phase error ever exceeds $\pm\pi$, there will be no deterministic restoring force tending to restore lock. We can show by an obvious generalization of the previous results that for an arbitrary odd error function (which need not be periodic), the mean time to loss of lock when the loop is initially in lock ($\phi_0=0$) is given by the expression [*cf.* (53)][1]

$$T(\phi) = \frac{1}{\gamma}\int_0^{\phi_l}\int_\phi^{\phi_l} \exp(-\alpha)[g(\phi)-g(x)]dxd\phi, \qquad (92)$$

where

$$\alpha = \frac{A^2}{N_0 B_L}, \qquad \gamma = \frac{4B_L}{\alpha}, \qquad B_L = AK/4,$$

and

$$g(\phi) = \int^\phi h(x)dx.$$

For the delay-lock tracking loop, we find by integrating the error function of Fig. 13 over a half period, that

$$-\alpha g(\phi) = \begin{cases} \dfrac{\alpha}{\pi}\left[\left(\dfrac{\pi}{2}\right)^2 - \phi^2\right] & 0 \le \phi \le \pi/2 \\[4mm] \dfrac{\alpha}{\pi}\left[-\left(\dfrac{\pi}{2}\right)^2 + (\phi-\pi)^2\right] & \pi/2 \le \phi \le \pi. \end{cases} \qquad (93)$$

We are most interested in the case $\phi_l=\pi$. However, the case $\phi_l=\pi/2$ is also worth consideration since it represents the mean time required for the loop to pass beyond the central linear region when initially $\phi=0$. The triangular domain of integration may be divided into three regions (Fig. 14) according to the regions of definition of $g(\phi)$ (93). Then referring to (93) and Fig. 14, we obtain from (92),

$$T(\pi) = A + B + C, \qquad (94)$$

where

$$A = T(\pi/2) = \frac{1}{\gamma} \int_0^{\pi/2} \int_\phi^{\pi/2} \exp\left(\frac{\alpha}{\pi}\right)(-\phi^2 + x^2)dxd\phi$$

$$B = \frac{1}{\gamma} \int_{\pi/2}^{\pi} \int_\phi^{\pi} \exp\left(\alpha/\pi\right)[(\phi - \pi)^2 - (x - \pi)^2]dxd\phi$$

$$C = \frac{1}{\gamma} \int_0^{\pi/2} \int_{\pi/2}^{\pi} \exp\left(\frac{\alpha}{\pi}\right)\left[\frac{\pi^2}{2} - \phi^2 - (x - \pi)^2\right]dxd\phi.$$

By making the proper changes of variables we can show that

$$C = \frac{\alpha \exp\left(\dfrac{\alpha\pi}{2}\right)}{4B_L}\left[\int_0^{\pi/2} \exp -\left(\frac{\alpha\phi^2}{\pi}\right)d\phi\right]^2, \qquad (95)$$

which is a tabulated integral, and

$$
\begin{aligned}
B = A &= T(\pi/2) \\
&= \frac{1}{\gamma} \int_0^{\pi/2} \int_0^x \exp\left(\frac{\alpha}{\pi}\right)(x^2 - \phi^2)d\phi dx \\
&= \frac{\alpha}{4B_L}\left[\frac{\pi\alpha}{4} + \sum_{n=2}^{\infty} \frac{(\pi\alpha/4)^n}{n!} \frac{\prod_{i=1}^{n-1} 2i}{\prod_{i=1}^{n}(2i - 1)}\right], \qquad (96)
\end{aligned}
$$

as can be shown by expanding the integrand in Taylor series. From (94), (95), and (96) we obtain the mean times $T(\pi)$ and $T(\pi/2)$ multiplied by B_L whose inverses are shown in Fig. 15. Similar results can be obtained for any odd error function by evaluating (92) for the proper $g(\phi)$.

Conclusions and Comparisons

This paper has dealt with the exact analysis of the nonlinear device described by the model of Fig. 2 or the nonlinear differential equation (10). The principal results, which can be generalized to any odd error function, have been:

1) The stationary (steady-state) probability density function, distribution, and variance for the first-order loop.
2) The expected time to loss of lock and frequency of skipping cycles for the first-order loop (this is particularly useful for constant-velocity Doppler tracking applications).
3) Approximate expressions for the stationary probability density of the second-order loop.
4) The partial differential (Fokker-Planck) equation for the probability density for higher-order loops,

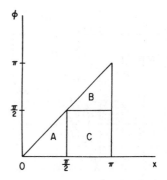

Fig. 14.—Domains of integrals of (94).

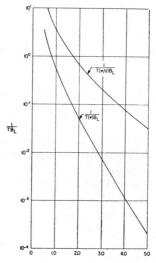

Fig. 15—Inverse-mean times to loss of lock and to first-passage from linear region (normalized by loop bandwidth).

including the case in which the signal is frequency modulated by a stationary Gaussian process.

The limitations of the method become evident when we attempt to solve the partial differential equation for the higher-order cases. Although the equations are linear in the probability densities, since the coefficients are nonlinear functions of the dependent variables, no exact solution seems possible. Thus for the higher-order cases and particularly for modulated signals, we must rely on the approximate models mentioned in the first section of this paper which lend themselves to more direct methods. However, our exact results for the first-order loop are quite useful in determining the validity threshold of a particular model, which we shall define loosely as the value of SNR below which the model no longer yields useful results.

The most obvious parameter to consider for this comparison is the variance of the first-order loop. We have already seen (Fig. 7) that for the linear model (wherein $\sin \phi$ is replaced by ϕ), the variance as determined from the model underestimates the actual variance by more than 20 per cent when the SNR α is less than 4 (or 6 db), so that $\alpha = 4$ may be taken as the validity thresh-

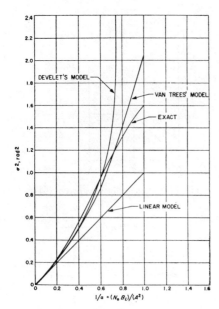

Fig. 16—Comparison of variance for first-order loop with results of approximate models.

old of the linear model. The variance for the exact model and the linear model are reproduced in Fig. 16 together with the variance obtained by using the models of Van Trees [8] and Develet [2]. Van Trees has shown that for the first-order loop the variance of the phase error can be written as a power series in $1/\alpha$. By calculating the first five terms of the Volterra functional expansion, Van Trees [8] found that the first three terms of the power series are

$$\sigma^2 = 1/\alpha + \frac{1}{2}(1/\alpha)^2 + \frac{13}{24}(1/\alpha)^3 \cdots \quad \text{(Van Trees)}.$$

This is shown in Fig. 16. Using the quasi-linearization technique Develet [2] replaced the sinusoidal nonlinearity by its average gain under the assumption that the input distribution is approximately Gaussian. Since the gain of a sinusoidal nonlinearity for an input value x is $A \cos x$, the average gain when the input is Gaussian of zero mean and variance σ^2 is

$$\int_{-\infty}^{\infty} A \cos x \exp\left(\frac{-x^2}{2\sigma^2}\right) dx = A \exp\left(-\frac{\sigma^2}{2}\right).$$

Replacing the nonlinear element of Fig. 2 by this gain, we obtain by the usual linear analysis the variance of the phase error for the first-order loop:

$$\sigma^2 = (1/\alpha) \exp(\sigma^2/2) \quad \text{(Develet)}.$$

The solution of this transcendental equation yields the value of the variance shown in Fig. 16. The maximum of $\sigma^2 \exp(-\sigma^2/2)$ is $2/e$ so that there can be no solution for $\alpha < e/2$, which means that the validity threshold of this model can be no lower than this value.

From Fig. 16 we note that the error in the Develet model is less than 10 per cent for $1/\alpha < 0.65$ or $\alpha > 1.54$,

while the Van Trees approximation involving the first five Volterra kernels yields results of this accuracy for $1/\alpha < 0.80$ or $\alpha > 1.25$. Of course, with sufficient effort one can compute arbitrarily many terms of the Volterra series and consequently obtain arbitrarily many terms of the power series expansion of σ^2 thus extending the validity threshold of the model as far as may be desired. However, for higher-order loops, Van Trees' method becomes exceedingly complex and tedious, while Develet's method remains simple for all loop filters and even for modulated signals. In fact, using this method he has obtained fairly general results on the threshold of the phase-locked loop as a frequency modulation discriminator [2].

While we have certainly not exhausted the realm of application of closed-loop tracking devices, the results of this paper may serve as a guide for the analysis of a large class of such systems.

Acknowledgment

The author is indebted to Prof. J. N. Franklin and Dr. E. C. Posner for several valuable discussions during the preparation of this manuscript.

References

[1] W. B. Davenport, Jr., and W. L. Root, "Random Signals and Noise," McGraw-Hill Book Co., Inc., New York, N. Y.; 1958.
[2] J. A. Develet, Jr., "A threshold criterion for phase-lock demodulation," Proc. IRE, vol. 51, pp. 349–356; February, 1956.
[3] W. J. Gruen, "Theory of AFC synchronization," Proc. IRE, vol. 41, pp. 1043–1048; August, 1953.
[4] A. J. Viterbi, "Acquisition and Tracking Behavior of Phase-Locked Loops," Proc. Symp. on Active Networks and Feedback Systems, Polytechnic Inst. of Brooklyn, N. Y., vol. 10, pp. 583–619; April, 1960.
[5] R. M. Jaffe and E. Rechtin, "Design and performance of phase-lock circuits capable of near-optimum performance over a wide range of input signal and noise levels," IRE Trans. on Information Theory, vol. IT-1, pp. 66–76; March, 1955.
[6] S. G. Margolis, "The response of a phase-locked loop to a sinusoid plus noise," IRE Trans. on Information Theory, vol. IT-3, pp. 135–144; March, 1957.
[7] R. C. Booton, Jr., "The Analysis of Nonlinear Control Systems with Random Inputs," Proc. Symp. on Nonlinear Circuit Analysis, Polytechnic Inst. of Brooklyn, N. Y., pp. 369–391; April, 1953.
[8] H. L. Van Trees, "Functional Techniques for the Analysis of the Nonlinear Behavior of Phase-Locked Loops," presented at WESCON, San Francisco, Calif.; August 20–23, 1963.
[9] V. I. Tikhonov, "The effects of noise on phase-lock oscillation operation," Automatika i Telemakhanika, vol. 22, no. 9; 1959.
[10] ——, "Phase-lock automatic frequency control application in the presence of noise," Automatika i Telemekhanika, vol. 23, no. 3; 1960.
[11] G. E. Uhlenbeck and L. S. Ornstein, "On the theory of Brownian motion," Phys. Rev., vol. 36, pp. 823–841; September, 1930.
[12] M. C. Wang and G. E. Uhlenbeck, "On the theory of Brownian motion II," Rev. Mod. Phys., vol. 17, pp. 323–342; April–July, 1945.
[13] A. A. Andronov, L. S. Pontryagin, and A. A. Witt, "On the statistical investigation of a dynamical system," J. Exp. Theoret. Phys. (USSR), vol. 3, p. 165; 1933.
[14] A. J. F. Siegert, "On the first passage time probability problem," Phys. Rev., vol. 81, pp. 617–623; February, 1951.
[15] D. A. Darling and A. J. F. Siegert, "The first passage time problem for a continuous Markov process," Ann. Math. Stat., vol. 24, pp. 624–639; 1953.
[16] J. J. Spilker, Jr., and D. T. Magill, "The delay-lock discriminator—an optimun tracking device," Proc. IRE, vol. 49, pp. 1403–1416; September, 1961.
[17] M. R. O'Sullivan, "Tracking system employing the delay-lock discriminator," IRE Trans. on Space Electronics and Telemetry, vol. SET-8, pp. 1–7; March, 1962.
[18] J. J. Spilker, Jr., "Delay-lock tracking of binary signals," IEEE Trans. on Space Electronics and Telemetry, vol. SET-9, pp. 1–8; March, 1963.

IV

Digital FM

PAPER NO. 16

Reprinted from *BSTJ*, Vol. 42, No. 5, pp. 2387–2426, Sept. 1963

Binary Data Transmission by FM over a Real Channel

By W. R. BENNETT and J. SALZ

(Manuscript received April 26, 1963)

Formulas are derived for probability of error in the detection of binary FM signals received from a channel characterized by arbitrary amplitude- and phase-vs-frequency distortion as well as additive Gaussian noise. The results depend on the signal sequence and can be presented in terms of averages over all signal sequences or as bounds for the most and least vulnerable ones. Illustrative examples evaluated include Sunde's method of suppressing intersymbol interference in band-limited FM. The effects of various representative channel filters are also analyzed. A solution is given for the problem of optimizing the receiving bandpass filter to minimize error probability at constant transmitted signal power. It is found that a performance from 3 to 4 db poorer than that theoretically attainable from binary PM is realizable over a variety of filtering situations.

I. INTRODUCTION

This paper undertakes to refine and extend the state of knowledge concerning performance of FM systems for binary data transmission over real-life channels. The particular aim is application to facilities such as exist in the telephone plant. Efficient use of the available channels constrains the bandwidth allowed for a given signaling speed. The luxury of a bandwidth sufficient to permit frequency transitions without amplitude variations and without dependence of present waveform on past signal history would in general imply an unjustifiably low information rate for the frequency range occupied. We therefore concentrate our attention on the band-limited channel with its inherent distortion of the FM data wave.

We assume a linear time-invariant transmission medium specified by its amplitude- and phase-vs-frequency functions and the statistics of its additive noise sources. The limiting noise environment in the telephone plant is typically nongaussian and not well defined even in a

349

statistical sense. Nevertheless, with the usual apology, we shall perform our analysis in terms of additive Gaussian noise. Justification of the relevancy is based on the following considerations:

(a) Laboratory tests on data transmission systems are made at present by adding Gaussian noise and counting errors. Good performance in terms of low error rate as a function of signal-to-noise ratio under such test conditions is found to be indicative of good performance on actual channels.

(b) Identification and removal of nongaussian disturbances is a feasible and continuing process which should eventually lead to a more nearly Gaussian description of the residue.

Our measure of performance is expressed in terms of error probability vs the ratio of average transmitted signal power to average Gaussian noise power. In most of the work we assume white Gaussian noise is added at the receiver input. A convenient reference is then the average noise power in a band of frequencies having width equal to the transmitted information rate in bits per second.

II. STATEMENT OF PROBLEM

A block diagram of the transmission system under study is shown in Fig. 1. The data source emits a sequence of binary symbols which for full information rate are independent of each other and have equal probability. The analysis can be generalized without analytical inconvenience to assign a probability m_1 to one of the two binary symbols and $1 - m_1$ to the other. In conventional binary notation the symbols are 1 and 0. It is convenient to express binary frequency modulation of

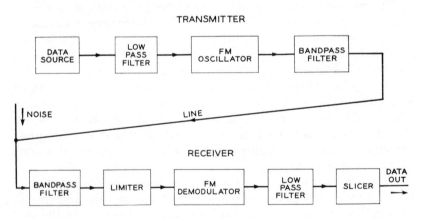

Fig. 1 — Binary FM transmission system.

2

an oscillator in terms of positive and negative frequency deviations. The combination of data source and low-pass filter is accordingly defined by the shaped baseband data wave train

$$s(t) = \sum_{n=-\infty}^{\infty} b_n g(t - nT) \qquad (1)$$

where

$$b_n = 2a_n - 1. \qquad (2)$$

The values of a_n represent the data sequence in binary notation. The probability is m_1 that the typical a_n is unity, and $1 - m_1$ that it is zero. The value of b_n is $+1$ if a_n is unity, and -1 if a_n is zero. The function $g(t)$ represents a standard pulse emitted by the low-pass filter for a signal element centered at $t = 0$.

Ideally, the oscillator frequency follows the baseband signal wave $s(t)$. This would imply an output voltage from the FM oscillator specified by

$$V(t) = A \cos\left[\omega_c t + \theta_0 + \mu \int_{t_0}^{t} s(\lambda)\, d\lambda \right]. \qquad (3)$$

Here, A is the carrier amplitude, ω_c is the frequency of the oscillator with no modulating signal applied, t_0 is an arbitrary reference time, θ_0 is the phase at $t = t_0$, and μ is a conversion factor relating frequency displacement to baseband signal voltage. The instantaneous frequency of the wave (3) is defined as the derivative of the argument of the cosine function. It can be written in the form $\omega_c + \omega_i$, where ω_i, the deviation from midband, is ideally expressed by

$$\omega_i = \mu s(t). \qquad (4)$$

In the practical case, the transmitting bandpass filter restricts the frequency-modulated wave to the range of frequencies passed by the channel. The purpose of this filter is to prevent both waste of transmitted power in components which will not reach the receiver and contamination of the line at frequencies assigned to other channels. The result is a transformation of the voltage wave (3) to a band-limited form, which must depart in more or less degree from the ideal conditions of constant amplitude and linear relationship between frequency and baseband signal. The line also inserts variations in amplitude- and phase-vs-frequency which cause further departures from the ideal. For our purposes it is sufficient to combine the line characteristics with those of the transmitting filter into a single com-

3

351

posite network function determining the wave presented to the receiving bandpass filter.

The receiving bandpass filter is necessary to exclude out-of-band noise and interference from the detector input. It also shapes the signal waveform and can include compensation for linear in-band distortion suffered in transmission. Two contradictory attributes are sought in the filter — a narrow band to reject noise and a wide band to supply a good signal wave to the detector. An opportunity for an optimum design thus exists and will be explored in this paper.

The frequency detector is assumed to differentiate the phase with respect to time. The post-detection filter can do further noise rejection and shaping in the baseband range, but its only function in our present analysis is to separate the wave representing the frequency variation from the higher-frequency detection products. The slicer delivers positive voltage when the detected frequency is above midband and negative voltage when the detected frequency is below midband. The slicer output is sampled at appropriate instants to recover the binary data sequence.

The noise-free input to the detector will be written in the form

$$V_r(t) = P(t) \cos (\omega_c t + \theta) - Q(t) \sin (\omega_c t + \theta). \qquad (5)$$

$P(t)$ and $Q(t)$ represent in-phase and quadrature signal modulation components respectively, which are associated with a carrier wave at the midband frequency ω_c with specified phase θ. Such a resolution can always be made, even though the details in actual examples may be burdensome. The added noise wave at the detector input is assumed to be Gaussian with zero mean and can likewise be written as

$$v(t) = x(t) \cos (\omega_c t + \theta) - y(t) \sin (\omega_c t + \theta). \qquad (6)$$

If $v(t)$ represents Gaussian noise band-limited to $\pm 2\omega_c$, $x(t)$ and $y(t)$ are also Gaussian and are band-limited to $\pm \omega_c$. If the spectral density of $v(t)$ is $w_v(\omega)$, the spectral densities of $x(t)$ and $y(t)$ are given by[1]

$$w_x(\omega) = w_y(\omega) = w_v(\omega_c + \omega) + w_v(\omega_c - \omega), \qquad |\omega| < \omega_c \quad (7)$$

In general, $x(t)$ and $y(t)$ are dependent, with cross-spectral density

$$w_{xy}(\omega) = j[w_v(\omega_c - \omega) - w_v(\omega_c + \omega)] \qquad (8)$$

and cross-correlation function expressed in terms of $R_v(\tau)$, the auto-correlation function of $v(t)$, by

$$R_{xy}(\tau) = -2R_v(\tau) \sin \omega_c \tau. \qquad (9)$$

4

The cross correlation vanishes at $\tau = 0$, and hence the joint distribution of $x(t)$, $y(t)$ at any specified t is that of two independent Gaussian variables.

We shall also require the joint distribution of x and y with their time derivatives $\dot{x}$ and $\dot{y}$. The latter are Gaussian with spectral densities

$$w_{\dot{x}}(\omega) = w_{\dot{y}}(\omega) = \omega^2 w_x(\omega). \tag{10}$$

The cross-spectral densities are

$$w_{x\dot{x}}(\omega) = w_{y\dot{y}}(\omega) = j\omega w_x(\omega) \tag{11}$$

$$w_{x\dot{y}}(\omega) = j\omega_{xy}(\omega) = \omega[w_v(\omega_c + \omega) - w_v(\omega_c - \omega)] = -w_{\dot{x}y}. \tag{12}$$

The cross correlations are

$$R_{x\dot{x}}(\tau) = \int_{-\infty}^{\infty} w_{x\dot{x}}(\omega)e^{j\tau\omega}\,d\omega = -\int_{-\infty}^{\infty} \omega w_x(\omega)\,\sin\tau\omega\,d\omega$$
$$= R_{y\dot{y}}(\tau) \tag{13}$$

$$R_{x\dot{y}}(\tau) = -R_{\dot{x}y}(\tau) = \int_{-\infty}^{\infty} w_{x\dot{y}}(\omega)e^{j\tau\omega}\,d\omega$$
$$= \int_{-\infty}^{\infty} \omega[w_v(\omega_c + \omega) - w_v(\omega_c - \omega)]\,\cos\tau\omega\,d\omega. \tag{14}$$

The cross correlation of x and $\dot{x}$ as well as of y and $\dot{y}$ vanish at $\tau = 0$, and hence at any instant $\dot{x}$ is independent of x, and $\dot{y}$ is independent of y. The cross correlations of x and $\dot{y}$, and of $\dot{x}$ and y, do not vanish in general, but do vanish in the special case in which

$$w_v(\omega_c + \omega) = w_v(\omega_c - \omega). \tag{15}$$

This is the case of a noise spectrum which is symmetrical with respect to the midband and represents a reasonable objective in system design. Since the simplification in computational details is quite considerable when the condition of symmetry is imposed, and since the departures caused by lack of symmetry are not of primary interest, we shall assume henceforth that (15) is satisfied. The four variables x, $\dot{x}$, y, and $\dot{y}$ are then independent and have the joint Gaussian probability density function

$$p(x, y, \dot{x}, \dot{y}) = \frac{1}{4\pi^2\sigma_0^2\sigma_1^2}\exp\left[-\frac{x^2 + y^2}{2\sigma_0^2} - \frac{\dot{x}^2 + \dot{y}^2}{2\sigma_1^2}\right] \tag{16}$$

$$\sigma_0^2 = \int_{-\infty}^{\infty} w_x(\omega)\,d\omega = 2\int_{-\infty}^{\infty} w_v(\omega_c + \omega)\,d\omega \tag{17}$$

5

$$\sigma_1{}^2 = \int_{-\infty}^{\infty} w_{\dot{x}}(\omega)\, d\omega = 2 \int_{-\infty}^{\infty} \omega^2 w_v(\omega_c + \omega)\, d\omega. \qquad (18)$$

The noise-free detector input wave (5) can be written in the equivalent form

$$V_r(t) = R(t) \cos\left[\omega_c t + \phi(t)\right] \qquad (19)$$

where

$$R^2(t) = P^2(t) + Q^2(t) \qquad (20)$$

$$\tan \phi(t) = Q(t)/P(t). \qquad (21)$$

The frequency detector and post-detection filter combine to deliver a wave proportional to the instantaneous frequency deviation from mid-band. Taking the constant of proportionality as unity, we write for the output wave

$$\phi'(t) = \frac{d}{dt} \arctan \frac{Q(t)}{P(t)} = \frac{P(t)Q'(t) - Q(t)P'(t)}{P^2(t) + Q^2(t)}. \qquad (22)$$

With the functional dependence on t understood, we write this equation in the form

$$\phi'(t) = \dot{\phi} = (P\dot{Q} - Q\dot{P})/R^2. \qquad (23)$$

When the noise is added, the detected frequency is changed to

$$\psi'(t) = \dot{\psi} = \frac{(P + x)(\dot{Q} + \dot{y}) - (Q + y)(\dot{P} + \dot{x})}{(P + x)^2 + (Q + y)^2}. \qquad (24)$$

Assuming that the system does not make errors in the absence of noise, we can express the probability of error in a given sample of instantaneous frequency taken at the time $t = nT$ as the probability that $\psi'(nT)$ is negative if $\phi'(nT)$ is positive or the probability that $\psi'(nT)$ is positive if $\phi'(nT)$ is negative. Since the system has memory, the values of P, Q, $\dot{P}$, and $\dot{Q}$ at any sampling instant depend on the entire signal sequence. Our procedure is first to show how the error probability can be evaluated at any sampling instant for any sequence. We then calculate error rates for specific sequences and establish bounds for most and least vulnerable sequences.

Since the denominators of (23) and (24) are inherently positive, the decisions are made entirely on the basis of the signs of the numerators. Therefore, we do not require the distribution function of the instantaneous frequency itself. In fact if we let

$$x + P = x_1, \qquad \dot{x} + \dot{P} = \dot{x}_1$$
$$y + Q = y_1, \qquad \dot{y} + \dot{Q} = \dot{y}_1 \qquad (25)$$

6

354

we require only one value of the distribution function of the variable z defined by

$$z = x_1\dot{y}_1 - y_1\dot{x}_1. \qquad (26)$$

The error probability is fully determined in any specific case either by the probability that z is negative or by the probability that z is positive. That is, if $F(z)$ is the distribution function of z, we only require the value of $F(0)$.

We shall derive a general expression for $F(0)$ in terms of a single definite integral. From this integral we shall then obtain definite integrals representing bounds for the error probability when arbitrary binary data sequences are transmitted. No restrictions on range of signal-to-noise ratios are made. The results will be applied to special cases of practical interest. One is Sunde's binary FM system which avoids intersymbol interference in a finite band in the absence of noise. When noise is added in this system, the detected samples become dependent on past signal history. It has been found possible to give a complete treatment of the Sunde method, including optimization of the receiving filter for minimum probability of error with fixed average transmitted signal power. The other cases analyzed in detail are based on design parameters actually in use on FM data transmission terminals.

III. GENERAL SOLUTION

Our first observation is that when x_1 and y_1 are fixed, the variable z of (26) is defined by a linear operation on the two independent Gaussian variables $\dot{x}_1$ and $\dot{y}_1$. Hence the conditional probability density function $p(z \mid x_1, y_1)$ of z when x_1 and y_1 are given is Gaussian with readily determined parameters. We accordingly write

$$p(z \mid x_1, y_1) = \frac{1}{\sigma\sqrt{2\pi}} \exp\left[-\frac{(z - z_0)^2}{2\sigma^2}\right]. \qquad (27)$$

The mean z_0 is the sum of the means of $x_1\dot{y}_1$ and $-y_1\dot{x}_1$, that is,

$$z_0 = x_1 \text{ av } \dot{y}_1 - y_1 \text{ av } \dot{x}_1 = x_1\dot{Q} - y_1\dot{P}. \qquad (28)$$

The variance σ^2 is the sum of the variances of $x_1\dot{y}_1$ and $y_1\dot{x}_1$; hence

$$\sigma^2 = (x_1^2 + y_1^2)\sigma_1^2. \qquad (29)$$

The complete probability density function $p(z)$ for z is obtained by averaging the conditional probability density function over x_1 and y_1. This is done by multiplying (27) by the joint probability density function of x_1 and y_1 and then integrating over all x_1 and y_1. Calling the

7

latter function $q(x_1, y_1)$, we can express its value by substituting the values of x and y from (25) in (16) and integrating out the $\dot{x}$ and $\dot{y}$ terms. The result is

$$q(x_1, y_1) = \frac{1}{2\pi\sigma_0{}^2} \exp\left[-\frac{(x_1 - P)^2 + (y_1 - Q)^2}{2\sigma_0{}^2}\right]. \quad (30)$$

Then

$$p(z) = \int_{-\infty}^{\infty} \int_{-\infty}^{\infty} p(z \mid x_1, y_1) q(x_1, y_1) \, dx_1 \, dy_1. \quad (31)$$

The probability of error when the noise-free sample of frequency deviation is positive is

$$P_+ = \int_{-\infty}^{0} p(z) \, dz = \int_{0}^{\infty} p(-z) \, dz. \quad (32)$$

Likewise, when the noise-free sample is negative, we obtain a probability of error

$$P_- = \int_{0}^{\infty} p(z) \, dz. \quad (33)$$

The problem is thus reduced to the evaluation of the triple integral obtained by combining (27), (30), and (31) with either (32) or (33). It is shown in Appendix A that the result of these operations can be expressed in the following form

$$P_+ = \frac{1}{2} \operatorname{erfc} \frac{R}{\sqrt{2}\sigma_0}$$
$$+ \frac{R}{2\sigma_0\sqrt{2\pi}} \int_{-1}^{1} \exp\left(-\frac{R^2 x^2}{2\sigma_0{}^2}\right) \operatorname{erfc} \frac{R\phi(1 - x^2)^{\frac{1}{2}} - \dot{R}x}{\sqrt{2}\sigma_1} \, dx. \quad (34)$$

The value of P_- is obtained by subtracting the right-hand member of (34) from unity. We note that ϕ is positive for P_+ and negative for P_-. The symbol $\dot{R}$ is used for dR/dt where R is given by (20). In a pure FM wave, $\dot{R} = 0$, but this condition cannot be maintained in a finite bandwidth.

Differentiating partially with respect to $\dot{R}$ and rearranging, we obtain

$$\frac{\partial P_+}{\partial \dot{R}} = \frac{R}{\pi\sigma_0\sigma_1} \int_{0}^{1} x \exp\left[-\frac{R^2 x^2}{2\sigma_0{}^2} - \frac{R^2\phi^2(1 - x^2) + \dot{R}^2 x^2}{2\sigma_1{}^2}\right]$$
$$\sinh \frac{R\dot{R}\phi x(1 - x^2)^{\frac{1}{2}}}{\sigma_1{}^2} \, dx. \quad (35)$$

8

356

We note that $\partial P_+/\partial \dot{R}$ vanishes when $\dot{R} = 0$ and at no other value of $\dot{R}$. The latter follows from the fact that the integrand of (35) cannot change sign in the interval of integration. We also find that $\partial^2 P_+/\partial \dot{R}^2$ is positive when $\dot{R} = 0$. We conclude that P_+ is minimum with respect to $\dot{R}$ when and only when $\dot{R} = 0$. A lower bound on the probability of error for any fixed R and ϕ is therefore obtained by setting $\dot{R} = 0$, giving

$$P_l = \frac{1}{2}\operatorname{erfc}\frac{R}{\sqrt{2}\sigma_0}$$
$$+ \frac{R}{2\sigma_0\sqrt{2\pi}}\int_{-1}^{1}\exp\left(-\frac{R^2 x^2}{2\sigma_0^2}\right)\operatorname{erfc}\frac{R\phi(1-x^2)^{\frac{1}{2}}}{\sqrt{2}\sigma_1}\,dx. \tag{36}$$

Also, since P_+ must be monotonic increasing with $|\dot{R}|$, the largest probability of error for any fixed R and ϕ occurs when $\dot{R}$ has its largest possible absolute value. These deductions are of aid in selecting the data sequences which have most and least probabilities of error.

It is shown in Appendix A that P_l can be written in the equivalent form

$$P_l = \frac{1}{\pi}\int_0^{\pi/2}\exp\left[-\frac{R^2\phi^2/(2\sigma_1^2)}{1+\left(\frac{\sigma_0^2\phi^2}{\sigma_1^2}-1\right)\cos^2\theta}\right]d\theta. \tag{37}$$

It is also shown that when $\phi < (\sigma_1/\sigma_0)$, the limiting form for large signal-to-noise ratio — i.e., R large compared with σ_0 — is given by

$$P_l \sim \frac{\sigma_0^2\phi}{R\sigma_1\sqrt{2\pi}}\left(\frac{\sigma_0^2\phi^2}{\sigma_1^2}-1\right)^{-\frac{1}{2}}\exp\left(-\frac{R^2}{2\sigma_0^2}\right). \tag{38}$$

When $\phi > (\sigma_1/\sigma_0)$, the limiting form becomes

$$P_l \sim \frac{\sigma_1}{R\phi\sqrt{2\pi}}\left(1-\frac{\sigma_0^2\phi^2}{\sigma_1^2}\right)^{-\frac{1}{2}}\exp\left(-\frac{R^2\phi^2}{2\sigma_1^2}\right). \tag{39}$$

When $\phi = \sigma_1/\sigma_0$, we have the exact result

$$P_l = \frac{1}{2}\exp\left(-\frac{R^2}{2\sigma_0^2}\right). \tag{40}$$

The general equation for error probability (34) can conveniently be expressed in terms of the following three parameters

$$\rho^2 = \frac{R^2}{2\sigma_0^2} \tag{41}$$

9

$$a^2 = \frac{\sigma_0^2 \dot{\phi}^2}{\sigma_1^2} \qquad (42)$$

$$b^2 = \frac{\dot{R}^2}{2\sigma_1^2}. \qquad (43)$$

Equation (34) then becomes

$$P_+ = \frac{1}{2} \operatorname{erfc} \rho + \frac{\rho}{2\sqrt{\pi}} \int_{-1}^{1} e^{-\rho^2 x^2} \operatorname{erfc} \left[a\rho(1 - x^2)^{\frac{1}{2}} - bx \right] dx. \quad (44)$$

Evaluation of this equation in terms of the three parameters ρ, a, and b gives the error probability for any of the FM systems considered.

IV. ERROR PROBABILITY VS SIGNAL-TO-NOISE RATIO

In analog systems the performance is often expressed in terms of signal-to-noise ratio in the receiver output. In the case of audio and video signals, where subjective judgments determine the requirements, the signal-to-noise ratio furnishes a good criterion. In the case of data signals, however, performance is judged in terms of errors made, and the errors cannot be predicted from the signal-to-noise ratio alone. The error rate depends in general on the distribution of the noise values. Furthermore, in good systems the errors are rare and hence are associated with infrequent noise conditions. The central part of the noise distribution is of less importance than the tails.

We illustrate the difference between a straight signal-to-noise ratio analysis and a direct error probability calculation in FM by a simple example. Consider the case of a long sequence of mark signals leading to a constant signal frequency $\omega_c + \omega_d$. The signal wave can then be written in the form

$$\begin{aligned} V(t) &= A \cos (\omega_c + \omega_d)t \\ &= A \cos \omega_d t \cos \omega_c t - A \sin \omega_d t \sin \omega_c t. \end{aligned} \qquad (45)$$

Comparing with (5) and noting that we are omitting the arbitrary phase angle θ, which is of trivial interest, we make the identifications

$$P(t) = A \cos \omega_d t \qquad Q(t) = A \sin \omega_d t. \qquad (46)$$

Then, by differentiation

$$P'(t) = -\omega_d A \sin \omega_d t \qquad Q'(t) = \omega_d A \cos \omega_d t. \qquad (47)$$

If a sample is taken at $t = 0$

$$P = A \qquad \dot{P} = 0 \qquad Q = 0 \qquad \dot{Q} = \omega_d A. \qquad (48)$$

10

358

Then from (24) the error $\dot{\psi} - \omega_d$ in the detected frequency deviation because of additive Gaussian noise is

$$\nu = \dot{\psi} - \omega_d = \frac{(A + x)(\omega_d A + \dot{y}) - y\dot{x}}{(A + x)^2 + y^2} - \omega_d. \qquad (49)$$

In a signal-to-noise ratio calculation for the case in which the signal amplitude is usually much larger than the noise on the line, (49) would be written in the form

$$\nu = \frac{\omega_d(1 + x/A) + \dot{y}/A + (x\dot{y} - y\dot{x})/A^2}{(1 + x/A)^2 + (y/A)^2} - \omega_d. \qquad (50)$$

If we then assume that A is large compared with x, y, $\dot{x}$, and $\dot{y}$, we retain only first-order terms in small quantities and construct the following approximate result, valid most of the time

$$\nu \approx \omega_d(1 + x/A) + \dot{y}/A - \omega_d(1 + 2x/A)$$
$$= (\dot{y} - \omega_d x)/A. \qquad (51)$$

The approximate spectral density of the frequency deviation error is then

$$w_\nu(\omega) \approx [w_{\dot{y}} + \omega_d^2 w_x(\omega)]/A^2$$
$$= 2(\omega^2 + \omega_d^2)w_v(\omega_c + \omega)/A^2. \qquad (52)$$

The approximate mean-square value of error can now be found by integrating the spectral density function $w_\nu(\omega)$ over all frequencies. However, we cannot obtain the probability of error from this value because we do not know the distribution function. A nonlinear operation has been performed on a Gaussian process, and the result must be non-gaussian. In this case Rice[2] has shown that the central part of the frequency error distribution is approximately Gaussian. His argument does not apply to the tail. When the signal exceeds the noise most of the time, it is only the tails of the distribution which are important in determining the probability that an error is made in distinguishing between mark and space frequencies.

Since there is no intersymbol interference in our example the exact expression for probability of error is given by (37) with $R = A$ and $\dot{\phi} = \omega_d$. It can be seen from the limiting forms for large signal-to-noise ratio, (38) through (40), that the Gaussian approximation from (52) cannot approach the correct result. The result obtained from (52) must contain both the original and differentiated noise spectra in the argument of the exponential part of the approximation at large signal-to-noise

11

359

ratios. In (38) and (39) the exponential depends on either σ_0 or σ_1 but not both.

As another example of the difference between inferences from signal-to-noise ratio and error probability, it is interesting to consider the case of differentially detected binary phase modulation. In this system the polarity of the present carrier wave is compared with the polarity one bit ago. The binary message is read as 1 for a phase reversal and 0 for no phase change. By intuitive reasoning one could easily conclude that there would be a 3-db penalty relative to synchronous detection with a noise-free period. Certainly, in the differential case noise is added to both waves under comparison, and the bit interval is usually long enough to make the two noise samples substantially independent of each other. Signal-to-noise ratio analysis supports the intuitive argument when the average noise power is small relative to the average signal power. A direct calculation of error probability, however, exposes the fallacy and reminds us sharply that the noise is not small compared with the signal when errors occur. If we focus attention on the large noise peaks which cause error, we can see that the simultaneous combination of disturbances on both waves does not imply the same probability of disaster as would follow from concentration of all the noise on one wave.

The differential binary PM problem can in fact be solved as a simple special case of the analysis we have developed for FM. The input wave to the detector can be written as

$$V_r(t) = [P(t) + x(t)] \cos \omega_c t - y(t) \sin \omega_c t. \tag{53}$$

The detector operates by multiplying $V_r(t)$ and $V_r(t - T)$, selecting the low-frequency components of the product, and sampling the output at intervals T apart. If we assume $\omega_c T$ is a multiple of 2π and identify quantities evaluated at $t - T$ by the subscript d, the binary decisions are based on the sign of the wave

$$V_a(t) = (P + x)(P_d + x_d) + yy_d . \tag{54}$$

When the correct binary decision is 0, the signs of P and P_d are the same, and an error occurs if the sampled value V_a is negative. When the correct binary decision is 1, the signs of P and P_d are opposite, and an error occurs if the sampled value of V_a is positive. The two cases are symmetric and an analysis of either suffices. For the case of the symbol 0, $P = P_d$, while for the case of 1, $P = -P_d$.

In calculating the signal-to-noise ratio for the case of a symbol 0, we would write

$$V_a = P \left(P + x + x_d + \frac{xx_d + yy_d}{P} \right). \tag{55}$$

12

Then if P is large compared with x, x_d, y, and y_d, we approach a condition in which the decisions are based on the sign of $P + x + x_d$. If x and x_d are independent, the sum $x + x_d$ represents samples from random noise with twice as much average power as the samples of either x or x_d alone. This tempting argument leads to the 3-db rule.

In a direct calculation of error probability, we recognize that the influence of xx_d and yy_d cannot be ignored at the tails of the noise distribution where the errors occur. In particular, if x and x_d are both very negative, tending to cause an error in a symbol 0, the value of xx_d is large and positive, tending to prevent the threatened damage.

To find the error probability, we compare (54) with (26), and note that we have a special case of the previous solution if we make the following identification

$$z \equiv V_a \qquad x_1 \equiv P + x \qquad \dot{y}_1 \equiv P_d + x_d$$
$$y_1 \equiv y \qquad \dot{x}_1 \equiv -y_d \, . \qquad\qquad (56)$$

The remainder of the solution proceeds as before if x, y, x_d, and y_d are independent Gaussian variables. The independence is guaranteed if the second-order correlation functions vanish at lag time T. One difference between this case and the earlier one is that the variables x, y, x_d, and y_d all have the same variance. This specialization can be made in the earlier work by setting $\sigma_0 = \sigma_1 = \sigma$. By comparing with (25), we further note that we can now set $Q = \dot{P} = 0$, $\dot{Q} = P_d = P$. Hence we also have $R = P$ and $\dot{R} = 0$. Corresponding to ϕ we insert the value which V_a/R^2 assumes in the absence of noise, namely $\phi \equiv P^2/P^2 = 1$. In terms of (41), (42), and (43) we then have

$$\rho^2 = \frac{P^2}{2\sigma^2} \qquad a^2 = 1 \qquad b^2 = 0. \qquad\qquad (57)$$

Hence the answer is given by (40), namely

$$P_+ = P_- = \tfrac{1}{2}e^{-\rho^2}. \qquad\qquad (58)$$

In the ideal case, a bandwidth f_0 is sufficient to send signals by binary PM at a rate f_0 bits per second without intersymbol interference. This allows for upper and lower sidebands with widths $f_0/2$. If the spectral density of the noise is ν_0 watts/cps, it follows that $\sigma^2 = \nu_0 f_0$. Then M, the ratio of average signal power to the average noise power in a band of width equal to the bit rate, is equal to the ratio of $P^2/2$ to $\nu_0 f_0$ and hence $M = \rho^2$. The formula for error probability is thus found to agree with the one given by Lawton.[3] Average signal power 0.9 db greater than the coherent case is required for an error probability of 10^{-4}. The

13

difference in performance between the differential and purely coherent cases approaches zero at very high signal-to-noise ratios.

E. D. Sunde[4] has described a binary FM system in which the intersymbol interference in the absence of noise can be made to vanish at the sampling instants, even when the bandwidth is limited to an extent comparable with that used in AM transmission. The method is remarkable in that a type of result similar to that given by Nyquist[5] for AM systems is obtained for all sequences in spite of the nonlinear FM detection process which invalidates the principle of superposition. The performance falls a little short of the corresponding AM case, in that some dependence on the message appears when noise is added.

Fig. 2 shows a diagram of Sunde's method. The binary message is sent by switching between two oscillators. The difference between the oscillator frequencies must be locked to the bit rate, and the oscillators must be so phased that the frequency transitions are accomplished with continuous phase. The combination of sending filter, line, and receiving filter modify the switched output to produce a spectrum at the input to the frequency detector with even symmetry about the midband and with Nyquist's vestigial symmetry about the marking and spacing fre-

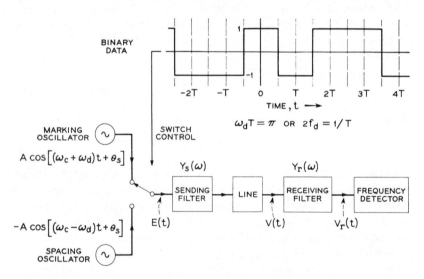

Fig. 2 — Sunde's band-limited binary FM system.

14

362

quencies. The latter must be high enough relative to the bit rate to prevent appreciable lower sideband foldover.

The output of the switch is represented by

$$E(t) = \frac{A}{2} [1 - s(t)] \cos [(\omega_c - \omega_d)t + \theta_s]$$

$$+ \frac{A}{2} [1 + s(t)] \cos [(\omega_c + \omega_d)t + \theta_m].$$

(59)

In (59) A represents the amplitude of the output and must be the same for each oscillator. The switching function $s(t)$ represents the baseband data wave of (1). When $s(t) = -1$, the first term has amplitude A and the second term vanishes. When $s(t) = +1$, the first term vanishes and the second has amplitude A. The center of the band is the frequency ω_c and the total frequency shift is $2\omega_d$. For minimum bandwidth the angular signaling frequency $\omega_0 = 2\pi/T$ must be equal to $2\omega_d$. One of the two phase angles θ_s and θ_m can be arbitrary, but the two angles must differ by 180 degrees. Under these restrictions, the value of $E(t)$ can be written as

$$E(t) = A \sin \omega_d t \sin (\omega_c t + \theta_s) - As(t) \cos \omega_d t \cos (\omega_c t + \theta_s). \quad (60)$$

Sunde requires that the input wave to the frequency detector can be written in the form

$$V_r(t) = A \sin \omega_d t \sin (\omega_c t + \theta_r) - As_1(t) \cos (\omega_c t + \theta_r) \quad (61)$$

where $s_1(t)$ represents the data sequence with $g(t)$ replaced by $g_1(t)$. The latter must be a pulse which gives no intersymbol interference when the data rate is $1/T$. That is,

$$s_1(t) = \sum_{n=-\infty}^{\infty} (-)^n b_n g_1(t - nT) \quad (62)$$

and $g_1(t)$ assumes the value unity at $t = 0$ and has nulls at all instants differing from $t = 0$ by multiples of T. In mathematical notation

$$g_1[(m - n)T] = \delta_{mn} \quad (63)$$

and

$$s_1(mT) = (-)^m b_m. \quad (64)$$

The requirement as actually stated by Sunde differs from (61) in that his analysis is based on a switching function which assumes the values 1 and 0 at the sampling instant rather than 1 and -1. The two expressions for the requirement can be shown to be equivalent. Equation

15

(61) has the advantage that the function $s_1(t)$ has the average value zero for a random data sequence with equal probability of the two binary symbols. This fact enables an easy separation of the spectral density of $V_r(t)$ into line spectra contributed by the first term of (61) and a continuous spectral density function for the second part.

Incidentally, it is clear from (61) that all the signal information is contained in the second term, and that the first term can be regarded as a pair of pilot tones at the marking and spacing frequencies $\omega_c \pm \omega_d$. The sole function of these pilot tones is to enable an FM detector to recover the message. The information carrying part of $V_r(t)$ can equally well be regarded as double-sideband suppressed-carrier binary AM or binary phase modulation, with the carrier frequency placed at ω_c. The ideal way of detecting such signals is by multiplication with a coherent carrier wave, which must be transmitted as part of the data wave in some way. Detection of $V_r(t)$ as FM has a practical advantage in that there is no carrier recovery problem; the wave is ready for the frequency detector with no further processing. The penalty for transmitting pure sine waves is a waste of signal power. As will be shown quantitatively later, such waste results in an unfavorable comparison with more nearly ideal systems.

To show that the stipulated conditions are sufficient to suppress intersymbol interference in the detected frequency of $V_r(t)$, we identify $P(t)$ and $Q(t)$ of (5) with the applicable terms of (61) as follows

$$P(t) = -As_1(t) \tag{65}$$

$$Q(t) = -A \sin \omega_d t. \tag{66}$$

We then calculate

$$P'(t) = -As_1'(t) \tag{67}$$

$$Q'(t) = -\omega_d A \cos \omega_d t. \tag{68}$$

If we take frequency samples at $t = mT$ we find that since $\omega_d T = \pi$

$$
\begin{aligned}
P(mT) &= (-)^{m+1} b_m \\
P'(mT) &= -As_1'(mT) \\
Q(mT) &= 0 \\
Q'(mT) &= (-)^{m+1} \omega_d A.
\end{aligned}
\tag{69}
$$

Hence in (23), evaluated at $t = mT$

$$\dot\phi = \dot Q / P = \omega_d / b_m = b_m \omega_d . \tag{70}$$

16

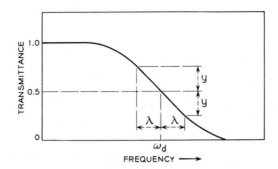

Fig. 3 — Nyquist's condition of vestigial symmetry.

The value of the instantaneous frequency deviation at the mth sampling point is, therefore, equal to ω_d if $s(mT) = 1$ and equal to $-\omega_d$ if $s(mT) = -1$. Freedom from intersymbol interference is thus obtained if (64) is satisfied.

As shown by Nyquist, a sufficient condition for obtaining (64) is that the standard pulse $g_1(t)$ is the impulse response of a network with transmittance $G_1(\omega)$ of the form shown in Fig. 3, described mathematically by

$$G_1(\pm\omega_d - \lambda) + G_1(\pm\omega_d + \lambda) = 2G_1(\omega_d) = T \qquad 0 < \lambda < \omega_d . \quad (71)$$

We say that a function satisfying (71) has vestigial symmetry about frequency ω_d because it has the type of symmetry called for in a vestigial sideband filter with the carrier at ω_d. We can think of the response at a frequency exceeding ω_d by an amount λ as exactly compensating the deficiency in the response at the frequency less than ω_d by the same amount λ. The ideal low-pass filter is a limiting special case occurring when the transmittance vanishes for $|\omega| > \omega_d$. The amplitude can be associated with linear phase shift, which changes only the origin of time. Unnecessary complication is avoided by carrying through the calculations with zero phase shift.

The conditions imposed on the filters and line to transform (60) to (61) can be expressed in terms of the Fourier transforms of $g(t) \cos \omega_d t$ and $g_1(t)$, which we represent respectively by $C(\omega)$ and $G_1(\omega)$. Both $C(\omega)$ and $G_1(\omega)$ are purely real and are given by

$$C(\omega) = \int_{-\infty}^{\infty} g(t) \cos \omega_d t \cos \omega t \, dt$$
$$= [G(\omega - \omega_d) + G(\omega + \omega_d)]/2 \quad (72)$$

17

365

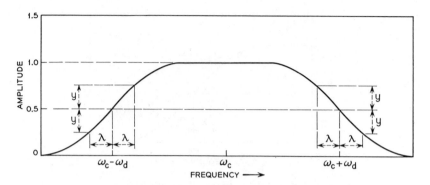

Fig. 4 — Spectrum at input to detector in Sunde's FM system.

$$G_1(\omega) = \int_{-\infty}^{\infty} g_1(t) \cos \omega t \, dt. \tag{73}$$

The result, obtained by multiplying $\cos (\omega_c t + \theta_s)$ by $g(t) \cos \omega_d t$ or $g_1(t)$, is to place upper and lower sidebands on the frequencies $\pm \omega_c$, as shown in Fig. 4, with spectra equal to $C(\omega - \omega_c)/2$ and $G_1(\omega - \omega_c)/2$ respectively on ω_c. The required transmittance function for the combination of sending filter, line, and receiving filter is then

$$Y(\omega) = \frac{G_1(\omega - \omega_c)}{C(\omega - \omega_c)}. \tag{74}$$

This function transforms the second term of (60) to the second term of (61). It is also necessary for the first term of (60) to remain unchanged. The first term can be written as the difference of sine waves of frequencies $\omega_c - \omega_d$ and $\omega_c + \omega_d$. These components will be unchanged by the operation $Y(\omega)$ if

$$C(\pm \omega_d) = G_1(\pm \omega_d) \quad \text{or} \quad Y(\omega_c \pm \omega_d) = 1. \tag{75}$$

It can readily be seen that the condition (71) required on $G_1(\omega)$ translates to the same condition for $G_1(u)$ where $u = \omega - \omega_c$.

The relations can be made clearer by working out an example. Suppose the switching is rectangular and there is no lost time between contacts. The function $g(t)$ is then defined by

$$g(t) = \begin{array}{ll} 1, & -T/2 < t < T/2 \\ 0, & |t| > T/2. \end{array} \tag{76}$$

Let the received signal $V_r(t)$ have a full raised cosine spectrum centered at ω_c, with vestigial symmetry about $\omega_c + \omega_d$ and $\omega_c - \omega_d$. We then write

18

$$G_1(u) = \begin{array}{ll} T\left(1 + \cos\dfrac{\pi u}{2\omega_d}\right)\Big/ 2 & |u| \leqq 2\omega_d \\ 0 & |u| > 2\omega_d. \end{array} \qquad (77)$$

We calculate

$$C(\omega) = 2\int_0^{T/2} \cos\omega_d t \cos\omega t\, dt = \frac{2\omega_d \cos(\omega T/2)}{\omega_d{}^2 - \omega^2} \qquad (78)$$

$$Y(\omega) = \frac{\pi(\omega_d{}^2 - u^2)\left(1 + \cos\dfrac{\pi u}{2\omega_d}\right)}{4\omega_d{}^2 \cos\dfrac{\pi u}{2\omega_d}} \qquad u = \omega_c - \omega. \qquad (79)$$

This function satisfies the required condition that $Y(\omega_c \pm \omega_d) = 1$.

In practice it is difficult to control two oscillators with the necessary precision to meet Sunde's requirements. One method of realizing the system approximately is to begin with two high-frequency crystal-controlled oscillators of frequencies $n(\omega_c - \omega_d)$ and $n(\omega_c + \omega_d)$, where n is a large integer. The phases of the two oscillators are not under control and are assumed to be θ_1 and θ_2, respectively. Frequency step-down circuits are introduced after each oscillator to give outputs of frequency $\omega_c - \omega_d$ and $\omega_c + \omega_d$ with respective phases θ_1/n and θ_2/n. By multiplying these two outputs and selecting the low-frequency component as shown in Fig. 5, we obtain a wave of frequency $2\omega_d$ and phase $(\theta_2 - \theta_1)/n$. This wave can be used to control the timing of the binary input symbols. For the switched marking and spacing frequency sources we use the stepped-down component of frequency $\omega_c - \omega_d$ directly and the component of frequency $\omega_c + \omega_d$ with reversed polarity. The required frequency and phase relations are then satisfied

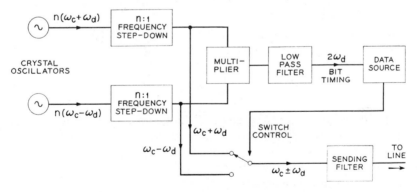

Fig. 5 — Practical realization of Sunde's system.

19

except for a slow drift in the time scale caused by the lack of perfect stability in the original oscillators.

To calculate the probability of error when Gaussian noise is added to Sunde's FM signal, we identify the values of $P(mT)$, $P'(mT)$, $Q(mT)$, and $Q'(mT)$ of (69) with P, $\dot{P}$, Q, and $\dot{Q}$ respectively. The general expression for the probability of error, (34), is expressed in terms of R and $\dot{R}$. We calculate

$$R = (P^2 + Q^2)^{\frac{1}{2}} = A \tag{80}$$

$$R'(t) = \frac{d}{dt} [P^2(t) + Q^2(t)]^{\frac{1}{2}}$$

$$= [P(t)P'(t) + Q(t)Q'(t)]/R(t) \tag{81}$$

$$\dot{R} = R'(mT) = (P\dot{P} + Q\dot{Q})/R = (-)^{m+1} A b_m s_1'(mT). \tag{82}$$

From (62)

$$s_1'(mT) = \sum_{n=-\infty}^{\infty} (-)^n b_n g_1'[(m - n)T]. \tag{83}$$

From (73) we verify

$$
\begin{aligned}
g_1(rT) &= \frac{1}{\pi} \int_0^{2\omega_d} G_1(\omega) \cos (\omega rT) \, d\omega \\
&= \frac{1}{\pi} \int_0^{\omega_d} G_1(\omega_d - \omega) \cos [rT(\omega_d - \omega)] \, d\omega \\
&\quad + \frac{1}{\pi} \int_0^{\omega_d} G_1(\omega_d + \omega) \cos [rT(\omega_d + \omega)] \, d\omega \\
&= \frac{2}{\pi} G_1(\omega_d) \cos r\pi \int_0^{\omega_d} \cos r\omega T \, d\omega = \delta_{r0}.
\end{aligned} \tag{84}
$$

This checks our previous requirements expressed by (63) and (64). By differentiating (73) and substituting $t = rT$, we find

$$g_1'(rT) = -\frac{1}{\pi} \int_0^{2\omega_d} \omega G_1(\omega) \sin \omega rT \, d\omega. \tag{85}$$

The value of this integral in general is not zero except when $r = 0$. It appears, therefore, that at any sampling instant $t = mT$ the value of $\dot{R}$ depends on all the values of b_n in the sequence except b_m.

For further progress we take a specific example, namely the full raised cosine spectrum for $G_1(\omega)$. We set

20

368

$$G_1(\omega) = T\left(1 + \cos\frac{\pi\omega}{2\omega_d}\right)\Big/2 \qquad |\omega| \leqq 2\omega_d. \tag{86}$$

Then

$$g_1'(rT) = -\frac{T}{2\pi}\int_0^{2\omega_d} \omega\left(1 + \cos\frac{\pi\omega}{2\omega_d}\right)\sin \omega rT\, d\omega$$

$$= \frac{f_0}{r(1 - 4r^2)} \qquad r \neq 0. \tag{87}$$

From (85), we noted that $g_1'(0) = 0$. The value of $\dot{R}$ can now be found from (82), thus

$$\dot{R} = (-)^{m+1}\, b_m f_0 A \left[\sum_{n=-\infty}^{m-1} + \sum_{n=m+1}^{\infty}\right]\frac{(-)^n b_n}{(m - n)[1 - 4(m - n)^2]} \tag{88}$$

$$= -b_m f_0 A \sum_{n=1}^{\infty} (-)^n \frac{b_{m+n} - b_{m-n}}{n(4n^2 - 1)}.$$

We observe from our previous study of the integral defining the probability of error that for fixed R the most vulnerable sequence is the one which has the largest absolute value of $\dot{R}$. The least vulnerable sequence is the one for which $\dot{R} = 0$, and this can be obtained by setting $b_{m+n} = b_{m-n}$ for all n. The maximum absolute value of $\dot{R}$ occurs when b_{m+n} and b_{m-n} have opposite signs and the signs are reversed when n changes by unity. The resulting value of $|\dot{R}|$ is[6]

$$\dot{R}_m = 2f_0 A \sum_{n=1}^{\infty} \frac{1}{n(4n^2 - 1)} \tag{89}$$

$$= 2f_0 A\, (\log_e 4 - 1) = 0.7726\, f_0 A.$$

The upper and lower bounds for the error probability are found by substituting $\dot{R}_m$ and 0 respectively for $\dot{R}$ in (34). By (80) the value of R is constant and equal to A. From (70), $\phi = b_m \omega_d$. It is important to note that while the intersymbol interference is suppressed in the absence of noise the error probability with noise present does depend on the signal sequence. This occurs because frequency detection is a nonlinear process, and the effect of noise cannot be found by merely adding a noise wave to the detected frequency output.

The actual spectral density of the noise facing the frequency detector is under the control of the system designer, since the selectivity of the receiving bandpass filter is not determined by the requirements thus far discussed. We have stated what the received signal spectrum at the

21

detector input should be, but this is a resultant of signal shaping at the transmitter, the transmitting filter selectivity, and the transmittance of the line, as well as receiving filter selectivity. The latter can be varied within reasonable limits if the others are adjusted in a complementary fashion to obtain the desired output response. In evaluating the merit of different receiving filter designs it is reasonable to compare them with the same average signal power on the line. We shall also assume that the line has been equalized for unity gain and linear phase over the band so that it can be considered as a transparent link in the system.

The average signal power on the line can be computed in terms of (a) the transmittance function $Y_r(\omega)$ of the receiving filter, (b) the required function $G_1(\omega)$ representing the spectrum of the modified switching function $g_1(t)$ at the detector input, and (c) the statistics of the data sequence. Details of the calculation are given in Appendix B. An interesting consequence of the assumptions that the FM wave has continuous phase and that the frequency shift is equal to the signaling rate is the appearance of discrete components on the line at the marking and spacing frequencies even when the data sequence is random. This means there are transmitted sine waves which consume power but carry no information. An optimization procedure aimed at conserving power would very nearly suppress these components at the transmitter by balance or by sharp antiresonances and restore them to their proper relative amplitudes by complementary narrow-band resonance peaks in the response of the receiving bandpass filter. The bandwidth used to augment these frequencies at the receiver could in theory be made so small that no appreciable effect on the accepted noise would result. The system would then only have to deliver the average power associated with the continuous part of the FM spectrum.

Actually, even a partial suppression of the steady-state components on the line would destroy much of the advantage of signaling by FM. The system would become more sensitive to gain changes and overload distortion. Accurate tracking of the suppression and recovery circuits for the marking and spacing frequencies would be difficult at best and would be practically impossible over a channel with carrier frequency offset. The narrow-band recovery circuits would contribute to a sluggish start-up time. In fact, about the only remaining resemblance to FM would be the use of an FM detector. If low-level tones can actually be recovered successfully from a received wave, it would be better to use them for synchronous PM detection, which is a linear method capable of attaining ideal performance in the presence of additive Gauss-

22

370

ian noise. It appears that Sunde's system should carry the power in the steady-state components in order to deserve the name of FM.

Standard variational procedures can be applied to find the shape of receiving filter selectivity which minimizes probability of error when the average signal power and the spectral density of added Gaussian noise on the line are specified. The solution of the optimization problem is given in Appendix B, and means are shown for completing the computation of the corresponding probabilities of error for the most and least vulnerable data sequences. In the case of white Gaussian noise on the line, the optimum receiving filter has very nearly the same cosine characteristic found by Sunde for optimum binary AM transmission. The bounds for error probability are plotted in Fig. 6 for both FM proper with no suppression of steady-state tones and the abnormal FM with marking and spacing frequencies suppressed. Also shown is the ideal curve representing what can be proved to give the best possible binary performance. The ideal curve can theoretically be obtained for example by coherent detection of binary phase modulation. Differentially detected phase modulation requires about 1 db more signal power than ideal at an error probability of 10^{-4}.

It is seen from Fig. 6 that when the suppression bands are inserted in Sunde's binary FM system, the theoretical performance is only about a half db poorer than ideal, but, as previously pointed out, this does not represent a true FM system. The more legitimate FM has error bounds from 3 to 3.5 db poorer than ideal. However, a penalty of this order of magnitude could be a fair trade in many cases for the advantages of a much simplified receiver relatively immune to many channel faults.

VI. APPLICATION TO DATA TERMINALS FOR USE ON TELEPHONE CHANNELS

We now apply our formulas to calculate error probabilities in binary FM transmission with terminals more closely resembling those actually in use on telephone channels. In the design of real-life terminals, the emphasis is placed on ruggedness and simplicity. The bit rate is not locked to the frequency deviation. The filters do not meet elaborate optimization requirements. The significant conclusion from our evaluation of error probabilities for the practical systems is that the degradation of performance compared with the ideal is actually very slight.

The probability of error as given in (44) is generally applicable to FM systems. There are three parameters, ρ, a, and b, given in (41) to (43). The first parameter ρ is a signal-to-noise ratio. It depends on the

23

371

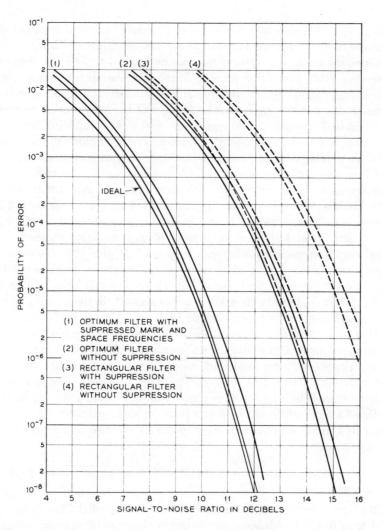

Fig. 6 — Error probabilities for Sunde's binary FM system with additive Gaussian noise. Bounds are for most and least vulnerable sequences. Noise reference is mean noise power in bandwidth equal to bit rate.

ratio of instantaneous envelope of the received signal to the rms noise voltage at the detector input. For any given front-end filter, this parameter can be expressed in terms of average signal-to-noise ratio at the input of the receiver. The parameter a depends on the ratio of instantaneous frequency displacement at the sampling time to the Gabor noise bandwidth, σ_1/σ_0, of the receiver. The third parameter b depends

24

on the derivative of the instantaneous envelope at the sampling time. For a given channel these parameters can be computed for any particular signaling sequence. The true probability of error could conceivably be obtained by averaging over all possible sequences, but this would be a formidable task. Instead we will give bounds on the probability of error for the most and least vulnerable sequences over a finite representative set of signaling intervals.

We first consider the system in Fig. 7, which has amplitude-vs-frequency "raised cosine" type roll-off but no phase distortion. Equal filtering takes place at the transmitter and receiver. The modulator applies a pure FM wave of constant envelope to the transmitting filter. In other words, the modulator and the demodulator are ideal. The data source is composed of rectangular pulses. The frequency deviation in cps is equal to half the bit rate. These rates and deviations are characteristic of practical systems.

With the aid of a digital computer, S. Habib has calculated the parameters given in (41) to (43) for 2^{10} sequences. From these calculations we have computed an upper and a lower bound on the probability of error. These results are shown in Fig. 8. The probability of error for all other sequences will fall between the two curves labeled "best" and "worst." Superimposed on the same graph is the ideal curve, which can only be achieved with ideal phase systems and coherent detection. The FM detection is, of course, incoherent.

Our next example applies the theory to a real bandpass filter used in an operational data set. Fig. 9 shows the system considered. The curve

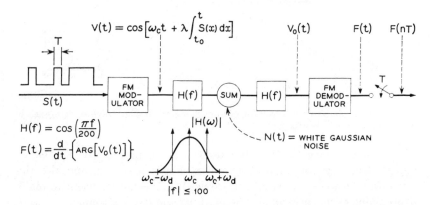

Fig. 7 — Ideal FM modulator and demodulator with transmitted and received signals equally shaped by "raised cosine" type roll-off amplitude characteristics and no phase distortion.

25

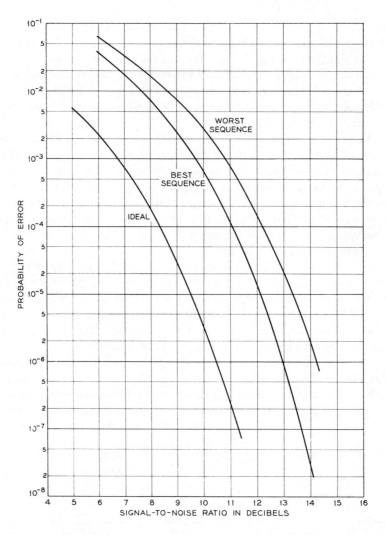

Fig. 8 — Probability of error for system depicted in Fig. 7. Noise reference is mean power in bandwidth equal to bit rate.

of loss vs frequency for the filter used is given in Fig. 10. The curve departs from the condition of symmetry about midband, and also the separation between the signal and carrier bands is not sufficient to make overlapping effects negligible. The marking and spacing frequencies were assumed to be 1200 and 2200 cps, respectively, and the signaling rate 1200 bits per second. As shown in Fig. 11, the calculated results are

26

374

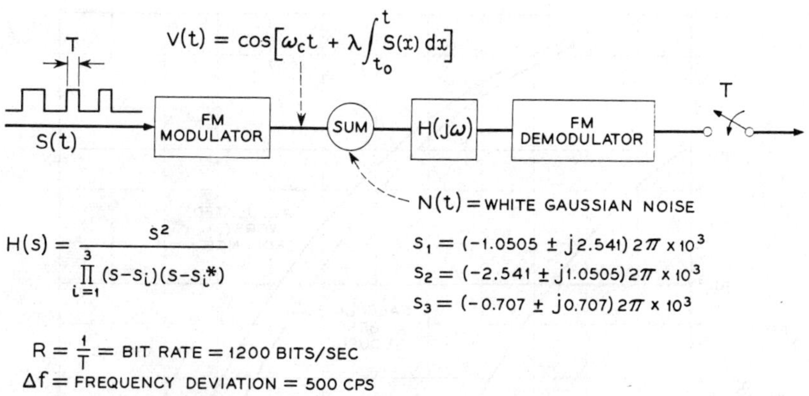

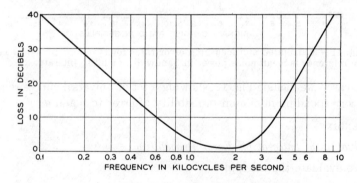

T
$V(t) = \cos\left[\omega_c t + \lambda \int_{t_0}^{t} S(x)\,dx\right]$
S(t)
FM MODULATOR
SUM
H(jω)
FM DEMODULATOR
T
N(t) = WHITE GAUSSIAN NOISE
$H(s) = \frac{s^2}{\prod_{i=1}^{3} (s-s_i)(s-s_i^*)}$
$s_1 = (-1.0505 \pm j2.541)\, 2\pi \times 10^3$
$s_2 = (-2.541 \pm j1.0505)\, 2\pi \times 10^3$
$s_3 = (-0.707 \pm j0.707)\, 2\pi \times 10^3$
$R = \frac{1}{T}$ = BIT RATE = 1200 BITS/SEC
Δf = FREQUENCY DEVIATION = 500 CPS

Fig. 9 — Ideal modulator and demodulator with received signal shaped by filter characteristics used in FM data set and shown in Fig. 10.

about 1 db better than the experimental results obtained with a random word generator, random noise generator, and error counter. The experimental system included an axis-crossing detector and post-detection low-pass filter, which do not correspond precisely with the theoretical model. In view of the differences cited, the agreement between calculated and experimental curves is good. The penalty suffered by the actual back-to-back channel compared with the best theoretical FM performance is between 2 and 3 db. Somewhat more optimistic estimates have been given in other published studies.[7,8] The effects of amplitude and delay-versus-frequency variation in the channel are calculable by use of the computer programs we have established.

It was shown in the previous sections that a lower bound on the probability of error occurs when the parameter b is set equal to zero. For

40
30
20
10
0
LOSS IN DECIBELS
0.1
0.2
0.3
0.4
0.6
0.8
1.0
2
3
4
5
6
8
10
FREQUENCY IN KILOCYCLES PER SECOND

Fig. 10 — Receiver bandpass filter loss vs frequency characteristic.

27

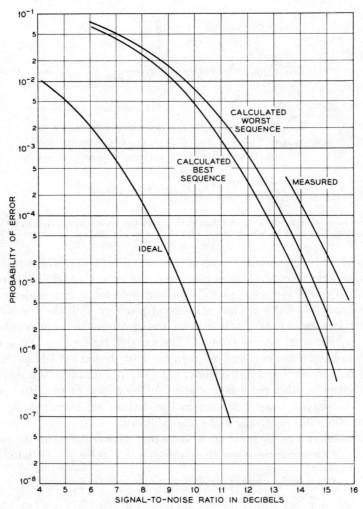

Fig. 11 — Calculated and measured error rates for system depicted in Fig. 9. Noise reference is mean noise power in bandwidth equal to bit rate.

this reason we include Fig. 12, showing a set of universal curves relating the corresponding minimum probability of error to ρ and a.

APPENDIX A

Evaluation of Integral for Error Probability

We evaluate the integral

$$P_+ = \int_0^\infty dz \int_{-\infty}^\infty \int_{-\infty}^\infty p(-z \mid x,y)q(x,y) \; dx \; dy \qquad (90)$$

28

376

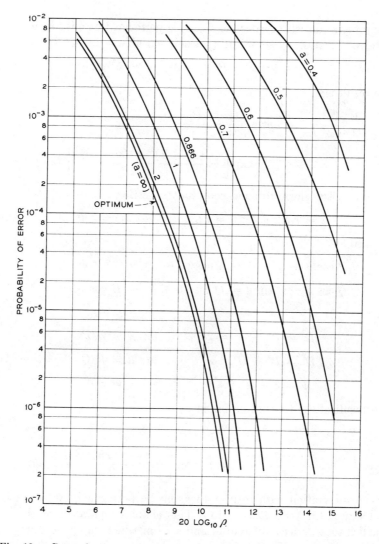

Fig. 12 — General curves of minimum probability of error vs ρ for different values of a in the range of interest.

where

$$p(-z \mid x,y) = \frac{1}{\sigma_1[2\pi(x^2 + y^2)]^{\frac{1}{2}}} \exp\left[-\frac{(z + \dot{Q}x - \dot{P}y)^2}{2\sigma_1^2(x^2 + y^2)}\right] \quad (91)$$

$$q(x,y) = \frac{1}{2\pi\sigma_0^2} \exp\left[-\frac{(x - P)^2 + (y - Q)^2}{2\sigma_0^2}\right]. \quad (92)$$

29

The integration with respect to z can be performed at once in terms of the error function by substituting a new variable u defined by

$$(z + \dot{Q}x - \dot{P}y)^2 = 2\sigma_1^2(x^2 + y^2)u^2. \tag{93}$$

The result is:

$$P_+ = \frac{1}{2}\frac{1}{4\pi\sigma_0^2}\int_{-\infty}^{\infty}\int_{-\infty}^{\infty} \text{erf}\,\frac{\dot{Q}x - \dot{P}y}{\sigma_1[2(x^2 + y^2)]^{\frac{1}{2}}}$$
$$\cdot\exp\left[-\frac{(x - P)^2 + (y - Q)^2}{2\sigma_0^2}\right] dx\,dy. \tag{94}$$

We now transform to polar coordinates, setting

$$x = r\cos\theta \qquad y = r\sin\theta \qquad dx\,dy = r\,dr\,d\theta \tag{95}$$

We also let

$$P\cos\theta + Q\sin\theta = R\cos(\theta - \alpha) = R\cos\psi$$
$$\dot{Q}\cos\theta - \dot{P}\sin\theta = D\cos(\theta + \beta) = D\cos(\psi + \gamma)$$

where

$$R^2 = P^2 + Q^2 = 2\sigma_0^2\rho^2 \qquad \tan\alpha = Q/P$$
$$D^2 = \dot{P}^2 + \dot{Q}^2 \qquad \tan\beta = \dot{P}/\dot{Q} \tag{96}$$
$$\psi = \theta - \alpha \qquad \gamma = \alpha + \beta.$$

The result of the transformation is

$$1 - 2P_+ =$$
$$\frac{e^{-\rho^2}}{2\pi\sigma_0^2}\int_{-\pi}^{\pi}\text{erf}\,\frac{D\cos(\psi + \gamma)}{\sqrt{2}\,\sigma_1}\,d\psi\int_0^{\infty}\exp\left[-\frac{r^2 - 2rR\cos\psi}{2\sigma_0^2}\right]r\,dr. \tag{97}$$

The integration with respect to r can be performed by subtracting and adding the term $R\cos\psi$ to r. This enables separation of the integrand into a perfect differential and a term which can be expressed as an error function. We thereby obtain

$$1 - 2P_+ = \frac{1}{2\pi}\int_{-\pi}^{\pi}\text{erf}\,\frac{D\cos(\psi + \gamma)}{\sqrt{2}\,\sigma_1}$$
$$\left[1 + \sqrt{2\pi}\,\frac{R}{2\sigma_0}\exp\left(-\frac{R^2}{2\sigma_0^2}\sin^2\psi\right)\cos\psi\left(1 - \text{erf}\,\frac{R\cos\psi}{\sqrt{2}\,\sigma_0}\right)\right] d\psi. \tag{98}$$

We note that both $\cos\psi$ and $\cos(\psi + \gamma)$ change sign when ψ is increased by π and that $\sin^2(\psi + \pi) = \sin^2\psi$. Furthermore, the inte-

30

378

gration in (98) is over one full period in ψ, and for every value of ψ in the left half of the period there is a corresponding value in the right half at $\psi + \pi$. Since the error function, erf z, is an odd function of z, a change in the sign of $\cos \psi$ or $\cos (\psi + \gamma)$ changes the sign of the corresponding error function in the integrand. If we multiply the first term under the integral sign by the terms within the bracket following, we see that there is only one product which does not change sign at points π apart. The integral of the other products must vanish. The integral of the one which does not change sign is twice the integral over a half period of ψ. Hence

$$1 - 2P_+ = \frac{R}{\sigma_0 \sqrt{2\pi}} \int_{-\pi/2}^{\pi/2} \exp\left(- \frac{R^2}{2\sigma_0^2} \sin^2 \psi\right)$$

$$\cdot \cos \psi \; \mathrm{erf} \; \frac{D \cos(\psi + \gamma)}{\sqrt{2}\,\sigma_1} \; d\psi. \tag{99}$$

From (96) and (23)

$$D \cos \gamma = D(\cos \alpha \cos \beta - \sin \alpha \sin \beta)$$

$$= D\left(\frac{P}{R}\frac{\dot{Q}}{D} - \frac{Q}{R}\frac{\dot{P}}{D}\right) = \frac{P\dot{Q} - Q\dot{P}}{R} = R\dot{\phi} \tag{100}$$

$$D \sin \gamma = D(\sin \alpha \cos \beta + \cos \alpha \sin \beta$$

$$= D\left(\frac{Q}{R}\frac{\dot{Q}}{D} + \frac{P}{R}\frac{\dot{P}}{D}\right) = \frac{Q\dot{Q} + P\dot{P}}{R} \tag{101}$$

$$= \frac{1}{2R}\frac{d}{dt}(R^2) = \frac{dR}{dt} = \dot{R}.$$

Therefore

$$D \cos (\psi + \gamma) = D \cos \gamma \cos \psi - D \sin \gamma \sin \psi$$

$$= R\dot{\phi} \cos \psi - \dot{R} \sin \psi. \tag{102}$$

Now substituting $x = \sin \psi$ in (99) we rearrange to obtain

$$P_+ = \frac{1}{2} - \frac{\rho}{2\sqrt{\pi}} \int_{-1}^{1} e^{-\rho^2 x^2} \; \mathrm{erf} \; \frac{R\dot{\phi}(1 - x^2)^{\frac{1}{2}} - \dot{R}x}{\sqrt{2}\,\sigma_1} \; dx. \tag{103}$$

Equation (34) of the main text is obtained from (103) by substituting the complementary function erfc $z = 1 - \mathrm{erf}\; z$.

The lower bound P_l on the probability of error for any fixed R and ϕ was shown in the text to be obtained by setting $\dot{R} = 0$. When this substitution is made in (103) and the definition of the error function

31

in terms of an integral is inserted, we obtain

$$P_l = \frac{1}{2} - \frac{2\rho}{\pi} \int_0^1 e^{-\rho^2 x^2} \, dx \int_0^{a\rho(1-x^2)^{\frac{1}{2}}} e^{-z^2} \, dz. \qquad (104)$$

The parameters a and ρ are defined by (41) and (42). If we substitute $\rho x = y$ the expression becomes

$$P_l = \frac{1}{2} - \frac{2}{\pi} \int_0^\rho \int_0^{a(\rho^2 - y^2)^{\frac{1}{2}}} e^{-y^2 - z^2} \, dy \, dz. \qquad (105)$$

The region of integration in the double integral consists of the first quadrant of the ellipse

$$z^2/(a\rho)^2 + y^2/\rho^2 = 1. \qquad (106)$$

After transforming to polar coordinates by setting $y = r \cos \theta$ and $z = r \sin \theta$, we can perform the integration with respect to r. The result is

$$P_l = \frac{1}{\pi} \int_0^{\pi/2} \exp\left[- \frac{a^2\rho^2}{\sin^2 \theta + a^2 \cos^2 \theta} \right] d\theta. \qquad (107)$$

This is equivalent to (37) of the main text.

The integral has a simple value when $a = 1$, which is equivalent to $\dot{\phi} = \sigma_1/\sigma_0$. For this case the integrand is seen to become a constant and (40) results. This coincides with a result given for a special case by Montgomery.[9] By a change in the meaning of the parameters it also gives the error probability for differential binary phase detection as discussed in Section IV. In the general case, the limiting form of P_l for large signal-to-noise ratio can be calculated by the method of steepest descents. Saddle points occur at $\theta = 0$ and $\theta = \pi/2$. When $a > 1$, the saddle point at $\theta = 0$ determines the asymptotic form of the integral for large ρ and (38) is obtained. When $a < 1$, the saddle point at $\theta = \pi/2$ is dominant and we obtain (39).

APPENDIX B

Optimization of Receiving Filter for Sunde's FM System

Our problem is to find the receiving filter characteristic which minimizes the probability of error in Sunde's FM system when the average transmitted signal power and the spectral density of the noise on the line are specified. In terms of Fig. 13 the transmittance function for the filter is $Y_r(\omega)$ and the output of the filter is $V_r(t)$ as defined by (61), (62), (71), and (73), namely

32

380

$$V_r(t) = A \sin \omega_d t \sin (\omega_c t + \theta_r) - A s_1(t) \cos (\omega_c t + \theta_r) \quad (108)$$

$$s_1(t) = \sum_{n=-\infty}^{\infty} (-)^n b_n g_1(t - nT) \quad (109)$$

$$G_1(\pm \omega_d - \lambda) + G_1(\pm \omega_d + \lambda) = 2G_1(\omega_d) = T \quad 0 < \lambda < \omega_d \quad (110)$$

$$G_1(\omega) = \int_{-\infty}^{\infty} g_1(t) \cos \omega t \, dt. \quad (111)$$

The input to the filter is the sum of the signal wave $V(t)$ plus the Gaussian noise wave $v_0(t)$. The wave $V(t)$ is defined as that function of time which when operated on by $Y_r(\omega)$ produces $V_r(t)$. The noise wave $v(t)$ at the input to the frequency detector has a spectral density equal to $| Y_r(\omega) |^2$ times that of $v_0(t)$.

We shall simplify our treatment by assuming a random sequence of data in which the two binary symbols are selected with equal probability. The probability is then equal to 0.5 that any particular b_n has the value $+1$ and also 0.5 that the value is -1. We regard $V_r(t)$ as a member of an ensemble of random functions with a distribution in the infinite number of independent random parameters b_n. The randomness appears entirely in the function $s_1(t)$. We can calculate the ensemble average of $s_1(t)$ at fixed t by adding the individual averages of the terms in the infinite series defining $s_1(t)$. When we do this we find that the only random variable in each term is b_n, which assumes the values ± 1 with equal likelihood and therefore has the average value zero. Hence the ensemble average of $s_1(t)$, which we shall designate by $\langle s_1(t) \rangle$, is zero for any fixed value of t. It follows that $s_1(t)$ can contain no periodic components, for the presence of any such components would give a non-zero average at some values of t. Therefore, the spectral density function of the second term in $V_r(t)$ must be a continuous function of frequency.

To calculate the average square of $s_1(t)$ over the ensemble, we note that $s_1(t)$ is the sum of an infinite number of independent random variables of form

$$z_n = (-)^n b_n g_1(t - nT). \quad (112)$$

The average value of each z_n is zero and the variance, or mean square minus the square of the mean, is equal to the square of $g_1(t - nT)$.

Fig. 13 — Function of receiving filter in Sunde's system.

33

Since the variance of the sum of independent variables is equal to the sum of the variances of the individual variables, we can write

$$\langle s_1^2(t) \rangle = \sum_{n=-\infty}^{\infty} g_1^2(t - nT). \tag{113}$$

The average in (113) is an ensemble average at fixed t. We can show that this average is periodic in t with period T by noting that

$$\langle s_1^2(t + T) \rangle = \sum_{n=-\infty}^{\infty} g_1^2(t + T - nT)$$

$$= \sum_{m=-\infty}^{\infty} g_1^2(t - mT) \tag{114}$$

$$= \langle s_1^2(t) \rangle.$$

Therefore the average over t can be computed by averaging over a single period from $t = 0$ to $t = T$. Hence the average over time which we shall designate by av is

$$\text{av } s_1^2(t) = \frac{1}{T} \int_0^T \langle s_1^2(t) \rangle \, dt = \frac{1}{T} \sum_{n=-\infty}^{\infty} \int_0^T g_1^2(t - nT) \, dt$$

$$= \frac{1}{T} \sum_{n=-\infty}^{\infty} \int_{-nT}^{-(n-1)T} g_1^2(\lambda) \, d\lambda = \frac{1}{T} \int_{-\infty}^{\infty} g_1^2(\lambda) \, d\lambda. \tag{115}$$

By application of Parseval's theorem

$$\text{av } s_1^2(t) = \frac{1}{2\pi T} \int_{-\infty}^{\infty} G_1^2(\omega) \, d\omega. \tag{116}$$

From (116) we deduce that the spectral density of $s_1(t)$ is given by

$$w_1(\omega) = \frac{G_1^2(\omega)}{2\pi T} = \frac{\omega_d G_1^2(\omega)}{2\pi^2}. \tag{117}$$

The spectral density of $V_r(t)$ can now be easily calculated. The first term can be expressed as the sum of sine waves of amplitude $A/2$ and frequencies $\omega_c + \omega_d$ and $\omega_c - \omega_d$. The first term therefore contributes line spectra with mean square $A^2/8$ at the marking and spacing frequencies. The average square of the second term can be written

$$\text{av } [A^2 s_1^2(t) \cos^2 (\omega_c t + \theta_r)] = \frac{A^2}{2} \text{ av } s_1^2(t). \tag{118}$$

The spectral components comprising $s_1(t) \cos (\omega_c t + \theta_r)$ are those of $s_1(t)$ shifted from their original positions to appear as sidebands around the frequencies $\pm \omega_c$. Hence $w_r(\omega)$, the spectral density of $V_r(t)$ with

34

all power assigned to positive frequencies, is given by

$$w_r(\omega) = \frac{A^2}{8} \delta(\omega - \omega_c + \omega_d) + \frac{A^2}{8} \delta(\omega - \omega_c - \omega_d)$$

$$+ \frac{\omega_d A^2 G_1^2(\omega - \omega_c)}{4\pi^2} \qquad \omega \geqq 0. \tag{119}$$

It is convenient to let $\omega - \omega_c = u$ and write for the transmittance of the filter

$$U(u) = Y_r(\omega - \omega_c). \tag{120}$$

We shall also designate the spectral density of $V(t)$ as $w(u)$. Since the linear operator $U(u)$ can be applied individually to the components which make up (119) we must have

$$w(u) = \frac{A^2 \delta(u + \omega_d)}{8 \, | \, U(-\omega_d) \, |^2} + \frac{A^2 \delta(u - \omega_d)}{8 \, | \, U(\omega_d) \, |^2} + \frac{\omega_d A^2 G_1^2(u)}{4\pi^2 \, | \, U(u) \, |^2}. \tag{121}$$

The average power on the line is proportional to W_0, the average square of $V(t)$, which is given by

$$W_0 = \int_{-2\omega_d}^{2\omega_d} w(u) \, du = \frac{A^2}{8 \, | \, U(-\omega_d) \, |^2} + \frac{A^2}{8 \, | \, U(\omega_d) \, |^2}$$

$$+ \frac{\omega_d A^2}{4\pi^2} \int_{-2\omega_d}^{2\omega_d} \frac{G_1^2(u)}{| \, U(u) \, |^2} \, du. \tag{122}$$

We make the reasonable assumption that $| \, U(u) \, |$ is an even function of u. Combined with the further assumption that the spectral density of the noise on the line is symmetrical about ω_c, this furnishes a convenient assurance of a symmetrical spectral density for the noise in the output of the receiving filter. Since $G_1(u)$ is also an even function of u, we can write (122) in the equivalent form

$$W_0 = \frac{A^2}{4X(\omega_d)} + \frac{\omega_d A^2}{2\pi^2} \int_0^{2\omega_d} \frac{G_1^2(u)}{X(u)} \, du \tag{123}$$

where

$$X(u) = | \, U(u) \, |^2. \tag{124}$$

The function $X(u)$ is to be chosen to minimize the probability of error under the constraint that W_0 is held constant. In calculating the optimum function, the signal power represented by the steady-state components can be ignored, since this power could be reduced to an arbitrarily small value by the use of narrow-band suppression tech-

35

niques. The constraint on the signal power is therefore that the integral in (123) is to be held constant.

Let $N(u)$ represent the spectral density of the Gaussian noise wave $v_0(t)$ on the line. Then the spectral density of $v(t)$, the noise in the output of the receiving filter, is $X(u)N(u)$. In terms of the spectral density $w_v(\omega)$ previously defined for $v(t)$ with values symmetrically distributed between positive and negative frequencies, we have

$$X(u)N(u) = 2w_v(u + \omega_c). \tag{125}$$

The values of σ_0 and σ_1 necessary to complete the calculation of the probability of error by (34) can now be found by substituting (125) in (17) and (18) giving the results

$$\sigma_0^2 = 2\int_0^{2\omega_d} X(u)N(u)\ du \tag{126}$$

$$\sigma_1^2 = 2\int_0^{2\omega_d} u^2 X(u)N(u)\ du. \tag{127}$$

If we substitute (126) and (127) into the general expression for error probability, (34), and attempt to formulate a variational problem, the expressions become unmanageable. Instead, we concentrate attention on the lower bound for error probability obtained by setting $\dot{R} = 0$, (36), in which it is evident that to make the error probability as small as possible both σ_0 and σ_1 should be made as small as possible. As shown by (126) and (127), σ_0 and σ_1 are not independent. The effect of the dependence can be taken into account by performing the minimization problem in two steps. First we minimize σ_0 with both σ_1 and W_0 held constant. After this solution is obtained, we find by trial the value of σ_1 which yields the lowest minimum probability of error.

Omitting inconsequential multiplying factors, we set the variational problem as

$$\delta\left[\int_0^{2\omega_d} X(u)N(u)\ du + \lambda\int_0^{2\omega_d} u^2 X(u)N(u)\ du \right.$$
$$\left. + \mu\int_0^{2\omega_d} \frac{|G_1(u)|^2}{X(u)}\ du\right] = 0 \tag{128}$$

where λ and μ are Lagrange multipliers and the function under variation is $X(u)$. The solution is

$$X(u) = \frac{\mu\,|G_1(u)|}{(1 + \lambda u^2)^{\frac{1}{2}}\,N^{\frac{1}{2}}(u)}. \tag{129}$$

36

It is straightforward to verify that this stationary value of $X(u)$ actually gives a minimum value of σ_0 and hence minimum probability of error for fixed values of σ_1 and W_0.

Substituting our partially optimized solution in (123), (126), and (127), we obtain

$$\sigma_0^2 = 2\mu \int_0^{2\omega_d} \frac{|G_1(u)| N^{\frac{1}{2}}(u)}{(1 + \lambda u^2)^{\frac{1}{2}}} \, du \qquad (130)$$

$$\sigma_1^2 = 2\mu \int_0^{2\omega_d} \frac{u^2 |G_1(u)| N^{\frac{1}{2}}(u)}{(1 + \lambda u^2)^{\frac{1}{2}}} \, du. \qquad (131)$$

$$W_o - W_s = \frac{\omega_d A^2}{2\pi^2 \mu} \int_0^{2\omega_d} |G_1(u)| N^{\frac{1}{2}}(u)(1 + \lambda u^2)^{\frac{1}{2}} \, du \qquad (132)$$

$$W_s = \frac{A^2 N^{\frac{1}{2}}(\omega_d)(1 + \lambda \omega_d^2)^{\frac{1}{2}}}{4\mu |G_1(\omega_d)|} \qquad (133)$$

$$\frac{1}{\rho^2} = \frac{2\sigma_0^2}{A^2} = \frac{2\omega_d I_1 I_2}{\pi^2(W_0 - W_s)} \qquad (134)$$

$$\frac{1}{a^2 \rho^2} = \frac{2\sigma_1^2}{A^2 \omega_d^2} = \frac{2 I_2 I_3}{\pi^2 \omega_d (W_0 - W_s)} \qquad (135)$$

where

$$I_1 = \int_0^{2\omega_d} \frac{|G_1(u)| N^{\frac{1}{2}}(u)}{(1 + \lambda u^2)^{\frac{1}{2}}} \, du \qquad (136)$$

$$I_2 = \int_0^{2\omega_d} |G_1(u)| N^{\frac{1}{2}}(u)(1 + \lambda u^2)^{\frac{1}{2}} \, du \qquad (137)$$

and

$$I_3 = \int_0^{2\omega_d} \frac{u^2 |G_1(u)| N^{\frac{1}{2}}(u)}{(1 + \lambda u^2)^{\frac{1}{2}}} \, du. \qquad (138)$$

These equations furnish a straightforward procedure for calculating the optimum filter characteristic. Each assumed value of λ determines a pair of values ρ and $a\rho$ from which the corresponding upper and lower bounds for the error probability can be evaluated by computer techniques. By successive trials the best value of λ can be approximated to any desired degree and substituted in (129) to obtain the best filter selectivity function. In actual examples tried, this procedure could be shortened because the error probability turned out to be very much more sensitive to the value of ρ than to the value of $a\rho$. If this were known beforehand, we would place no constraint on σ_1 in the minimiza-

37

tion of σ_0. This is equivalent to setting $\lambda = 0$, leading to the simpler formulas

$$\frac{1}{\rho^2} = \frac{2\omega_d}{\pi^2(W_0 - W_s)}\left[\int_0^{2\omega_d} |G_1(u)| N^{\frac{1}{2}}(u) \ du\right]^2 \qquad (139)$$

$$\frac{1}{a^2\rho^2} = \frac{2}{\pi^2\omega_d(W_0 - W_s)}\int_0^{2\omega_d} u^2 |G_1(u)| N^{\frac{1}{2}}(u) \ du$$

$$\cdot\int_0^{2\omega_d} |G_1(u)| N^{\frac{1}{2}}(u) \ du. \qquad (140)$$

By applying Schwarz' inequality to the products of integrals in (134) and (135), we verify that the case of $\lambda = 0$ gives the maximum value of ρ, but that the maximum value of $a\rho$ occurs when $\lambda = \infty$. It seems therefore that an intermediate nonzero value of λ would be best, but in the cases computed the improvement obtainable in this way turned out to be negligibly small.

As an example, consider the raised cosine signal spectrum in which $G_1(u)$ is given by (77). We also assume a white noise spectrum in which $N(u)$ is equal to a constant N_0. It is convenient to introduce as a signal-to-noise ratio the quantity M defined by

$$M = \frac{W_0}{N_0\omega_0} = \frac{W_0}{2N_0\omega_d}. \qquad (141)$$

This is the ratio of average transmitted signal power to the average noise power in a band of frequencies of width equal to the bit rate. Computer results show that the case of $\lambda = 0$ is practically indistinguishable from the optimum λ. Hence we set $\lambda = 0$ and calculate for the optimum filter

$$U(u) = X^{\frac{1}{2}}(u) = \left(\frac{\mu\pi}{\omega_d N_0}\right)^{\frac{1}{2}} \cos\frac{\pi u}{4\omega_d} \qquad |u| < 2\omega_d. \qquad (142)$$

This is the same cosine filter characteristic found by Sunde to be optimum for binary AM with synchronous detection. From (132) and (133) we find that with $\lambda = 0$

$$W_0 - W_s = \frac{A^2\omega_d N_0^{\frac{1}{2}}}{2\pi\mu} = W_s. \qquad (143)$$

Hence

$$W_s = W_0/2 \quad \text{and} \quad W_0 - W_s = W_0/2. \qquad (144)$$

From (139), (140), and (141) we then calculate

38

$$\rho^2 = \frac{W_0}{4\omega_d N_0} = \frac{M}{2} \tag{145}$$

$$a^2\rho^2 = \frac{3\pi^2 W_0}{16\omega_d N_0(\pi^2 - 6)}$$
$$= \frac{3\pi^2 M}{8(\pi^2 - 6)} = 0.956M. \tag{146}$$

If the steady-state components were suppressed, we would set $W_s = 0$ and would then obtain $\rho^2 = M$, $a^2\rho^2 = 1.913M$. This would correspond to a 3-db shift in the direction of lower signal-to-noise ratio when the error probability curves are plotted against $10 \log_{10} M$.

The curves of Fig. 6, showing the upper and lower bounds for error probability when Sunde's FM system is optimized, were calculated by S. Habib on the digital computer. The case of a nonoptimum receiving filter is illustrated by the corresponding curves for a rectangular band defined by

$$X(u) = X_0 \qquad |u| < 2\omega_d . \tag{147}$$

For this case we compute from (126) and (127)

$$\sigma_0^2 = 2\int_0^{2\omega_d} X_0 N_0 \, du = 4\omega_d X_0 N_0 \tag{148}$$

$$\sigma_1^2 = 2\int_0^{2\omega_d} u^2 X_0 N_0 \, du = \frac{16\omega_d^3 X_0 N_0}{3} . \tag{149}$$

From (123)

$$W_0 = \frac{A^2}{4X_0} + \frac{A^2}{8\omega_d X_0}\int_0^{2\omega_d} \left(1 + \cos\frac{\pi u}{2\omega_d}\right)^2 du = \frac{5A^2}{8X_0}. \tag{150}$$

We then calculate

$$\rho^2 = 2M/5 \qquad a^2\rho^2 = 3M/10. \tag{151}$$

If the steady-state components are suppressed, the average transmitted power could be reduced to $(\frac{5}{8} - \frac{1}{4})/(\frac{5}{8}) = \frac{3}{5}$ of the previously determined value, which is a saving of 2.2 db.

REFERENCES

1. Bennett, W. R., *Electrical Noise*, McGraw-Hill Book Co., Inc., New York, 1960, pp. 234–238.
2. Rice, S. O., Statistical Properties of a Sine Wave Plus Random Noise, B.S.T.J., **27**, Jan., 1948, pp. 109–157.
3. Lawton, J. G., Theoretical Error Rates of "Differentially Coherent" Binary

39

and "Kineplex" Data Transmission Systems, Proc. I.R.E., **47**, Feb., 1959, pp. 333–334.

4. Sunde, E. D., Ideal Pulses Transmitted by AM and FM, B.S.T.J., **38**, Nov., 1959, pp. 1357–1426.

5. Nyquist, H., Certain Topics in Telegraph Transmission Theory, Trans. A.I.E.E., **47**, April, 1928, pp. 617–644.

6. Knopp, K., *Theory and Applications of Infinite Series*, Blackie and Son, Ltd., London, 1928, p. 269.

7. Meyerhoff, A. A. and Mazer, W. M., Optimum Binary FM Reception Using Discriminator Detection and I-F Shaping, RCA Review, **22**, Dec., 1961, pp. 698–728.

8. Smith, E. F., Attainable Error Probabilities in Demodulation of Random Binary PCM/FM Waveforms, I.R.E. Trans. on Space Elec. and Telemetry, **SET-8**, Dec., 1962, pp. 290–297.

9. Montgomery, G. F., A Comparison of Amplitude and Angle Modulation for Narrow-Band Communication of Binary-Coded Messages in Fluctuation Noise, Proc. I.R.E., **42**, Feb., 1954, pp. 447–454.

40

PAPER NO. 17

Reprinted from *RCA Rev.*, Vol. 27, No. 2, pp. 226–244, June 1966

DEMODULATOR THRESHOLD PERFORMANCE AND ERROR RATES IN ANGLE-MODULATED DIGITAL SIGNALS*

By

J. KLAPPER

RCA Communications Systems Division
New York, N. Y.

Summary—A theory is developed that provides a link between the threshold mechanism in analog FM signals and error rates in digital FM reception. A main result is a formula that predicts error rates for binary FM signals with limiter–discriminator reception and integrate–dump decision. Unlike the work of earlier investigators, this formula is also applicable to large deviation indices, is relatively simple, and facilitates the inclusion of center-frequency shifts. Several important illustrations are included.

INTRODUCTION

AVAILABLE formulas for predicting error rates in digital FM reception do not include the effect of post-detection processing of the digital signals.[1-3] The difficulty is due to the non-gaussian distribution of the post-detection noise. However, some form of post-detection low-pass filtering is essential for all except small deviation indices that, in turn, are associated with appropriately narrow-band predetection filtering. System constraints, such as those due to Doppler shifts and frequency instabilities, often dictate the use of predetection bandwidths that are substantially wider than the bit rate. A main result in this paper is a formula that permits the prediction of the probability of error in such systems. In addition, this formula is simple to use and permits the ready inclusion of center-frequency shifts, such as those due to the Doppler effect.

Additive gaussian (not necessarily white) noise is assumed to be the sole source of interference. The derivation is based on the work

---

* The work reported herein was sponsored by the AF Avionics Laboratory, Wright Patterson AFB, Ohio, under Contract AF 33 (615) 2426.

[1] A. A. Meyerhoff and W. M. Mazer, "Optimum Binary FM Reception Using Discriminator Detection and IF Shaping," *RCA Review*, Vol. 22, p. 698, Dec. 1961.

[2] P. D. Shaft, "Error Rate of PCM-FM Using Discriminator Detection," *Trans. IEEE PGSET*, Vol. SET-9, p. 131, Dec. 1963.

[3] W. R. Bennett and J. Salz, "Binary Data Transmission by FM over a Real Channel," *Bell Syst. Tech. Jour.*, Vol. 42, p. 2387, Sept. 1963.

of Rice[4] and on experimental evidence that the noise at the output of a limiter–discriminator for additive gaussian-noise contamination of the input is comprised mainly of small, nearly gaussian, noise with superimposed, randomly occurring threshold impulses. These impulses are all approximately of unity area in the output-versus-time plane if the ordinate is calibrated in terms of frequency deviation. For the limiter–discriminator, however, they are not necessarily of identical shape. Rice has postulated that an impulse is generated each time the vector resulting from the addition of instantaneous noise to the instantaneous signal encircles the origin. The speed of encirclement is a function of the instantaneous signal and noise conditions, and generally differs from case to case. Thus the shapes of the impulses are not expected to be identical. The encirclement produces a 2π jump in the phase-versus-time plane, and differentiation with respect to time yields a spike of unity area in the frequency-versus-time plane. Since these spikes are all of essentially the same area and are relatively sharp, they are referred to herein as impulses. According to present theory these impulses are the cause of the well-known FM threshold phenomenon for analog signals.[4] When the rate at which these impulses occur is small, their effect on the output signal-to-noise power ratio is negligible. The analog FM threshold is caused by a certain rate of these impulses, and thus they are referred to herein as threshold impulses (TI).

In angular feedback demodulators, the TI rate is considerably reduced by the compressive and filtering actions of the demodulator. However, a different phenomenon occurs, "loss-of-lock", that also results in 2π phase jumps and impulsive post-detection noise. This noise is referred to as loss-of-lock impulses (LLI). In a well-designed angular feedback demodulator, the combined effect of TI and LLI is less than that of TI in a limiter–discriminator (LD). Thus, the noise picture analysis for the phase-locked loop and FM feedback types of low-threshold demodulators is similar, although a new mechanism of impulse creation appears. The LLI of the phase-locked demodulator (PLD) are more uniform in that they depend mainly on the relock mechanism, and can be made sharper than those of the discriminator by circuit design. The extension of Rice's noise analysis to the PLD was proposed by Schilling et al,[5] who used it in the prediction of error

---

[4] S. O. Rice, "Noise in FM Receivers," Chapter 25, *Time Series Analysis,* John Wiley, 1963.

[5] D. L. Schilling, J. Billig and D. Kermish, "Error Rates in FSK Using the Phase Locked Loop Demodulator," 1st IEEE Annual Communications Convention, Boulder, Colorado, June 1965; also, Research Report No. PIB MRI-1254-65, Polytechnic Institute of Brooklyn.

rates. Their results differ from those in this paper in that their work accounts only for the effect of a single loss-of-lock impulse and is thus applicable only over a certain range of deviation indices. Also, their system uses a simple low-pass filter instead of the integrate–dump circuit used here and thus requires a different analytical approach. In the system described here, the probability of error is shown to be mainly a function of the input-carrier-to-noise ratio and the deviation

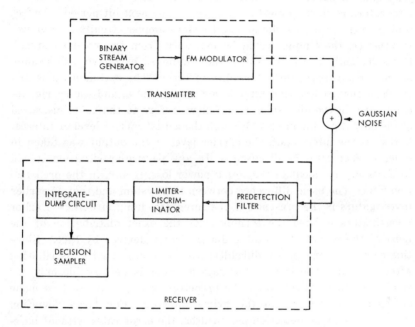

Fig. 1—Communication system.

index, where the deviation index is defined as the ratio of the peak-to-peak deviation to the bit rate;

$$D = \frac{2\Delta f}{B_R} \quad \text{cylces per bit}, \tag{1}$$

where Δf is the peak deviation in cps and B_R is the bit rate in bits/sec.

THE SYSTEM

Figure 1 is a block diagram of the system under consideration. At the transmitter end, a binary stream of essentially rectangular transitions frequency modulates a carrier. The only admitted contamination by the channel is additive gaussian noise. The receiver input signal is filtered to reject as much noise as possible consistent with the bandwidth requirements of the signal. The filter output is fed to a limiter–discriminator for demodulation. The integrate–dump circuit integrates the output of the limiter–discriminator for the duration of the bit. At the end of each bit, the decision sampler decides whether a mark or a space was received, depending upon the polarity of the integrator output at that instant. The contents of the integrator are then dumped to prepare for the next bit period. Perfect timing in the integrate–dump and decision-sampler circuits is assumed. In practice, the timing would be extracted from the received signal. For sufficiently low error rates, the timing so extracted will usually not materially affect the probability of error. The optimum shape and width of the predetection filter is not treated here, and the carrier-to-noise ratio in the discussion that follows refers to the value measured at the output of this filter. Although the actual carrier level of interest is that at the filter input, the carrier level at the output was taken in order to facilitate the discussion. In any particular situation it will be necessary to calculate the signal power loss, if any, in the predetection filter. The basic difference between this system and that of earlier investigators is the presence of the integrate–dump circuit.[1-3] Earlier investigators utilized the formulas for the exact distribution of the noise at the output of a limiter–discriminator derived by Rice.[6] Since this noise is nongaussian, difficulty arose in obtaining its distribution after filtering. The lack of low-pass filtering is permissible only for small deviation indices, since the predetection filter can then be made sufficiently narrow to do the prime filtering. For large deviation indices or large predetection bandwidths, the error rates without baseband filtering become prohibitive.

The integrate–dump circuit results in improved system performance[7] and, in addition, makes possible the "equivalent area" analytical approach, which facilitates the inclusion of the effect of a plurality of threshold impulses.

Error Rates Due to Gaussian Noise

Consider the post-detection noise with the threshold impulses removed. According to Rice,[4] this noise has a nearly gaussian distribution, an assumption that works well for analog signal S/N ratio calculations. With digital signals, the tails of the distribution (or the

---

[6] S. O. Rice, "Statistical Properties of a Sine-Wave Plus Noise," *Bell Syst. Tech. Jour.*, Vol. 27, Jan. 1948.

[7] J. Klapper, "The Effect of the Integrator-Dump Circuit on PCM/FM Error Rates," *Trans. IEEE PGCT*, June 1966.

infrequent events) are of primary interest, and therefore the assumption that this noise is gaussian needs further evaluation. It will be shown that if this noise can be assumed to be gaussian, then it is negligible as a cause of errors except for very small deviation indices, provided that appropriate filtering follows the discriminator.

The integrate–dump circuit is a sin x/x type of filter for gaussian noise. The formula for the probability of error for rectangular binary video signals immersed in additive gaussian noise is given by[8]

$$P_e = \frac{1}{2} \operatorname{erfc} \frac{A/2}{\sqrt{2}\,\sigma}, \qquad (2)$$

where A is the peak-to-peak amplitude of the video pulse, σ^2 is the power of the gaussian noise, and

$$\operatorname{erfc}(x) = \frac{2}{\sqrt{\pi}} \int_x^\infty e^{-y^2}\, dy. \qquad (3)$$

Utilizing the FM improvement factor and writing in terms of the carrier-to-noise ratio at the limiter–discriminator input, ρ, one obtains[5]

$$P_e = \frac{1}{2} \operatorname{erfc} \sqrt{3\left(\frac{\Delta f}{B_v}\right)^2 \left(\frac{B_i/2}{B_v}\right) \rho}, \qquad (4)$$

where Δf is the peak deviation of the input signal from the center frequency, B_v is the video bandwidth, and B_i is the input predetection bandwidth.

Since rectangular filters are assumed in Equation (4) and since the integrate–dump circuit is a sin x/x type of filter, the results obtained are not exactly applicable. Assuming that the relation between the input predetection and the video bandwidths is given by Carson's rule,

$$\frac{B_i}{2} = (1 + D)B_v, \qquad (5)$$

the error rate can be written

---

[8] M. Schwartz, *Information Transmission, Modulation, and Noise*, McGraw-Hill, New York, 1959.

$$P_e = \frac{1}{2} \operatorname{erfc} \sqrt{3D^2 \frac{E}{N_0}}, \tag{6}$$

where E is the input energy per bit, N_0 is the input noise power density, and D is defined by Equation (1). If the video bandwidth is taken as one-half the bit rate, which is consistent with the Nyquist signalling rate,[9] then

$$\rho = \frac{E}{N_0} \frac{1}{B_i T}, \tag{7}$$

where T is the bit duration in seconds.

A practical value of E/N_0 is 12 db. The error probability is plotted in Figure 2 as a function of the deviation index, D, for this value of E/N_0. Two points are marked on the curve; one is the minimum error rate obtained by Meyerhoff and Mazer[1] for any deviation index, and the other is the minimum error rate available in a coherent phase-shift-keyed (PSK) system. It is evident that the contribution of this gaussian noise is negligible except for very small deviation indices.

THE RATE OF OPPOSING AND AIDING THRESHOLD IMPULSES

This section shows that threshold impulses appear mainly in the direction opposing the frequency deviation of the signal from the center. When the carrier is at the center frequency of a symmetrical noise-power density, then, as expected, the probability of threshold impulses of either polarity is the same.

Let N_+ be the number of aiding threshold impulses per second and similarly, let N_- be the rate of opposing threshold impulses. Rice's formulas[*] may be put in the form

$$N_+ = \frac{\Delta f}{2} \left\{ \left[1 + \left(\frac{r}{\Delta f} \right)^2 \right]^{1/2} \operatorname{erfc} \left[\rho \left(1 + \left(\frac{\Delta f}{r} \right)^2 \right) \right]^{1/2} \right.$$
$$\left. - e^{-\rho} \operatorname{erfc} \left[\frac{\Delta f}{r} \sqrt{\rho} \right] \right\} \tag{8}$$

and

$$N_- = N_+ + \Delta f e^{-\rho}, \tag{9}$$

---

[9] H. Nyquist, "Certain Topics in Telegraph Transmission Theory," *Trans. AIEE*, Vol. 47, p. 617, April 1928.

[*] See Equation (71) in Reference (4).

where r is the radius of gyration of the power spectrum of the noise. Two basic assumptions were made in the derivation of Equations (8) and (9). First, the noise power density is assumed to have arithmetic

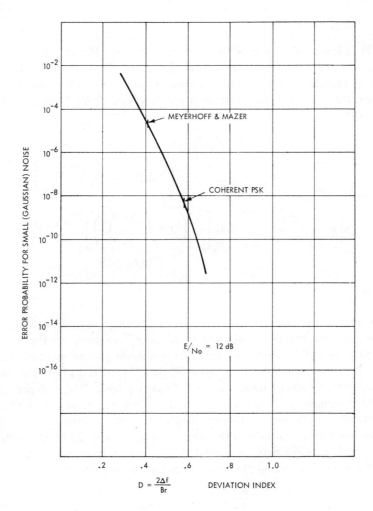

Fig. 2—Error probability for small noise.

symmetry about its center frequency. This assumption was also made by Meyerhoff and Mazer[1] and by Shaft.[2] Second, the signal is assumed to be a sine wave existing for all time. The frequency-shift-keyed signal is a wave in which the frequency is being switched. For this analysis, however, it is assumed (as it was by Shaft[2]) that the signal may be represented as being in steady state.

For large values of its argument, the complementary error function can be closely approximated by[4]

$$\text{erfc}(x) \cong \frac{e^{-x^2}}{\sqrt{\pi}\,x}. \tag{10}$$

It is interesting to note that, when Equation (10) is inserted into Equation (8), N_+ becomes identically zero. How readily can the aiding threshold impulses be ignored in practice? The region over which the approximation can be made is given, from Equations (8) and (9), by

$$\frac{N_+}{\Delta f} \ll \frac{N_-}{\Delta f}, \tag{11}$$

or

$$\frac{1}{2}\left[\left(\frac{r}{\Delta f}\right)^2 + 1\right]^{1/2} \text{erfc}\left[\rho\left(1 + \left(\frac{\Delta f}{r}\right)^2\right)\right]^{1/2}$$
$$- e^{-\rho}\,\text{erfc}\left[\frac{\Delta f}{r}\sqrt{\rho}\right] \ll e^{-\rho}. \tag{12}$$

The approximation is dependent on two parameters, $\Delta f/r$ and $\sqrt{\rho}$. The first quantity is usually less than unity; the value of the second quantity is mainly dependent upon the quality of service and the deviation ratio. One cannot say, in general, that Equation (10) holds; thus a test on the basis of Relationship (11) is required. Figure 3 presents plots of comparative values of N_+ and N_- for practical values of the parameters $\Delta f/r$ and ρ. One concludes from the curves that for the usual operating values of the normalized deviation, $\Delta f/r$, the aiding threshold impulses may be neglected. In what follows we shall therefore assume the rate of threshold impulses to be given by

$$N_+ = 0 \tag{13}$$

and

$$N_- = \Delta f e^{-\rho} = N. \tag{14}$$

This fact, which has been shown experimentally, results in important simplifications in the prediction of error rates. For example, in the usual case where the mark and space frequencies are spaced symmetrically about the center, one can ignore the effect of the "aiding" threshold impulses. Also, for an extreme shift in the signal frequencies (due to the Doppler effect, oscillator instabilities, etc.), there will be very few errors in the symbol that is closest to the center frequency.

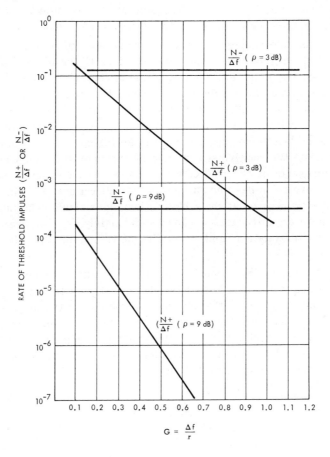

Fig. 3—Comparative rates of threshold impulses.

THE ERROR-RATE FORMULA

The assumed decision method is basic to our discussion. The output of the demodulator is integrated over a bit period. At the end of the integration period the output of the integrator is sampled and a binary decision is made on the basis of the polarity of the output. After the sampling, the integrator content is rapidly dumped and a new integration period begins. The binary decision is thus based on the polarity of the area under the demodulator output wave, comprising both signal and noise, over a bit period. It is crucial to note here that the time integral of frequency is phase, and, therefore, the decision method stated above can be restated in terms of phase as follows— the binary decision is based on the polarity of the difference between the final and initial phases of the resultant; the resultant is the sum of the signal and noise waves at the demodulator input.

The noise superimposed on the carrier causes the resultant's phase to deviate from that of the noise-free carrier. If the phase difference between the beginning and the end of the bit period introduced by the noise cancels the phase difference introduced by the modulation, then an error in decision takes place. Note that in this decision the phase modulations within the bit do not matter; only the phase difference between the beginning and the end of the bit counts. We can begin to appreciate the relative significance of the impulsive and nonimpulsive components of the post-detection noise on the error rates. The nonimpulsive noise has undulations in both directions, and the probability of a large value at the sampling instant is small. The impulsive noise, however, introduces a phase step, mainly in the direction to oppose the modulation, and generally no return of this step takes place.

There is yet another important limitation on the nonimpulsive noise. It was stated previously, from Rice's theory,[4] that a threshold impulse occurs nearly every time the additional angle due to the noise exceeds $\pm\pi$. In this manner, the nonimpulsive noise can contribute phase undulations no greater than 2π. We will now compare this effect to the angle introduced by the modulation. The angle due to the modulation at the sampling instant, θ_s, is given by

$$\theta_s = 2\pi \int_0^T (f - f_0)\, dt$$

where $f - f_0$ is the instantaneous frequency deviation and T is the bit duration. For a frequency step $f - f_0 = \Delta f$, and from the definition of the deviation index, Equation (1),

$$\theta_s = 2\pi \Delta f T = \pi D.$$

Thus, for deviation indices greater than two, an error must be accompanied by the occurrence of a threshold or loss-of-lock impulse. Intersymbol interference is neglected in this development. The major cause of errors seems to be the impulses, whether they are of the threshold (phase jumps in the received wave) or loss-of-lock type (phase jumps in the voltage-controlled-oscillator wave). The nonimpulsive noise appears to be of importance for very small deviation indices or as a correction term at the higher deviation indices. Only the impulsive noise will be considered in the development of the error-rate formula that follows.

In the plane of frequency deviation versus time, the uncorrupted signal area is[7] $\Delta f T$. For the same scaling, the threshold impulse area is —1. Thus an error occurs in a particular bit if

$$i \geqq \Delta f T, \tag{15}$$

where i is the number of opposing threshold impulses in the bit. It is assumed that the threshold impulses are so narrow in the time scale that only an integral number of these appear in a bit. In other words, the area of a threshold impulse is never shared by adjoining bits but is fully contained in a single bit. Since T is the reciprocal of the bit rate, Equation (15) can be written in terms of the deviation index (Equation (1)) as

$$i \geqq \frac{D}{2}. \tag{16}$$

It is assumed that the threshold impulses have a Poisson distribution. Such a distribution was assumed by Rice[4] and was later verified experimentally.[10] The probability of exactly k threshold impulses in a time interval T is given by[11]

$$P(k) = \frac{(NT)^k e^{-NT}}{k!}, \tag{17}$$

where N is the number of threshold impulses per second given by Equation (14). Since the events are mutually exclusive, the probability of k or more threshold impulses in T seconds is given by the sum

---

[10] I. Ringdahl and D. L. Schilling, "On the Distribution of the Spikes Seen at the Output of an FM Discriminator Below Threshold," *Proc. IEEE*, Vol. 52, p. 1756, Dec. 1964.

[11] E. Parzen, *Modern Probability Theory and Its Applications*, John Wiley and Sons, New York, 1960.

$$P(i \geq k) = \sum_{i=k}^{\infty} \frac{(NT)^i e^{-NT}}{i!} = 1 - \sum_{i=0}^{k-1} \frac{(NT)^i e^{-NT}}{i!}. \quad (18)$$

For our application, Equation (17) yields a sufficiently close value for the probability of k or more threshold impulses in time T. It can

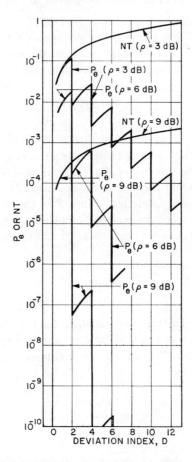

Fig. 4—P_e or NT versus deviation index.

be readily shown that

$$\frac{P(k+1)}{P(k)} = \frac{NT}{k+1}. \quad (19)$$

Typical values of NT are given in Figure 4. NT is small for small deviation indices where k is not large, and may be near unity for large deviation indices where k is large. In either case, the ratio $NT/(k+1)$ is small compared to unity. Since, in addition, the Poisson distribution monotonically decreases in the region of our interest, the assertion that Equation (17) is sufficiently close in value to Equation (18) is justified. Thus

$$P(i \geqslant k) \cong \frac{(NT)^k e^{-NT}}{k!} = P(k).$$ (20)

In the same manner we shall ignore the "or greater" sign and write Equation (16) as

$$k = D/2 \quad \text{(next higher integer)}.$$ (21)

It takes at least one opposing threshold impulse to cause an error for deviation indices of less than two. Two or more opposing threshold impulses are required to cause an error if the deviation index is between two and four, three or more are required between four and six, and so on. This sudden jump in the required number of threshold impulses at deviation indices that are multiples of two yields some interesting changes in slope in the curves of error rates versus deviation index.

Thus, the probability of error, P_e, is given from Equations (20), (1) and (14) by the relatively simple expressions

$$P_e = \frac{I^k e^{-I}}{k!},$$ (22)

where

$$I = NT = \frac{D}{2} e^{-\rho} \quad \text{impulses per bit.}$$ (23)

As written in Equations (22) and (23), the probability of error is only a function of the deviation index and the carrier-to-noise ratio (CNR) at the limiter input. The average number of threshold impulses per bit is directly proportional to the deviation index and decreases exponentially with increasing CNR.

It is of interest to amend the error-rate formulas so that center frequency shifts of the signal can be readily incorporated. It is assumed that the d-c value associated with the center frequency shift is cancelled out at the output of the discriminator. Since the area of the signal bit has not changed, Equation (21) still gives the number of threshold impulses that cause an error. Also, since the Poisson distribution still holds, Equation (22) is correct. The only equation that will differ is Equation (23), since the threshold impulse rate for the mark and space symbols will be different. Let d be the shift in the signal center frequency in cycles per second and let

$$\delta = \frac{d}{\Delta f}. \tag{24}$$

Equation (23) can now be generalized to our case using Equations (1) and (14):

$$I = \frac{D}{2}(1 \pm \delta)e^{-\rho}. \tag{25}$$

Equation (25) would indicate that there are no threshold impulses when the signal frequency is at the center. This is clearly incorrect and is due to the failure of Relationship (11). The errors increase for the symbol whose effective deviation from the center has increased, and vice versa.

ILLUSTRATIONS

The foregoing formulas will now be used in some important examples that will illustrate properties of the binary FM receiver, some of which were unknown heretofore. The data for the plots were obtained on a digital computer.

Error Rates for Constant Carrier-to-Noise Ratio

Suppose we are constrained to keep the predetection bandwidth at a certain value. Practical reasons for this may be the required accommodation of a Doppler shift, oscillator frequency uncertainty, or the lack of sufficiently narrow-band filters. With the noise-power density, N_0, a constant, the noise-power level at the input cannot be reduced, and for a certain carrier power the maximum CNR is fixed. Suppose further that we are free to choose the deviation index in order to achieve a specified probability of error. Since there is a limit to the value of the frequency deviation before excessive signal power losses are suffered in the predetection filter, the maximum information rate is also limited. In any case, the appropriate plots of error rates (P_e) versus deviation index are given in Figure 4 for CNR of 3, 6, and 9 db using Equations (22) and (23).

402

The general trend of the curves is, as expected, towards increased reliability of transmission with an increasing deviation index or, equivalently, with a decreased information rate. The surprising part, however, is the zigzag behavior of the plot, which indicates that a number of relative minima are available to the designer. This behavior is due to the consideration of only an integral number of threshold impulses per bit and the neglect of the nonimpulsive component of the noise. For example, at $D = 2$ we jump from the probability of one threshold impulse per bit to the probability of two, resulting in a large jump in error rates.

The curves marked NT represent the average number of threshold impulses per bit. For deviation indices of less than two, the probability of a threshold impulse per bit, the average number of threshold impulses per bit, and the probability of an error are numerically nearly the same. NT is actually increasing linearly with the deviation index, as predicted by Equation (23).

Error Rates for a Constant Energy Ratio

We shall again predict the probability of an error as a function of the deviation index. However, the parameter will be E/N_0, i.e., the ratio of the energy per bit to the noise power in a 1-cps bandwidth. This parameter is used extensively in the literature in judging the quality of a digital communication system.

To use Equations (22) and (23), we must convert from ρ to E/N_0 using Equation (7). Also, $B_i T$ must be replaced by known parameters. An optimum relation for digital FM is not known, so we shall use Carson's rule (see Equation (5)). Then, from Equations (5) and (7),

$$\rho = \frac{E}{N_0} \frac{1}{1 + D}. \qquad (26)$$

The probability of error obtained using Equations (22), (23) and (26) is plotted in Figure 5 for E/N_0 of 9 and 15 db. The general behavior is similar to that in Figure 4, except that here the probability of error increases with D, verifying the generally known fact that, if the system has no constraints preventing such operation, the deviation index should be kept small. There are, however, systems where a small predetection bandwidth is not possible (e.g., due to Doppler shifts), and for such cases these curves indicate the error rates obtained and penalty suffered. It is observed from Equation (7) that ρ and E/N_0 are the same for $B_i T = 1$.

It would appear in Figure 5 that the error rates decrease limitlessly as the deviation index is decreased. However, as discussed earlier, the nonimpulsive noise begins to predominate at small deviation indices. If the small noise were truly gaussian, its effect would be described by Equation (4) (also see Figure 2). Since it deviates from the

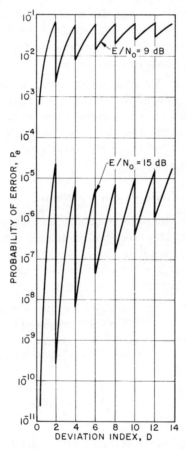

Fig. 5—Probability of error versus deviation index.

gaussian distribution, the exact value of D at which it becomes important is not known. At deviation indices of 0.7 it is still sufficient to consider only the threshold impulses, since the results are the same as those obtained by other analytical methods and experimentally.[2]

Center Frequency Shift

To illustrate the variation in error rates as the signal center frequency is shifted away from the center frequency of the noise-power density, we utilize Equations (25) and (22) and obtain the curves given in Figure 6. The CNR is kept at 6 db and the curves are plotted for two deviation indices: 0.75 and 3. Three curves are given for each deviation index:

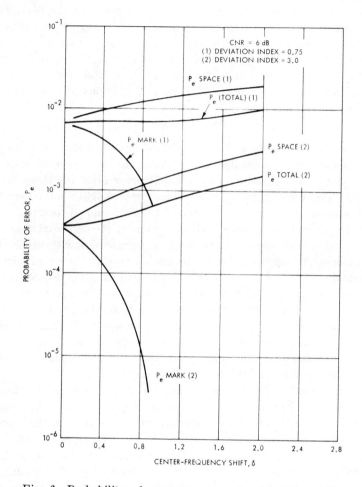

Fig. 6—Probability of error versus center-frequency shift.

P_e(space), which is the probability of error for the symbol being moved further away from the center frequency of the filter;

P_e(mark), which is the probability of error for the symbol being brought closer to the center frequency of the filter; and

P_e(total), which is the probability of error per bit transmitted. In general,

$$P_e(\text{total}) = \frac{1}{2}\left[\,P_e(\text{space}) + P_e(\text{mark})\,\right].\qquad(27)$$

The shift in the d-c level at the discriminator output due to this center frequency shift is cancelled out, as explained in the derivation of Equation (25). An important conclusion from Figure 6 is that, if the signal frequency is shifted, the symbol that is further removed from the filter center frequency carries the brunt of errors, and careful centering is required for equal error rates. The total error rate, however, does not change appreciably. These conclusions were also found experimentally.

CONCLUSION

In addition to giving further insight into the error-generation process, the theory presented here results in an error-rate formula with the following advantages: it predicts error rates for large deviation indices; it gives the effect of video processing by an integrate–dump circuit; it is relatively simple to use; and it permits easy inclusion of center-frequency shifts. Further, the essential features of the theory also apply to the recently developed low-threshold demodulators. The only additional information required is the prediction or measurement of the rate of the TI and LLI of these demodulators. The theory can also be extended to multilevel FM reception. The system designer will note the various tradeoffs involved, especially when doppler shifts or other constraints do not permit a narrow-band predetection filter and small deviation indices. When narrow-band operation is possible, the theory verifies the conclusions of earlier researchers that small deviation indices are desirable.

Experimentally, it was found that the error rates of a limiter–discriminator receiver are in reasonable agreement with the theory, but the zigzag behavior is not pronounced. Experimental data available on the phase-locked demodulators do show these predicted rapid changes in slope, and the lack of the zigzag pattern for the limiter–discriminator is believed to be due to the consideration of only integral numbers of threshold impulses and the neglect of the non-impulsive noise. In view of the greater sharpness of these impulses in the phase-locked demodulators, the shape of the error-rate curve versus deviation index is closer to that predicted.

Figure 6 is of interest in showing the behavior of the system when the signal and noise center frequencies do not coincide. While the probability of error on the total-number-of-bits basis does not change considerably with a center-frequency shift, one of the two symbols bears the brunt of the errors.

It is assumed in Figure 2 that the small noise is gaussian. Since the distribution is not exactly gaussian, the effect of the nonimpulsive noise will be felt at different deviation indices than those shown in Figure 2. It appears, however, from a comparison of our results with those of earlier investigators and from available experimental data, that the nonimpulsive noise can still be ignored at $D = 0.7$.

It is worthwhile to list the assumptions made in the development:

(1) If post-detection filtering (e.g., integrate–dump circuit) is used, the threshold impulses are the main source of digital FM errors. This assumption was justified analytically for the case where the nonimpulsive noise can be assumed to be gaussian.

(2) The signal is of rectangular shape.

(3) The signal dwells sufficiently long at each frequency so that an analysis based on a cw signal applies.

(4) The input noise is gaussian.

(5) The noise-power density is of arithmetic symmetry.

(6) The parameters of the system are such that the aiding threshold impulses can be ignored.

(7) Only an integral number of threshold impulses appear in a bit.

(8) The probability of k or more threshold impulses per bit is the same as the probability of k impulses per bit.

(9) All threshold impulses are of unity area.

(10) The threshold impulses follow a Poisson distribution in time.

(11) Intersymbol interference can be ignored.

Some of these assumptions are not essential, and were made for the sake of analytical simplicity.

The error rates predicted herein do not give any bound on optimum demodulation, but they are indicative of what can be obtained with the system described. For example, it may be possible to obtain lower error rates even with a limiter–discriminator if nonrectangular bits or different filtering is used.

Acknowledgment

The author is indebted to G. Aaronson, J. Frankle, M. Masonson, S. J. Mehlman, and A. Newton for helpful comments and discussions.

PAPER NO. 18

Reprinted from *IEEE Trans.*, Vol. SET-10, No. 1, pp. 39–44, March 1964

The Autocorrelation Function and Power Spectrum of PCM/FM with Random Binary Modulating Waveforms

M. G. PELCHAT, MEMBER, IEEE

Summary—This paper presents the derivation of exact expressions for the autocorrelation function and power spectrum of PCM/FM or FSK when the frequency modulating waveform is a random sequence of binary pulses of length T. The problem treated is that of true frequency modulation of an oscillator, a process which, except for a few special cases, generates waveforms and spectra different from those produced by sequentially switching between the outputs of two continuously running oscillators. The final expression for the power spectrum is simple and written in closed form.

The power spectrum of PCM/FM is dependent on the bit rate f_B, usually defined as the reciprocal of T, and the deviation ratio D, defined as the difference between the two possible values of the instantaneous frequency divided by the bit rate. For small values of $D(D<.5)$ the spectrum of PCM/FM has a shape resembling a high-Q resonance curve with a 3-db bandwidth given by $\frac{1}{2}\pi\, f_B D^2$. As D increases, f_B being fixed, the resonance curve becomes a poorer approximation, and with D in the vicinity of 0.7 the spectral density is nearly flat across a frequency band equal to the bit rate and drops abruptly on either side of this frequency band. It is interesting to note that a value of D near 0.7 has been found to yield minimum probability of error for given received power and receiver noise temperature. As D increases from 0.7 to 1.0, pronounced peaks in the spectrum develop about each of the two possible values of the instantaneous frequency. With D equal to 1, 2, 3, ... half of the total power is concentrated in two spectral lines, one at each possible value of the instantaneous frequency.

The power spectrum of PCM/FM is quite different from the power spectrum of the waveform generated by alternately switching between two continuously running oscillators. It has been shown in the literature that in this last case the power spectrum of the waveform contributed by one oscillator consists of a discrete line at the oscillator frequency and a continuous spectrum of the form $\sin^2 \pi f T/(\pi f T)^2$ centered on this line. The power spectrum of the complete waveform is obtained as the sum[1] of the individual power spectra. The tails of the power spectrum of PCM/FM behave as f^{-4}, whereas the tails of the power spectrum of the output of the switched oscillators behave as f^{-2}, f being the frequency measured from the center of the spectrum. This is an important consideration with regard to adjacent channel interference.

The Inter-Range Instrumentation Group (IRIG) Standards recommend the use of premodulation filtering to reduce adjacent channel interference. The computation of the exact effect of a premodulation filter on the spectrum of PCM/FM is a very difficult problem. From experimental results obtained for $D = 0.8$, it appears that a premodulation filter, as recommended by IRIG, does not materially affect the shape of the spectrum in the frequency band containing the bulk of the total power. Also, the effects of a premodulation filter on the tails of the power spectrum can be estimated by the very simple and effective procedure of Watt, Zurick and Coon.

Manuscript received September 20, 1963.
The author is with Radiation Incorporated, Melbourne, Fla.
[1] This simple addition is not generally valid for D equal to an integer including zero since this situation corresponds to cross-spectra different from zero.

INTRODUCTION

THIS PAPER presents the derivation of an expression for the power spectrum of PCM/FM, or FSK, when the frequency modulating waveform is a random binary sequence. The general approach taken is conventional and consists in computing the autocorrelation function from which the power spectrum is obtained by application of the Weiner-Khintchine theorem.

Some of the gross characteristics of the power spectrum of PCM/FM are as follows. If the deviation ratio D is much smaller than unity, the bulk of the power is contained in a bandwidth small compared to the bit rate and centered on the average frequency. Also, if D is equal to an integer, the power spectrum contains two discrete spectral lines. In those cases the PCM/FM waveform can be generated by switching between two equal amplitude- and continuously-running oscillators phased such that the instantaneous phase of the resulting waveform is continuous at the switching instant. Finally, the tails of the spectrum, for any value of D, decreases as f^{-4}, f being the frequency measured from the average frequency.[2]

Comparison of theoretical and experimental results[3] for $D = 0.8$ indicates that the presence of a premodulation filter, as recommended for the purpose of interference suppression, does not materially affect the spectrum in the high-power density region (a bandwidth equal to the bit rate and centered on the average frequency). In frequency regions of low spectral density, the effect of a premodulation filter on the power spectrum can be estimated by a simple procedure discussed by Watt, Zurick and Coon[5] for the case of square-wave frequency modulation.

COMPUTATION OF AUTOCORRELATION FUNCTION

A common expression for the output voltage of a

[2] This result is normally true when the frequency-modulating waveform contains abrupt steps. There are exceptions, however, such as when the steps are constrained to occur at the peaks of the FM wave, in which case the tails of the power spectrum behave as f^{-6}.
[3] With PCM/FM a value of D around 0.7 has been shown to yield minimum probability of error for a given received power.[4]
[4] E. L. Smith, "Attainable error probabilities in demodulation of random binary PCM/FM waveforms," IRE TRANS. ON SPACE ELECTRONICS AND TELEMETRY, vol. SET-8, pp. 290–297; December, 1962.
[5] A. D. Watt, V. J. Zurick and R. M. Coon, "Reduction of adjacent-channel interference components from frequency shift-keyed carriers," IRE TRANS. ON COMMUNICATIONS SYSTEMS, vol. CS-6, pp. 39–47; December, 1958.

frequency-modulated oscillator is written as

$$e(t) = \sqrt{2} \cos\left[\omega_c t + \Delta\omega \int_{t_0}^{t} V(t)\, dt' + \theta\right], \qquad (1)$$

where

- ω_c is the carrier angular frequency
- $V(t)$ is the frequency modulating waveform
- $\Delta\omega$ is the constant which fixes the degree of modulation
- t is time measured from some reference time t_0
- θ is a phase angle.

In the problem under discussion $V(t)$ is a binary waveform which can assume only two values, $+1$ and -1, both with probability 0.5. Transitions in $V(t)$ are instantaneous and are separated in time by an amount qT, q being an integer greater or equal to one, and T denotes the bit length or the reciprocal of the bit rate f_B. A sequence $V(t)$ can be generated by tossing a true coin every T seconds and writing $+1$ for heads and -1 for tails.

If $\psi(t)$ is defined as

$$\psi(t) \triangleq \Delta\omega \int_{t_0}^{t} V(t')\, dt' + \theta \qquad (2)$$

(1) can be rewritten as

$$e(t) = \sqrt{2} \cos\left[\omega_c t + \psi(t)\right]. \qquad (3)$$

It has been shown that $k(\tau)$, the autocorrelation function of $e(t)$, can be written as

$$k(\tau) \triangleq E\{e(t)\cdot e(t+\tau)\} \qquad (4)$$

$$= \text{Real part } [e^{i\omega_c\tau}\cdot E\{e^{i[\psi(\tau)-\psi(0)]}\}], \qquad (5)$$

provided that $\psi(t+\tau) - \psi(t)$ is stationary.[6] In the present case $\psi(\tau) - \psi(0)$ is evenly distributed about zero and (5) simplifies to

$$k(\tau) = \cos\omega_c\tau\cdot E\{e^{i[\psi(\tau)-\psi(0)]}\}. \qquad (6)$$

For the purpose of evaluating $k(\tau)$, it is convenient to define a new random process $\phi(t)$ as

$$\phi(t+\delta) \triangleq \psi(t), \qquad (7)$$

with $-\delta$ denoting the value of t in the interval $-T < t \leq 0$ at which a bit transition is possible. δ is a random variable with uniform probability density $p(\delta)$, such that

$$p(\delta) = \frac{1}{T}, \qquad 0 \leq \delta < T,$$
$$\qquad\qquad\qquad\qquad\qquad (8)$$
$$p(\delta) = 0, \qquad \delta < 0, \qquad \delta \geq T.$$

Fig. 1 shows the possible values for $\phi(\tau) - \phi(0)$ as a function of τ. In terms of $\phi(t)$ and δ, $k(\tau)$, given by (6), can be written as

$$k(\tau) = \cos\omega_c\tau\cdot E\{e^{i[\phi(\tau+\delta)-\phi(\delta)]}\} \qquad (9)$$

$$\triangleq \cos\omega_c\tau\cdot R(\tau).$$

---

[6] In (4) and (5) and in what follows $E[X]$ denotes the statistical average of the random variable X. Eq. (5) is exact only if θ is independent of $\psi(t)$ and uniformly distributed between 0 and 2π. It is approximately correct, regardless of the statistics of θ, if $e(t)$ is narrowband. See D. Middleton, "The distribution of energy in randomly modulated waves," *Phil. Mag.*, Ser. 7, vol. XLII, p. 689; July, 1951.

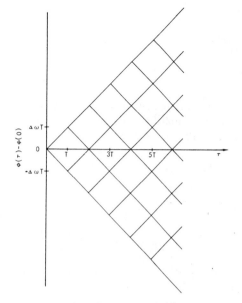

POSSIBLE VALUES $\phi(\tau)-\phi(0)$

Fig. 1—Possible values $\phi(\tau) - \phi(0)$.

The evaluation of $R(\tau)$, which is defined in (9), will be carried out in two steps: $R(\tau)$ will be computed for $0 \leq \tau < T$, and then for $nT \leq \tau < (n+1)T$, $n = 1, 2, 3, \cdots$. For the purpose of computing $R(\tau)$ for $0 \leq \tau < T$, define set A as consisting of the set of all $\phi(t+\delta)$ for which δ satisfies $0 \leq \delta < T - \tau$, and let set B consist of the set of all $\phi(t+\delta)$ for which δ satisfies $T - \tau \leq \delta < T$. Accordingly, $P(A)$, the probability of set A and $P(B)$, the probability of set B, are given by

$$P(A) = \frac{T - \tau}{T},$$

$$P(B) = \frac{\tau}{T}.$$

Consideration of Fig. 1 shows that over set A $\phi(\tau + \delta) - \phi(\delta)$ is given by

$$\phi_A(\tau + \delta) - \phi_A(\delta) = k_1\Delta\omega\tau, \qquad k_1 = \pm 1,$$

and $R_A(\tau)$, the contribution of set A to $R(\tau)$, is

$$R_A(\tau) = P(A)\cdot E[e^{ik_1\Delta\omega\tau}]$$

$$R_A(\tau) = \frac{T-\tau}{T}\cdot\frac{e^{i\Delta\omega\tau} + e^{-i\Delta\omega\tau}}{2} \qquad (11)$$

$$R_A(\tau) = \frac{T-\tau}{T}\cdot\cos\Delta\omega\tau.$$

Over set B $\phi(\tau + \delta) - \phi(\delta)$ can be written as

$$\phi_B(\tau + \delta) - \phi_B(\delta) = k_1\Delta\omega(T - \delta) + k_2\Delta\omega(\tau + \delta - T),$$

$$k_1 = \pm 1, \qquad k_2 = \pm 1,$$

and $R_B(\tau)$, the contribution of set B to $R(\tau)$, is

$$R_B(\tau) = P(B)\cdot E[e^{i[k_1\Delta\omega(T-\delta)+k_2\Delta\omega(\tau+\delta-T)]}]. \qquad (12)$$

Averaging (12) over k_1 and k_2 gives

$$R_B(\tau) = \frac{P(B)}{4}$$

$$\cdot E[e^{j\Delta\omega\tau} + e^{-j\Delta\omega\tau} + e^{j\Delta\omega(\tau+2\delta-2T)} + e^{-j\Delta\omega(\tau+2\delta-2T)}]. \quad (13)$$

In (13) δ is uniformly distributed in the interval $T - \tau \le \delta < T$ and has a probability density $1/\tau$ over this interval. Accordingly, (13) can be written as

$$R_B(\tau) = \frac{P(B)}{4}$$

$$\cdot \int_{T-\tau}^{T} [e^{j\Delta\omega\tau} + e^{-j\Delta\omega\tau} + e^{j\Delta\omega(\tau-2\delta-2T)} + e^{-j\Delta\omega(\tau-2\delta-2T)}] \frac{1}{\tau} d\delta$$

$$R_B(\tau) = \frac{\tau}{2T} \cos \Delta\omega\tau + \frac{\tau}{2T} \frac{\sin \Delta\omega\tau}{\Delta\omega\tau}. \quad (14)$$

$R(\tau)$, for $0 \le \tau < T$ can now be obtained by adding (11) and (14)

$$R(\tau) = \frac{2T - \tau}{2T} \cos \Delta\omega\tau + \frac{\tau}{2T} \frac{\sin \Delta\omega\tau}{\Delta\omega\tau}, \quad 0 \le \tau < T. \quad (15)$$

Now consider $R(\tau)$ for τ in the interval $nT \le \tau < (n + 1)T$, $n = 1, 2, 3, \cdots$ and let set C consist of all $\phi(\tau + \delta)$ for which $0 \le (\tau - nT) + \delta < T$, and, similarly, let set D consist of all $\phi(\tau + \delta)$ for which $T \le (\tau - nT) + \delta < 2T$.

If τ_s is defined as

$$\tau_s \triangleq \tau - nT, \quad (16)$$

we have

$$P(C) = \frac{T - \tau_s}{T},$$
$$\quad (17)$$
$$P(D) = \frac{\tau_s}{T}.$$

Over set C, $\phi(\tau + \delta) - \phi(\delta)$ can be written as

$$\phi_C(\tau + \delta) - \phi_C(\delta)$$

$$= k_1\Delta\omega(T - \delta) + p\Delta\omega T + k_2\Delta\omega(\tau_s + \delta), \quad (18)$$

$$k_1 = \pm 1, \quad k_2 = \pm 1,$$

where p denotes the number of $+1$'s minus the number of -1's in the bit pattern $\phi(\tau) - \phi(0)$ over the interval $T \le \tau \le nT$. Thus, p can assume the $(n - 1)$ values: $(n - 1)$, $(n - 3) \cdots$, $-(n - 1)$. $R_C(\tau)$, the contribution of set C to $R(\tau)$ is

$$R_C(\tau) = P(C) \cdot E[e^{j\Delta\omega[k_1(T-\delta)+pT+k_2(\tau_s+\delta)]}], \quad (19)$$

and because the random variable p is independent of k_1, k_2 and δ, (19) can be written as

$$R_C(\tau) = P(C) \cdot E[e^{j\Delta\omega[k_1(T-\delta)+k_2(\tau_s+\delta)]}] \cdot Ee^{j\Delta\omega pT} \quad (20)$$

Noting that over set C, δ is uniformly distributed between $0 \le \delta < T - \tau_s$, it is easily verified that the first average in (20) is

$$E[e^{j\Delta\omega[k_1(T-\delta)+k_2(\tau_s+\delta)]}] = \tfrac{1}{2} \cos \Delta\omega(T + \tau_s)$$

$$+ \frac{1}{2} \frac{1}{2(T - \tau_s)\Delta\omega} \sin \Delta\omega(T - \tau_s). \quad (21)$$

The second average in (20) involves p which has a Bernoulli distribution, and it follows that

$$E[e^{j\Delta\omega pT}] = \cos^{(n-1)} \Delta\omega T. \quad (22)$$

This result can be obtained directly from the characteristic function for the Bernoulli distribution or by application of the binomial expansion theorem. Substituting (17), (21) and (22) into (20) gives

$$R_C(\tau) = \frac{\cos^{(n-1)} \Delta\omega T}{2T}$$

$$\cdot \left\{ (T - \tau_s) \cos \Delta\omega(T + \tau_s) + \frac{1}{\Delta\omega} \sin \Delta\omega(T - \tau_s) \right\},$$

$$nT \le \tau < (n + 1)T. \quad (23)$$

Over set D, $\phi(\tau + \delta) - \phi(\delta)$ can be written as

$$\phi_D(\tau + \delta) - \phi(\delta)$$

$$= k_1\Delta\omega(T - \delta) + p\Delta\omega T + k_2\Delta\omega(\tau_s + \delta - T), \quad (24)$$

and $R_D(\tau)$, the contribution of set D to $R(\tau)$, is

$$R_D(\tau) = P(D) \cdot E\{e^{j\Delta\omega[(k_1 T-\delta)+pT+k_2(\tau_s+\delta-T)]}\}. \quad (25)$$

In (24) p is defined as in (18), but now p takes on the values n, $n - 2$, $\cdots - n$. The average value of (25) is obtained by retracing the steps followed to obtain the average value of (19). $R_D(\tau)$ is found to be

$$R_D(\tau) = \frac{\cos^n \Delta\omega T}{2T} \left\{ \tau_s \cos \Delta\omega\tau_s + \frac{1}{\Delta\omega} \sin \Delta\omega\tau_s \right\},$$

$$nT \le \tau < (n + 1)T. \quad (26)$$

Adding (23) and (24) gives $R(\tau)$ which can be written as

$$R(\tau) = R_C(\tau) + R_D(\tau) = \frac{\cos^{n-1} \Delta\omega T}{2} \left\{ \left(\cos \Delta\omega T + \frac{\sin \Delta\omega T}{T} \right) \right.$$

$$\left. \cdot \cos \Delta\omega\tau_s - \frac{T - \tau_s}{T} \cdot \sin \Delta\omega T \cdot \sin \Delta\omega\tau_s \right\},$$

$$nT \le \tau < (n + 1)T. \quad (27)$$

Eqs. (27) and (15) describe $R(\tau)$ for $\tau \ge 0$. $R(\tau)$ is automatically known for $\tau < 0$, because $R(\tau)$ must be an even function of τ.

COMPUTATION OF POWER SPECTRUM

In the last section an expression for the autocorrelation function was computed and the result written in the form

$$k(\tau) = \cos \omega_c \tau \cdot R(\tau).$$

The desired power spectrum $S(f)$ is given by the Fourier transform of $k(\tau)$

$$S(f) = \int_{-\infty}^{\infty} \cos \omega_c \tau \cdot R(\tau) e^{-j\omega\tau} \, d\tau$$

$$= \tfrac{1}{2}[\delta(f_c) + \delta(-f_c)]*G(f) \tag{28}$$

$$\rightleftharpoons \tfrac{1}{2}[S'(f - f_c) + S'(f + f_c)]$$

In (28) the star denotes convolution, $\tfrac{1}{2}[\delta(f_c) + \delta(-f_c)]$ is the power spectrum corresponding to $\cos \omega_c \tau$, and $G(f)$ is the power spectrum corresponding to $R(\tau)$, i.e.,

$$G(f) = \int_{-\infty}^{\infty} R(\tau)^{-j\omega\tau} \, d\tau = 2 \int_0^{\infty} R(\tau) \cos \omega\tau \, d\tau. \tag{29}$$

In view of the form in which $R(\tau)$ was obtained, (29) is rewritten as

$$G(f) = \sum_{n=0}^{n=\infty} G_n(f) = 2 \sum_{n=0}^{n=\infty} \int_{nT}^{(n+1)T} R_n(\tau) \cos \omega\tau \, d\tau, \tag{30}$$

where $R_n(\tau)$ is the expression for $R(\tau)$ applicable for $nT \leq \tau < (n+1)T$ as given by (15) and (27). It turns out to be convenient to evaluate $G_0(f)$ separately and $\sum_1^{\infty} G_n(f)$ as a group. Consider $G_0(f)$ which is given by

$$G_0(f) = 2 \int_0^T \left[\left(1 - \frac{\tau}{2T} \right) \cdot \cos \Delta\omega\tau \right.$$

$$\left. + \frac{1}{2 \Delta\omega T} \cdot \sin \Delta\omega\tau \right] \cos \omega\tau \, d\tau.$$

Writing θ for $\Delta\omega T$, and ϕ for ωT, the above integral can be written as

$$G_0(f) = \frac{T}{2}\left[\frac{\sin(\theta - \phi)}{(\theta - \phi)} + \frac{\sin(\theta + \phi)}{(\theta + \phi)} + \frac{2\sin^2\left(\frac{\theta - \phi}{2}\right)}{(\theta - \phi)^2}\right.$$

$$\left. \cdot \left(\frac{2\theta - \phi}{\theta} \right) + \frac{2\sin^2\left(\frac{\theta + \phi}{2}\right)}{(\theta + \phi)^2} \frac{(2\theta + \phi)}{\theta}\right]. \tag{31}$$

Now consider

$$G_r(f) = 2 \sum_{n=1}^{\infty} \int_{nT}^{(n+1)T} R_n(\tau) \cos \omega\tau \, d\tau,$$

which upon substitution of z for $\tau - nT$ becomes

$$G_r(f) = 2 \sum_{n=1}^{\infty} \int_0^T R_n(z + nT) \cos \omega(z + nT) \, dz$$

$$= 2 \sum_{n=1}^{\infty} \cos \omega nT \int_0^T R_n(z + nT) \cos \omega z \, dz$$

$$- 2 \sum_{n=1}^{\infty} \sin \omega nT \cdot \int_0^T R_n(z + nT) \sin \omega z \, dz. \tag{32}$$

Note that $R_n(z + nT)$ given by (27) can be written as

$$R_n(z + nT) = \frac{\cos^{n-1}\theta}{2} \cdot f(z), \tag{33}$$

with

$$f(z) = \left\{ \left[\cos\theta + \frac{\sin\theta}{\theta} \right] \cos \Delta\omega z - \sin\theta \cdot \left(1 - \frac{z}{T} \right) \sin \Delta\omega z \right\}.$$

Substituting (33) into (32) gives

$$G_r(f) = \sum_{n=1}^{\infty} \cos nT \cdot \cos^{n-1}\theta \cdot \int_0^T f(z) \cos \omega z \, dz$$

$$- \sum_{n=1}^{\infty} \sin n\omega T \cdot \cos^{n-1}\theta \cdot \int_0^T f(z) \sin \omega z \, dz. \tag{34}$$

With the help of the identities[7]

$$\sum_0^{\infty} a^n \cos n\omega T = \frac{1 - a \cos \omega T}{1 - 2a \cos \omega T + a^2}, \qquad |a| < 1,$$

$$\sum_1^{\infty} a^{n+1} \sin n\omega T = \frac{a \sin \omega T}{1 - 2a \cos \omega T + a^2}, \qquad |a| < 1,$$

(34) can be written in closed form as

$$G_r(f) = \frac{1}{1 - 2\cos\theta \cos\omega T + \cos^2\theta}\left[(\cos\omega T - \cos\theta) \right.$$

$$\left. \cdot \int_0^T f(z) \cos \omega z \, dz - \sin\omega T \int_0^T f(z) \sin \omega z \, dz \right]. \tag{35}$$

The two integrals in (35) are elementary and are readily evaluated. The result is

$$\int_0^T f(z) \cos \omega z \cdot dz = \left(\cos\theta + \frac{\sin\theta}{\theta} \right) \int_0^T \cos \Delta\omega z \cdot \cos \omega z \, dz$$

$$- \sin\theta \int_0^T \left(1 - \frac{z}{T} \right) \sin \Delta\omega z \cos \omega z \, dz$$

$$= T\left[\left(\cos\theta + \frac{\sin\theta}{\theta} \right)\left(\frac{\theta\sin\theta \cos\phi - \phi\sin\phi \cos\phi}{\theta^2 - \phi^2} \right) \right.$$

$$- \sin\theta\left(\frac{\theta}{\theta^2 - \phi^2} \right)$$

$$\left. - \frac{(\theta^2 + \phi^2)\sin\phi \sin\theta - 2\phi\theta \cos\theta \sin\phi}{(\theta^2 - \phi^2)^2} \right], \tag{36}$$

and

$$\int_0^T f(z) \sin \omega z \, dz = \left[\cos\theta + \frac{\sin\theta}{\theta} \right] \cdot \left[\int_0^T \cos \Delta\omega z \cdot \sin \omega z \, dz \right.$$

$$\left. - \sin\theta \int_0^T \left(1 - \frac{z}{T} \right) \sin \Delta\omega z \cdot \sin \omega z \, dz \right] \tag{37}$$

$$= T\left[\left(\cos\theta + \frac{\sin\theta}{\theta} \right)\left(\frac{\theta\sin\theta \sin\phi + \phi\cos\theta \cos\phi - \phi}{\theta^2 - \phi^2} \right) \right.$$

$$\left. - \sin\theta\left(\frac{2\phi\theta - (\theta^2 + \phi^2)\sin\phi \sin\theta - 2\phi\theta \cos\theta \cos\phi}{(\theta^2 - \phi^2)^2} \right) \right]. \tag{38}$$

Substituting (38) and (36) into (35) and adding the result to (31) gives $G(f)$. Fortunately, many terms cancel one another out leaving the following simple equation

$$G(f) = T\frac{4\theta^2}{(\theta^2 - \phi^2)^2}\frac{(\cos\theta - \cos\phi)^2}{(1 - 2\cos\theta \cos\phi + \cos^2\theta)}, \tag{39}$$

$$|\cos\theta| < 1.$$

The complete power spectrum can be obtained by substituting (39) into (28). The final result is simpler, however, if we change the frequency variable to $X =$

[7] H. B. Dwight, "Tables of Integrals and Other Mathematical Data," The Macmillan Co., New York, N. Y., p. 85; 1957.

$2(f - f_c)/f_B$. Thus, $X = 1$ corresponds to a frequency half the bit rate higher than the carrier frequency. It is also convenient to express θ in terms of the deviation ratio, D, which is usually defined as $D = 2\Delta f/f_B = \theta/\pi$. In terms of X and D the desired power spectrum is written as

$$S'(f) = \frac{4}{f_B}\left[\frac{D}{\pi(D^2 - X^2)}\right]^2$$

$$\cdot\frac{(\cos \pi D - \cos \pi X)^2}{(1 - 2\cos \pi D \cdot \cos \pi X + \cos^2 \pi D)}, \quad |\cos \pi D| < 1. \quad (40)$$

If $|\cos \pi D| = 1$, (40) does not apply. In this case, however, $\sin \theta = 0$ and $R(\tau)$ as given by (27) can be simplified to the following form

$$R(\tau) = \tfrac{1}{2}\cos \Delta\omega\tau + f(\tau), \quad 0 \leq \tau,$$

$$f(\tau) = \frac{T - \tau}{2T}\cos \Delta\omega\tau + \frac{1}{2\Delta\omega T}\cdot\sin \Delta\omega\tau, \quad 0 \leq \tau \leq T$$

$$= 0, \quad \tau > T.$$

The power spectrum corresponding to $\tfrac{1}{2}\cos \Delta\omega\tau$ is clearly that of two sine waves, one at frequency $(\omega_c - \Delta\omega)$, the other at frequency $(\omega_c + \Delta\omega)$. Each sine wave has a power of 0.25 watt. In addition to the discrete spectrum, there is a continuous spectrum with a total power of 0.5 watt which is obtained by taking the Fourier transform of $f(\tau)$. It is easily verified that the continuous portion of this spectrum can be written as

$$T\frac{2\theta^2}{(\theta^2 - \phi^2)^2} \quad (1 - \cos \theta \cos \phi).$$

In terms of X and D, defined as in (40), the total power spectrum can be written as

$$S'(f) = \tfrac{1}{4}\delta(X + D) + \tfrac{1}{4}\delta(X - D)$$

$$+ \frac{2}{f_B}\left[\frac{D}{\pi(D^2 - X^2)}\right]^2\cdot(1 - \cos \pi D \cos \pi X),$$

$$|\cos \pi D| = 1, \quad D \neq 0. \quad (41)$$

DISCUSSION OF RESULTS

The principal results of this paper are given by (15), (27), (40), and (41). Eqs. (15) and (27) give the envelope of the autocorrelation function of a PCM/FM wave for random binary modulation, and (40) and (41) give the corresponding power spectrum.

Fig. 2 shows plots obtained from (40) and (41) for deviation ratios of 0.5, 0.7, 0.8, and 1.0. Only one half of $S'(f)$ is shown since $S'(f)$ is symmetrical about f_c. It is interesting to note that the power spectrum corresponding to $D = 0.7$ gives almost constant spectral density over a bandwidth equal to the bit rate, and that the spectral density dies down abruptly outside this frequency band. It is shown by Smith[4] that $D = 0.715$ yields minimum probability of error with optimum demodulation tech-

[8] "Telemetry System Study," vol. 2, Aeronutronic Sys., Inc., Newport Beach, Calif.; and "Experimental Evaluation Program," U. S. Army Signal Research and Dev. Labs., Ft. Monmouth, N. J., Contract No. DA-36-039, SC-73182, December, 1959.

nique; also, experimental work referenced by him indicates that a value of D near 0.7 yields minimum probability of error for demodulation with conventional frequency discriminators.

In studying the behavior of the power spectrum, it is helpful to note that the power spectrum, as expressed by (40), consists of the product of an envelope term

$$\frac{4}{\pi 2 f_B}\left[\frac{D}{D^2 - X^2}\right]^2, \quad (42)$$

and a term

$$\frac{(\cos \pi D - \cos \pi X)^2}{1 - 2\cos \pi D \cos \pi X + \cos^2 \pi D}, \quad (43)$$

which is periodic in X. Fig. 3 shows a geometrical interpretation of (43) which clearly displays the behavior of (43) with variations in X or D.

For small values of D the power spectrum is crowded in the region of small X. If $D \ll 1$, (36) can be rewritten as

$$S'(f) \doteq \frac{4}{f_B\pi^2 D^2}\frac{1}{\left[1 + \left(\frac{2X}{\pi D^2}\right)^2\right]}, \quad D \ll 1, \quad X < 0.5. \quad (44)$$

Thus, for very small values of D the carrier is smeared into a spectrum having the shape of a high-Q resonance curve centered on the carrier frequency and having a 3-db bandwidth of $\tfrac{1}{2}\pi f_B D^2$.

Fig. 4 shows the theoretical power spectrum for $D = 0.8$ and the measured power spectrum for $D = 0.8$ and a 6-pole Butterworth premodulation filter having a 3-db bandwidth equal to the bit rate. The experimental result is Fig. II-3-15.[8] It is seen that the primary effect of the premodulation filter is to attenuate the tails of the power spectrum and to smooth the fine structural details in the spectrum.

Fig. 5 shows the same experimental result as Fig. 4 in addition to an approximate computed result which accounts for premodulation filtering by the method of Watt, Zurick and Coon.[3] Clearly, the theoretical result is sufficiently accurate to yield useful information with regard to adjacent channel interference problems. The procedure for obtaining the theoretical result of Fig. 5 may be described as follows. The power spectrum of PCM/FM before filtering is written as $S(X)$, and the frequency response of the premodulation filter, translated about the carrier (or average) frequency is written as $F(X)$. $S_f(X)$, the power spectrum with premodulation filtering, is then written as

$$S_f(X) \doteq S(X), \quad 0 < X < D$$

$$S_f(X) \doteq S(X)F(X - D), \quad D < X$$

$$S_f(-X) = S_f(X), \quad \text{all values of } X.$$

In the case of Fig. 5, $S(X)$ is available from Fig. 4 or (40), and $F(X)$ for a 6-pole Butterworth filter with a cutoff

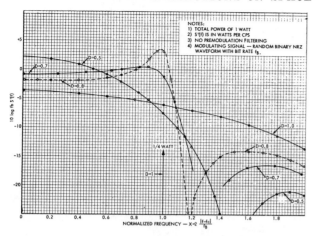

Fig. 2—Power spectrum of PCM/FM with deviation ratio as a parameter.

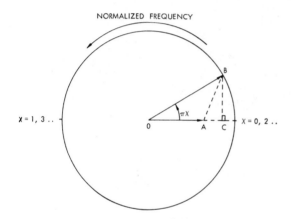

$$\overline{OA} = \cos \pi D \text{ (DRAWN FOR } \cos \pi D = +0.6$$

$$\overline{AC}^2 = (\cos \pi D - \cos \pi X)^2$$

$$\overline{AB}^2 = 1 - 2 \cos \pi D \cos \pi X + \cos^2 \pi D$$

$$\text{EQUATION 43} = \frac{\overline{AC}^2}{\overline{AB}^2}$$

Fig. 3—Geometrical interpretation of (43).

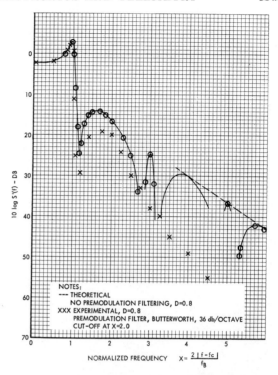

Fig. 4—Spectrum of PCM/FM.

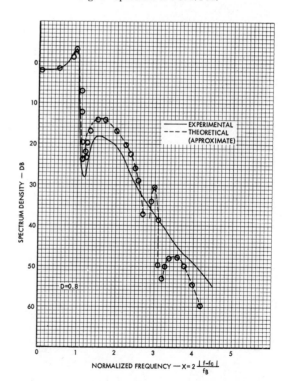

Fig. 5—Power spectrum of PCM/FM with premodulation filtering.

frequency equal to the bit rate is

$$F(X) = \frac{1}{1 + \left(\dfrac{X}{2}\right)^{12}}.$$

The procedure just described has also been used successfully to predict adjacent channel interference levels for a 3-pole Butterworth filter having a cutoff frequency equal to half the bit rate and for a 5-pole Bessel filter with cutoff frequency equal to the bit rate. This simple technique appears to yield useful results in cases of practical interest. It should be remembered, however, that there are cases in which it fails completely, *e.g.*, the ideal sharp cutoff premodulation filter.

ACKNOWLEDGMENT

The author wishes to thank Dr. C. J. Palermo for his helpful discussions and suggestions and H. E. O'Kelley for his support.

SELECTED BIBLIOGRAPHY FOR FURTHER READING

I. General FM Theory and Basic Experiments

No. 1 is the paper in which Carson proves that an FM wave occupies more spectrum than its AM counterpart. No. 2 treats the interference to FM reception by impulsive noise. Its postulation of "pop" and "click" noise, depending on whether or not a cycle is slipped, is interesting. Nos. 3, 4, 5, 6, 9, and 10 give further data on the output noise of an FM demodulator: No. 3 investigates the effect of limiting; No. 4 assumes no limiting; No. 5 is a detailed analysis of the spectrum; No. 6 investigates minimum signal detectability; No. 9 proposes the concept of "pop" noise as the mechanism of FM threshold and errors in digital FM. Later investigators have calculated in this manner the threshold noise (text paper 7) and probability of error (text paper 17); No. 10 is primarily a tutorial paper on the subject of discriminator output noise. Nos. 7 and 8 give the RF spectrum of a carrier frequency-modulated by random noise, using different approximations.

1. J. R. Carson, Notes on the Theory of Modulation, *Proc. IRE*, Vol. 10, No. 1, Feb. 1922, pp. 57–64.
2. D. B. Smith and W. E. Bradley, The Theory of Impulse Noise in Ideal Frequency Modulation Receivers, *Proc. IRE*, Vol. 34, No. 10, Oct. 1946, pp. 743–751.
3. D. Middleton, On Theoretical Signal-to-Noise Ratios in FM Receivers: A Comparison with Amplitude Modulation, *J. Appl. Phys.*, Vol. 20, No. 4, April 1949, pp. 334–351.
4. N. M. Blachman, The Demodulation of a Frequency-Modulated Carrier and Random Noise by a Discriminator, *J. Appl. Phys.*, Vol. 20, No. 10, Oct. 1949, pp. 976–983.
5. D. Middleton, The Spectrum of Frequency-Modulated Waves After Reception in Random Noise I, II, *Quart. Appl. Math.*, Vol. 7, 1949, pp. 129–174; and Vol. 8, 1950, pp. 59–80.
6. M. C. Wang, Chapter 13 in *Threshold Signals* by J. L. Lawson and G. E. Uhlenbeck, MIT Rad. Lab. Ser., Vol. 24, 1950, pp. 365–383.
7. J. L. Stewart, The Power Spectrum of a Carrier Frequency Modulated by Gaussian Noise, *Proc. IRE*, Vol. 42, Oct. 1954, pp. 1539–1542. *See also* J. A. Mullen and D. Middleton, Limiting Forms of FM Noise Spectra, *Proc. IRE*, Vol. 45, June 1957, pp. 874–877.
8. R. G. Medhurst, RF Bandwidth of Frequency-Division Multiplex Systems Using Frequency Modulation, *Proc. IRE*, Vol. 44, Feb. 1956, pp. 189–199.
9. J. Cohn, A New Approach to the Analysis of FM Threshold Reception, *Proc. Nat. Elect. Conf.*, Vol. 12, Oct. 1956, pp. 221–236.
10. M. Schwartz, Signal-to-Noise Effects and Threshold Effects in FM, *Proc. Nat. Elect. Conf.*, Vol. 18, Nov. 1962, pp. 59–71.

II. FM Circuit Theory

Nos. 11, 14, and 16 treat the response of circuits to carriers frequency-modulated by transient signals: No. 11 uses contour integration in the complex plane; No. 14 analyzes the discriminator response; No. 16 uses decomposition into convenient waveforms. Nos. 12 and 13 give design information for discriminators: No. 12 on the Phase Discriminator; No. 13 on the Ratio Detector. No.

15 presents an approximate method for transient FM signals. Nos. 17, 18, 19, and 20 discuss the quasi-stationary response: No. 17 unifies and extends the theories of Carson and Fry (text paper No. 8) and van der Pol and Stumpers (text papers Nos. 9 and 10); No. 18 keeps a tight bound on the error term; No. 19 gives experimental values of distortion for single-tuned and double-tuned circuits; No. 20 is a discussion of possible misapplication of some approximations in FM circuit theory.

11. H. Salinger, Transients in Frequency Modulation, *Proc. IRE*, Vol. 30, Aug. 1942, pp. 378–383.
12. K. R. Sturley, The Phase Discriminator, *Wireless Engr.* (London), Vol. 21, Feb. 1944, pp. 72–78.
13. S. M. Seeley and J. Avins, The Ratio Detector, *RCA Rev.*, Vol. 8, June 1947, pp. 201–236.
14. M. K. Zinn, Transient Response of an FM Receiver, *Bell System Tech. J.*, Vol. 27, Oct. 1948, pp. 714–731.
15. T. T. N. Bucher, Network Response to Transient Frequency Modulation Inputs, *Transactions AIEE, Communications and Electronics*, Vol. 78, Part 1, Jan. 1960, pp. 1017–1022.
16. D. A. Linden, Transient Response in FM, *Proc. IRE*, Vol. 45, July 1957, pp. 1017–1018.
17. E. J. Baghdadi, Theory of Low-Distortion Reproduction of FM Signals in Linear Systems, *IRE Trans.*, Vol. CT-5, Sept. 1958, pp. 202–214. *See also* H. E. Rowe, Distortion of Angle-Modulated Waves by Linear Networks, *IRE Trans.*, Vol. CT-9, Sept. 1962, pp. 286–290.
18. D. T. Hess, Transmission of FM Signals through Linear Filters, *Proc. NEC*, Vol. 18, Oct. 1962, pp. 469–476.
19. J. B. Izatt, The Distortion Produced When Frequency-Modulated Signals Pass Through Certain Networks, *Proc. IEE* (British), Vol. 110, No. 1, Jan. 1963, pp. 149–156.
20. A. S. Gladwin, R. G. Medhurst, L. H. Enloe, and C. L. Ruthroff, A Common Error in FM Distortion Theory, *Proc. IEEE*, Vol. 51, p. 846, May 1963 and Vol. 52, Feb. 1964, pp. 186–189.

III. FM Threshold Reduction

No. 21 presents early work done in Japan on threshold reduction. No. 22 gives a synthesis procedure for an FM feedback demodulator. Nos. 23, 24, 25, 26, 27, and 28 give alternate threshold analyses. No. 23 uses a quasi-linearization technique of Booton (text paper No. 14); No. 24 analyzes a band-divided discriminator; No. 25 uses a Volterra technique for the phase-locked loop; No. 26 discusses a regenerative factor in the discriminator; No. 27 analyzes the PLL by means of loss-of-lock pulses; No. 28 discusses the discriminator and the FMFB in terms of the distribution of the noise envelope. No. 29 gives the performance of demodulators having lower thresholds than the FM feedback or phase-locked loop types. They utilize combinations of FM feedback or phase-locked loops, or phase detectors with an extended monotonic range. No. 30 presents a method and the improved performance characteristics for a phase-locked loop with an extended range phase detector.

21. M. Morita and S. Ito, High Sensitivity Receiving System for Frequency Modulated Wave, *IRE Conv. Record*, Vol. 8, Pt. 5, March 1960, pp. 228–237.
22. L. H. Enloe, The Synthesis of Frequency Feedback Demodulators, *Nat. Electr. Conf.*, Oct. 1962, pp. 477–497.
23. J. A. Develet, Statistical Design and Performance of High-Sensitivity Frequency-Feedback Receivers, *IEEE Trans. on Military Electronics*, Oct. 1963, pp. 281–284.
24. S. Darlington, Demodulation of Wideband Low-Power FM Signals, *BSTJ*, Vol. XLIII, No. 1, Part 2, Jan. 1964, pp. 339–374.
25. H. L. Van Trees, Functional Techniques for the Analysis of the Nonlinear Behavior of Phase-Locked Loops, *Proc. IEEE*, Vol. 52, No. 8, Aug. 1964, pp. 894–911.
26. B. R. Davis, Factors Affecting the Threshold of Feedback FM Detectors, *IEEE Trans.*, Vol. SET-10, Sept. 1964, pp. 90–94.
27. D. L. Schilling and J. Billig, A Comparison of the Threshold Performance of the Frequency Demodulator Using Feedback and the Phase-Locked Loop, *Record of 1965, Int. Space Electr. Symp.*, Nov. 1965, pp. 3E1–3E9.
28. P. Frutiger, Noise in FM Receivers with Negative Frequency Feedback, *Proc. IEEE*, Vol. 54, No. 11, Nov. 1966, pp. 1506–1520.
29. J. Frankle, Threshold Performance of Analog

FM Demodulators, *RCA Rev.*, Vol. 27, No. 4, Dec. 1966, pp. 521–562.

30. A. Acampora and A. Newton, Use of Phase Subtraction to Extend the Range of a Phase-Locked Demodulator, *RCA Rev.*, Vol. 27, No. 4, Dec. 1966, pp. 577–599.

IV. Digital FM

No. 31 optimizes for impulsive or Gaussian-pulse FM signaling. No. 32 finds the best performance of rectangular binary FM. No. 33 is a discussion of the spectrum of rectangular binary FM with and without phase continuity. No. 34 extends No. 31 to rectangular binary FM. No. 35 considers the effect of selective fading. No. 36 gives an alternate binary FM detection scheme. No. 37 treats the power spectrum of non-rectangular binary FM. No. 38 obtains lower error rates with transmitter filtering. Nos. 39 and 40 are further work on the role of cycle slipping in error rates (text paper 17): No. 39 also deals with multilevel FM; No. 40 uses a low-pass filter in place of integrate-dump decision.

31. A. A. Meyerhoff and W. M. Mazer, Optimum Binary FM Reception Using Discriminator Detection and IF Shaping, *RCA Rev.*, Vol. 22, No. 4, Dec. 1961, pp. 698–728.

32. E. F. Smith, Attainable Error Probabilities in Demodulation of Random Binary PCM/FM Waveforms, *IRE Trans.*, Vol. SET-8, Dec. 1962, pp. 290–297.

33. W. R. Bennett and S. O. Rice, Spectral Density and Autocorrelation Functions Associated with Binary Frequency-Shift Keying, *BSTJ*, Vol. 42, No. 5, Sept. 1963, pp. 2355–2386.

34. P. D. Shaft, Error Rate of PCM-FM Using Discriminator Detection, *IEEE Trans.*, Vol. SET-9, Dec. 1963, pp. 131–137.

35. E. D. Sunde, Digital Troposcatter Transmission and Modulation Theory, *BSTJ*, Vol. 43, Jan. 1964, pp. 143–214.

36. R. R. Anderson, W. R. Bennett, J. R. Davey and J. Salz, Differential Detection of Binary FM, *BSTJ*, Vol. 44, Jan. 1965, pp. 111–159.

37. M. G. Pelchat, Power Spectrum of PAM/FM and PAM/PM, *IRE Trans.*, Vol. SET-11, June 1965, pp. 70–77.

38. J. Salz, Performance of Multilevel Narrow-Band FM Digital Communication Systems, *IEEE Trans.*, Vol. Com-13, No. 4, Dec. 1965, pp. 420–424. *See also* J. Salz and V. G. Koll, An Experimental Digital Multilevel FM Modem, *IEEE Trans.*, Vol. COM-14, No. 3, June 1966, pp. 259–265.

39. J. E. Mazo and J. Salz, Theory of Error Rates for Digital FM, *BSTJ*, Vol. XLV, No. 9, Nov. 1966, pp. 1511–1535.

40. D. L. Schilling and E. Hoffman, Demodulation of Digital Signals Using an FM Discriminator, *Nat. Electronics Conf.*, Vol. 22, Oct. 1966, pp. 369–374.

SOME DOVER SCIENCE BOOKS

SOME DOVER SCIENCE BOOKS

WHAT IS SCIENCE?,
Norman Campbell
This excellent introduction explains scientific method, role of mathematics, types of scientific laws. Contents: 2 aspects of science, science & nature, laws of science, discovery of laws, explanation of laws, measurement & numerical laws, applications of science. 192pp. 5⅜ x 8.   60043-2 Paperbound $1.25

FADS AND FALLACIES IN THE NAME OF SCIENCE,
Martin Gardner
Examines various cults, quack systems, frauds, delusions which at various times have masqueraded as science. Accounts of hollow-earth fanatics like Symmes; Velikovsky and wandering planets; Hoerbiger; Bellamy and the theory of multiple moons; Charles Fort; dowsing, pseudoscientific methods for finding water, ores, oil. Sections on naturopathy, iridiagnosis, zone therapy, food fads, etc. Analytical accounts of Wilhelm Reich and orgone sex energy; L. Ron Hubbard and Dianetics; A. Korzybski and General Semantics; many others. Brought up to date to include Bridey Murphy, others. Not just a collection of anecdotes, but a fair, reasoned appraisal of eccentric theory. Formerly titled *In the Name of Science*. Preface. Index. x + 384pp. 5⅜ x 8.
20394-8 Paperbound $2.00

PHYSICS, THE PIONEER SCIENCE,
L. W. Taylor
First thorough text to place all important physical phenomena in cultural-historical framework; remains best work of its kind. Exposition of physical laws, theories developed chronologically, with great historical, illustrative experiments diagrammed, described, worked out mathematically. Excellent physics text for self-study as well as class work. Vol. 1: Heat, Sound: motion, acceleration, gravitation, conservation of energy, heat engines, rotation, heat, mechanical energy, etc. 211 illus. 407pp. 5⅜ x 8. Vol. 2: Light, Electricity: images, lenses, prisms, magnetism, Ohm's law, dynamos, telegraph, quantum theory, decline of mechanical view of nature, etc. Bibliography. 13 table appendix. Index. 551 illus. 2 color plates. 508pp. 5⅜ x 8.
60565-5, 60566-3 Two volume set, paperbound $5.50

THE EVOLUTION OF SCIENTIFIC THOUGHT FROM NEWTON TO EINSTEIN,
A. d'Abro
Einstein's special and general theories of relativity, with their historical implications, are analyzed in non-technical terms. Excellent accounts of the contributions of Newton, Riemann, Weyl, Planck, Eddington, Maxwell, Lorentz and others are treated in terms of space and time, equations of electromagnetics, finiteness of the universe, methodology of science. 21 diagrams. 482pp. 5⅜ x 8.
20002-7 Paperbound $2.50

CHANCE, LUCK AND STATISTICS: THE SCIENCE OF CHANCE,
Horace C. Levinson
Theory of probability and science of statistics in simple, non-technical language.
Part I deals with theory of probability, covering odd superstitions in regard to
"luck," the meaning of betting odds, the law of mathematical expectation,
gambling, and applications in poker, roulette, lotteries, dice, bridge, and other
games of chance. Part II discusses the misuse of statistics, the concept of statistical probabilities, normal and skew frequency distributions, and statistics applied to various fields—birth rates, stock speculation, insurance rates, advertis-
ing, etc. "Presented in an easy humorous style which I consider the best kind of
expository writing," Prof. A. C. Cohen, Industry Quality Control. Enlarged
revised edition. Formerly titled *The Science of Chance*. Preface and two new
appendices by the author. xiv + 365pp. 5⅜ x 8. 21007-3 Paperbound $2.00

BASIC ELECTRONICS,
prepared by the U.S. Navy Training Publications Center
A thorough and comprehensive manual on the fundamentals of electronics.
Written clearly, it is equally useful for self-study or course work for those with
a knowledge of the principles of basic electricity. Partial contents: Operating
Principles of the Electron Tube; Introduction to Transistors; Power Supplies
for Electronic Equipment; Tuned Circuits; Electron-Tube Amplifiers; Audio
Power Amplifiers; Oscillators; Transmitters; Transmission Lines; Antennas and
Propagation; Introduction to Computers; and related topics. Appendix. Index.
Hundreds of illustrations and diagrams. vi + 471pp. 6½ x 9¼.
61076-4 Paperbound $2.95

BASIC THEORY AND APPLICATION OF TRANSISTORS,
prepared by the U.S. Department of the Army
An introductory manual prepared for an army training program. One of the
finest available surveys of theory and application of transistor design and
operation. Minimal knowledge of physics and theory of electron tubes required.
Suitable for textbook use, course supplement, or home study. Chapters: Introduction; fundamental theory of transistors; transistor amplifier fundamentals;
parameters, equivalent circuits, and characteristic curves; bias stabilization;
transistor analysis and comparison using characteristic curves and charts; audio
amplifiers; tuned amplifiers; wide-band amplifiers; oscillators; pulse and switching circuits; modulation, mixing, and demodulation; and additional semi-
conductor devices. Unabridged, corrected edition. 240 schematic drawings,
photographs, wiring diagrams, etc. 2 Appendices. Glossary. Index. 263pp.
6½ x 9¼. 60380-6 Paperbound $1.75

GUIDE TO THE LITERATURE OF MATHEMATICS AND PHYSICS,
N. G. Parke III
Over 5000 entries included under approximately 120 major subject headings of
selected most important books, monographs, periodicals, articles in English,
plus important works in German, French, Italian, Spanish, Russian (many
recently available works). Covers every branch of physics, math, related engineering. Includes author, title, edition, publisher, place, date, number of
volumes, number of pages. A 40-page introduction on the basic problems of
research and study provides useful information on the organization and use of
libraries, the psychology of learning, etc. This reference work will save you
hours of time. 2nd revised edition. Indices of authors, subjects, 464pp. 5⅜ x 8.
60447-0 Paperbound $2.75

THE RISE OF THE NEW PHYSICS (formerly THE DECLINE OF MECHANISM),
A. d'Abro
This authoritative and comprehensive 2-volume exposition is unique in scientific publishing. Written for intelligent readers not familiar with higher mathematics, it is the only thorough explanation in non-technical language of modern mathematical-physical theory. Combining both history and exposition, it ranges from classical Newtonian concepts up through the electronic theories of Dirac and Heisenberg, the statistical mechanics of Fermi, and Einstein's relativity theories. "A must for anyone doing serious study in the physical sciences," *J. of Franklin Inst.* 97 illustrations. 991pp. 2 volumes.
20003-5, 20004-3 Two volume set, paperbound $5.50

THE STRANGE STORY OF THE QUANTUM, AN ACCOUNT FOR THE GENERAL READER OF THE GROWTH OF IDEAS UNDERLYING OUR PRESENT ATOMIC KNOWLEDGE, *B. Hoffmann*
Presents lucidly and expertly, with barest amount of mathematics, the problems and theories which led to modern quantum physics. Dr. Hoffmann begins with the closing years of the 19th century, when certain trifling discrepancies were noticed, and with illuminating analogies and examples takes you through the brilliant concepts of Planck, Einstein, Pauli, de Broglie, Bohr, Schroedinger, Heisenberg, Dirac, Sommerfeld, Feynman, etc. This edition includes a new, long postscript carrying the story through 1958. "Of the books attempting an account of the history and contents of our modern atomic physics which have come to my attention, this is the best," H. Margenau, Yale University, in *American Journal of Physics*. 32 tables and line illustrations. Index. 275pp. 5⅜ x 8.
20518-5 Paperbound $2.00

GREAT IDEAS AND THEORIES OF MODERN COSMOLOGY,
Jagjit Singh
The theories of Jeans, Eddington, Milne, Kant, Bondi, Gold, Newton, Einstein, Gamow, Hoyle, Dirac, Kuiper, Hubble, Weizsäcker and many others on such cosmological questions as the origin of the universe, space and time, planet formation, "continuous creation," the birth, life, and death of the stars, the origin of the galaxies, etc. By the author of the popular *Great Ideas of Modern Mathematics*. A gifted popularizer of science, he makes the most difficult abstractions crystal-clear even to the most non-mathematical reader. Index.
xii + 276pp. 5⅜ x 8½. 20925-3 Paperbound $2.50

GREAT IDEAS OF MODERN MATHEMATICS: THEIR NATURE AND USE,
Jagjit Singh
Reader with only high school math will understand main mathematical ideas of modern physics, astronomy, genetics, psychology, evolution, etc., better than many who use them as tools, but comprehend little of their basic structure. Author uses his wide knowledge of non-mathematical fields in brilliant exposition of differential equations, matrices, group theory, logic, statistics, problems of mathematical foundations, imaginary numbers, vectors, etc. Original publications, appendices. indexes. 65 illustr. 322pp. 5⅜ x 8. 20587-8 Paperbound $2.25

THE MATHEMATICS OF GREAT AMATEURS, *Julian L. Coolidge*
Great discoveries made by poets, theologians, philosophers, artists and other non-mathematicians: Omar Khayyam, Leonardo da Vinci, Albrecht Dürer, John Napier, Pascal, Diderot, Bolzano, etc. Surprising accounts of what can result from a non-professional preoccupation with the oldest of sciences. 56 figures. viii + 211pp. 5⅜ x 8½. 61009-8 Paperbound $2.00

COLLEGE ALGEBRA, *H. B. Fine*

Standard college text that gives a systematic and deductive structure to algebra; comprehensive, connected, with emphasis on theory. Discusses the commutative, associative, and distributive laws of number in unusual detail, and goes on with undetermined coefficients, quadratic equations, progressions, logarithms, permutations, probability, power series, and much more. Still most valuable elementary-intermediate text on the science and structure of algebra. Index. 1560 problems, all with answers. x + 631pp. 5⅜ x 8. 60211-7 Paperbound $2.75

HIGHER MATHEMATICS FOR STUDENTS OF CHEMISTRY AND PHYSICS, *J. W. Mellor*

Not abstract, but practical, building its problems out of familiar laboratory material, this covers differential calculus, coordinate, analytical geometry, functions, integral calculus, infinite series, numerical equations, differential equations, Fourier's theorem, probability, theory of errors, calculus of variations, determinants. "If the reader is not familiar with this book, it will repay him to examine it," *Chem. & Engineering News.* 800 problems. 189 figures. Bibliography. xxi + 641pp. 5⅜ x 8. 60193-5 Paperbound $3.50

TRIGONOMETRY REFRESHER FOR TECHNICAL MEN, *A. A. Klaf*

A modern question and answer text on plane and spherical trigonometry. Part I covers plane trigonometry: angles, quadrants, trigonometrical functions, graphical representation, interpolation, equations, logarithms, solution of triangles, slide rules, etc. Part II discusses applications to navigation, surveying, elasticity, architecture, and engineering. Small angles, periodic functions, vectors, polar coordinates, De Moivre's theorem, fully covered. Part III is devoted to spherical trigonometry and the solution of spherical triangles, with applications to terrestrial and astronomical problems. Special time-savers for numerical calculation. 913 questions answered for you! 1738 problems; answers to odd numbers. 494 figures. 14 pages of functions, formulae. Index. x + 629pp. 5⅜ x 8. 20371-9 Paperbound $3.00

CALCULUS REFRESHER FOR TECHNICAL MEN, *A. A. Klaf*

Not an ordinary textbook but a unique refresher for engineers, technicians, and students. An examination of the most important aspects of differential and integral calculus by means of 756 key questions. Part I covers simple differential calculus: constants, variables, functions, increments, derivatives, logarithms, curvature, etc. Part II treats fundamental concepts of integration: inspection, substitution, transformation, reduction, areas and volumes, mean value, successive and partial integration, double and triple integration. Stresses practical aspects! A 50 page section gives applications to civil and nautical engineering, electricity, stress and strain, elasticity, industrial engineering, and similar fields. 756 questions answered. 556 problems; solutions to odd numbers. 36 pages of constants, formulae. Index. v + 431pp. 5⅜ x 8. 20370-0 Paperbound $2.25

INTRODUCTION TO THE THEORY OF GROUPS OF FINITE ORDER, *R. Carmichael*

Examines fundamental theorems and their application. Beginning with sets, systems, permutations, etc., it progresses in easy stages through important types of groups: Abelian, prime power, permutation, etc. Except 1 chapter where matrices are desirable, no higher math needed. 783 exercises, problems. Index. xvi + 447pp. 5⅜ x 8. 60300-8 Paperbound $3.00

FIVE VOLUME "THEORY OF FUNCTIONS" SET BY KONRAD KNOPP

This five-volume set, prepared by Konrad Knopp, provides a complete and readily followed account of theory of functions. Proofs are given concisely, yet without sacrifice of completeness or rigor. These volumes are used as texts by such universities as M.I.T., University of Chicago, N. Y. City College, and many others. "Excellent introduction . . . remarkably readable, concise, clear, rigorous," *Journal of the American Statistical Association.*

ELEMENTS OF THE THEORY OF FUNCTIONS,
Konrad Knopp
This book provides the student with background for further volumes in this set, or texts on a similar level. Partial contents: foundations, system of complex numbers and the Gaussian plane of numbers, Riemann sphere of numbers, mapping by linear functions, normal forms, the logarithm, the cyclometric functions and binomial series. "Not only for the young student, but also for the student who knows all about what is in it," *Mathematical Journal.* Bibliography. Index. 140pp. 5⅜ x 8.   60154-4 Paperbound $1.50

THEORY OF FUNCTIONS, PART I,
Konrad Knopp
With volume II, this book provides coverage of basic concepts and theorems. Partial contents: numbers and points, functions of a complex variable, integral of a continuous function, Cauchy's integral theorem, Cauchy's integral formulae, series with variable terms, expansion of analytic functions in power series, analytic continuation and complete definition of analytic functions, entire transcendental functions, Laurent expansion, types of singularities. Bibliography. Index. vii + 146pp. 5⅜ x 8.   60156-0 Paperbound $1.50

THEORY OF FUNCTIONS, PART II,
Konrad Knopp
Application and further development of general theory, special topics. Single valued functions. Entire, Weierstrass, Meromorphic functions. Riemann surfaces. Algebraic functions. Analytical configuration, Riemann surface. Bibliography. Index. x + 150pp. 5⅜ x 8.   60157-9 Paperbound $1.50 ·

PROBLEM BOOK IN THE THEORY OF FUNCTIONS, VOLUME 1.
Konrad Knopp
Problems in elementary theory, for use with Knopp's *Theory of Functions,* or any other text, arranged according to increasing difficulty. Fundamental concepts, sequences of numbers and infinite series, complex variable, integral theorems, development in series, conformal mapping. 182 problems. Answers. viii + 126pp. 5⅜ x 8.   60158-7 Paperbound $1.50

PROBLEM BOOK IN THE THEORY OF FUNCTIONS, VOLUME 2,
Konrad Knopp
Advanced theory of functions, to be used either with Knopp's *Theory of Functions,* or any other comparable text. Singularities, entire & meromorphic functions, periodic, analytic, continuation, multiple-valued functions, Riemann surfaces, conformal mapping. Includes a section of additional elementary problems. "The difficult task of selecting from the immense material of the modern theory of functions the problems just within the reach of the beginner is here masterfully accomplished," *Am. Math. Soc.* Answers. 138pp. 5⅜ x 8.
60159-5 Paperbound $1.50

NUMERICAL SOLUTIONS OF DIFFERENTIAL EQUATIONS,
H. Levy & E. A. Baggott
Comprehensive collection of methods for solving ordinary differential equations
of first and higher order. All must pass 2 requirements: easy to grasp and
practical, more rapid than school methods. Partial contents: graphical integration of differential equations, graphical methods for detailed solution. Numerical solution. Simultaneous equations and equations of 2nd and higher orders.
"Should be in the hands of all in research in applied mathematics, teaching,"
Nature. 21 figures. viii + 238pp. 5⅜ x 8.   60168-4 Paperbound $1.85

ELEMENTARY STATISTICS, WITH APPLICATIONS IN MEDICINE AND THE
BIOLOGICAL SCIENCES, *F. E. Croxton*
A sound introduction to statistics for anyone in the physical sciences, assuming no prior acquaintance and requiring only a modest knowledge of math.
All basic formulas carefully explained and illustrated; all necessary reference
tables included. From basic terms and concepts, the study proceeds to frequency
distribution, linear, non-linear, and multiple correlation, skewness, kurtosis,
etc. A large section deals with reliability and significance of statistical methods.
Containing concrete examples from medicine and biology, this book will prove
unusually helpful to workers in those fields who increasingly must evaluate,
check, and interpret statistics. Formerly titled "Elementary Statistics with Applications in Medicine." 101 charts. 57 tables. 14 appendices. Index. vi +
376pp. 5⅜ x 8.   60506-X Paperbound $2.25

INTRODUCTION TO SYMBOLIC LOGIC,
S. Langer
No special knowledge of math required — probably the clearest book ever
written on symbolic logic, suitable for the layman, general scientist, and philos-
opher. You start with simple symbols and advance to a knowledge of the
Boole-Schroeder and Russell-Whitehead systems. Forms, logical structure, classes,
the calculus of propositions, logic of the syllogism, etc. are all covered. "One
of the clearest and simplest introductions," *Mathematics Gazette.* Second enlarged, revised edition. 368pp. 5⅜ x 8.   60164-1 Paperbound $2.25

A SHORT ACCOUNT OF THE HISTORY OF MATHEMATICS,
W. W. R. Ball
Most readable non-technical history of mathematics treats lives, discoveries of
every important figure from Egyptian, Phoenician, mathematicians to late 19th
century. Discusses schools of Ionia, Pythagoras, Athens, Cyzicus, Alexandria,
Byzantium, systems of numeration; primitive arithmetic; Middle Ages, Renais-
sance, including Arabs, Bacon, Regiomontanus, Tartaglia, Cardan, Stevinus,
Galileo, Kepler; modern mathematics of Descartes, Pascal, Wallis, Huygens,
Newton, Leibnitz, d'Alembert, Euler, Lambert, Laplace, Legendre, Gauss,
Hermite, Weierstrass, scores more. Index. 25 figures. 546pp. 5⅜ x 8.
20630-0 Paperbound $2.75

INTRODUCTION TO NONLINEAR DIFFERENTIAL AND INTEGRAL EQUATIONS,
Harold T. Davis
Aspects of the problem of nonlinear equations, transformations that lead to
equations solvable by classical means, results in special cases, and useful
generalizations. Thorough, but easily followed by mathematically sophisticated
reader who knows little about non-linear equations. 137 problems for student
to solve. xv + 566pp. 5⅜ x 8½.   60971-5 Paperbound $2.75

An Introduction to the Geometry of N Dimensions,
D. H. Y. Sommerville
An introduction presupposing no prior knowledge of the field, the only book in English devoted exclusively to higher dimensional geometry. Discusses fundamental ideas of incidence, parallelism, perpendicularity, angles between linear space; enumerative geometry; analytical geometry from projective and metric points of view; polytopes; elementary ideas in analysis situs; content of hyper-spacial figures. Bibliography. Index. 60 diagrams. 196pp. 5⅜ x 8.

60494-2 Paperbound $1.50

Elementary Concepts of Topology, *P. Alexandroff*
First English translation of the famous brief introduction to topology for the beginner or for the mathematician not undertaking extensive study. This unusually useful intuitive approach deals primarily with the concepts of complex, cycle, and homology, and is wholly consistent with current investigations. Ranges from basic concepts of set-theoretic topology to the concept of Betti groups. "Glowing example of harmony between intuition and thought," David Hilbert. Translated by A. E. Farley. Introduction by D. Hilbert. Index. 25 figures. 73pp. 5⅜ x 8.

60747-X Paperbound $1.25

Elements of Non-Euclidean Geometry,
D. M. Y. Sommerville
Unique in proceeding step-by-step, in the manner of traditional geometry. Enables the student with only a good knowledge of high school algebra and geometry to grasp elementary hyperbolic, elliptic, analytic non-Euclidean geometries; space curvature and its philosophical implications; theory of radical axes; homothetic centres and systems of circles; parataxy and parallelism; absolute measure; Gauss' proof of the defect area theorem; geodesic representation; much more, all with exceptional clarity. 126 problems at chapter endings provide progressive practice and familiarity. 133 figures. Index. xvi + 274pp. 5⅜ x 8.

60460-8 Paperbound $2.00

Introduction to the Theory of Numbers, *L. E. Dickson*
Thorough, comprehensive approach with adequate coverage of classical literature, an introductory volume beginners can follow. Chapters on divisibility, congruences, quadratic residues & reciprocity. Diophantine equations, etc. Full treatment of binary quadratic forms without usual restriction to integral coefficients. Covers infinitude of primes, least residues. Fermat's theorem. Euler's phi function, Legendre's symbol, Gauss's lemma, automorphs, reduced forms, recent theorems of Thue & Siegel, many more. Much material not readily available elsewhere. 239 problems. Index. I figure. viii + 183pp. 5⅜ x 8.

60342-3 Paperbound $1.75

Mathematical Tables and Formulas,
compiled by Robert D. Carmichael and Edwin R. Smith
Valuable collection for students, etc. Contains all tables necessary in college algebra and trigonometry, such as five-place common logarithms, logarithmic sines and tangents of small angles, logarithmic trigonometric functions, natural trigonometric functions, four-place antilogarithms, tables for changing from sexagesimal to circular and from circular to sexagesimal measure of angles, etc. Also many tables and formulas not ordinarily accessible, including powers, roots, and reciprocals, exponential and hyperbolic functions, ten-place logarithms of prime numbers, and formulas and theorems from analytical and elementary geometry and from calculus. Explanatory introduction. viii + 269pp. 5⅜ x 8½.

60111-0 Paperbound $1.50

A SOURCE BOOK IN MATHEMATICS,
D. E. Smith
Great discoveries in math, from Renaissance to end of 19th century, in English translation. Read announcements by Dedekind, Gauss, Delamain, Pascal, Fermat, Newton, Abel, Lobachevsky, Bolyai, Riemann, De Moivre, Legendre, Laplace, others of discoveries about imaginary numbers, number congruence, slide rule, equations, symbolism, cubic algebraic equations, non-Euclidean forms of geometry, calculus, function theory, quaternions, etc. Succinct selections from 125 different treatises, articles, most unavailable elsewhere in English. Each article preceded by biographical introduction. Vol. I: Fields of Number, Algebra. Index. 32 illus. 338pp. 5⅜ x 8. Vol. II: Fields of Geometry, Probability, Calculus, Functions, Quaternions. 83 illus. 432pp. 5⅜ x 8.

60552-3, 60553-1 Two volume set, paperbound $5.00

FOUNDATIONS OF PHYSICS,
R. B. Lindsay & H. Margenau
Excellent bridge between semi-popular works & technical treatises. A discussion of methods of physical description, construction of theory; valuable for physicist with elementary calculus who is interested in ideas that give meaning to data, tools of modern physics. Contents include symbolism; mathematical equations; space & time foundations of mechanics; probability; physics & continua; electron theory; special & general relativity; quantum mechanics; causality. "Thorough and yet not overdetailed. Unreservedly recommended," *Nature* (London). Unabridged, corrected edition. List of recommended readings. 35 illustrations. xi + 537pp. 5⅜ x 8.

60377-6 Paperbound $3.50

FUNDAMENTAL FORMULAS OF PHYSICS,
ed. by D. H. Menzel
High useful, full, inexpensive reference and study text, ranging from simple to highly sophisticated operations. Mathematics integrated into text—each chapter stands as short textbook of field represented. Vol. 1: Statistics, Physical Constants, Special Theory of Relativity, Hydrodynamics, Aerodynamics, Boundary Value Problems in Math, Physics, Viscosity, Electromagnetic Theory, etc. Vol. 2: Sound, Acoustics, Geometrical Optics, Electron Optics, High-Energy Phenomena, Magnetism, Biophysics, much more. Index. Total of 800pp. 5⅜ x 8.

60595-7, 60596-5 Two volume set, paperbound $4.75

THEORETICAL PHYSICS,
A. S. Kompaneyets
One of the very few thorough studies of the subject in this price range. Provides advanced students with a comprehensive theoretical background. Especially strong on recent experimentation and developments in quantum theory. Contents: Mechanics (Generalized Coordinates, Lagrange's Equation, Collision of Particles, etc.), Electrodynamics (Vector Analysis, Maxwell's equations, Transmission of Signals, Theory of Relativity, etc.), Quantum Mechanics (the Inadequacy of Classical Mechanics, the Wave Equation, Motion in a Central Field, Quantum Theory of Radiation, Quantum Theories of Dispersion and Scattering, etc.), and Statistical Physics (Equilibrium Distribution of Molecules in an Ideal Gas, Boltzmann Statistics, Bose and Fermi Distribution. Thermodynamic Quantities, etc.). Revised to 1961. Translated by George Yankovsky, authorized by Kompaneyets. 137 exercises. 56 figures. 529pp. 5⅜ x 8½.

60972-3 Paperbound $3.50

MATHEMATICAL PHYSICS, *D. H. Menzel*
Thorough one-volume treatment of the mathematical techniques vital for classical mechanics, electromagnetic theory, quantum theory, and relativity. Written by the Harvard Professor of Astrophysics for junior, senior, and graduate courses, it gives clear explanations of all those aspects of function theory, vectors, matrices, dyadics, tensors, partial differential equations, etc., necessary for the understanding of the various physical theories. Electron theory, relativity, and other topics seldom presented appear here in considerable detail. Scores of definition, conversion factors, dimensional constants, etc. "More detailed than normal for an advanced text . . . excellent set of sections on Dyadics, Matrices, and Tensors," *Journal of the Franklin Institute*. Index. 193 problems, with answers. x + 412pp. 5⅜ x 8.   60056-4 Paperbound $2.50

THE THEORY OF SOUND, *Lord Rayleigh*
Most vibrating systems likely to be encountered in practice can be tackled successfully by the methods set forth by the great Nobel laureate, Lord Rayleigh. Complete coverage of experimental, mathematical aspects of sound theory. Partial contents: Harmonic motions, vibrating systems in general, lateral vibrations of bars, curved plates or shells, applications of Laplace's functions to acoustical problems, fluid friction, plane vortex-sheet, vibrations of solid bodies, etc. This is the first inexpensive edition of this great reference and study work. Bibliography, Historical introduction by R. B. Lindsay. Total of 1040pp. 97 figures. 5⅜ x 8.   60292-3, 60293-1 Two volume set, paperbound $6.00

HYDRODYNAMICS, *Horace Lamb*
Internationally famous complete coverage of standard reference work on dynamics of liquids & gases. Fundamental theorems, equations, methods, solutions, background, for classical hydrodynamics. Chapters include Equations of Motion, Integration of Equations in Special Gases, Irrotational Motion, Motion of Liquid in 2 Dimensions, Motion of Solids through Liquid-Dynamical Theory, Vortex Motion, Tidal Waves, Surface Waves, Waves of Expansion, Viscosity, Rotating Masses of Liquids. Excellently planned, arranged; clear, lucid presentation. 6th enlarged, revised edition. Index. Over 900 footnotes, mostly bibliographical. 119 figures. xv + 738pp. 6⅛ x 9¼.   60256-7 Paperbound $4.00

DYNAMICAL THEORY OF GASES, *James Jeans*
Divided into mathematical and physical chapters for the convenience of those not expert in mathematics, this volume discusses the mathematical theory of gas in a steady state, thermodynamics, Boltzmann and Maxwell, kinetic theory, quantum theory, exponentials, etc. 4th enlarged edition, with new material on quantum theory, quantum dynamics, etc. Indexes. 28 figures. 444pp. 6⅛ x 9¼.
          60136-6 Paperbound $2.75

THERMODYNAMICS, *Enrico Fermi*
Unabridged reproduction of 1937 edition. Elementary in treatment; remarkable for clarity, organization. Requires no knowledge of advanced math beyond calculus, only familiarity with fundamentals of thermometry, calorimetry. Partial Contents: Thermodynamic systems; First & Second laws of thermodynamics; Entropy; Thermodynamic potentials: phase rule, reversible electric cell; Gaseous reactions: van't Hoff reaction box, principle of LeChatelier; Thermodynamics of dilute solutions: osmotic & vapor pressures, boiling & freezing points; Entropy constant. Index. 25 problems. 24 illustrations. x + 160pp. 5⅜ x 8.   60361-X Paperbound $2.00

CELESTIAL OBJECTS FOR COMMON TELESCOPES,
Rev. T. W. Webb
Classic handbook for the use and pleasure of the amateur astronomer. Of inestimable aid in locating and identifying thousands of celestial objects. Vol I, The Solar System: discussions of the principle and operation of the telescope, procedures of observations and telescope-photography, spectroscopy, etc., precise location information of sun, moon, planets, meteors. Vol. II, The Stars: alphabetical listing of constellations, information on double stars, clusters, stars with unusual spectra, variables, and nebulae, etc. Nearly 4,000 objects noted. Edited and extensively revised by Margaret W. Mayall, director of the American Assn. of Variable Star Observers. New Index by Mrs. Mayall giving the location of all objects mentioned in the text for Epoch 2000. New Precession Table added. New appendices on the planetary satellites, constellation names and abbreviations, and solar system data. Total of 46 illustrations. Total of xxxix + 606pp. 5⅜ x 8.   20917-2, 20918-0 Two volume set, paperbound $5.00

PLANETARY THEORY,
E. W. Brown and C. A. Shook
Provides a clear presentation of basic methods for calculating planetary orbits for today's astronomer. Begins with a careful exposition of specialized mathematical topics essential for handling perturbation theory and then goes on to indicate how most of the previous methods reduce ultimately to two general calculation methods: obtaining expressions either for the coordinates of planetary positions or for the elements which determine the perturbed paths. An example of each is given and worked in detail. Corrected edition. Preface. Appendix. Index. xii + 302pp. 5⅜ x 8½.   61133-7 Paperbound $2.25

STAR NAMES AND THEIR MEANINGS,
Richard Hinckley Allen
An unusual book documenting the various attributions of names to the individual stars over the centuries. Here is a treasure-house of information on a topic not normally delved into even by professional astronomers; provides a fascinating background to the stars in folk-lore, literary references, ancient writings, star catalogs and maps over the centuries. Constellation-by-constellation analysis covers hundreds of stars and other asterisms, including the Pleiades, Hyades, Andromedan Nebula, etc. Introduction. Indices. List of authors and authorities. xx + 563pp. 5⅜ x 8½.   21079-0 Paperbound $3.00

A SHORT HISTORY OF ASTRONOMY, *A. Berry*
Popular standard work for over 50 years, this thorough and accurate volume covers the science from primitive times to the end of the 19th century. After the Greeks and the Middle Ages, individual chapters analyze Copernicus, Brahe, Galileo, Kepler, and Newton, and the mixed reception of their discoveries. Post-Newtonian achievements are then discussed in unusual detail: Halley, Bradley, Lagrange, Laplace, Herschel, Bessel, etc. 2 Indexes. 104 illustrations, 9 portraits. xxxi + 440pp. 5⅜ x 8.   20210-0 Paperbound $2.75

SOME THEORY OF SAMPLING, *W. E. Deming*
The purpose of this book is to make sampling techniques understandable to and useable by social scientists, industrial managers, and natural scientists who are finding statistics increasingly part of their work. Over 200 exercises, plus dozens of actual applications. 61 tables. 90 figs. xix + 602pp. 5⅜ x 8½.
61755-6 Paperbound $3.50

PRINCIPLES OF STRATIGRAPHY,
A. W. Grabau
Classic of 20th century geology, unmatched in scope and comprehensiveness.
Nearly 600 pages cover the structure and origins of every kind of sedimentary,
hydrogenic, oceanic, pyroclastic, atmoclastic, hydroclastic, marine hydroclastic,
and bioclastic rock; metamorphism; erosion; etc. Includes also the constitution
of the atmosphere; morphology of oceans, rivers, glaciers; volcanic activities;
faults and earthquakes; and fundamental principles of paleontology (nearly 200
pages). New introduction by Prof. M. Kay, Columbia U. 1277 bibliographical
entries. 264 diagrams. Tables, maps, etc. Two volume set. Total of xxxii $+$
1185pp. 5⅜ x 8.   60686-4, 60687-2 Two volume set, paperbound $6.25

SNOW CRYSTALS, *W. A. Bentley and W. J. Humphreys*
Over 200 pages of Bentley's famous microphotographs of snow flakes—the product of painstaking, methodical work at his Jericho, Vermont studio. The
pictures, which also include plates of frost, glaze and dew on vegetation, spider
webs, windowpanes; sleet; graupel or soft hail, were chosen both for their
scientific interest and their aesthetic qualities. The wonder of nature's diversity
is exhibited in the intricate, beautiful patterns of the snow flakes. Introductory
text by W. J. Humphreys. Selected bibliography. 2,453 illustrations. 224pp.
8 x 10¼.   20287-9 Paperbound $3.25

THE BIRTH AND DEVELOPMENT OF THE GEOLOGICAL SCIENCES,
F. D. Adams
Most thorough history of the earth sciences ever written. Geological thought
from earliest times to the end of the 19th century, covering over 300 early
thinkers & systems: fossils & their explanation, vulcanists vs. neptunists, figured
stones & paleontology, generation of stones, dozens of similar topics. 91 illustrations, including medieval, renaissance woodcuts, etc. Index. 632 footnotes,
mostly bibliographical. 511pp. 5⅜ x 8.   20005-1 Paperbound $2.75

ORGANIC CHEMISTRY, *F. C. Whitmore*
The entire subject of organic chemistry for the practicing chemist and the
advanced student. Storehouse of facts, theories, processes found elsewhere only
in specialized journals. Covers aliphatic compounds (500 pages on the properties and synthetic preparation of hydrocarbons, halides, proteins, ketones,
etc.), alicyclic compounds, aromatic compounds, heterocyclic compounds, or-
ganophosphorus and organometallic compounds. Methods of synthetic preparation analyzed critically throughout. Includes much of biochemical interest.
"The scope of this volume is astonishing," *Industrial and Engineering
Chemistry*. 12,000-reference index. 2387-item bibliography. Total of x $+$
1005pp. 5⅜ x 8.   60700-3, 60701-1 Two volume set, paperbound $4.50

THE PHASE RULE AND ITS APPLICATION,
Alexander Findlay
Covering chemical phenomena of 1, 2, 3, 4, and multiple component systems,
this "standard work on the subject" (*Nature*, London), has been completely
revised and brought up to date by A. N. Campbell and N. O. Smith. Brand
new material has been added on such matters as binary, tertiary liquid
equilibria, solid solutions in ternary systems, quinary systems of salts and
water. Completely revised to triangular coordinates in ternary systems, clarified
graphic representation, solid models, etc. 9th revised edition. Author, subject
indexes. 236 figures. 505 footnotes, mostly bibliographic. xii $+$ 494pp. 5⅜ x 8.
60091-2 Paperbound $2.75

THE PRINCIPLES OF ELECTROCHEMISTRY,
D. A. MacInnes
Basic equations for almost every subfield of electrochemistry from first principles, referring at all times to the soundest and most recent theories and results; unusually useful as text or as reference. Covers coulometers and Faraday's Law, electrolytic conductance, the Debye-Hueckel method for the theoretical calculation of activity coefficients, concentration cells, standard electrode potentials, thermodynamic ionization constants, pH, potentiometric titrations, irreversible phenomena. Planck's equation, and much more. 2 indices. Appendix. 585-item bibliography. 137 figures. 94 tables. ii + 478pp. 5⅜ x 8⅜.
60052-1 Paperbound $3.00

MATHEMATICS OF MODERN ENGINEERING,
E. G. Keller and R. E. Doherty
Written for the Advanced Course in Engineering of the General Electric Corporation, deals with the engineering use of determinants, tensors, the Heaviside operational calculus, dyadics, the calculus of variations, etc. Presents underlying principles fully, but emphasis is on the perennial engineering attack of set-up and solve. Indexes. Over 185 figures and tables. Hundreds of exercises, problems, and worked-out examples. References. Total of xxxiii + 623pp. 5⅜ x 8.
60734-8, 60735-6 Two volume set, paperbound $3.70

AERODYNAMIC THEORY: A GENERAL REVIEW OF PROGRESS,
William F. Durand, editor-in-chief
A monumental joint effort by the world's leading authorities prepared under a grant of the Guggenheim Fund for the Promotion of Aeronautics. Never equalled for breadth, depth, reliability. Contains discussions of special mathematical topics not usually taught in the engineering or technical courses. Also: an extended two-part treatise on Fluid Mechanics, discussions of aerodynamics of perfect fluids, analyses of experiments with wind tunnels, applied airfoil theory, the nonlifting system of the airplane, the air propeller, hydrodynamics of boats and floats, the aerodynamics of cooling, etc. Contributing experts include Munk, Giacomelli, Prandtl, Toussaint, Von Karman, Klemperer, among others. Unabridged republication. 6 volumes. Total of 1,012 figures, 12 plates, 2,186pp. Bibliographies. Notes. Indices. 5⅜ x 8½. 61709-2, 61710-6, 61711-4, 61712-2, 61713-0, 61715-9 Six volume set, paperbound $13.50

FUNDAMENTALS OF HYDRO- AND AEROMECHANICS,
L. Prandtl and O. G. Tietjens
The well-known standard work based upon Prandtl's lectures at Goettingen. Wherever possible hydrodynamics theory is referred to practical considerations in hydraulics, with the view of unifying theory and experience. Presentation is extremely clear and though primarily physical, mathematical proofs are rigorous and use vector analysis to a considerable extent. An Engineering Society Monograph, 1934. 186 figures. Index. xvi + 270pp. 5⅜ x 8.
60374-1 Paperbound $2.25

APPLIED HYDRO- AND AEROMECHANICS,
L. Prandtl and O. G. Tietjens
Presents for the most part methods which will be valuable to engineers. Covers flow in pipes, boundary layers, airfoil theory, entry conditions, turbulent flow in pipes, and the boundary layer, determining drag from measurements of pressure and velocity, etc. Unabridged, unaltered. An Engineering Society Monograph. 1934. Index. 226 figures, 28 photographic plates illustrating flow patterns. xvi + 311pp. 5⅜ x 8.
60375-X Paperbound $2.50

PRINCIPLES OF ART HISTORY,
H. Wölfflin
Analyzing such terms as "baroque," "classic," "neoclassic," "primitive," "picturesque," and 164 different works by artists like Botticelli, van Cleve, Dürer, Hobbema, Holbein, Hals, Rembrandt, Titian, Brueghel, Vermeer, and many others, the author establishes the classifications of art history and style on a firm, concrete basis. This classic of art criticism shows what really occurred between the 14th-century primitives and the sophistication of the 18th century in terms of basic attitudes and philosophies. "A remarkable lesson in the art of seeing," *Sat. Rev. of Literature*. Translated from the 7th German edition. 150 illustrations. 254pp. 6⅛ x 9¼. 20276-3 Paperbound $2.25

PRIMITIVE ART,
Franz Boas
This authoritative and exhaustive work by a great American anthropologist covers the entire gamut of primitive art. Pottery, leatherwork, metal work, stone work, wood, basketry, are treated in detail. Theories of primitive art, historical depth in art history, technical virtuosity, unconscious levels of patterning, symbolism, styles, literature, music, dance, etc. A must book for the interested layman, the anthropologist, artist, handicrafter (hundreds of unusual motifs), and the historian. Over 900 illustrations (50 ceramic vessels, 12 totem poles, etc.). 376pp. 5⅜ x 8.   20025-6 Paperbound $2.50

THE GENTLEMAN AND CABINET MAKER'S DIRECTOR,
Thomas Chippendale
A reprint of the 1762 catalogue of furniture designs that went on to influence generations of English and Colonial and Early Republic American furniture makers. The 200 plates, most of them full-page sized, show Chippendale's designs for French (Louis XV), Gothic, and Chinese-manner chairs, sofas, canopy and dome beds, cornices, chamber organs, cabinets, shaving tables, commodes, picture frames, frets, candle stands, chimney pieces, decorations, etc. The drawings are all elegant and highly detailed; many include construction diagrams and elevations. A supplement of 24 photographs shows surviving pieces of original and Chippendale-style pieces of furniture. Brief biography of Chippendale by N. I. Bienenstock, editor of *Furniture World*. Reproduced from the 1762 edition. 200 plates, plus 19 photographic plates. vi + 249pp. 9⅛ x 12¼.   21601-2 Paperbound $3.50

AMERICAN ANTIQUE FURNITURE: A BOOK FOR AMATEURS,
Edgar G. Miller, Jr.
Standard introduction and practical guide to identification of valuable American antique furniture. 2115 illustrations, mostly photographs taken by the author in 148 private homes, are arranged in chronological order in extensive chapters on chairs, sofas, chests, desks, bedsteads, mirrors, tables, clocks, and other articles. Focus is on furniture accessible to the collector, including simpler pieces and a larger than usual coverage of Empire style. Introductory chapters identify structural elements, characteristics of various styles, how to avoid fakes, etc. "We are frequently asked to name some book on American furniture that will meet the requirements of the novice collector, the beginning dealer, and . . . the general public. . . . We believe Mr. Miller's two volumes more completely satisfy this specification than any other work," *Antiques*. Appendix. Index. Total of vi + 1106pp. 7⅞ x 10¾.
21599-7, 21600-4 Two volume set, paperbound $7.50

THE PRINCIPLES OF PSYCHOLOGY,
William James

The full long-course, unabridged, of one of the great classics of Western literature and science. Wonderfully lucid descriptions of human mental activity, the stream of thought, consciousness, time perception, memory, imagination, emotions, reason, abnormal phenomena, and similar topics. Original contributions are integrated with the work of such men as Berkeley, Binet, Mills, Darwin, Hume, Kant, Royce, Schopenhauer, Spinoza, Locke, Descartes, Galton, Wundt, Lotze, Herbart, Fechner, and scores of others. All contrasting interpretations of mental phenomena are examined in detail—introspective analysis, philosophical interpretation, and experimental research. "A classic," *Journal of Consulting Psychology*. "The main lines are as valid as ever," *Psychoanalytical Quarterly*. "Standard reading . . . a classic of interpretation," *Psychiatric Quarterly*. 94 illustrations. 1408pp. 5⅜ x 8.

20381-6, 20382-4 Two volume set, paperbound $6.00

VISUAL ILLUSIONS: THEIR CAUSES, CHARACTERISTICS AND APPLICATIONS,
M. Luckiesh

"Seeing is deceiving," asserts the author of this introduction to virtually every type of optical illusion known. The text both describes and explains the principles involved in color illusions, figure-ground, distance illusions, etc. 100 photographs, drawings and diagrams prove how easy it is to fool the sense: circles that aren't round, parallel lines that seem to bend, stationary figures that seem to move as you stare at them — illustration after illustration strains our credulity at what we see. Fascinating book from many points of view, from applications for artists, in camouflage, etc. to the psychology of vision. New introduction by William Ittleson, Dept. of Psychology, Queens College. Index. Bibliography. xxi + 252pp. 5⅜ x 8½.

21530-X Paperbound $1.50

FADS AND FALLACIES IN THE NAME OF SCIENCE,
Martin Gardner

This is the standard account of various cults, quack systems, and delusions which have masqueraded as science: hollow earth fanatics. Reich and orgone sex energy, dianetics, Atlantis, multiple moons, Forteanism, flying saucers, medical fallacies like iridiagnosis, zone therapy, etc. A new chapter has been added on Bridey Murphy, psionics, and other recent manifestations in this field. This is a fair, reasoned appraisal of eccentric theory which provides excellent inoculation against cleverly masked nonsense. "Should be read by everyone, scientist and non-scientist alike," R. T. Birge, Prof. Emeritus of Physics, Univ. of California; Former President, American Physical Society. Index. x + 365pp. 5⅜ x 8.

20394-8 Paperbound $2.00

ILLUSIONS AND DELUSIONS OF THE SUPERNATURAL AND THE OCCULT,
D. H. Rawcliffe

Holds up to rational examination hundreds of persistent delusions including crystal gazing, automatic writing, table turning, mediumistic trances, mental healing, stigmata, lycanthropy, live burial, the Indian Rope Trick, spiritualism, dowsing, telepathy, clairvoyance, ghosts, ESP, etc. The author explains and exposes the mental and physical deceptions involved, making this not only an exposé of supernatural phenomena, but a valuable exposition of characteristic types of abnormal psychology. Originally titled "The Psychology of the Occult." 14 illustrations. Index. 551pp. 5⅜ x 8. 20503-7 Paperbound $3.50

THE MUSIC OF THE SPHERES: THE MATERIAL UNIVERSE — FROM ATOM TO QUASAR, SIMPLY EXPLAINED, *Guy Murchie*
Vast compendium of fact, modern concept and theory, observed and calculated data, historical background guides intelligent layman through the material universe. Brilliant exposition of earth's construction, explanations for moon's craters, atmospheric components of Venus and Mars (with data from recent fly-by's), sun spots, sequences of star birth and death, neighboring galaxies, contributions of Galileo, Tycho Brahe, Kepler, etc.; and (Vol. 2) construction of the atom (describing newly discovered sigma and xi subatomic particles), theories of sound, color and light, space and time, including relativity theory, quantum theory, wave theory, probability theory, work of Newton, Maxwell, Faraday, Einstein, de Broglie, etc. "Best presentation yet offered to the intelligent general reader," *Saturday Review*. Revised (1967). Index. 319 illustrations by the author. Total of xx + 644pp. 5⅜ x 8½.
21809-0, 21810-4 Two volume set, paperbound $5.00

FOUR LECTURES ON RELATIVITY AND SPACE, *Charles Proteus Steinmetz*
Lecture series, given by great mathematician and electrical engineer, generally considered one of the best popular-level expositions of special and general relativity theories and related questions. Steinmetz translates complex mathematical reasoning into language accessible to laymen through analogy, example and comparison. Among topics covered are relativity of motion, location, time; of mass; acceleration; 4-dimensional time-space; geometry of the gravitational field; curvature and bending of space; non-Euclidean geometry. Index. 40 illustrations. x + 142pp. 5⅜ x 8½.   61771-8 Paperbound $1.35

HOW TO KNOW THE WILD FLOWERS, *Mrs. William Starr Dana*
Classic nature book that has introduced thousands to wonders of American wild flowers. Color-season principle of organization is easy to use, even by those with no botanical training, and the genial, refreshing discussions of history, folklore, uses of over 1,000 native and escape flowers, foliage plants are informative as well as fun to read. Over 170 full-page plates, collected from several editions, may be colored in to make permanent records of finds. Revised to conform with 1950 edition of Gray's Manual of Botany. xlii + 438pp. 5⅜ x 8½.   20332-8 Paperbound $2.50

MANUAL OF THE TREES OF NORTH AMERICA, *Charles Sprague Sargent*
Still unsurpassed as most comprehensive, reliable study of North American tree characteristics, precise locations and distribution. By dean of American dendrologists. Every tree native to U.S., Canada, Alaska; 185 genera, 717 species, described in detail—leaves, flowers, fruit, winterbuds, bark, wood, growth habits, etc. plus discussion of varieties and local variants, immaturity variations. Over 100 keys, including unusual 11-page analytical key to genera, aid in identification. 783 clear illustrations of flowers, fruit, leaves. An unmatched permanent reference work for all nature lovers. Second enlarged (1926) edition. Synopsis of families. Analytical key to genera. Glossary of technical terms. Index. 783 illustrations, 1 map. Total of 982pp. 5⅜ x 8.
20277-1, 20278-X Two volume set, paperbound $6.00

IT'S FUN TO MAKE THINGS FROM SCRAP MATERIALS,
Evelyn Glantz Hershoff
What use are empty spools, tin cans, bottle tops? What can be made from rubber bands, clothes pins, paper clips, and buttons? This book provides simply worded instructions and large diagrams showing you how to make cookie cutters, toy trucks, paper turkeys, Halloween masks, telephone sets, aprons, linoleum block- and spatter prints — in all 399 projects! Many are easy enough for young children to figure out for themselves; some challenging enough to entertain adults; all are remarkably ingenious ways to make things from materials that cost pennies or less! Formerly "Scrap Fun for Everyone." Index. 214 illustrations. 373pp. 5⅜ x 8½.   21251-3 Paperbound $1.75

SYMBOLIC LOGIC and THE GAME OF LOGIC, *Lewis Carroll*
"Symbolic Logic" is not concerned with modern symbolic logic, but is instead a collection of over 380 problems posed with charm and imagination, using the syllogism and a fascinating diagrammatic method of drawing conclusions. In "The Game of Logic" Carroll's whimsical imagination devises a logical game played with 2 diagrams and counters (included) to manipulate hundreds of tricky syllogisms. The final section, "Hit or Miss" is a lagniappe of 101 additional puzzles in the delightful Carroll manner. Until this reprint edition, both of these books were rarities costing up to $15 each. Symbolic Logic: Index. xxxi + 199pp. The Game of Logic: 96pp. 2 vols. bound as one. 5⅜ x 8.
20492-8 Paperbound $2.50

MATHEMATICAL PUZZLES OF SAM LOYD, PART I
selected and edited by M. Gardner
Choice puzzles by the greatest American puzzle creator and innovator. Selected from his famous collection, "Cyclopedia of Puzzles," they retain the unique style and historical flavor of the originals. There are posers based on arithmetic, algebra, probability, game theory, route tracing, topology, counter and sliding block, operations research, geometrical dissection. Includes the famous "14-15" puzzle which was a national craze, and his "Horse of a Different Color" which sold millions of copies. 117 of his most ingenious puzzles in all. 120 line drawings and diagrams. Solutions. Selected references. xx + 167pp. 5⅜ x 8.
20498-7 Paperbound $1.35

STRING FIGURES AND HOW TO MAKE THEM, *Caroline Furness Jayne*
107 string figures plus variations selected from the best primitive and modern examples developed by Navajo, Apache, pygmies of Africa, Eskimo, in Europe, Australia, China, etc. The most readily understandable, easy-to-follow book in English on perennially popular recreation. Crystal-clear exposition; step-by-step diagrams. Everyone from kindergarten children to adults looking for unusual diversion will be endlessly amused. Index. Bibliography. Introduction by A. C. Haddon. 17 full-page plates, 960 illustrations. xxiii + 401pp. 5⅜ x 8½.
20152-X Paperbound $2.25

PAPER FOLDING FOR BEGINNERS, *W. D. Murray and F. J. Rigney*
A delightful introduction to the varied and entertaining Japanese art of origami (paper folding), with a full, crystal-clear text that anticipates every difficulty; over 275 clearly labeled diagrams of all important stages in creation. You get results at each stage, since complex figures are logically developed from simpler ones. 43 different pieces are explained: sailboats, frogs, roosters, etc. 6 photographic plates. 279 diagrams. 95pp. 5⅝ x 8⅜.
20713-7 Paperbound $1.00

A Course in Mathematical Analysis,
Edouard Goursat

Trans. by E. R. Hedrick, O. Dunkel, H. G. Bergmann. Classic study of fundamental material thoroughly treated. Extremely lucid exposition of wide range of subject matter for student with one year of calculus. Vol. 1: Derivatives and differentials,· definite integrals, expansions in series, applications to geometry. 52 figures, 556pp. 60554-X Paperbound $3.00. Vol. 2, Part I: Functions of a complex variable, conformal representations, doubly periodic functions, natural boundaries, etc. 38 figures, 269pp. 60555-8 Paperbound $2.25. Vol. 2, Part II: Differential equations, Cauchy-Lipschitz method, nonlinear differential equations, simultaneous equations, etc. 308pp. 60556-6 Paperbound $2.50. Vol. 3, Part I: Variation of solutions, partial differential equations of the second order. 15 figures, 339pp. 61176-0 Paperbound $3.00. Vol. 3, Part II: Integral equations, calculus of variations. 13 figures, 389pp. 61177-9 Paperbound $3.00 60554-X, 60555-8, 60556-6 61176-0, 61177-9 Six volume set,

paperbound $13.75

Planets, Stars and Galaxies,
A. E. Fanning

Descriptive astronomy for beginners: the solar system; neighboring galaxies; seasons; quasars; fly-by results from Mars, Venus, Moon; radio astronomy; etc. all simply explained. Revised up to 1966 by author and Prof. D. H. Menzel, former Director, Harvard College Observatory. 29 photos, 16 figures. 189pp. 5⅜ x 8½.   21680-2 Paperbound $1.50

Great Ideas in Information Theory, Language and Cybernetics,
Jagjit Singh

Winner of Unesco's Kalinga Prize covers language, metalanguages, analog and digital computers, neural systems, work of McCulloch, Pitts, von Neumann, Turing, other important topics. No advanced mathematics needed, yet a full discussion without compromise or distortion. 118 figures. ix + 338pp. 5⅜ x 8½.
21694-2 Paperbound $2.25

Geometric Exercises in Paper Folding,
T. Sundara Row

Regular polygons, circles and other curves can be folded or pricked on paper, then used to demonstrate geometric propositions, work out proofs, set up well-known problems. 89 illustrations, photographs of actually folded sheets. xii + 148pp. 5⅜ x 8½.   21594-6 Paperbound $1.00

Visual Illusions, Their Causes, Characteristics and Applications,
M. Luckiesh

The visual process, the structure of the eye, geometric, perspective illusions, influence of angles, illusions of depth and distance, color illusions, lighting effects, illusions in nature, special uses in painting, decoration, architecture, magic, camouflage. New introduction by W. H. Ittleson covers modern developments in this area. 100 illustrations. xxi + 252pp. 5⅜ x 8.
21530-X Paperbound $1.50

Atoms and Molecules Simply Explained,
B. C. Saunders and R. E. D. Clark

Introduction to chemical phenomena and their applications: cohesion, particles, crystals, tailoring big molecules, chemist as architect, with applications in radioactivity, color photography, synthetics, biochemistry, polymers, and many other important areas. Non technical. 95 figures. x + 299pp. 5⅜ x 8½.
21282-3 Paperbound $1.50

Applied Optics and Optical Design,
A. E. Conrady
With publication of vol. 2, standard work for designers in optics is now
complete for first time. Only work of its kind in English; only detailed work
for practical designer and self-taught. Requires, for bulk of work, no math
above trig. Step-by-step exposition, from fundamental concepts of geometrical,
physical optics, to systematic study, design, of almost all types of optical
systems. Vol. 1: all ordinary ray-tracing methods; primary aberrations; necessary higher aberration for design of telescopes, low-power microscopes, photographic equipment. Vol. 2: (Completed from author's notes by R. Kingslake,
Dir. Optical Design, Eastman Kodak.) Special attention to high-power micro-
scope, anastigmatic photographic objectives. "An indispensable work," *J., Optical Soc. of Amer.* Index. Bibliography. 193 diagrams. 852pp. 6⅛ x 9¼.
60611-2, 60612-0 Two volume set, paperbound $8.00

Mechanics of the Gyroscope, the Dynamics of Rotation,
R. F. Deimel, Professor of Mechanical Engineering at Stevens Institute of
Technology
Elementary general treatment of dynamics of rotation, with special application
of gyroscopic phenomena. No knowledge of vectors needed. Velocity of a moving
curve, acceleration to a point, general equations of motion, gyroscopic horizon,
free gyro, motion of discs, the damped gyro, 103 similar topics. Exercises.
75 figures. 208pp. 5⅜ x 8.   60066-1 Paperbound $1.75

Strength of Materials,
J. P. Den Hartog
Full, clear treatment of elementary material (tension, torsion, bending, com-
pound stresses, deflection of beams, etc.), plus much advanced material on
engineering methods of great practical value: full treatment of the Mohr circle,
lucid elementary discussions of the theory of the center of shear and the
"Myosotis" method of calculating beam deflections, reinforced concrete, plastic
deformations, photoelasticity, etc. In all sections, both general principles and
concrete applications are given. Index. 186 figures (160 others in problem
section). 350 problems, all with answers. List of formulas. viii + 323pp. 5⅜ x 8.
60755-0 Paperbound $2.50

Hydraulic Transients,
G. R. Rich
The best text in hydraulics ever printed in English . . . by former Chief Design
Engineer for T.V.A. Provides a transition from the basic differential equations
of hydraulic transient theory to the arithmetic integration computation re-
quired by practicing engineers. Sections cover Water Hammer, Turbine Speed
Regulation, Stability of Governing, Water-Hammer Pressures in Pump Dis-
charge Lines, The Differential and Restricted Orifice Surge Tanks, The
Normalized Surge Tank Charts of Calame and Gaden, Navigation Locks,
Surges in Power Canals—Tidal Harmonics, etc. Revised and enlarged. Author's
prefaces. Index. xiv + 409pp. 5⅜ x 8½.   60116-1 Paperbound $2.50

Prices subject to change without notice.

Available at your book dealer or write for free catalogue to Dept. Adsci,
Dover Publications, Inc., 180 Varick St., N.Y., N.Y. 10014. Dover publishes more
than 150 books each year on science, elementary and advanced mathematics,
biology, music, art, literary history, social sciences and other areas.